AGRICULTURAL ENTOMOLOGY

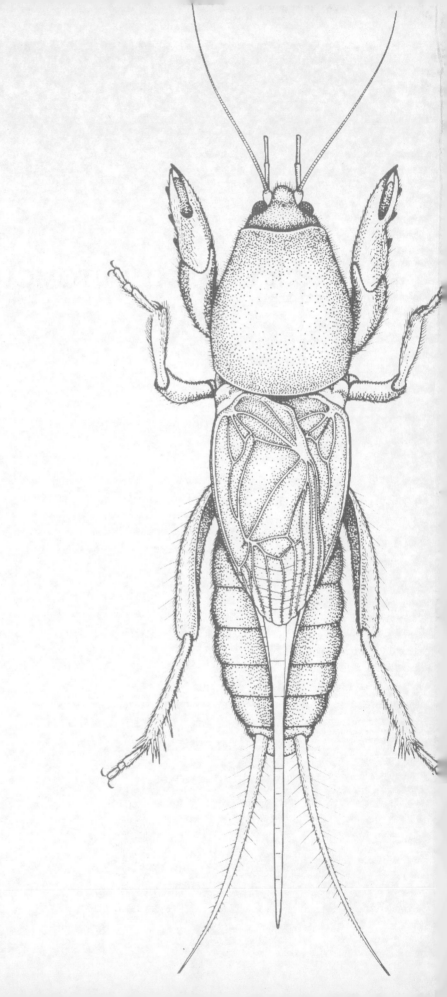

AGRICULTURAL ENTOMOLOGY

by

Dennis S. Hill

M.Sc., Ph.D., F.L.S., C. Biol., M.I. Biol., F.R.E.S.

Consultant in Entomology, Ecology, and Plant Protection
Formerly Professor of Crop Protection,
Alemaya University of Agriculture, Ethiopia

Senior Lecturer in Entomology and Ecology,
Department of Zoology, University of Hong Kong

Assisted by
Jeremy D. Hill, B.Sc., M.Sc.

TIMBER PRESS
Portland, Oregon

Illustrations courtesy Dr. Dennis S. Hill;
reprint permission required

ISBN 0-88192-223-4
Printed in Hong Kong

TIMBER PRESS, INC.
The Haseltine Building
133 S.W. Second Ave., Suite 450
Portland, Oregon 97204-3527, U.S.A.

Library of Congress Cataloging-in-Publication Data

Hill, Dennis S., 1934-
    Agricultural entomology : by Dennis S. Hill ; assisted by Jeremy
D. Hill.
        p.   cm.
    Includes bibliographical references and index.
    ISBN 0-88192-223-4
    1. Insect pests.   2. Beneficial insects.   3. Plant mites.
4. Agricultural pests.   I. Hill, Jeremy D.   II. Title.
    SB931.H448   1994
    632'.7--dc20
                                                    91-39244
                                                      CIP

CONTENTS

This book is dedicated to
the late Professors O. W. Richards, FRS,
and John G. Phillips, FRS,
to whom I shall always be indebted.

PREFACE

In the literature, there is a vacant niche for a book such as this. At the time of writing, there is no publication that takes a global overall view of the agricultural scene with regard to entomology. There are excellent books on basic entomology, such as those by Richards & Davies (1977) and by Borror & Delong (1971), but the former has a bias towards Europe, while the latter is restricted to North America. The regional books (C.S.I.R.O., 1969; Kalshoven, 1981; etc.) are good, but limited in approach. The books on crop pests can usually only concentrate on the more important pest species, and are often overly concerned with their control. There are many publications of high academic merit dealing with the pests of specific crops, and, similarly, some of the medical and veterinary parasitology texts are excellent, but very detailed and specialized. The ones being produced in France under the general editorial guidance of Balachowsky (1966) are massive, expensive, and in French. Two of the books that do attempt a systematic view of agricultural insects are those by Smith (1951) and Caswell (1962), but the treatment is very superficial and there are few or no illustrations. For specialized courses at B.Sc. level, and for M.Sc. courses, it does appear that a general text is needed that looks at agricultural entomology in the widest sense, especially one that is well illustrated, for few students have any experience of seeing these insects in the field.

The experience on which this book is based stems from professional activities and residence in the U.K. for 12 years, Hong Kong for 13 years, Uganda for 2½, and Ethiopia for 4 years; and during this period of time there were visits for entomological and other purposes to the following countries for spells of weeks or months—Malaysia, Thailand, India, Seychelles, Malawi, Libya, Japan (Kyoto), U.S.A., and Trinidad. In all these countries data were collected on local insect pests and photographs taken. The photographs used in this book were taken by the author, with one or two exceptions. The insect drawings were mostly make by Karen Phillipps and Hilary Broad with a few by Alan Forster and others.

Over the years, most of the specimens illustrated were identified by staff of the Commonwealth Institute of Entomology International, or the Natural History Museum formerly known as the British Museum (Natural History). Some specimens of insect pests were borrowed from the Natural History Museum for drawing purposes, through the Keepers of Entomology (Dr. P. Freeman, then Dr. L. A. Mound), and the Trustees are accordingly thanked.

A major contribution to this project has been made by Jeremy D. Hill with both insect collecting and the rearing of many species, especially the Lepidoptera, in Uganda, Cambridge (U.K.) and Hong Kong.

Since 1960, I have received help in many forms from many colleagues, friends and neighbors to whom I owe my sincere thanks. The people who made specific contributions are listed below: Dr. D. V. Alford, Dr. H. Banziger, Dr. and Mrs. M. Bascombe, Dr. D. K. Butani, Dr. J. M. Cherrett, Mr. T. J. Crowe, Mr. T. Dunn, Dr. V. F. Eastop, Dr. D. Evans, Susan D. Feakin, Dr. D. J. Greathead, Mr. R. Gair, Dr. K. M. Harris, Mr. W. R. Ingram, Mrs. G. Johnston, Dr. R. H. Le Pellay, Dr. Li Li-ying, Dr. Lee Hay-yue, Mr. R. J. A. W. Lever, Dr. L. A. Mound, Mr. D. N. McNutt, Dr. J. L. Nickel, Dr. K. A. Spencer, Mr. H. Stroyan, Prof. Dr. J. T. Wiebes, Dr. D. J. Williams, Mr. R. Winney, and Mr. Robin Wong.

While at Makerere University, Uganda, I received support and encouragement from both Dr. J. L. Nickel and Prof. K. Oland to whom I shall always be grateful, and the original project on tropical crop pests was funded by a grant from the Rockefeller Foundation (part of the aid vote to Makerere University).

In the Department of Zoology at Hong Kong University I was given both support and excellent facilities, including technical help from the Faculty photographers, and I would like to thank Professors David Barker, John G. Phillips, and Brian Lofts. Dr. J. H. Sudd kindly made facilities available at Hull University in 1983. This was later followed by four years (1986–88; 1990–92) at Alemaya University of Agriculture, Ethiopia, whose assistance is gratefully acknowledged.

My month-long visit to Malaya in 1976 to entomological and agricultural research centers was made possible by a grant from the Commonwealth Fund for Technical Cooperation, London, whose assistance was both timely and generous.

Dennis S. Hill
"Haydn House"
20 Saxby Avenue
Skegness
Lincs.
ENGLAND

Part A

INTRODUCTION

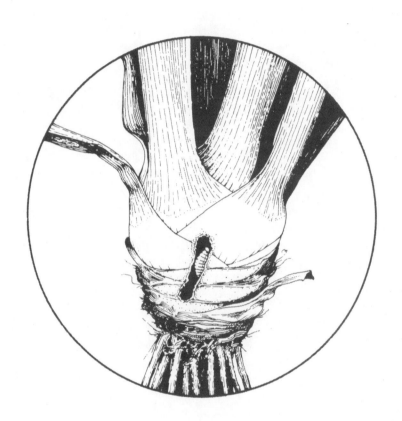

The object of this book is to look at the Class Insecta (and some Arachnida) systematically and to indicate which are pest species and which are species beneficial in one way or another. Agricultural pest species are regarded in the widest possible sense, to include veterinary pests as well as crop pests, and pests of forestry, of ornamental plants, and pests of stored products (grain, foodstuffs, etc.). Medical pests are also included in so far as they affect farmers and farm workers. However, the major effort is on crop pests, and for obvious reasons the veterinary, medical and forestry pests are not dealt with in great detail. All the main insect groups and many of the more important genera of insects and mites are illustrated, either by black and white photographs or line drawings to show their salient features.

The book is intended as a tertiary level textbook for courses in "Agricultural Entomology", "Applied Entomology", and the like, preferably at M.Sc. level, and also as a source of reference for B.Sc. level courses and for field workers and agriculturalists.

Most texts on crop pests are, for several reasons, restricted to either one country or one region, or at most refer to the tropics (e.g., Hill, 1983) or to temperate regions of the world (e.g., Hill, 1987). The distinction between tropical and temperate is not easy climatically/geographically, and is even more difficult agriculturally as new varieties of crops are bred that have greater tolerance ranges. Almost all countries in the world are undergoing agricultural diversification programs with the introduction of new and exotic crops. In many tropical countries there are truly tropical crops grown in the lowlands, but wheat, potatoes, and other "temperate" crops are grown in the cooler highlands. Most training courses are nowadays orientated narrowly to the local situation; the old days of the D.T.A. ((Diploma in Tropical Agriculture, Trinidad) and the training of entomologists for work anywhere in the "British Empire" have long since gone, but there would seem to be a place for one book that looks at the global agricultural scene with regard to entomology. One of the main aims of the book is to indicate the extent of each insect family, both geographically and biologically, and to demonstrate similarities from continent to continent, and occasionally to stress major differences. Ideally, any working entomologist should have knowledge extending beyond the immediate province and should be aware of important species that might well spread or be distributed. An entomologist should also be aware of the existence of all the other closely related species that might occur along with the main pest species, and should be wary of unsubstantiated identifications. One generally accepted premise is that, worldwide, there is a total of about 1000 species of serious pests of crops and cultivated plants, and probably some 30,000 species of minor pests.

The text is based upon the 10th edition of the standard British work by Imms, the last revision by Richards & Davies (1977) and retitled, *Imms' general textbook of entomology.* The somewhat simplified systematic treatment afforded by this work is most suitable for student use, and has been followed in almost all cases. The previously very small introductory text colloquially referred to by generations of students as "Little Imms" has now been extensively revised by Professor Davies, after the death of Professor Richards, and in its new form (Davies, 1988) is an extremely good "outline" of entomology and to be highly recommended as a basic entomology text. It is presupposed that students have a working knowledge of basic entomology before using this text. Additional material, concerning both pest species and beneficial ones, has been collated from many of the sources listed in the References and from 30 years of personal observations and collecting, in England, eastern Africa and tropical Asia, and in some other places.

Common names used in the text are either those used on the C.A.B.I.I.E. pest distribution maps, or else names used frequently in international entomological literature, or if they are local species, then the names suggested by local entomological societies, etc. The difficulty is to find common names that are both meaningful and acceptable worldwide. In some instances both British and American approved common names are given in the text.

Every reasonable effort has been made to check the scientific names used but this is difficult, very time-consuming, and, in the final analysis, there is often a lack of agreement from one country to another, first concerning the respective taxa employed and secondly the precise limits of various genera and specific interpretation. There is also the problem of genuine name changes which are occurring all the time, so that a book manuscript can never be completely up to date in this respect. There have been a few name changes that I have been reluctant to use, not that their authenticity is in doubt, but they would replace names that have been widely accepted and used for decades and are firmly entrenched in the entomological literature. As a teacher I am well aware that in many regions students are using older editions of standard works and such name changes would be very confusing at the present time. In such cases the "new" name is given in parentheses. To the taxonomic purists I apologize and plead expediency, and the problem of working in remote locations where the main problem is isolation and lack of communication. The main criterion I feel is that the identity of the insect concerned should not be in doubt, and by using a combination of scientific name and common name, this is hopefully achieved.

The C.A.B.I.I.E. have recognized this problem and to help alleviate matters have now produced a checklist of insect and mite pest species names (Wood, 1989; *Insects of economic importance: a checklist of preferred names;* pp. 150). Regrettably, this publication was obtained too late to permit detailed changes, and also at this stage it cannot be known how widely the checklist will be accepted internationally; further, it does not include common names.

For several reasons it was decided to omit the name of the taxonomic authority after each species name. This is not a book on taxonomy, although it has a taxonomic approach, and the number of species mentioned is very large.

The basic sources of data have been many and diversified and, in a few cases, it was clear that a misidentification locally had been made. When this was apparent the record was obviously discounted, but there were doubtless other misidentifications (that a specialist might have detected) that were not obvious and these records have been accepted at face value. It is regrettably inevitable that, for these and other reasons, there will be nomenclatural errors and some misidentifications evident to the taxonomic specialists; for these I apologize, but most were largely unavoidable.

One problem with being an agricultural entomologist is that, to be most effective, you need a great breadth of knowledge and experience, in that you need first to know which insects are pests and which are not, and which might be potential pests. Thus you need to be able to recognize the more important pests at sight, and to be conversant with their biology, life cycle, and the damage they inflict on the host organism, and finally you need to have some knowledge as to the most appropriate method of control (details of the latter can always be looked up back at the laboratory). You also need to be able to recognize most of the minor pests of the local crops, etc., and also the other species that are beneficial and should not be harmed. There will be many insects whose presence on a plant is purely fortuitous, so it is really necessary to be able to recognize insects generally to family level. Of course, it basically takes years to accumulate the experience required, and it is in the hope of speeding up the learning process that this book is so detailed and has such a large number of illustrations of the more serious and widespread pest species. The extensive illustrations should materially assist the students to recognize the different groups of insects generally, as well as the more important species.

CURRENT PROBLEMS FOR AGRICULTURE

For some years now it has been assumed quite widely that agricultural science has the answers to the problem of increasing agricultural production to meet both the existing need for food and to provide for the ever-increasing world population. It was thought that the only problem remaining was to get the new techniques adopted. In the temperate regions this situation seems to have been true, in that many new varieties of crops are high-yielding and disease-resistant and harvests are good (ever-increasing in fact). At the same time, in these regions, population increase is slowing dramatically as the birthrate falls, so many temperate countries are now producing considerable crop surpluses, particularly of grains and basic foodstuffs such as butter, milk, eggs, and oils. The production surpluses are now becoming an economic embarrassment and, in recent years, the U.S.A. has been starting an agricultural "set-aside" scheme, in 1988 followed by the U.K., whereby a set percentage (10–20% in some areas) of poorer quality agricultural land is left fallow for a period of 5–10 years and the farmers are given financial compensation. The idea is to reduce overall production levels of grain to the required levels for the country, and at the same time allow the poorer quality land to recover natural fertility to some extent.

In Europe and North America there is an ever-increasing minority of people becoming vociferously concerned with the environment—some of these belong to the so-called "green" political parties. One of their beliefs is that more food should be produced without the use of artificial fertilizers and pesticides. Organic farming is feasible, this has now clearly been demonstrated, but yields are lower and pest damage levels are higher, and the whole production system is less cost-effective. Modern agriculture as practiced in North America and Europe, etc., is very efficient using new improved crop varieties, but it is very input-intensive in terms of equipment, fertilizers, pesticides, etc. Widespread adoption of "organic" methods of agriculture would result in substantial price rises for most types of food, but in the Western world this is a growing trend and there might be radical changes on the way.

Throughout the tropics the situation is less satisfactory. However, the "green revolution" in India was very successful and did transform the country into a grain-exporter. The new varieties of crops do, however, require high levels of fertilizer use and plenty of water, and, in most of India, it was possible to provide these conditions. But in many other tropical regions there have been repeated harvest failures and subsequent widespread famine. This was brought very forcibly to the attention of the world in 1984 with a series of documentary films made in northern Africa showing the effects of the widespread famine. This publicity resulted in considerable rethinking of agricultural research priorities of the international aid programs. Clearly the main problems are in tropical Africa, but Asia and South America are not exempt. The chief problems appear to be those listed below, and, of course, they are usually acting in concert so that the overall situation is aggravated.

1. General climatic changes; a warming effect with increasing drought and desertification.

2. Poverty.

3. Emphasis of research on cash crops, and neglect of food crops.

4. Land shortage, and erosion problems.

5. General overpopulation.

6. Population drift from the countryside into towns and cities.

CLIMATIC CHANGES

In view of the geological history of the earth, with the recurring ice ages and warm interglacial periods, it should be expected that long-term climatic trends are likely to be experienced. One such trend is the reduction of rainfall in parts of tropical Africa, Australia, and South America, and in some areas an overall rise in temperature. In these areas, deserts have been increasing in size and this led to the term "desertification" being coined. It has been clear for the last 20 years or more that the Sahara Desert in North Africa is extending its southern boundaries. Actually, it has been known historically for many years, since Roman days when much of the staple grain that fed the legions was cultivated on the plains of North Africa bordering the Mediterranean. The southern fringe of the Sahara is geographically known as the Sahel Region, and it has been popu-lated by a mixture of nomadic tribes, essentially pastoral with an economy based on flocks of goats and sheep, or herds of camels or cattle, and often a crop of grain grown during a lengthy bivouac. The other inhabitants of the southern Sahel are dryland farmers who grow millets and sorghum, partly as their subsistence food and partly for sale to the nomads. During the past 20 years or so in the Sahel the desert has been extending its southern boundary by 5 km (3 miles) per year, and this annual encroach-ment has been accompanied by reduced rainfall and loss of endemic vegetation. Rain-fall reduction in this region amounts to a loss of 50–66% in the arid (northern) zone, and 30–35% less in the semiarid (southern) zone, over the last 20 years.

The two rainfall problems in this region are, first the overall reduction in precipita-tion, and second its erratic nature. With reduced and unpredictable rainfall, crop produc-tion is seriously affected and yields swiftly diminish until a point is reached when avail-able water is insufficient to permit the crop to mature, and total harvest failure results. Across the southern fringe of the Sahara, from Mali in the west to Ethiopia in the east, the population of this semiarid area amounts to many millions. In fact the semiarid area extends southwards through Somalia and northern Kenya, to parts of Tanzania and Mozambique. Repeated crop failures result in either starvation of the rural population or else causes mass migration, either into the cities and towns, or into other rural areas usually already fully populated.

Then the famine victims need urgent international aid relief in the form of grain and other foodstuffs. The food provided by international relief organizations is of course only a palliative, and the basic problem remains, that is to increase crop production of staples to a level of national self-sufficiency (and to minimize losses in storage). The people who die in famines are the subsistence farmers and the pastoralists, and they die not because the country has run out of food but because they cannot produce a surplus sufficient to carry them through bad years when harvests fail.

As already mentioned, in these regions the effect of increasing desertification and drought is exacerbated by the other factors, namely shortage of cultivatable land, over-population, poverty, and lack of knowledge concerning staple food crops.

POVERTY

. Most smallholder farmers have no money savings, and many see little actual cash at all. Some have livestock which in emergency are sold for cash, and of course the pastoralists have their livestock as their assets. For the farmers there may be new high-yielding crop varieties available but to benefit from the promised increase in yield they need fertilizers and pesticides and often the farmers just do not have the money or the credit with which to buy these required inputs. Money is needed in most countries for tax payment, school fees, clothing, medicine, etc., and often the food staples have to be sold to realize the cash. Foodstuffs provided by international relief agencies in times of famine are often sold locally because of the urgent need of cash.

EMPHASIS ON CASH CROPS

Most tropical countries are still totally (or almost so) dependent on agriculture as their prime economic resource, and the national importance of the local cash crops cannot be overemphasized. Since the cash crops are the source of vitally needed foreign exchange it is understandable that, in the past, emphasis has been placed on research into these crops, so that a great deal is known about coffee, sugarcane, tea, cotton, cocoa, oil palm, etc. What is not understandable is that research into staple foodstuffs has been until relatively recently almost totally neglected in most countries. Fortunately this is not entirely true in that for some years now the Consultative Group for International Agricultural Research (C.G.I.A.R.) have funded research stations throughout the world such as the Centre for International Tropical Agriculture (Colombia) (C.I.A.T.), the International Rice Research Institute (Philippines) (I.R.R.I.), the International Institute of Tropical Agriculture (Nigeria) (I.I.T.A.), the International Maize and Wheat Improvement Centre (Mexico) (C.I.M.M.Y.T.), the International Potato Centre (Peru) (C.I.P.), the International Crops Research Institute for the Semi-Arid Tropics (India) (I.C.R.I.S.A.T.), etc., where research is aimed at local basic food crops such as rice, cassava, wheat, maize, sorghum and the millets, etc. Now the research findings are being applied to local smallholder food production with considerable success in some regions—the work done at I.R.R.I. in the Philippines and other parts of S.E. Asia is well known and highly regarded. But, in general, research has not been aimed at improvements for the subsistence farmers in most countries, although the situation is improving. In most tropical countries the bulk of the national production, even of most of the cash crops, is still in the hands of the smallholders and peasant farmers with their shambas of 1–2 ha each. In East Africa generally for example, more than 80% of coffee production is by smallholders, despite the impressive coffee plantations that can be seen. Even tea in Malawi is now being grown by more and more smallholders. Examination of international aid programs in recent years will reveal relatively enormous sums being spent on research into coffee, cocoa, tea, tobacco, etc., whereas very little was being spent on bananas, cassava, sweet potato, yams, millets, sorghum, etc. It was published in *New Scientist* (April, 1985) that in 1976 total international research effort on maize amounted to U.S. $23 million, whereas in the same year $39 million was spent on coffee, and virtually nothing spent in Africa on sorghum and millets. However, the situation appears to be changing, and in a recent F.A.O. (U.N.D.P.) program—the "Action Programme for Improved Plant Protection" (for Africa)—the need for emphasis on staple food crops was stressed and there has been reaction from the international aid agencies.

The "Green Revolution" of the 1970s largely failed to improve the lot of the subsistence farmers, sad to say; although, as a result of this time period, India did become self-sufficient in grains, and started to export wheat and other cereals. Generally, on some large-scale cultivation schemes significant crop yield increases of both quantity and quality were achieved. Spectacular improvements were possible, but only with an increased input of technology and money. New high-yielding rice varieties in Asia required considerable application of fertilizer and pesticides in order to achieve the high yield possible. Similarly, new maize varieties in Africa require an increased agricultural input and, after harvesting, these relatively soft grains are extremely susceptible to insect pest damage in storage. The locally evolved ("semi-indigenous") maize varieties in

Africa have a "flint" grain that is difficult to grind and process but is highly resistant to the attack of stored products insects so storage losses are seldom very high; soft grains are often devastated in storage!

In parts of S.E. Asia and South America there are national agricultural programs to increase local acreage of plantation cash crops in order to significantly increase national productivity. What is happening is that vast tracts of natural rain forest are being razed, the trees sold as timber (for an immediate cash profit), and the exposed ground being planted with oil palm or rubber trees, or coconut palms in coastal regions. The long-term effects of the ecological devastation may well prove to be horrendous. Unless the ground is rapidly carpeted with the crop plants and suitable cover crops, soil erosion can be expected to be a serious problem.

Cassava is one of the main crops being studied at C.I.A.T. in Colombia, and also at I.I.T.A. in Nigeria, and local tuber production is being greatly enhanced. At A.V.R.S. in Taiwan, sweet potato is receiving similar intensive attention. With the establishment of I.C.R.I.S.A.T. in India to study the crops of semiarid areas, it is to be hoped that, very soon, crop production of sorghum and the millets will be greatly improved. In the last couple of years international effort has been directed to the problem of banana decline in East Africa.

On the other hand however, it can be a mistake to go to the other extreme and neglect the important cash crops; over the last 20–30 years there was a major decline in research effort on coffee in eastern Africa, largely following independence of the main countries, and by 1986/87 it was clear that the coffee industry of this region had seriously declined, with both quantity and quality of yield greatly diminished. It should perhaps be mentioned that some of the reasons for this decline were governmental rather than agricultural. In 1987 a major E.E.C. aid program was being extended throughout eastern Africa to try to reestablish the coffee industry (it had been started in Uganda in 1982) but it will not be an easy matter.

LAND SHORTAGE AND EROSION

In most countries in the world all suitable agricultural land is already being cultivated. And, as already mentioned, the new land "set-aside" scheme being practiced in the U.S.A. and now in the U.K. is designed to take fields of marginal fertility out of agricultural production. In some of the developed countries, prime agricultural land is being lost annually by the encroachment of urban conclaves, road building, new airports, dam construction, and, of course, by soil exhaustion (especially the highly organic fen and peat or muck soils). Population growth rates may be declining appreciably but the total population is still in most cases increasing, which demands greater urbanization. Most tropical countries however still have rapidly expanding populations and, in most countries, all the cultivable land has been in use for some time. In early agriculture the traditional approach was the "slash and burn" method. Forest or scrub was cleared by a combination of cutting and burning and the exposed land was then planted with food crops. Usually each plot was productive for 2–3 years but nutrient levels fall rapidly and it becomes necessary to move the family or the village on to a new location. This approach sufficed until the population became large in relation to available land, and then villages became more static. With the static villages there was initially still sufficient arable land to allow areas to lie fallow for several years between cultivations, but as the populations increase the land eventually comes under continuous cultivation. Many traditional inheritance systems work on the basis of the family farm being split up between the children of each successive generation. Now, in many countries in Africa and other regions, the hereditary subplots are too small to provide food for a family. Nowadays, marginal land is being planted but often with minimal success. Cultivation of hillsides and slopes, unless carefully terraced, inevitably leads to surface (soil) erosion, which then spreads down into the valleys and plains. In the countries where cultivable land is in such short supply, especially in most African countries, crop production can be increased by very careful use of suitable techniques. But this would require strong governmental support (which is often lacking), and it may involve considerable local changes in respect to life-styles and agricultural practices, and the concept of "appropriate technology" is important here.

In many tropical countries the erratic rainfall, when it does come, is likely to be torrential, with the sad result that most of the water is lost as runoff, taking with it tons of topsoil in surface erosion. Rain runoff and soil surface erosion must both be curtailed by appropriate cultivation methods.

Many Third World countries face an acute shortage of firewood, charcoal, kerosene, coal, or whatever is used to fuel cooking fires, and one result is that all available wood is being collected, cut, and felled for cooking purposes or for sale. Hillsides are being denuded and animal browsing (especially by goats) is preventing any natural reforestation. This loss of plant cover is exposing hillsides to erosion, which in some countries is quite serious. In parts of tropical Asia and in South America there is a tendency to encourage deforestation, partly for the immediate cash advantage of the timber, and also to provide open land for farmers to utilize. But this forest devastation often results in severe surface soil erosion long before any serious attempt is made to farm the area. When the rains fall, the rivers flow with almost solid red silt. Thousands of square kilometers of forest are, at the present time, being destroyed annually and there is a great destruction of the rather delicate lateritic tropical soils. The prognosis for such ill-advised activity is gloomy to say the least. In the long term, agriculture will inevitably deteriorate and the ecological damage to the environment is totally incalculable.

The shortage of agricultural land in the world is inevitable and, in order to maintain or increase crop production, it is necessary that agriculture becomes more efficient, and that protection measures are applied for post-harvest storage. The role of plant protection and produce protection in storage, are both vital parts of the improved systems of agriculture required.

OVERPOPULATION

In the early days of settled agriculture, the having of a large family was a means of providing labor for cultivation and maintenance purposes, and this practice once established is proving difficult to counter. In the past, high levels of infant mortality have generally prevailed, so that even more children were born in order to provide the labor force. One result has been that, in many parts of the tropical world, having a large family has been synonymous with prosperity for so long that present birth control programs are having little effect. With improved medical services, the first result invariably is a dramatic increase in infant survival, and this results in a population explosion that aggravates the situation as far as food supplies are concerned. In countries such as Ethiopia it is difficult to see how agricultural production can ever hope to feed the population until the birth rate is controlled. It is sad to hear that, according to the latest news broadcasts, the initial success being achieved in China in birth control is not being maintained.

POPULATION DRIFT

As rural communities become more urbanized and more sophisticated, which is inevitable, there is a drift of young people from the villages into nearby towns and cities, where they work in factories instead of on the fields of their ancestors. On the whole this trend is, generally, not too disadvantageous, for the new urban industries require laborers, and the more efficient agriculture being practiced rurally soon requires less labor as mechanization increases. In the U.S.A. a century ago each farmer supported only two people, whereas now one farmer supports some 50–60 people. But if the drift of the young people into towns occurs before the farms are adequately mechanized, agricultural production will suffer, and there may not be jobs in the towns for them to go to.

STUDENT TRAINING

Some of the problems prevalent at the present time can be gradually countered by education, and by the dissemination of information to the rural communities. But this requires an increased output of graduates from both the universities and the colleges in most tropical countries, in all subjects pertaining to agriculture in the broadest sense. An important aspect is the further training of specialists (in this case in Agricultural Entomology, and in Crop Protection) and a large cadre of Agricultural Officers and Extension staff is needed as well as the research workers and the lecturers.

On a worldwide basis, there is now a tremendous quantity of information available both in basic entomology and the many aspects of applied entomology and pest control, etc. This book is designed to provide information to help train agricultural entomologists and crop protection specialists in universities and colleges of agriculture in the English-speaking world. Ideally it should be used in conjunction with similar books relating to crop pests and their control (e.g. Hill, 1983, 1987), and those dealing with pests of specific crops, such as Le Pelley (1968), or Gair, Jenkins & Lester (1983), etc. As a textbook it would be most useful for students on M.Sc. courses or specialized B.Sc. courses; but as a source of reference it should be applicable to many B.Sc. level courses where a basic knowledge of applied entomology is required. It is also hoped to be useful to extension workers and to some agriculturalists as a reference book.

1

Insect Distributions and Ecology

The basic approach of this book is a systematic review, group by group, describing briefly the agricultural importance in the widest possible sense. But the scope is world-wide and, since many pests are regionally distributed, it is important to stress distributional differences and to mention the factors that are thought to be responsible. Insect/host relations are not as static as they might seem at first sight, although changes are not usually very obvious and may be quite few. But changes are occurring all the time and occasionally a major change takes place with disconcerting results. There is a great need to understand more fully the complex interrelationships in the joint ecology of the insects and their host organisms. So this small introductory section is used to stress the need for a greater accumulation of ecological knowledge about the insects concerned.

INSECT DISTRIBUTIONS

Most of the insects of concern here are phytophagous and feed on plants, so the study of natural insect distributions is closely linked with the distribution of plants throughout the world. As will be discussed in the next section, it is well known that insects are greatly influenced by ambient temperature and humidity conditions, so the basic controlling factors affecting the natural distribution of most insects are a combination of the natural vegetation and ambient climatic conditions.

Studies by geographers and biologists on the world distribution of flora and fauna are known as biogeography for the overall science, or phytogeography for the distribution of plants, and zoogeography for the distribution of animals. Global maps are also constructed to show climatic zones, ecological zones, zones based on soil types or substrate types, and so on.

BIOGEOGRAPHY

Because of the pattern of the Earth's rotation around the Sun, there are vast differences in the amount of solar radiation falling upon different parts of the globe, with the resulting pronounced differences in climate and the associated seasonal variations. As a result of these differences, the geographical globe is traditionally divided into broad climatic belts, as shown in Fig. 1.1.

The distinction between what is generally regarded as a tropical flora (and fauna) and a temperate flora in some parts of the world is not at all clear, for at latitude 23.5° there are deserts and mountain ranges to confuse the boundaries as well as continental effects.

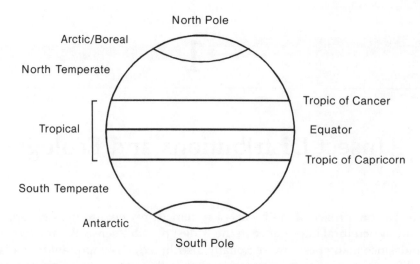

Fig. 1.1   The geographical globe (Earth) divided into the main broad climactic belts.

Arctic/boreal—the extreme northern region above latitude 65°N where climatic conditions are rigorous, with a long, cold, dark winter. There is a short summer period of continuous daylight when the surface soil and water thaws and conditions are warm. Tree species here are prostrate mostly. The population of aquatic insects is very high, especially Diptera (mosquitoes, midges, black flies, etc.). Under the surface soil the permafrost remains frozen, so plant growth is superficial.

North temperate—a broad zone that contains several distinct ecological (plant) biota. Floristically the northern belt is the *taiga,* a region of coniferous (needle-leaved) forest; this is also called the *Cold temperate region* with a long, cold winter and cool summer.

Below this belt is the broad-leaved deciduous forest; an extensive zone also called the (true) *Temperate region,* dominated by oaks and the family Fagaceae, with a temperate winter and cool summer.

The lowest zone is the *Southern temperate* and includes most of the Mediterranean Basin in Europe. It contains a mixture of summergreen deciduous forest and evergreen scrub, including evergreen scrub oaks in Europe, and other similar types in the U.S.A. The climate has a mild winter, and a summer that is dry but also cool in some areas.

Subtropical—most maps do not include such a zone, but the term is commonly used, and in the dictionary by Lincoln, Boxshall & Clark (1982) they define this region as that lying between the Tropic (23.5°) and 34.0° in either hemisphere. The climate is usually a mild winter with a hot, wet summer, and the vegetation is broad-leaved evergreen forest, or scrub in places. This is the zone exploited for intense fruit production with such species as citrus, peach, avocado, pineapple, etc. These are often referred to as the "subtropical fruits."

Tropical—the central region of the globe extending from the Tropic of Cancer in the north and the Tropic of Capricorn in the south. Much of the tropical zone has a dry and hot winter with summer rains, but the central strip is designated the Equatorial Belt and is characterized by evergreen rain forest and continual rainfall—but to some extent there may be two short, dry seasons and two long, rainy seasons, although sometimes the so-called dry seasons may be scarcely apparent.

World vegetation maps are produced by phytogeographers in bewildering detail (see Good, 1953, etc.) but the very basic plant types and their world distribution are

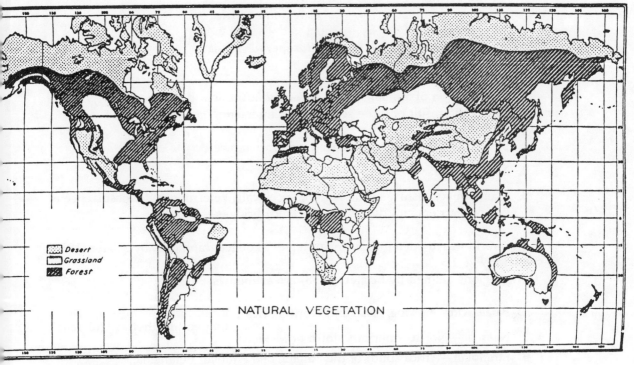

Fig. 1.2 Generalized world distribution of the three major divisions of the natural vegetation. (After Henry *et al.*) Taken from Klages, K. H. W. (1942), *Ecological crop geography* (Macmillan: New York), page 302.

highly relevant to agriculture and agricultural entomology. In Klages (1942) there is a map showing a generalized world distribution of the three major divisions of natural vegetation—namely forest, grassland, and desert (Fig. 1.2).

The main controlling factor that keeps these natural climax vegetation types stable is rainfall:

1. Forest—trees predominant—high rainfall.

2. Grassland—grasses (Gramineae) basically—low rainfall.

3. Desert—scanty xerophytic vegetation (low productivity)—very low rainfall.

Human evolution was very closely associated with the natural grassland areas, often near the edges of forest regions. Cereal crops were developed and evolved in the grasslands, each of the main cereals evolving in one of the main continental grassland areas. People became dependent upon the local cereal as the staple food crop, whose cultivation permitted the development of human populations which in turn lead to the main civilizations. The origins of the main cereals are as follows:

Maize	—New World
Rice	—S.E. Asia
Wheat	—temperate Asia
Barley	—Near East/Ethiopia
Oats	—northern parts of Europe and Asia
Rye	—northernmost parts of Europe and Asia
Sorghum	—drier areas of Africa

Finger millet	—East Africa
Bulrush millet	—Sahel region of Africa
Foxtail millet	—China
Common millet	—east Europe/west Asia
Tef	—Ethiopia

The distribution of animal species throughout the world (zoogeography) at the present time is the result of three separate but interacting factors:

1. Basic climate—mostly influenced by temperature.

2. Natural flora.

3. Evolutionary history of the region; specifically referring to the occurrence of natural barriers and distributional (land) bridges.

The world was divided into six major regions zoogeographically by Sclater (1858) as shown in Fig. 1.3.

The regions defined by Sclater (1858) are as follows:

1. Palaearctic—Europe, Asia, and North Africa ⎱
2. Nearctic—North American, north of Mexico ⎰ lumped together as Holarctic.

3. Ethiopian—Africa south of the Sahara, and Arabian Peninsula.

4. Oriental—India and S.E. Asia.

5. Australasian—Australia, New Zealand, Papua New Guinea, and the Oceanic Islands.

6. Neotropical—Central and South America.

The Palaearctic and Nearctic are very closely related faunistically and are often referred to collectively as the Holarctic region. Some insects are Holarctic in distribution, and others have the genus Holarctic but with separate species in the Palaearctic and

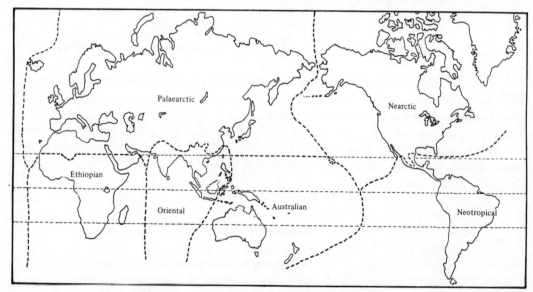

Fig. 1.3 Zoogeographical regions of the world.

in the Nearctic regions. These six regions are separated by oceans, deserts and mountain ranges and so they are quite distinct biologically; the one area lacking a major natural barrier to dispersal is in South China where there is overlap between Oriental and West Palaearctic faunas (and floras).

It is not surprising that the major botanical divisions of the world made by Good in 1953 are essentially similar to the zoogeographical regions, since the underlying controlling factors are basically the same.

Further subdivisions of the world, either faunistically or floristically, are not relevant here.

TEMPERATURE

This really belongs as an ecological factor but it is of such importance in the lives of insects, so directly influences their distribution, and is so easily measured, that it is included here. Insects are poikilothermous (ectothermous) and have only a little control over their body temperature (as distinct from the homeothermous or endothermous mammals and birds); they are directly influenced by ambient conditions. They do have some control in that metabolic heat is generated, and, under crowded or restricted conditions, total heat production can be considerable. For example, honeybees in a hive may collectively beat their wings on cold days to generate heat; bluebottle maggots in a sheep carcass in Australia can raise the corpse temperature to 30°C (86°F). Grasshoppers, at first light, will sit in prominent locations to catch the first rays of early morning sun to warm up their flight muscles for activity—at lower temperatures the grasshoppers are able to walk but neither to leap nor to fly.

Insects (and other poikilothermous animals) can be divided into three broad categories on the basis of their reaction to temperature. It should be remembered that the *preferred temperature* for a species is that range within which the insect thrives and at each end of the temperature range the insect becomes inactive and passes into torpor—both higher and lower than this is the thermal death point at which the insect dies. It is well known that animals can withstand cold relatively better than heat. The three categories are based on the insect's cold death point, and are as follows:

Tropical insects—cold death point at about 10–15°C (50–60°F).
Temperate insects—death occurs at 0°C (32°F), usually due to ice crystal formation.
Arctic/boreal insects—can withstand freezing and survive; body when supercooled turns to "glass" with a noncrystalline nature; blood chemicals (glycols, etc.) permit tissue supercooling.

These definitions apply to insect survival and not necessarily to activity—no insect will be active at 0°C (32°F), but it may well be alive (in torpor) and revivable, and many temperate insects are sluggishly active at 5–10°C (40–50°F). Being poikilothermous their activity will be broadly correlated with ambient temperature, but the tropical species require correspondingly higher temperatures for activity—many tropical species become inactive at temperatures about 18°C (64°F).

Some species of insect are *eurythermal* and will function over a relatively wide range of temperature, whereas *stenothermal* species will only be active over a narrow range of temperature, although this range can be either high or low, as evidenced strikingly by some freshwater insects in thermal springs and others in glacier streams.

Some widespread or cosmopolitan species of insects appear to occur as different physiological races in different parts of the world (although they may be morphologically identical), so that individuals in the tropics may be multivoltine and continuously active, whereas others inhabiting temperate regions may be uni- or bivoltine, with either facultative or even obligatory diapause, at any instar, for over-winter survival. Some of the cosmopolitan crop pests (some aphids, other bugs, some Lepidoptera) fall into this category, which, in part, accounts for their success as major international crop pests.

DISTRIBUTION

Many pest organisms are species that are biologically aggressive and with an opportunist nature in relations to food, and hosts, and generally their distribution tends to be ever-increasing to the limits of suitable environmental conditions. Over recent years most countries have been engaged in agricultural diversification programs so that many more crops are being more widely cultivated than before, and new cultivars and varieties with greater tolerances are enabling even more widespread cultivation. The distributional limits of an insect are usually a combination of food availability and suitable climatic conditions. Thus, at the present time, many insect pest species (and some beneficial species) are not yet spread into all the areas suitable for their habitation and humans have to be careful to curtail their dispersal. International phytosanitary regulations, quarantine regulations, and other legislation, both internationally and nationally, are designed to curb the spread of noxious pest organisms.

Most adult insects have wings and can fly and so are able to surmount many natural barriers to dispersal. Species which are parthenogenic clearly start new populations more easily than most. Adults that feed and are long-lived are likewise more likely to be dispersed more effectively. Larger insects may rely more on their own flying ability for dispersal, but even locusts usually fly downwind once they are airborne. Aphids, thrips, flies, and the like can be carried on air currents for great distances, and several important pest species have very recently been transported across the Atlantic Ocean, presumably on the prevailing wind systems.

In the general vocabulary for the description of distributions of animal and plant organisms are a number of general geographical terms based on the main geographical and biological regions of the world, as shown on the map (Fig. 1.4).

Most of the categories are fairly obvious and clear-cut, but a few are used inconsistently with imprecise limits. One major problem is the usage of Near East and Middle East. The Far East (China, Japan, etc.) is usually synonymous with the Orient.

Many synanthropic species, and some medical and veterinary pests, have all been globally distributed by human activity, as have a number of crop pests, all within the broad climatic tolerances of the particular insect species concerned. But, for many phytophagous insects, their distribution is often closely correlated with that of their natural host plants, and so the basic phytogeographical (Good, 1953) and zoogeographical divisions are relevant to insect studies. In addition to these floral and faunal global subdivisions, a number of other distributional terms are widely used and require brief definition, as follows:

1. Cosmopolitan—worldwide; occurring on several continents (at least two).

2. Pantropical (= tropicopolitan)—throughout the tropical regions of the world (both Old World and New World tropics).

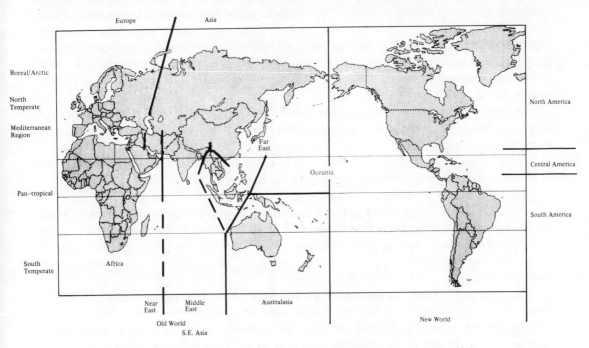

Fig. 1.4 General geographical/biological subdivisions of the world.

3. Arctic/alpine—remnants from the Ice Age glaciations; found either in extreme northern parts, or on mountain tops; tundra species.

4. Boreal (= northern)—cold adapted species of northern latitudes; extreme northern species may be called "Arctic."

5. Temperate—organisms occurring between the Tropic of Cancer and the Arctic Circle, and the converse.

6. Indigenous (= native; autochthonous)—of old stock locally; originating locally so far as is known.

7. Naturalized—introduced by humans and breeding locally.

8. Domesticated—introduced by humans, but only breeding if protected.

9. Introduced (= exotic; allochthonous)—brought into the country by humans; originating elsewhere.

DISPERSAL

Most populations spread away from their source (origin) at times of high population density. With plants, dispersal is important and special dispersal structures (diaspores) are produced in the forms of spores, seeds, fruits, etc.; since plants are immobile they rely on their diaspores mainly for dispersal. With insects, dispersal is often associated with overcrowding and a reduction in the availability of suitable food (such as the progressive drying of leaves or grains), or in response to dwindling food supplies, or sometimes it is a behavioral oddity coinciding with certain weather conditions, or often it might be quite accidental and the insects are carried by air currents.

Dispersal generally appears to be of great importance for the overall survival of the species because new available habitats are colonized, diminished populations can be replenished, and the spread of genetic material through the whole population is advantageous. All animal and plant populations have an innate tendency to increase, and they also have a similar tendency to dispersal. A recent short review on the topic is available (Stinner *et al.,* 1983).

Migration. This refers to the movements of individuals or populations from one area to another that are more definite than just a general dispersal. There are three different types of movement:

Immigration—movement of animals into an area.

Emigration—movement of animals out of an area.

Migration—in the strict sense this is a double journey, first going out of a region to another with the population survivors later returning to the original area. True migration is common with birds and some mammals, and some fish, but is rare in the Insecta.

Migration is typical of colder temperate regions with pronounced seasonal variations, where animals move away from colder northern parts before the onset of the long, cold winter when food is scarce; but there are some spectacular tropical migrations, usually the movement being correlated with local food shortage and abundance elsewhere (African armyworm, desert locust, etc.). Most flying insects are carried passively by winds and air currents, and so the regular annual movements of insects from the U.S.A. into Canada, from South China into Japan, and from the Mediterranean region up into western and northern Europe, are the result of the regular summer air streams in these regions.

In Canada, many insect pests cannot survive the long, cold winter; they arrive from the U.S.A. in early summer, and breed during the hot Canadian summer but die in late fall. A number of rice pests move up the coast of east Asia in large numbers in the spring and are annual immigrants into Japan—arriving in Kyushu in the period mid June to mid July. The brown planthopper of rice (*Nilaparvata lugens*), the white-backed planthopper (*Sogatella furcifera*), and some green rice leafhoppers (*Nephotettix* spp.) (Fig. 8.18) invade Japan annually, and the planthoppers regularly have four generations each summer on the rice, before the populations die out in the fall after the rice harvest. In Europe, moths such as *Helicoverpa armigera* and *Spodoptera littoralis* regularly invade Europe from North Africa and the Mediterranean region, and the polyphagous larvae can be found feeding on a wide range of cultivated plants, but pupae do not survive a normal winter.

Many migratory species spread regularly from area to area using prevailing wind systems, but some tropical species have distinct, gregarious phases when the basic physiology changes and the social-phase insects swarm and disperse collectively, sometimes in vast numbers. African armyworm (*Spodoptera exempta*) and desert locust (*Schistocerca gregaria*) are probably the two best-known species that have migrating swarms. They breed in areas of grassland associated with desert and only irrupt sporadically when weather conditions are conducive to rapid population growth. Flying adults are dispersed by prevailing wind systems, but the larvae march from one area of food to another. In 1988, desert locust populations were carried by high altitude wind systems from North Africa across the Atlantic Ocean to the West Indies—several different groups of individuals were reported as making the journey.

Changing distributions (and new pests). With the spread of agriculture and in particular with the cultivation of new crops, there is inevitably also a spread of some of the insect pests of these plants into the new regions. In many parts of the world suitable climatic conditions for many pests prevail and if the host plant is taken there, then food for the insects is available so it is to be expected that many pests will follow in the wake of the crop introduction, although some will follow quite belatedly. In the early days of world *Citrus* cultivation (in about 1900), the spread of plants, cuttings, rootstocks, and fruits to the U.S.A. (California, Florida, Hawaii), Australia, South Africa and the Mediterranean region included many of the scale insect pests (Coccoidea). But nowadays the dangers of such transportation are understood and international phytosanitary regulations prevent similar occurrences. But such is the extent of international trade in agricultural produce and in tourism that eventually most of the major pests will follow the distribution of the plants. As already mentioned, two basic qualities possessed by living organisms are the tendency for population increase and for population dispersal, so it is to be expected that many pest species will show signs of extending their geographical ranges. There appear to be four main ways in which distribution extensions are effected:

1. Natural dispersal of the pest species.

2. Sudden but natural invasion of a pest species population(s).

3. Deliberate introduction by humans.

4. Cargo contamination/accidental introduction by humans

Within the limits imposed by climatic conditions and food availability, most pest species will slowly spread and extend their distributional range within a country, and will spread from country to country, but this is typically a gradual process. A vivid example of this was seen with a bird species in the U.K. and Europe—the collared dove (*Streptopelia decaocto*) that spread from eastern Europe over a 30–40 year period. The reason for this change in distribution has never been discovered.

Occasionally a pest suddenly appears in a new location in a population large enough for establishment (survival) and procreation; sometimes these insects are carried on a storm wind (hurricane or typhoon), sometimes the invasion is the result of the insect's own powers of dispersal and there often appears to be something of a change in the insect's basic physiology (a new, more aggressive race has developed) but often the insect or animal just appears in the new location with no obvious mechanism of dispersal being responsible. Two recent and very striking examples are to be seen in the U.K. In 1981 the American lupine aphid (*Macrosiphum albifrons*) was recorded in London. Statutory eradication measures were taken but the insect withstood these actions and established itself locally, and then started to spread. By 1986 it had spread to much of England and Wales, as far north as Yorkshire. As far as is known, this aphid only attacks lupines (Fig. 8.50); but lupines are now being grown for seed (they have a very high protein content) and also as a break crop in cereals, and some perennial species are being used as pioneer colonizers on open cast mining reclamation sites. The aphids are very large and very abundant and can kill the host plant, so they are of considerable importance. As yet this pest has not spread to Europe. There are no clues as to how it arrived in the U.K. from the U.S.A., but it seems most likely to have been windborne. More recently, the Western flower thrips (*Frankliniella occidentalis*) has invaded greenhouses in the U.K.—it is a polyphagous pest of flowers (and maize) from America that has successfully resisted all attempts at extermination and is now spreading in the U.K. from greenhouse to greenhouse. Presumably, this species invaded the U.K. on infested plant material (flowers) exported from the U.S.A. The recent advent of palm thrips (a North African species) in the Caribbean (in about 1985) and of desert locust on the West Indies islands (1988) are clearly cases where the insects were transported by high altitude airstreams.

Deliberate introductions by humans were more typical of the last century, when there was little understanding of the far-reaching consequences of such actions. The gypsy moth was imported into the U.S.A. in about 1870 by an entomologist as a possible source of silk. Rabbit and *Opuntia* were taken into Australia quite deliberately, and there are many other examples listed by Simmonds and Greathead (in Cherrett & Sagar, 1977). The recent introduction of the African race of honeybee into South America was an attempt to crossbreed its desirable qualities into the South American race; escapes were inevitable and now the dreaded African honeybee is invading the U.S.A. Accidental introductions through human agency are legion, and this category is by far the largest. One of the earlier classic cases was that of the Colorado beetle taken to Europe (France) from the U.S.A. in a cargo of potato tubers in 1921; it is still spreading steadily throughout Europe and most years it invades the U.K. (see Hill, 1987; p. 45), but it has now been removed from Sweden after a rigorous campaign. This case was followed by a steady succession of similar agricultural accidents over the years and, sad to say, they continue to the present day. Cotton boll weevil has now become established in Brazil and is posing a threat to cotton cultivation throughout South America. In 1989, it was reported that the New World screwworm (*Cochliomyia hominivorax*) was established in North Africa following importation of infested goats from the U.S.A. The adult flies emerged in their new habitat and found climatic conditions ideal; the species is now prospering in the new location and threatening to spread. It could spread throughout the entire continent of Africa. Latest reports (1992) suggest that it has been exterminated after a rigorous international campaign.

Three other recent cases of accidental introduction in Africa have secured their places in the entomological literature as classic examples of agricultural carelessness. Cassava is native to Central and South America, and is reported to have been cultivated in Peru for 4000 years, and in Mexico for 2000 years. It was first taken to Africa (Congo) in about 1570 and its cultivation spread in the 19th century; by the 20th century it had become important as a crop. Now, more cassava is grown in Africa than anywhere else—grown purely for local consumption. Some of the pests of cassava in Africa are African insects that have adapted to the new host plant. Some New World species did make the journey to Africa, but most did not. Generally, in its new location (Africa) cassava has not been subjected to heavy pest attack. But in recent years, three pests of cassava were accidentally shipped from the New World to Africa where they have found conditions ideal—plenty of food, suitable climate, and absence of their usual American natural enemies. Population explosions were inevitable, and these three species have become very serious pests of great economic concern, even to the extent of threatening the continued cultivation of cassava in tropical Africa.

Cassava green (spider) mite (*Mononychellus tanajoa*—**Acarina; Tetranychidae**). This South American species was accidentally introduced into Uganda in 1971, it is thought on cassava stem cuttings imported for propagation. It rapidly established itself for there is much cassava cultivated in East Africa, and spread effectively—mites can be carried by the wind on silken threads. By 1982 it had almost completely colonized the entire cassava belt of tropical Africa (see Fig. 1.5 from I.I.T.A.). The feeding mites bronze the cassava leaves which wilt and later fall, young shoots can be killed, so the end result is often a leafless, stunted plant with a greatly reduced yield of tubers. In 1982, surveys in Uganda revealed that in many locations cassava crop losses of 50% or more were being suffered due to this pest.

Cassava mealybug (*Phenacoccus manihoti*—**Homoptera; Pseudococcidae**). Another South American species accidentally introduced into Africa in about 1973, in the region between Angola and Gabon on the west coast. Now it occurs throughout West and Central Africa through to Uganda, having spread rapidly across the continent

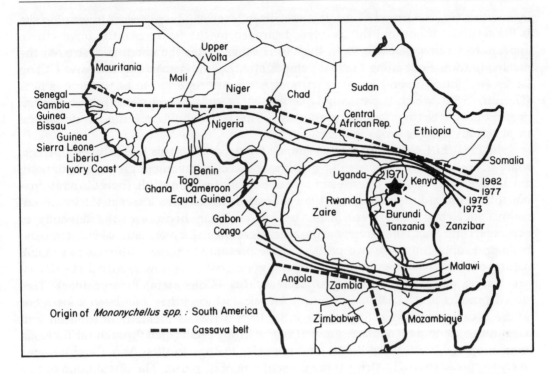

Fig. 1.5 Spread of cassava green spider mite (*Mononychellus* spp.) between 1971 and 1982 (I.I.T.A., 1982).

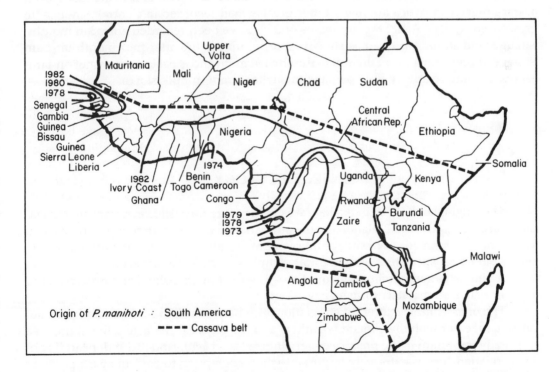

Fig. 1.6 Spread of cassava mealybug (*Phenacoccus manihoti*) between 1973 and 1982 (I.I.T.A., 1982).

in ten years (Fig. 1.6 from I.I.T.A.). It is specific to *Manihot*. The preferred infestation site is the terminal shoot and this is often destroyed by the feeding of the bugs—there appears to be toxic substances in the insect saliva. After the shoot is destroyed the mealybugs move out along the leaf petioles and the undersurface of the leaves. Crop losses (weight of tubers) in many parts of Africa are reported to be as high as 80% (Herren, 1981), which is particularly serious since cassava is a staple foodstuff often grown in dry areas as a famine reserve crop. This is now being regarded as one of the most serious pests of crops in Africa.

In 1981 (I.I.T.A. Report) it was estimated that in the cassava belt of tropical Africa, some 4.5 million ha were infested with either cassava mealybug or green mite, or both, and the annual crop losses were estimated to be $1800 million. C.I.E. (now CAB International Institute of Entomology) in 1986 estimated that cassava losses in Africa due to cassava mealybug alone were $1000 million per year. Because of the difficulty of pesticidal control and the poverty of the area, there is a major international effort into the biological control of these two pests, and the present situation is hopeful as suitable natural enemies have been introduced, and good control is now reported (1992).

Larger grain borer (*Prostephanus truncatus*—Coleoptera; Bostrychidae). This stored product pest is native to Central America, where it has long been known but where it causes relatively little damage. In 1975 it was accidentally introduced into Tanzania in a shipment of maize grain. It is now firmly established throughout Tanzania and has spread to southern Kenya, and now also to Togo in West Africa, and it seems likely to spread through Africa at the present rate of dispersal. The distribution of this pest in the New World includes the native Central America, southern U.S.A., and Brazil, and there are records of interceptions in infested produce throughout the U.S.A., Canada, Europe, and the Near East. A devastating pest of stored maize (Fig. 10.45), it will also feed on dried cassava tubers, and has been recorded damaging (boring) rice, beans, groundnuts and other dried foodstuffs, as well as construction timbers in the food stores. Damage levels in Africa are many times greater than recorded in Central America; in maize cribs in East Africa weight losses of 35% have been recorded after 3–6 months' storage, and stored cassava has shown losses of up to 70% after four month on-farm storage. It prefers maize on the cob to the shelled grain and is basically a pest of on-farm stores, causing relatively little damage to shelled grain in bulk stores.

INSECT ECOLOGY

The original definition of ecology was made more than 100 years ago, and can be summarized as, "the total relations of the animals, and plants, both to their organic and inorganic environment." Nowadays there are many different definitions made by different biological specialists who naturally emphasize different aspects.

A simple general definition would be, "The total relationships of the animals and plants of an area (habitat) to their environment."

The often complex, interlinked relationships between all the animals and plants of an area together with the climate (weather) and the physical characteristics of the area are vitally important in the practice of agriculture (*sensu lato*). And the understanding of these relationships is extremely important when attempting to control insect pests, or alternatively attempting to encourage or introduce beneficial insect species. It should be remembered that basically agriculture is the practice of applied ecology.

ENVIRONMENTAL FACTORS

The environment in respect to animals was defined by Andrewartha & Birch (1961) as being composed of four main factors:

1. Weather (or climate)—temperature, humidity, water, wind, light, etc.

2. Food—organic remains, plant materials, animal materials.

3. Other animals and plants—competition, predation, parasitism, etc.

4. Shelter (= a place in which to live).

It has long been customary on occasions, for convenience, to lump together environmental factors into two broad categories: biotic or organic factors, and physical factors, also called abiotic or inorganic; but this was probably started by botanists for whom the distinctions are more appropriate. For entomologists this categorization is less useful; weather is clearly a physical factor; food and other animals and plants are equally clearly biotic factors; but shelter may come into either category according to circumstances.

Weather (and climate)

Weather refers to the day-by-day conditions of temperature, humidity, rainfall, etc., which affect individual insects, whereas climate refers to average or mean conditions of an area over a long period of time, and affects insect populations and species. As already mentioned, the first three broad climatic divisions are boreal, temperate, and tropical, and there can follow an almost endless series of subdivisions based on different criteria, most of which are not relevant here.

There are also three other basic interpretations of climate of particular interest to entomologists and agriculturalists, as follows:

Macroclimate—general mean climate over a large area.

Mesoclimate—climate within a limited area, often clearly defined, such as a valley, sheltered hillside, nullah, etc., which differs significantly from the mean; such areas are often exploited agriculturally, especially in temperate regions: often both hot and sheltered.

Microclimate—conditions within a small, or very small, area (microhabitat), generally with little fluctuation in relation to the macroclimate; usually with moderate temperature, high humidity, no wind, and low light intensity (shade). Often of great importance for the survival of individuals and for small populations of insects and plants generally, and include such locations as a compost heap, a rotting log, under a large stone, a tree hole and other phytotelemata, etc. The area under the leaves of a plant can be regarded as a microhabitat.

The distinction between a mesohabitat and a microhabitat is often not clear since they intergrade and intermediate situations occur, but the larger and smaller habitats are clearly distinct. Protected cultivation is a means whereby an artificial microhabitat is constructed on quite a large scale for enhanced plant growth.

Temperature. This has already been mentioned because of its prime importance in the lives of most insects, and the ease with which it is measured.

Humidity. Air moisture vapor levels range generally from moist (wet), through moderate, to dry. Humidity plays a vital role in the lives of small terrestrial invertebrates such as insects. Most animals require their body water content to be kept within quite narrow limits, and small animals have a relatively large surface area for water evaporation but a relatively small volume of internal water reserves. Moisture evaporation (dissipation) occurs as a result of respiration, and evaporation is very important under hot (tropical) conditions as the main means of lowering body temperature. With a rise in ambient temperature there is increased water loss, the precise rate of loss being correlated with the level of ambient humidity. In practice it is usually not possible to separate temperature effects from the effects of changes in relative humidity. Most insects require a relative humidity of 70% or more for comfortable living.

In the drier parts of the tropics, or in regions with a pronounced dry season, or during a hot summer in the temperate regions, many insects live in constant daily threat of death by desiccation, induced usually by solar radiation and high air temperatures. The insects usually only survive by remaining in suitable microhabitats with favorable microclimates—thus many are only active at night (nocturnal) or in the early morning and evening periods (crepuscular), and during the heat of the day they hide to protect their bodies from exposure to the sun. To some extent they are also hiding from natural enemies.

Prolonged periods of drought, and also cold, are avoided by various means, such as life cycle synchronization (only eggs or pupae present), migration away from the area, diapause or aestivation while concealed in a microhabitat shelter. Thus, in most parts of the world, insect abundance is correlated with a warm, rainy season, this being especially the case for phytophagous species which require fresh tender plant foliage for larval food.

Most insects do not seem to be affected by very high humidity, but are very susceptible to low humidities because the consequent rapid loss of water leads to death by dehydration. But a few desert species (including some museum beetles—Dermestidae) are actually killed if subjected to quite moderate levels of humidity (40–50% R.H.). Insect species are regarded as *euryhygric* if they can live under a wide range of humidity conditions and *stenohygric* if they survive only under a narrow range of humidity, either high or low.

Water. This refers to free or available water (not atmospheric moisture) and includes ponds, groundwater, and precipitation in the form of rainfall, dew, etc. Whereas atmospheric moisture is vitally important for most insects, free water is less directly important. For mammals, birds, plants, available water is most important, and in many habitats it is often the major limiting factor. Many insects do like to drink from drops of dew in the early morning, but mostly they obtain fluid in their food (plant sap, blood, tissues, etc.), and some manage to suffice with metabolic water. A large number of adult insects do not feed (utilizing larval body fat reserves) but they do like to drink, often from dew drops. A few social species, such as honeybee, do need considerable water during dry seasons both for social consumption and also for nest cooling (*Vespa* etc.).

Soil or groundwater may be of importance in the lives of some insects in that they will drown in waterlogged soils. In parts of the tropics, one method of killing soil insect pests is called the flooded-fallow technique, effective even if only flooded for a few days or a week or two.

Aquatic insects, of course, require free water, particularly for the larval stages, most of which can only survive if immersed; insects that live in temporary pools in the tropics have adopted a wide range of special survival techniques—the chironomid larvae that can be dehydrated are quite exceptional. Saline water is generally inimical to insect life, although there are a number of *euryhaline* species, especially larvae of Diptera. Most of the species to be regarded as "marine" have mechanisms, structures, and behavior that allow them to inhabit salt-marshes, mangroves, and sand dunes, saline or brackish pools with minimal actual contact with the sea water.

Wind. For exposed small animals, such as insects, wind has a strong, drying effect, which is another reason why so many insects seek out suitable microhabitats. The cooling effect of wind is seldom of importance in the lives of insects. However, wind is very important for the dispersal of aeroplankton (or anemoplankton), consisting of tiny insects such as aphids, thrips, small flies, mites on silken strands, and some small caterpillars on silken threads, and it also carries the larger migratory insects such as moths and locusts. The annual summer dispersal of aphids and thrips ("thunderflies") are temperate phenomena characteristic of the cooler regions, such as the U.K., western Europe, and parts of the U.S.A. The spread of insect species is largely due to prevailing wind systems—one of the most striking recent examples is the recolonization of Krakatoa Island after the spectacular volcanic eruption in 1900 when the entire island was devastated and denuded of life. As already mentioned, the recent spread of desert locust and palm thrips from North Africa to the West Indies were clearly cases of wind dispersal; and the regular migrations of insects already mentioned clearly depend upon the prevailing wind systems that occur in most parts of the world in different seasons.

Wind is an important factor affecting local insect dispersal—for example, in the U.K. in the summer, the prevailing wind system is from the southwest, so, when insects invade crop fields, the influx is invariably from the southwest headland. Crop sampling for pests is always based on the presupposition that the immigrants are most highly concentrated in the southwestern corner of the field.

Light. One of the most important factors for plants, this is of little direct importance to most animals, so long as there is enough light for them to see by. The main effect of light on animals is for synchronization of lifecycles to the seasons, especially to ensure that breeding only takes place at the appropriate time. But of all the major groups of animals, the Insecta seem to be the least affected by this aspect of light. Most phytophagous insects are most influenced by factors such as overcrowding, drying of host-plant foliage, and other changes in the host plants, but of course these changes are usually seasonal and will thus be associated with a particular photoperiod (daylength). Insects that are active at night are termed *nocturnal;* species active during the day are *diurnal,* and the twilight species active at dawn and at dusk are *crepuscular.*

Solar radiation (insolation) can be of importance as a source of heat. In temperate regions, early morning in parts of the tropics or at high altitudes, insects have a particular need for heat from the Sun. But very high levels of insolation are both physically damaging (the ultraviolet radiations mostly), and the heat can lead to body dehydration. This is why many tropical crop pests are nocturnal—especially the soft-bodied larvae of Lepidoptera.

It seems that insects active at night are either temporary ectoparasites (micropredators) that are feeding on sleeping vertebrate hosts (bedbugs, assassin bugs, mosquitoes), synanthropic species such as cockroaches avoiding human presence, or species that are avoiding the more rigorous climatic conditions of high temperature, high insolation, lower humidity, and wind, that often prevail during daylight hours. A few species, such as slow-flying large moths and large beetles, are probably avoiding the more abundant diurnal predators. It is not thought that many nocturnal insects are actually avoiding daylight as such.

Food

Being animals, the Insecta require prepared organic food materials. Some are predacious or parasitic and feed on animal flesh or blood and are *carnivorous.* The majority of insect species, however, are *phytophagous* and eat plant materials, either solid or fluid, but some are less choosy and will eat both animal and plant material, and these catholic species are termed *omnivorous*. Quite a large number of insect species feed on decomposing organic materials and can be called *saprophagous* or *scavengers* or *detritivores,* and a number feed on fungi and are termed *mycetophagous* or *fungivorous.* It is now known that many species that feed on decaying vegetation are actually utilizing the fungal mycelium (and the bacteria) and not the actual cellulose.

Some insects are extremely food specific (or host specific) in their diet and are referred to as *monophagous.* Species which will feed on a wide range of foods (host plants) are called *polyphagous,* and those intermediate cases where the food consists of a few closely related types are *oligophagous.* Quality of food is often of importance to insects. Although many insects feed on plant materials such as cell sap, leaf tissues, or nectar, obtaining carbohydrates from these sources, they do require protein, in particular, for body growth and egg development. Some of the best-known examples are female mosquitoes, which happily feed on plant sap but are unable to develop an egg brood without a blood meal first, as a source of protein.

The Insecta are distinctive in the animal kingdom in that the adults of many species do not feed at all—they rely entirely on the extensive body fat (fat-body) laid down by the larval stage which does little but feed. Only some of the reserves are utilized during pupation and the remainder are kept for the adult stage. Some adults drink water (dew drops), and some do not and manage by using metabolic water. Nonfeeding adults are usually short-lived and the oviposition period is short and concentrated, and, of course, such adults do not disperse too far, and the adults rate less important as pests. Some of the adult insect pests that feed and drink can be very long-lived and be very effective dispersal agents. Insects are also very adaptable as far as food is concerned and most species are able to metamorphose (pupate) if food is scarce when little more than half normal size, which is why it is not uncommon to find tiny individuals in insect population studies. Caterpillars and beetle larvae (and others) may have fewer larval instars if food is in short supply, which enables them to pupate sooner. Some species that are often gregarious when the larvae are small become solitary as the larvae develop, presumably to help avoid overexploitation of local food sources and if overcrowded, many Noctuidae practice cannibalism to ensure the survival of some. In some situations there will be competition for food among insects, especially in agricultural situations with some crop pests. Ecologists stress that in most situations herbivores do not use up more than a fraction of their available food, but carnivores more often compete for food, although this seems to be less common with the Insecta than with groups such as fish, birds, Carnivora, etc.

As mentioned already, there are three different basic types of food material available to insects, as presented below:

Organic remains. The insects that feed on decaying animal and plant remains are saprophagous, or may be called detritivores or scavengers. Insects that feed on animal corpses are actually eating protein in the process of decay or drying, or else eating keratin if feeding on the skin, feathers, fur or horns. If feeding on dead insects they are eating chitin which is chemically very similar to keratin. Feeding on keratin and chitin is difficult in that it is a very resistant compound protein and only a very few insect groups have evolved enzymes capable of keratin degradation (Mantidae, some Dermestidae, Tineidae). Insects feeding on decaying plant foliage are still faced with the problem that bedevils the herbivores—namely that they are unable to digest cellulose. Cellulose is a complex polysaccharide that can be degraded into a mixture of starches and sugars. No animals, so far as is known (except for a few Protozoa) possess the cellulases that are required to break down cellulose, and the herbivores that feed on plant foliage require intermediary activity by a range of microorganisms—often a mixture of fungi, bacteria and some protozoans. Most detritivorous insects appear to be of the type that are actually feeding on the bacteria and fungal mycelium that is breaking down the dead cellulose plant material, including both the terrestrial detritivores and the aquatic forms such as the larvae of some Trichoptera and some Ephemeroptera. Usually the insects actually swallow the fragments of vegetable matter, digesting the bacteria and fungi and any suitable breakdown materials, but intact cellulose passes through the intestine unchanged to be voided as feces, to be reinfected with microorganisms, and probably later to be ingested again by another feeding insect.

Many aquatic insects (some adults and many larvae) practice either filter-feeding from the water, or browsing with their mandibles over the surface of stones, twigs and aquatic vegetation. Some of the smallest filter-feeders are clearly only collecting minute particles whereas the larger species will collect particles of a wide range in size and will probably include zooplankton (Protozoa, crustacean and molluscan larvae, etc.) as well as phytoplankton (diatoms, desmids, algae), microorganisms (bacteria, some fungi, etc.) as well as pieces of organic debris. It appears that in many cases particle size is the critical factor in food selection—mattering little whether it is plant or animal, alive or dead. Similarly, most browsers will be scraping off algae, bacterial mats, aquatic fungi, sponges, hydroids, and other small encrusting animals. So, in these cases, it is not possible to make a meaningful distinction between herbivorous and carnivorous feeding.

A number of soil insects, and larvae feeding on tree bark, and in leaf litter, show a similar lack of selectivity in their feeding. A large number of Coleoptera and Diptera have specialized in inhabiting animal dung (cow dung especially). They are found naturally in the savannas of Africa where they utilize the dung from wild ungulates, but some species are now of agricultural importance. In parts of eastern Africa it is reported that a mound of elephant dung can be dispersed by Coprinid beetles (and others) in less than two hours. The breakdown of dung and its incorporation into the topsoil is very important in the maintenance of soil fertility, and it minimizes the utilization of the dung by synanthropic flies. Australia has a major animal-ranching industry (cattle and sheep) but no endemic dung beetles, and one result of the vast quantities of dung on the ground surface is the tremendous fly problem in parts of the Outback. The adult flies are as irritating as the face flies of Africa and can be a great nuisance to humans and livestock alike. Many of the muscoid flies will lay eggs on the dung-soiled wool of the sheep and the hatching larvae cause "sheep-strike" when they attack the flesh—larval myiasis is responsible for serious economic losses. C.S.I.R.O. are now attempting to establish various species of Coprinae in Australia to reduce the severity of this problem.

A few species of insect have specialized in feeding on certain types of animal corpses, and they are included under the heading of feeding on animal materials.

Plant material. The great majority of insects are phytophagous (herbivorous) and feed on plants, one way or another. Technically the fungi belong to the plant kingdom, as also do algae and bacteria, but since these microorganisms are closely associated with litter and humus formation they are usually included with the category of decaying organic matter and detritivores, so there is some overlap in the use of these categories. The main types of plant eaters are briefly reviewed below.

Sap suckers include the Hemiptera, Thysanoptera, certain mites and the larvae of some Diptera; they have mouthparts modified for piercing tissues and sucking out fluids. Some groups and some species have toxic saliva which causes growth distortion or necrosis and death of plant tissues. It appears that all insects have certain enzymes in their saliva, and the sap suckers (and fluid feeders) usually inject saliva into the host during the feeding process. In some cases it seems that enzymes from the insect stomach may also be regurgitated into the host. The end result is that most members of some families cause drastic host reaction when feeding, with tissue death and necrosis being the norm (Coreidae, Miridae, Pentatomidae, etc.). Members of groups such as Aphididae and Pseudococcidae usually do not upset the host plant when feeding, but in either family there are a few species whose saliva clearly contains toxic enzymes and their feeding kills the shoot on which they are located, and the feeding of some Cicadellidae and Delphacidae, etc., produce host symptoms called "hopperburn."

All parts of the plant body may be attacked—leaves most often, but also shoots, stems, roots, and fruits. Insects that suck sap from seeds have a diet rich in protein, but mostly sap suckers imbibe large amounts of water and dissolved sugars. Xylem vessels and phloem systems are often tapped, but many insects and mites just take cell contents; these sap feeders (especially Homoptera) usually have a modified alimentary canal with a short circuit to remove excess water and sugar solution. Feeding cicadas can be seen to eject streams of liquid, but aphids and Coccoidea excrete the fluid as drops of saturated sugar solution (honeydew).

The Hemiptera, as a group, have evolved several different forms of a sucking proboscis from the basic and primitive biting and chewing mouthparts as shown in the Orthoptera/Dictyoptera. Some larvae of Diptera use their mouthparts or mouth-hooks (Muscoidea) to rasp away plant tissues so that they may suck up the exuding sap.

Leaf eaters include the majority of phytophagous insects with unspecialized biting and chewing mouthparts, such as grasshoppers, caterpillars, and beetles. Mostly the insects are actually feeding on the cell sap extracted from the chewed tissues and the cellulose material is excreted as feces—this is why feeding caterpillars produce such a vast amount of fecal matter. Some species specialize in the eating of buds or flowers. A large number of Lepidoptera have larvae that are leaf miners, as do some Diptera, and the tiny maggots have either small biting mandibles or the scraping mouth-hook structures with which they tunnel.

Nectar feeders have the advantage of a rich diet but it is mostly sugar. Both floral and extrafloral nectaries (Fig. 2.1) are involved, and the insects are mostly adults of Lepidoptera, some flies, some beetles, bees, some ants, some wasps, and others. Some insects have their mouthparts modified into a "tongue" (bees), or a suctorial proboscis (Lepidoptera) sometimes to a great length to reach down a long corolla to the floral nectaries (Sphingidae). Some of the nectar feeders may also frequent Homopteran infestations for the honeydew that is excreted. Nectar sources are very important as food for adult insects of many groups for their longevity, and indirectly for their dispersal. Floral nectaries are the main inducement for pollinating insects, and so are of great importance both ecologically and agriculturally.

Fruits and seed feeders. Fruits are typically fleshy and succulent, usually containing sugary fluids (but little protein) and used to attract animals (mammals and birds) in order to disperse the plant seeds. Most of the insect fruit feeders are fly larvae and caterpillars, and a few beetle larvae, but some adult insects will feed on ripe fruits, presumably for the sugar. Seeds (including grains) include the plant embryo and the food reserves, so they contain starch (endosperm), some proteins, minerals, and vitamins, and are an excellent source of a balanced diet. A number of insects feed selectively on the germinal region of seeds which gives them a superior diet, and makes them more important as pests. Most of the insects that feed on seeds are beetle larvae, and caterpillars, but a few are larvae of Diptera; some Homoptera (Aphididae) and many Heteroptera suck sap from seed grains at the unripe "milky" stage.

Wood borers have a feeding problem in that wood has little nutritive value directly, unless the cellulose and hemicelluloses can be digested. The beetle larvae and the few caterpillars and sawfly larvae that bore tree trunks, develop very slowly on this poor diet, taking one to several years for larval development. Usually, all that is digested is the cell sap from the lignified tissues, so a lot of larval effort is expended for a little food—but presumably there is sap seepage when the insect is near vascular tissues. A few timber beetles, however, have symbiotic microorganisms in their intestines which break down the celluloses into sugars which can be utilized by the insect.

Microorganisms are used by a number of major insect pests to break down plant cellulose into carbohydrates suitable for insect digestion—several species of fungi and bacteria are commonly involved. The two main groups of pest insects concerned are the mound-building termites (Termitidae) and the leaf-cutting or fungus ants of the New World (Formicidae; Attini) with their underground fungus gardens. The other termites, and a few timber beetles, have a symbiotic intestinal microflora for direct breakdown of celluloses. Many beetles, Psocoptera, mites, fly larvae, etc. are *fungivorous* and feed on fungal mycelium, and sometimes lichens. A number of stored-products beetles and mites are, in fact, typically feeding on mouldy produce and primarily eating the fungal mycelium, and not the grain, etc. Some caterpillars and fly larvae specialize in inhabiting fungal bodies, either mushrooms or arboreal bracket fungi.

As already mentioned, some soil and litter detritivores, and some aquatic insect larvae (Trichoptera, Ephemeroptera, etc.) ingest decomposing plant materials but only digest the accompanying microorganisms.

Animal material. Carnivorous insects may eat a certain amount of plant material from time to time (often seasonally), but, if any appreciable amount of plant material is eaten, then these species are called *omnivorous*. Carnivores are subdivided into three main categories on the basis of how they obtain their food.

Scavengers are saprozoic feeders, eating dead animals and cadavers, either rotting or dried (according to the climate). Burying beetles (*Nicrophorous* spp.) and the other carrion beetles (Silphidae) are mostly Holarctic in distribution and they specialize in burying small animal corpses. Muscoid fly larvae are commonly found in fresh corpses, hence the expression that meat is "fly-blown". These insects perform a useful function in the rapid disposal of (large) animal corpses, especially in areas where hyenas, jackals, vultures, etc., are absent. One problem is that some of these fly species go a step farther and will invade living tissues of large animals and cause myiasis. The leather/hide beetles (Dermestidae) eat dried skin, hooves and horns, and the clothes moths (Tineidae) specialize in eating hair and wool as well as dried skin and horns (Fig. 12.3). Some termites regularly eat dried bones in parts of Africa.

Predators are by definition carnivorous animals that kill other animals in order to feed on their bodies. The predator is usually larger than the prey which is killed, but there are exceptions. Some parasitologists regard bloodsucking insects such as mosquitoes and other biting flies and bugs as *"micro-predators"* rather than temporary ectoparasites, but this is just a matter of definition. Some mandibulate predatory insects such as Neuroptera have hollow toothed mandibles through which they suck blood and body fluids from their prey. Other liquid feeders (Heteroptera) have digestive enzymes in their saliva and they digest their prey internally before sucking out the liquid mass, in the same manner as feeding spiders. Predacious beetles and others with biting mouth-parts tend to eat out the soft parts of the prey and to discard the remainder. Mantidae are unusual in that they secrete chitinases and so are able to digest the entire body of the insect prey—they typically eat everything but the wings!

Parasites are generally species smaller than their prey (host) and usually do not kill the host, typically just taking some blood. This is true for insect parasites that attack birds and mammals, but insects that parasitize other insects often kill them. Because of the wide range of parasitic insects it is usual to distinguish between several different types, as follows:

- *Ectoparasites* are species that live on the exterior body surface of the host, such as lice, and adult fleas; some are permanent and some temporary (visiting).

- *Endoparasites* live inside the host body, and, more typically, the term refers to parasitic Protozoa, nematodes and tapeworms, but some insects live inside the host body, especially fly maggots in vertebrate hosts, and the Hymenoptera Parasitica whose larvae are found inside caterpillars, bugs and beetles, etc. The adult female jigger flea lives inside the tissues of the human foot at the side of the toenails.

- *Parasitoids* is the term coined to describe the Hymenoptera Parasitica and the Tachinidae (Diptera) where only the larvae are parasitic; it refers to organisms that in their life cycle are alternatively parasitic and free-living.

Other categories used descriptively include permanent parasites, temporary or visiting; obligatory, facultative; cyclical, etc.

Other animals and plants (= the community)

Any one insect individual or species is a member of a large, and often complex, community comprising many different species of animals and plants interacting to a greater or lesser degree, within a common habitat. The usual way in which these interactions are classified is as follows:

Competition. This occurs when two or more individuals or species are sharing a limited resource—the resource may be food, nesting sites, or just space. Andrewartha & Birch (1961) maintained that the term competition be restricted to cases where food or space is limited in relation to the numbers of animals utilizing it, or when one species physically harms the other.

There are two different types of competition: intraspecific competition occurring between individuals of the same species, and interspecific occurring between different species.

- *Intraspecific competition* occurs between individuals of the same species, and often results from the fact that all species and populations have an innate capacity for population increase. Large populations typically suffer crowding, and crowding results in shortage of space and of food. In temperate regions this is part of a regular behavior pattern apparently designed to ensure survival and success of the species. During the short, warm early summer period many insect populations (especially aphids, thrips, etc.) build up to high levels, suffering severe crowding, which culminate in extensive dispersal or migration flights. At the time of dispersal, not only are the individuals physically crowded but often the host plant is drying and foliage dying as the seeds (grain) ripen. In the tropics such annual migrations are far less common, but certain pest species such as locusts and armyworms do sporadically build up dense populations which then have to move off foraging for new food sources. In these cases they use the crowding effect to stimulate a phase of gregarious behavior and then they swarm, but such gregariousness is uncommon in the Insecta. Crowding in a population generally results in reduced adult longevity, fewer eggs laid, and individuals show increased susceptibility to disease (leading to epizootics); in some species cannibalism may occur, and there may be territorial fighting but this is rare with insects. These events are generally linked with an underlying

associated shortage of food, and in some cases seem to have evolved to antici-
pate a probable shortage of food.

   Competition for mates may often be serious with vertebrate animals but
seems unimportant in the Insecta; some of the most successful pest species have
evolved parthenogenesis.

- *Interspecific competition* is between different species, and it occurs sometimes in
   phytophagous species sharing the same food (host) plant, in saprozoic fly larvae
   in animal corpses, and with some insect parasitoids. Usually such competition is
   kept to a minimum by different species occupying different ecological niches.
   For example, most insect parasitoids can recognize an insect host that has
   already been parasitized and will refrain from ovipositing in that host. But a cultivated cabbage plant, for example, may be attacked and eaten by a dozen different animals at the same time and its destruction/consumption may result in
   starvation for some of the less mobile pests. Sheep in pastures tread on insects in
   the sward (turf) and kill them with their sharp-edged hooves, which is an
   indirect form of competition.

   The three main forms of competition between different species of animals
   are predation, parasitism and disease. As already defined, the differences
   between predation and parasitism may in practice be rather obscure. Together,
   these two groups of animals are collectively referred to as *natural enemies.* In
   natural (wild) situations the predators and parasites of a species play a very
   important role in population control. For insect pest species, in particular for
   agricultural pests, the population controlling influence of predators and
   parasites is referred to as "*natural control*" and is of inestimable value. Most
   modern pest control strategies are very much concerned with not upsetting the
   existing natural control, and so far as possible to enhance or augment this effect.
   Sometimes natural control is augmented by the deliberate introduction of
   suitable predators or parasites (or disease-causing pathogens) in what is termed
   biological control (see Chapter 2, page 79).

Pathogens cause *diseases,* and when extensive are referred to as *epidemics* in human
populations, *epizootics* in animal populations, and *epiphytotics* in plants. Most of the
organisms concerned are technically plants, being fungi, bacteria and viruses, but some
Protozoa and other primitive forms of animal life are involved. Diseases typically spread
most effectively in dense populations under conditions of crowding, and they usually
spread in waves, each outbreak leaving a few resistant individuals so the outbreak is
followed by a period of population immunity. The resistant individuals breed up a new
population that is largely resistant but, in the meantime the pathogen will often have
undergone minor mutation and may be once again pathogenic to the animal population. Because of the vast number of diaspores produced, most pathogens have a high rate
of natural mutation. Pathogens are spread by many different methods; by the use of
wind, water, rainsplash, touch, contamination, and through insect and other vectors and
intermediate hosts, etc. Many insect populations are checked by disease outbreaks, especially when the population level is very high. Insects are very important, however, as
vectors of many pathogens causing diseases in human and animal populations, and in
plant populations. Viruses causing plant diseases are all insect-borne (plus a few
nematode vectors)—mostly by members of the Homoptera. Some fungal spores are
transmitted biologically by Heteroptera and by thrips, and many are carried
mechanically by both bugs and beetles. Parasitic hematozoa, many nematodes
(microfilariae, etc.), and a wide range of pathogenic microorganisms are transmitted by

biting flies to both human and animal populations, and also lice, fleas, some Heteroptera, some mites and many ticks. Selections of different pathogens transmitted by different insects and Acarina are listed in Chapter 3, Tables 3.1, 3.2, and 3.3.

Shelter (= habitat; a place in which to live)

The precise habitat of an insect may be very broad, as with some polyphagous grassland grasshoppers (Acrididae), or it may be extremely specific, as with a fig leaf-gall psyllid which makes (and inhabits) a leaf gall on only one species of *Ficus* in parts of S.E. Asia. The habitat of each insect species generally has to be investigated separately, for the total range of available habitats is far too broad to permit generalization. An animal that inhabits a wide range of habitats is termed *eurytopic* (cf. *stenotopic*) and is generally very tolerant physiologically. In most families of Insecta there are some species that are very eurytopic and are cosmopolitan in a wide range of habitats (a good example would be *Myzus persicae*) and, at the other extreme, some species are only found on one species of plant on part of one continent; and of course there are many intermediate cases.

But some generalizations can be made concerning insect habitats, and some definitions need to be given, so in the following pages the main habitat types generally found in most parts of the world are listed.

Community requires definition. The habitat is the physical area inhabited by a collection of different animal and plant species, all in their respective *niches,* and all these species together constitute the community. The niche is defined as the basic biotic position of the insect in the community, in other words its lifestyle or way of life. Categorization starts very broadly, for example, a herbivore, or a parasite, and the definition can be made in further detail at several levels until a specific interpretation is reached; for example, from the level of "herbivore" down to "a caterpillar leaf folder (eater) on *Microstegium* grass in the ground flora of rain forest in South China." Or from "parasite" down to "mallophagan ectoparasite in the secondary wing feathers of Corvidae in Europe." The basic plan for all natural communities is similar, in that there will always be a range of plant types, and animal herbivores, detritivores, carnivores, parasites, etc., but the types of animals (and species) will differ from one community to another. For example, the top carnivore in the community foodweb might be a mammal, a bird, a fish, a crustacean, or an insect, etc., according to the community—it could even be a hydroid or some form of large protozoan. An ecological community can be analogized to a human township, which has the same basic structure in all parts of the world. Most habitats have a distinctive natural boundary, such as a lake or a copse, and, in the agricultural context, each separate field is a habitat containing its own particular community of plants and animals; of course, most agricultural communities are characterized by their basic simplicity.

HABITATS

The first basic division of the Earth's surface into major habitat types, called *biota*, is the separation into terrestrial, freshwater, and marine biota, and these can be subdivided almost endlessly, but the initial subdivisions will suffice for the present purpose.

Terrestrial habitats
—soil (including litter)
—grassland
—scrub
—forest
—desert
—agricultural
—urban

Freshwater habitats	—lotic habitats (flowing water—rivers, streams, ditches, etc.)
	—lentic habitats (standing water—lakes, ponds, meres, paddy fields, etc.)
Marine habitats	—littoral (seashores)
	—sublittoral (offshore shallow water)
	—pelagic (oceanic)
	—benthic (sea bottom)

There are basic differences between these habitat types that are important and influence the insect fauna so that certain groups are characteristic of each, as indicated below:

Terrestrial habitats

Soil habitats. These generally include surface litter, topsoil, and subsoil; the deep substrate is of prime importance, whether it be rock, sand, clay, for it determines the nature of the subsoil and also topsoil, and if impervious it impedes plant root penetration. The main soil features as a habitat will also vary according to the vegetation, but there are features of all soils that are important both agriculturally and as habitats for insects. The important soil features include: depth, texture, particle size, drainage, aeration, humus content, fertility, and also slope, aspect, groundwater, pH, etc. Temperate soils usually have a deep litter layer on the surface, with leafmold underneath. Most tropical soils tend to have a rapid litter degradation because of the high temperatures and usually high rainfall, and there is less leaf-fall as most trees are evergreen, so, in general, there is little litter accumulation, and little leafmold. Mostly, the topical soils are moist, warm, rather acid, with little humus, and, in many places where deforestation is in progress, surface erosion during the rainy season may be serious. Tropical soil insect fauna is usually not as rich as in many temperate situations.

The insect species concerned are mostly Collembola, Thysanura, crickets, cockroaches, termites, ants, beetle larvae and pupae, fly larvae and pupae, and a few lepidopterous larvae but many pupae. The soil faunal composition will, to quite a large extent, be determined by the type of vegetation growing on the surface. Thus grassland soils often have enormous numbers of white-grubs (larvae of Scarabaeidae), wireworms (larvae of Elateridae), dung beetles, and other beetle larvae, and there will be no fruit flies (Tephritidae) but there will be some pupae of shoot flies. Soil fauna can be categorized ecologically: some species only live in the litter, some make nests underground, and at the other extreme some live permanently underground (some Collembola, etc.), and many use the soil as a suitable microhabitat for pupation (Lepidoptera, Diptera, Coleoptera) or for egg-laying. Many of the nest-makers (crickets, ants, termites) forage on the surface. Detritus-feeders predominate in soil and litter, but many soil insects feed on plant roots; they are accompanied by many predators such as Carabidae and Staphylinidae.

Soil insects are classified according to size:

- *macrofauna*—body size 2–20 mm (0.08–0.8 in.); they are active burrowers.

- *meiofauna*—body size 2 mm–200 μ; mostly pass between soil particles without disturbing them.

- *microfauna*—body size less than 200 μ; mostly Protozoa, swim in soil particle water films.

Litter species are mostly flattened in form, thigmotactic, cryptozoic and nocturnal in behavior, but the proper burrowers show various anatomical modifications for their fossorial life, as exemplified by the mole crickets (*Gryllotalpa*) (Fig. 7.7).

Grassland habitats. Grasslands are characterized by vegetation composed principally of grasses and other graminoid plants, sometimes with a few scattered shrubs. This is a climatic climax controlled largely by a low rainfall with a long, dry season, and/or shallow soil. The vertebrate fauna of grasslands is quite characteristic, but with insects it is far less distinctive. Predominant groups are short-horned grasshoppers (Acrididae), some termites (Termitidae), leafhoppers (Cicadellidae), planthoppers (Delphacidae), Cercopidae, certain species of Lepidoptera (Hesperiidae) and Diptera which feed inside grass shoots and stems, and beetles of the family Scarabaeidae whose larvae live in the soil and feed on grass roots. If there is animal dung then there is usually a large population of dung beetles. If the sward includes clovers and other forage legumes there will be blue butterflies (Lycaenidae).

Most of the insects that live in these habitats are diurnal in habit, and many are saltatorial and can jump strongly; sight is an important sense.

Scrubland habitats. These are rather difficult habitats to define, for in most situations, they are essentially transitional seral stages between grassland and forest. But in certain areas, often with shallow sandy or chalky soil, and low rainfall, this may be a natural climatic climax. On a large scale, it is more typical of the southern temperate regions; some of the best-known examples are in southern Europe and parts of the U.S.A. The vegetation is usually xerophytic, and there is often a characteristic vertebrate fauna, but the insect fauna is seldom distinctive.

Forest habitats. Over much of the world's land surface the natural climax vegetation is forest; an area inhabited more or less continuously with trees, often with interlocking canopies. In the Taiga, the forest is needle-leaved evergreen (usually), the canopy designed to withstand the mantle of snow likely to persist throughout the long cold winter. In temperate regions, the forest is broad-leaved deciduous, mostly of oaks and their relatives, with beech in areas having a calcareous substrate, and birch in wet acid areas; there are usually only a few dominant tree species in temperate forests. The understory shrubs, climbers, and the ground layer herbaceous plants, are important members of the plant community especially as host plants for phytophagous insects. But certain minimal levels of rainfall and soil depth are required to support tree life-forms. Over a large part of the tropics the vegetation is tropical rain forest with its tall, luxuriant growth, characterized by a high continuous canopy, very pronounced vertical stratification, and a floral composition of great diversity having some 10–20 co-dominant tree species. The insect species in rain forests are greatly diversified, some being adapted to live on or in the dark forest floor (mostly detritivores), others live on, or tunnel into, the tree trunks and branches or fly between the trees in the deep shade. The great majority however, live in the canopy where the leaves, flowers, fruits of the trees, climbers and epiphytes provide the main sources of food. Here are found the insect herbivores in great profusion, and, of course, their predators and parasites too. Many of the canopy insects will never in life descend to the darker lower levels of the forest, but remain at a height of some 30–50 m (100–160 ft.). However, some of the spectacular moths and beetles may pupate in the forest soil, and the noisy giant cicadas have subterranean fossorial nymphs.

Forest insects show a mixture of adaptations to their mode of life; some are nocturnal and some diurnal, many are cryptic and hide in the foliage or on the tree trunks; they are basically climbers.

Desert habitats. Cold deserts are characteristic of northern regions and the extreme south, and some are found at high altitudes; and, as with other deserts, very low rainfall is the main controlling factor but, with low ambient temperatures both flora and fauna are even more sparse than in hot deserts. But in the Arctic tundra during the very short summer conditions are warm enough to melt the snow and surface ice, and many Diptera breed in these shallow pools in enormous numbers.

Hot deserts are found in the tropics, characterized by very low rainfall, little standing water or, at best, only occasionally temporary pools, a sparse xerophytic flora, and an equally impoverished fauna with pronounced behavioral and physiological modifications which permit their survival in this hostile habitat. A totally dry desert obviously has no life-forms whatsoever, but most deserts will have occasional rain most (or some) years which permit a temporary blossoming of life. Desert insects are mostly inactive during daylight hours and shelter under grass tussocks or shrubs, or burrow into the sandy substrate to avoid the high daytime surface temperatures (which would be lethal). At night the animals emerge to forage and feed. The insects are able to survive usually without actual drinking, metabolic water being sufficient for their needs, but many desert areas have dew precipitation at night from which the insects can drink. The insects living here generally have few anatomical modifications for their particular mode of life—the necessary modifications are mostly physiological and behavioral. Many of the plants here are annuals that flower briefly when rain does fall, and some insects are adapted to feed on seeds.

Some arid areas have temporary pools of fresh or brackish water which may have a sparse but specialized insect and crustacean fauna with adaptations for survival over long, dry periods. The most remarkable are probably the chironomid larvae that can withstand complete dehydration and remain viable.

Agricultural habitats. Using the term "agricultural" in the wide sense refers to the cultivation of plants and the rearing of animals for food, for specific products such as timber and wood, cotton, rubber, coffee, eggs, milk, skins, furs, etc., and for ornamental purposes. The main different aspects are:

Agriculture (*sensu stricta*)—cultivation of field crops.
Horticulture—cultivation of fruits, vegetables, plantation crops (coffee, tea, sugarcane, coconut, etc.), and ornamentals; protected cultivation.
Forestry—trees grown for timber and fuel wood; afforested areas often include important watersheds and may surround reservoirs.
Animal husbandry—rearing of cattle, sheep, goats, pigs and poultry, etc.; also includes apiculture, sericulture, etc.
Fish pond culture—in both freshwater and brackish inland ponds, both natural and constructed; mostly for fish of the carp family, but also shrimps and other crustaceans; fish rearing in paddy fields and irrigation ponds and ditches.

In each of these categories, humans are aiming at extensive monocultures of plants and animal communities under conditions as near ecologically optimum as possible, for maximum sustained yields, and this induces explosive population growth of many insect species—mostly these will be pest species, but there may also be some beneficial species. The most serious entomological aspects are the population increases of phytophagous insects that feed upon crops and cultivated plants, and of parasites of domesticated animals. The more widespread and intensive agriculture becomes, so the more serious the pests become as they are provided with conditions suitable for unlimited population increase. It can be said that virtually all agricultural pests are the product of agricultural practices and thus are self-induced.

During the development of agriculture worldwide, forests were destroyed to provide arable land and over the millennia many of the forest insects adapted to live on the cultivated relatives of their former wild hosts.

Urban habitats. Two of the main results of the continuing human population explosion are the increase in the extent of agriculture, already mentioned, and the concomitant increase in urbanization for the accommodation of all the extra people. The loss of natural habitats to urbanization will cause some ecological disruption and, in the vicinity of urban conurbations, foraging for fuel wood will strip the hillsides of woody vegetation. Urban animals, such as rats and insects, will increase as their habitats increase. All the disturbed ground produced by urbanization, and the practice of agriculture, is typically colonized by *ruderal* plants (sometimes referred to as *weeds*), which are in turn colonized by certain species of insects, which can be expected to continue to increase in numbers. Urban insects either share human food, or else they suck the blood of people and domestic animals, or else they feed on rubbish. A few feed on clothing and possessions, and a few on the materials used in building. There is normally no way in which an urban species of insect can be recognized as such by either anatomy or behavior; they are opportunist species that have adapted to this ever-increasing habitat. The main groups of insects found in human dwelling places include crickets, cockroaches, some bloodsucking bugs, lice, fleas, stored-products beetles and moths, mosquitoes, house flies, ants, wasps. There will be other species in the gardens, ponds, roadsides, parks, and other man-made (urban) habitats.

Freshwater habitats. The main groups of freshwater insects are, of course, the orders Ephemeroptera, Odonata and Trichoptera, and some families of Heteroptera, Coleoptera and Diptera, but with some species from almost all orders of insects. In some cases all instars are aquatic, but with the Diptera for example, only the larvae and some pupae are to be found in the water. Some are herbivorous, some are detritivores and many are predacious and feed on other insects, other invertebrates and some prey on small fish and amphibia.

The main method of group characterization for freshwater insects is based upon their life-style, as follows:

Neuston—walking on the water surface, or just under the surface film.
Plankton—small forms floating freely in the water.
Nekton—free-swimming active forms.
Benthos—bottom-dwelling forms:
        On the surface of the substrate.
        Under stones and boulders.
        Burrowing in the substrate.
Periphyton—attached to objects (plants, stones) submerged in the water.

The greatest diversity of forms and numbers of species, as well as numbers of individuals, are to be found in ponds, and a lesser number in lakes; the numbers in rivers are generally rather low because of the problem of resisting the current. But a few species have adapted totally to riverine life and are only found in these habitats—the most extreme cases are the "web-trap" caddis larvae and the *Simulium* larvae attached to rocks in stream torrents, and some species of Ephemeroptera and Odonata have species adapted for life in flowing water.

Marine habitats. The seas are basically a habitat inimical to insect life, and are the evolutionary domain of the Crustacea. But a few species of insects have adapted to live in brackish water, saline rock pools, in the substrate (mud) of mangroves and salt-marshes, in the sand of sandy and boulder shores, in seaweeds, and associated with some marine animals. Most of these species, however, manage to keep actual contact with the seawater to a minimal level. The sea skaters are interesting in that they can be seen on the sea surface miles from shore, but most are coastal and they do not normally submerge.

Symbiotic associations

Although this is not a habitat as such, many insects live in association with other organisms and a few definitions should be made as to their life-styles. In recent years the term symbiosis has become accepted as the generic term that covers the four more specific categories of relationship. These are the cases where an insect lives in close and intimate association with a plant, another animal, or another insect, in a special partnership that is usually beneficial to at least one of them. Four main types of relationship are usually recognized.

Commensalism is when one organism lives with another and shares its food; the one species thus benefiting from the association but the other not being harmed in any way. The most common examples of this type of association are vertebrates and some invertebrates other than insects.

Inquilinism is regarded as a form of commensalism where one species lives in the home or the nest of another, and obtains a share of food. The term was coined to describe the relationships that occur in ant and termite nests. It is now also used where insects "share" plant galls induced by some Diptera and Hymenoptera Parasitica. Certain cockroaches, crickets, termites and ants (and a few lepidopterous larvae) are only to be found living inside the nest colonies of ants and termites.

Parasitism refers to an organism living on or within another to its own advantage in terms of shelter and food. Previously, it was usual to add that the host animal (or plant) was injured in some way by the parasite, but recent studies are indicating that many well-established parasites do not in fact cause detectable harm to the host and, in some cases, the host appears to utilize various substances produced by the parasite. Of course, the less well-adapted parasites do still cause their host damage and in extreme cases death, but it can be argued that the evolutionary culmination of a symbiotic (parasitic) relationship is mutualism. Relative body size is a major factor in relation to harm, and it seems that insects that parasitize other insects cannot be easily compared to vertebrates carrying protozoa and nemathelminthes, or even insects. In an insect/insect relationship, because of the similarity in size, it is almost inevitable that the host will be killed.

Mutualism is a relationship between two organisms in which both parties derive a definite advantage; in its extreme cases there may be complete physiological interdependence, as with fig wasps (Agaonidae) and plants of *Ficus* (Moraceae).

In a few old texts, it may be found that phytophagous crop pests are sometimes referred to as being parasites of crop plants; but the usual eating of plant foliage by insects is better regarded as being "grazing." If a gall is induced on the plant however, this is usually referred to as parasitism.

INSECT POPULATIONS

In field studies of insect pest species, very often the single most important aspect is the size of the population, or the numbers of that species in any particular area. Clearly, the size of the pest population is the prime factor in determining the extent of damage to the host crop (or livestock).

All plant and animal populations show an innate capacity for increase, and in pest species this capacity is particularly pronounced. The study of animal populations is often referred to as *population dynamics* in order to emphasize the basic dynamic nature of populations. Few populations are ever static—they are either increasing or decreasing, sometimes with a cyclical rhythm. With human populations and trees, fluctuations cannot be easily observed because of the time factor—the life cycle is protracted, body size is large, body growth is slow, and so many years are required to observe direct population changes. But with insects which are small and develop rapidly, population changes can be observed in a matter of weeks or months.

A very simple population expression is as follows:

$$P_2 = P_1 + N - M \pm D$$

P_2 = the final population size
P_1 = the initial population size
N = natality; the population increase factor; birthrate
M = mortality; the population decrease factor; deathrate
D = dispersal; immigration—movement into the population, emigration—movement out of the population

In the study of pest populations it is necessary to be able to assess whether the population is increasing or whether it might be decreasing, for this may determine whether or not any control measure is applied. One of the aims of agricultural entomology is to be able to reduce insect pest populations most effectively, to levels at which the damage done is unimportant. The ultimate aim of course is often to *avoid* pest infestations arising in the first place. But when infestations cannot be avoided, then the object is population management. The three main ways of population reduction are clearly: to reduce the birthrate; to increase mortality; and to either reduce immigration or to encourage emigration, or both.

A major point of importance is that any pest (species) is only really a pest (of economic consequence) at or above the particular population density when economic damage is done. And, typically, the control measures employed against it are designed only to reduce the population below the level at which the insect is regarded as an economic pest. Only occasionally is complete population eradication aimed at (and even less often is it achieved!). In Chapter 3 these basic points are defined. Allee *et al.* (1955) produced a simple schematic representation of the growth of a population which follows (Fig. 1.7).

Several important features are evident from the graph. The initial growth curve (Stage I) of a population is always *sigmoid* (referred to as a *logistic curve*) with the central part exponential. When the initial growth of the population ceases there is a period of equilibrium, of numerical stability, termed the *asymptote*. However, this period of time is usually momentary and is followed by a series of population changes, either more or less regular oscillations, or more irregular fluctuations, which often continue indefinitely in many tropical situations. In temperate countries, because of the onset of the cold rigorous winter period, most insect populations show the full range of development depicted in the graph. But most populations do not become extinct, active forms mostly die off, and the population survives as diapausing eggs or pupae. All practicing entomologists have experience of a sudden population of a particular species of insect (usually a pest) appearing locally as an enormous infestation, and suddenly after one or two generations it just as quickly disappears. One example was seen in Hong Kong in 1976 in a row of five *Bauhinia* trees which suddenly developed a total leaf infestation of blackfly (*Aleurolobus murlatti*); the infestation lasted almost six months and then disappeared completely, and was not repeated during the following four years. Thus it is

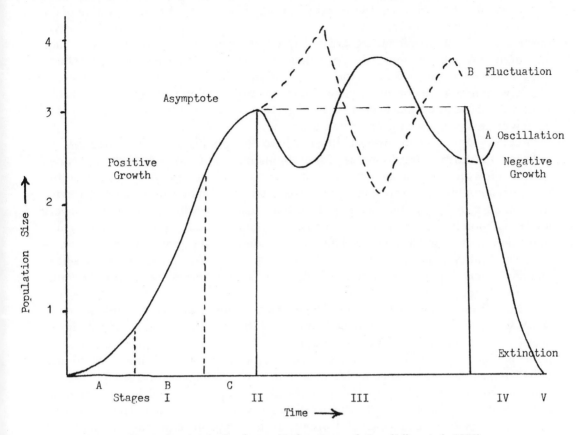

Fig. 1.7 The stylized growth of an animal pest population (Allee *et al.,* 1955)

Stage   I Period of positive, sigmoid growth; population increasing.
        A Establishment of population.
        B Period of rapid growth (exponential growth).
        C Growth rate slowing; period of numerical stability (asymptote) approaching.
    II Equilibrium position (asymptote); numerical stability.
   III Period of oscillations and/or population fluctuations.
   IV Period of population decline; negative growth.
    V Extinction

Hypothetical Pest Population Densities (at level when economic damage is done):
1. Antestia bugs on arabica coffee—threshold of two bugs per unsprayed bush.
2. A fairly serious pest such as *Helicoverpa armigera* on many different crops.
3. A pest occurring in larger numbers—leafhoppers or aphids.
4. An irregular, sporadic pest—desert locust; African armyworm; cutworms.

clear that insect populations do not remain static, they are constantly either increasing or decreasing in size. This basic understanding is vitally important in the study of insect pests and pest control.

The population fluctuations that are so characteristic of most insect infestations, and which bedevil control programs, are usually of unknown cause and are really little understood. There has been a great deal of research done in the last 50 years in several regions of the world, on a range of fundamental aspects of animal populations. Much of the work has been on insect populations because of the convenience of their rapid development and brief life cycles. The laboratory experiments have provided very useful data on how some insect populations develop under some controlled conditions. But, under field conditions there are apparently so many interacting factors involved in insect population fluctuations that our understanding is still very limited.

It has long been convenient to group important environmental factors into two categories: *density independent* factors which operate on populations irrespective of the size of the population, and aspects of weather (or climate) are the main factors. In parts of the tropics and subtropics there are not infrequently periods of drought and high temperature which devastate insect populations and, in these situations weather conditions are often the overriding factor in population control. In the U.K. in the late fall, clearly weather is of major importance in that the onset of winter kills all exposed active insects, but this is an annual event of great regularity and most life cycles have evolved to cope with this period of inclement weather.

Density dependent factors are those where the population size influences the outcome and further development of the populations, and the usual factors grouped here are crowding, food availability, predation, parasitism, disease outbreaks, and other forms of competition. Most of these effects are pronounced at high densities but, alternatively, some factors operate when populations are low or very low. For example, at very low population levels sexual isolation can be pronounced and most very small populations die out of their own accord. The literature on animal population studies is now quite vast and often bewildering in its detail, but few underlying basic principles have emerged from this plethora of publications. It would not be relevant to delve deeper into this topic here.

Population dynamics theory has changed drastically over recent years, especially since the recruitment of several well-known mathematicians to the field of insect ecology. A recent summarization by Southwood (1977) points out that although some insect populations may change dramatically in numbers, most of the time most populations are relatively stable in comparison with their prodigious powers of increase. It is commented that this spectrum of population change can be viewed in relation to the duration of habitat stability and has been expressed as the *r-K continuum*. The r-strategists are insect species that are basically opportunists, living in temporary habitats, and adapted to obtain maximum food intake in a short time; they are usually small in size, mobile, often migratory, and have a short generation time. K-strategists live in stable habitats, often under quite crowded conditions, and with their population near the carrying capacity of the habitat; they are usually larger insects, with less tendency to migrate, and with a longer generation time: usually living in equilibrium with their habitat resources. Obviously these two types represent extremes of this continuum and many insect species occupy an intermediate position with regard to these characteristics. Further information on these concepts of population dynamics, and also how they relate to aspects of pest control, can be found in the publications by May (1976) and Southwood (1977). The study of insect populations has now become very mathematical and recent advances have been made in application of computer models to simulate insect population dynamics.

Recent population studies have developed the theory of chaos in biological systems. This refers to the element of unpredictability inherent in many biological systems, and it is sufficiently apparent, it seems, to have to be taken into account in many pest population studies.

MIXED SPECIES POPULATIONS (species diversity)

In many teaching establishments there has been a tendency, when dealing with animal (insect) populations, to give undue concern to the idea of *competitive exclusion* and to stress that no two species can occupy the same ecological niche without one species (the "stronger") replacing the other (the "weaker"). This appears to be true enough but we tend to view the niche occupied by the animal far too broadly, which has been a drawback in the teaching. Thus when a particular crop pest species is found *in situ* on a host plant, it is too often assumed that the population consists only of that species. In practice, gross definitions of niche are usually clear (e.g. leafminer, fruitborer, etc.) but precise niche limitations appear to be very fine and subtle, and often not at all obvious. Field experience shows that some insect populations are not of just a single species; often several genera may be involved and they are usually quite distinct. In the U.K. in the fall, foliage of *Rosa* and *Rubus* carries large leafhopper populations (mostly *Empoasca, Edwardsiana, Zygina*)—the genera are distinct because of color differences, but there may be several species of some of these genera involved and at first sight they are not at all obvious. When the insect pest population is composed of several closely related species (of the same genus), presumably in their separate, closely related niches, this diversity can be very difficult to see unless a very careful study is made. In field studies it should be expected that some natural insect populations are likely to be composed of more than one species; these species may be closely related so that they are similar in appearance and habits and it may be quite difficult to distinguish them. In extreme cases the species can be indistinguishable morphologically, but will differ in genital structure, or behavior, or physiology. In the past, many ecological studies have failed because the recorder was unable to recognize a mixed-species population.

Closely related species represent a major taxonomic problem for field entomologists. Taxonomic characters are sometimes quite esoteric, usually comparative, and the structures involved vary from group to group. Some species can be recognized in the field, using a hand lens, but others need manipulation under a stereo-microscope in the laboratory, and some require microdissection and slide mounts.

Apart from the closely related species that may be present on the host (plant) there is always a problem with other species of the same genus, native to the area, that usually occur on other plants. Light-trapping, sweeping, and other forms of population sampling will invariably result in the collection of some of these species, and some of them will probably stray onto the crop hosts from time to time. In the U.K. there has been a population boom in the blossom beetle (*Meligethes*) since the increased acreage of rape crops; agriculturally usually only two species are mentioned (*M. aeneus* and *M. viridiscens*) but there are 33 species of *Meligethes* recorded in the U.K.! Similarly, with the increase in production of pot plants in the U.K., the occurrence of *Otiorhynchus* weevils, both as larvae in the pot soil, and as emerging adults, has greatly increased; there are 19 species of *Otiorhynchus* concerned, with larvae that are virtually indistinguishable although some adults are somewhat distinctive.

Considering the host organism as a whole, it is common to find different members of the same family of insects on different parts of the host body. Clearly this can be regarded as a population in the broad sense. Citrus plants often have aphid colonies on their roots, others on flush leaves and soft young shoots, and sometimes others on older leaves. Worldwide, 14 species of Aphididae are recorded on *Citrus.* Similarly, Coccidae may be present on the fruits, leaves, and green twigs—several genera are usually involved. But usually the closely related species will be inhabiting the same locus on the host. Some ectoparasites show a similar zonation on the host; many birds (e.g. pigeons) have several different Mallophaga in their plumage—one genus in the flight feathers, another on the head, and another one or two on hind parts of the body.

LIFE-TABLES

One of the more useful tools in studying insect pest infestations is the compilation of the life-table. This separates the population into the different age-group components (i.e., egg, larva, pupa, adult) and life expectancy can be studied in relation to various sources of mortality. The growth of an insect population, involving recruitment and the survival of the different stages, appears to vary considerably according to the type of insect concerned, and thus there are several different approaches that can be taken (see chapters 10 and 11 in Southwood, 1978).

The life-table concept was developed by demographers initially to study human longevity and the changes in populations due to births and deaths from different causes, and it was introduced into the ecological literature in 1947. A very recent paper by Carey (1989) looks at cause-of-death analysis in ecology and includes a very useful list of references.

2

Beneficial Insects

Most of the time entomology is concerned with the damage caused by insects and their inconvenience to humans in many ways, but it cannot be denied that in several ways they are distinctly useful in human society. In the early days of human evolution, it seems clear that insects were a major part of the diet, especially as a source of animal protein and fat, and vestiges of such habits are still to be seen in most parts of the tropical world. And, of course, the importance of insects in natural food chains and food webs in all terrestrial and freshwater habitats cannot be overemphasized.

For present purposes there are four main ways in which insects are beneficial, and these are reviewed briefly below, and also a fifth category of less important but interesting miscellany.

POLLINATION

In practical terms it is sometimes difficult to dissociate pollination from apiculture since the two are practiced more or less simultaneously, but for teaching purposes they are best viewed separately.

Most plants need pollination to produce seed—in other words, pollen from one plant has to be deposited on the receptive stigma of another plant in order to develop embryos for the seeds for procreation. A few plants practice self-pollination (pea, groundnut, arabica coffee, many Solanaceae, etc.) and some rely on vegetative means of propagation (sugarcane, etc.), and some use parthenocarpy to set fruit (banana, etc.), but most plants need cross-pollination.

The Gramineae (cereals, grasses) and many trees (especially Gymnospermae) are wind-pollinated and termed *anemophilous*. On days when pollen is shed by these plants the air is yellow with the pollen dust and the foliage of trees takes on a yellowish hue; most species have synchronized pollen release controlled by local weather conditions and, for a few days, the whole area may be quite yellow. Synchronized pollen release increases the chances of successful pollination when the wind is the only agency involved.

The plants pollinated by insects are called *entomophilous* and include most dicotyledonous plants and specifically most nonconiferous trees, most shrubs, most vegetables and herbaceous plants. This means that fruit trees and bushes, most vegetables, flowers, some legumes, cocoa, tea, cotton, etc. are insect-pollinated. A few crops are partly entomophilous and partly anemophilous, and there is a surprisingly large number of plants where there is uncertainty as to the precise manner of natural fertilization.

The value of insect pollination to human society is clearly vast and not really to be calculated, but it was estimated in the U.S.A. in 1986 that the value of insect-pollinated crops was in the region of U.S. $17,000 million. This was for a total of about 90 crops. If seed production in forage legumes grown for livestock food is included, then about one-third of the U.S. food supply depends directly or indirectly on pollination by insects.

With the ever-increasing spread of agriculture, destruction of hedgerows, increased urbanization, and similar activities that are taking place worldwide, there is a constant reduction in the extent of natural habitats; many insects are becoming less abundant as a result, and some species are becoming quite scarce. At the same time the acreage of some crops has increased considerably as more and more monocultures are established. Fortunately, the vast fields of cereals that in some regions stretch from hori-zon to horizon do not have to rely on insects for pollination—just wind! But when a crop suddenly leaps into prominence, as happened with oil palm in S.E. Asia, and rape in the U.K., and, in the past, with apples in many countries, then in some areas there may be pollination problems.

Most pesticides (possibly all) are harmful to bees, and many are particularly toxic to them, both to honeybees and wild bees. In the early days of the organochlorine boom (post-1940) the susceptibility of bees to these poisons was seriously underestimated and, before science had caught up with the basic implications, there was widespread destruction of pollinators in many regions. The scientists and farmers regard a range of insecticides as presenting minimal hazard to bees—when used as directed. But some of these less toxic chemicals, which do not actually kill the bees, may be carried back to the nest on the insect body, and some will cause the guard bees at the nest entrance to repel the returning foragers for they have lost the characteristic smell of the colony. Such fighting usually results in the death of the foragers and some of the guard workers, and losses can be high. Commercial beehives can be kept closed while nearby crops in flower are sprayed (with the less toxic insecticides) but, of course, the bumblebees and solitary bees are not aware of spray warnings and mortality rates can be very high.

The consensus of opinion is that, in many parts of the world, many crops are regularly underpopulated, and yields are far lower than could be expected. Some classic studies were done in the U.S.A. with red clover and showed that seed production could easily be quadrupled by increasing the local bee population, and in a special experi-mental plot with maximum pollination, the yield increase was more than tenfold! Forage legumes grown for seed are crops most likely to benefit from increased levels of pollina-tion, but many other crops would also show significant yield increases.

But of course not all entomophilous plants lend themselves to such yield increases. Some crops (apples, coconut, cotton, etc.) typically overproduce flowers so that there is a natural fruit-fall of young immature fruits once or even twice (apples) during the growing season; clearly, extra pollination of these crops would not necessarily increase the final yield of fruit. Apple belongs to this category, but it is well known that in some regions (Japan) there is a shortage of pollinators, and in many regions, in the spring of some years, weather conditions inhibit pollination and the reduced fruit-set results in a poor crop. Apple trees with an extra large number of fruits may give a harvest of many small apples, whereas a smaller crop of larger fruits would be more valuable commer-cially. So, with some crops the pollination position may be quite complicated. It appears that with rape crops in the U.K., increased pollination does not usually increase the yield but, with beehives present the crops mature uniformly and the final harvest quality is improved. Rape flowers are popular with pollinators and are a good source of nectar, but it seems that wind may be effecting some of the pollination of the crop. Apart from the case of Japan (see below) with top fruit generally supplementary pollination gives a sig-nificant improvement in quality of the fruit at harvest, especially shape and size. It seems evident that, with the practice of agriculture, there is a continual threat of destruction to the wild pollinators and potential crop reductions could be serious.

In Japan, bee destruction was very serious after World War II because of the very intensive nature of agriculture there. By 1980, in many orchards growers were having to resort to hand-pollination, and it was estimated hat about 25% of labor-time in fruit orchards was being spent on hand-pollination of the flowers. For some years there had been interest in the management of solitary bees in orchards, and Maeta & Kitamura (1980) described an experiment using the mason bee *Osmia cornifrons* (Japanese horn-faced bee). Artificial nest sites were made using small bundles of hollow reeds or narrow plastic tubes 5–6 mm (0.2–0.25 in.) diameter, and these were located in sheltered sites in the orchards where apples were grown. The bees readily colonized these nest sites. *Osmia* bees are good pollinators and in Japan they foraged from 08.00 h to 18.00 h daily, visiting some 15 flowers per minute. A local population of 500–600 bees per hectare gave sufficient pollination of the crop without recourse to outside pollinators, and gave a 50% fruit-set within 65 m (212 ft.) of the nest site. This species has now been introduced into the U.S.A. for the testing of its adaptability in the New World. One of the few draw-backs with protected cultivation of plants is that easy access to insect pollinators is denied and there has long been a tradition of hand-pollination for some plants grown in greenhouses. But, in recent years some flowers grown in greenhouses for seed are being pollinated by flies (blow flies mostly) and now (1989) there are commercial projects aiming at the use of bumblebee colonies in greenhouses both in the U.K. and in Europe.

A recent spectacular success with crop pollination is attributable to C.A.B. International. Oil palm, native to West Africa, is now a major crop in S.E. Asia, where it actually grows more successfully, with greater yields, because of the higher levels of sunshine. But in Malaysia and Papua New Guinea natural pollination was inadequate and so hand-pollination was necessary, adding greatly to cultivation costs (U.S. $11 million/year). In Peninsular Malaya pollination was better but still not adequate, and the main pollinators were thrips, but they were absent from Borneo. In West Africa, however, natural pollination had always been adequate, although most farmers did not know how the flowers were actually pollinated. In 1978, C.I.B.C. sent staff to West Africa, and after extensive field work they found a complex of insects to be responsible for oil palm flower pollination; most were beetles, and they were mainly members of the genus *Eiaedobius* (Col.; Curculionidae). After three years of careful study, *E. kamerunicus* was imported from Cameroon into Peninsular Malaysia where it is now well established and flourishing and natural pollination in the oil palm plantations is satisfactory. It is reported that the annual value of this beetle to the oil palm industry is about U.S. $115 million.

Insects visit flowers for various reasons: most are seeking nectar as a food material, particularly rich in carbohydrates—sugars; some are collecting pollen, largely proteinaceous; other destroy the anthers and other flower parts by their feeding activities and some of these may lay an egg for larval development in the developing fruit. The nectar gatherers usually carry pollen grains, more or less accidentally, from flower to flower and so effect pollination; during the course of evolution the relationship between flowers and insects has often culminated in a state of mutualism and interdependence. Most pollen eaters usually pollinate while they are collecting their food. The flower eaters may also pollinate but if their numbers are too high, their flower destruction can be serious. The insects that regularly visit flowers are termed anthophilous. A few predacious forms are included under this heading for they lurk in blossoms to catch prey that are attracted to the flowers.

Nectar sources are mostly floral and the nectaries are situated at the base of the flower petals (or the corolla tube). Usually the nectar-seeking anthophilus insect has to brush against the anther sacs in order to reach the floral nectaries and so pollen grains become dislodged and then some stick to the setae on the insect body. But extrafloral nectaries are found on a wide range of plants and are responsible for the attraction of some insects to the plants. Whether these nectaries indirectly make any real contribution to the overall pollination situation is seldom clear. The most common site for the extrafloral nectaries is either at the base of the leaf, on the ventral surface, or else at the distal tip of the leaf petiole (as in many forms of cherry), and the insects seen feeding are most often ants.

According to Proctor & Yeo (1973) some insects from most of the 29 orders are recorded visiting flowers and picking up pollen grains. Some of these probably did effect pollination occasionally, but the numbers of insects that are important in plant pollination are more limited. The importance of honeybees as crop plant pollinators, or rather as plant pollinators, on a worldwide basis is often overestimated by many biologists/agriculturalists. It is clear that where honeybees are extensively managed in association with widespread crop monocultures, they usually are the most important pollinators. But current thinking inclines to the view that, worldwide, the many solitary bee species are collectively the most important pollinators, with the bumblebees and other social bees next, and the honeybee third. Then, of course, there are the flies (Diptera), moths (Lepidoptera), beetles (Coleoptera), thrips (Thysanoptera), etc. The more important groups of insects in this respect are listed in the following pages, in order of importance.

From an agricultural point of view pollination efficiency is very important, in order to maximize crop yields consistently. Generally, pollination efficiency depends upon a combination of several factors, as listed below.

1. Tongue length—flowers with a long corolla are only attended by insects with a long "tongue" (bees mostly, and moths) for only they can reach the floral nectaries at the base.

2. Foraging distance—some bees will search for greater distances than others, maybe up to 10 km (6 miles).

3. Speed of foraging—measured as the number of flowers visited per minute.

4. Foraging period—the daily period of food seeking and flower visiting varies somewhat with the insect species.

5. All-weather foraging—typically honeybees do not fly in the rain or at low air temperatures, whereas some *Bombus* will forage.

6. Pollen transport mechanism—some species are able to transport greater quantities of pollen than others, and some deposit more pollen on to the stigma than others.

To examine these points more closely, the importance of bee "tongue" length is very evident in relation to the Leguminosae. Many legumes have a flower where a certain minimum body weight is needed to open the entrance, and then a long "tongue" is needed to reach the nectaries at the bottom of the corolla tube. Honeybees have a medium-length tongue (sometimes called "short"); bumblebees fall into two categories, some are long-tongued, such as *B. hortorum,* and others are short-tongued (*B. leucorum*) (Alford, 1975)—the latter are unable to obtain nectar from red clover and alfalfa although they can take pollen. Moths and butterflies have adapted to feed from flowers with a long corolla tube—the extreme cases include the Sphingidae with tongues up to 25 cm (10 in.) that feed from spider lilies most spectacularly. At the other extreme, blow flies can only obtain nectar from flowers that are quite open—such as the Cruciferae.

Foraging distance applies more to the social Hymenoptera which have a nest to return to, but dispersal distances for other insects are of importance in relation to their place of origin. Early work with hives of honeybees suggested that most foraging was within a couple of kilometers of the hive, but some work with wild honeybees has indicated that they might travel up to ten kilometers distance from the nest. And, clearly, the species that forage faster and for longer hours, and in worse weather, are the most effective at both collecting nectar and effecting pollination of the crop flowers. Many bees show a definite preference for certain flowers, and it was pointed out by Corbet (1987) that, on a crop preferred by bumblebees, a colony of 100 bumblebees could be as effective as a hive with 10,000 honeybees of which only a few worked the crop. In many respects bumblebees are more efficient foragers/pollinators than are honeybees.

In the more unspecialized cases, insects carry pollen grains stuck on to the hairs of the body and, if they rub against the stigma of another flower, they effect pollination quite accidentally. But many of the anthophilous insects have developed means of carrying the pollen they collect; some carry the pollen in the crop (*Hylaeus*) and it is later regurgitated but, in the more advanced cases, a special pollen-carrying structure (corbicle) has developed on either the body, or on a leg, of the bee. Very often the pollen in the corbicula is not available for pollination purposes, but it is an important part of the insect diet.

THE MAIN GROUPS OF INSECT POLLINATORS

Hymenoptera

APOIDEA   *20,000 spp.*

Adult females (workers mostly) forage and collect nectar and pollen as food for the larvae in the nest. More than 85% of the species are solitary (Batra, 1984) with a small nest of about ten cells, usually underground; gregarious nesting is common and there may be thousands of nests in a small area.

Colletidae (plaster bees; membrane bees). Primitive bees nesting in holes; nest cavity lined with impervious layer; bees short-tongued.
> *Colletes* spp. Holarctic in origin but now widespread; solitary bees.
> *Hylaeus* spp. Holarctic; many species.

Halictidae (mining bees; sweat bees). Some species solitary, some subsocial, some social in small colonies; nests are underground; some species short-tongued, some long-tongued; a group of some diversity worldwide.
> *Halictus* spp.—more temperate in distribution.
> *Nomia* spp.—more tropical; especially in the New World.
>> *Nomia melander* (alkali bee)—this American species is now successfully managed for pollination purposes in parts of the U.S.A.

Andrenidae (burrowing bees; digger bees). Nest underground; solitary but often nesting in colonies.
> *Andrena* spp.—the main genus; holarctic but some species elsewhere; most are short-tongued but a few are longer.

Megachilidae (leaf-cutting bees and mason bees). Most are solitary; a very large group, found worldwide; nest underground or in dead wood, or in hollows; long-tongued bees.
> *Anthidium* spp.—nest in ready-made cavities, in warmer countries.
> *Chalicodoma* spp. (mason bees, etc.)—nest stuck on to substrate.

Megachile spp. (leaf-cutting bees)—worldwide; nest in tunnel in soil or wood; cut leaf pieces for nest construction.

> *Megachile rotundata* (European alfalfa leaf cutter)—this is now established in the U.S.A. and managed for alfalfa pollination.

Osmia spp. (mason bees; hornfaced bees)—worldwide; mostly nest in ready-made cavities.

> *Osmia cornifrons* (Japanese hornfaced bee)—established in bee shelters in Japanese apple orchards and now introduced into the U.S.A. for experimental purposes by U.S.D.A.

Anthophoridae (cuckoo bees; mining bees, etc.). Solitary, but may nest gregariously, in soil burrows; long-tongued; many species parasitic.

> *Anthophora* spp.—large genus; stout hairy bees; some species in Europe and Asia start foraging very early in the season (on crocuses, etc.).

> *Hemesia* spp.—this is the main genus in the U.S.A.; large hairy bees.

Xylocopidae (carpenter bees). Solitary; timber/wood tunnelers, or nesting in plant stems; mostly tropical; long-tongued; many very large in size.

> *Allodape* spp.—many species; smaller; nest in plant stems, etc., in warmer climates.

> *Ceratina* spp.—many species; small; nest in pith of plant stems; many in Africa and elsewhere, and some in Europe.

> *Xylocopa* spp.—large, hairy tropical bees (one in southern Europe; also in U.S.A.); nest tunnel in timber, dead wood, bamboo stems, etc.

Apidae (social bees). Large family; worldwide; shows gradation from solitary species, through subsocial to social species with large nest colonies; very important pollinators; some species can be managed successfully.

Euglossinae (orchid bees). South America; large solitary bees with small nests; renowned as pollinators of certain orchids.

Bombinae (bumblebees). Holarctic; social bees with small nests; workers long-tongued or short-tongued.

> *Bombus* spp.—variation in body size and tongue length; some species very important pollinators of clovers, alfalfa, etc.; some introduced into New Zealand for red clover pollination; several species being tested for management possibilities, especially in glasshouses.

Meliponinae (250 spp.) (mosquito bees). Small, social bees, that do not sting; tropical in distribution.

> *Melipona* spp.—South America; quite large in size.

> *Trigona* spp.—Africa, Asia, northern Australia; tiny (3–4 mm/0.12–0.16 in.) bees; nest colony quite large.

Apinae (honeybees; social bees). Social species; moderate in size; colonies large; temperate and tropical species.

> *Apis mellifera* (honeybee)—Eurasia, but now worldwide; many subspecies and races; extensively reared and managed worldwide (apiculture).

> *Apis cerana* (eastern honeybee)—India, S.E. Asia; a few races; *A. c. indica* is the Indian honeybee; mostly wild.

> *Apis dorsata* (giant rock honeybee)—India; probably two subspecies; make large open nests; large colonies; very aggressive; most in Himalaya regions.

> *Apis florea* (little honeybee)—India; probably two subspecies.

CHALCIDOIDEA

Agaonidae (fig-wasps). Pantropical; live in a state of mutualism with plants of the genus *Ficus* (Moraceae).

> *Agaon, Blastophaga, Ceratosolen,* etc., some of the main genera responsible for pollination of the ovules inside the fig syconia.

Diptera

Many of the families are recorded visiting flowers and carrying pollen grains, but only the adults; pollination is accidental; nectar is imbibed as food. In a few families most species are anthophilous as adults, as listed below.

Bombyliidae (bee flies). Adults with a long proboscis for nectar collection.
Syrphidae (hover flies). Many species; worldwide; very important pollinators.
Muscidae (house flies, etc.). Both these groups are very large; worldwide.
Calliphoridae (blow flies; bluebottles, etc.). Many synanthropic; very important for pollination of open-flowered plants.

Lepidoptera

Butterflies are diurnal; most moths nocturnal; all adults have a long suctorial proboscis adapted to collecting nectar from flowers and other fluids; many will accidentally transport pollen from one flower to another; proboscis length is a major factor in choosing host plants to be visited; many tongues very long. Nearly 100 families are known and most are involved in plant pollination to varying degrees, none is markedly more important than the others.

Coleoptera

The largest Order of Insecta, with some 95 families and more than 330,000 species named; most adults have biting and chewing mouthparts, and the group shows tremendous diversity of habits. Many beetles are anthophilous, some manage to reach the nectar sources, some eat pollen, and a number have been recorded accidentally pollinating various flowers. Collectively their contribution to pollination may be appreciable, but few families are of particular importance.

Melyridae. A large family, worldwide; most adults frequent flowers.
Cantharidae (soldier beetles). Some species regularly found in large numbers on flowers of Umbelliferae, Compositae, etc. (only adults).
Chrysomelidae (leaf beetles). A very large family, but only some of the adults are found on flower heads.
Nitidulidae (sap beetles). Some known as blossom beetles (*Meligethes* spp.) are bred in flowers, but the feeding adults do pollinate as well as eating the anthers and pollen.
Curculionidae (weevils). *Elaeidobius* spp. (oil palm beetles) pollinate oil palm flowers; native to West Africa, now established in S.E. Asia.

Thysanoptera

Both adults and nymphs are regularly found inhabiting flowers and their feeding does damage to some extent, but they are recorded successfully pollinating a number of different flowers. Many thrips normally found in the flowers of legumes (Order Leguminales).

Thripidae. One of the largest families of thrips; very abundant; worldwide.

Frankliniella spp. (100+ spp.) (flower thrips, etc.)—most are called "flower thrips" as they are found inhabiting flowers of many types; some regularly pollinate some plants and can be of some importance.

Other Orders

As mentioned already, in the book by Proctor & Yeo (1973) many examples are given from most of the other insect orders.

To recapitulate, in the areas where honeybees are reared and managed they probably are the most important crop pollinators. But honeybees avoid some flowers and are not effective on others, for their tongue may not be long enough, and sometimes they forage slowly, and there are other reasons for poor pollination of some plants. Wild and feral colonies occur in some temperate countries but in Europe and the U.S.A. (including Canada) they are few; however, in most warmer countries wild populations are apparently numerous. The view, now widely held, is that in most parts of the world it is very probable that the local complex of solitary bees, together with the bumblebees (Bombinae) (and in the Southern Hemisphere the Euglossinae and Meliponinae), are the main pollinators. Other Hymenoptera, moths, flies, beetles, and thrips all contribute to the overall pollination effort. Each different type of insect, with its different mouthparts and different food preferences, will be important in the pollination of certain plants, but not others. Some solitary bee species collect pollen from only a few closely related plants and are called *oligolectic;* a few only visit one species of flower and these are *monolectic*. Honeybees, with a more or less "permanent" colony, have a long season of foraging activity and take pollen from many different plants and are termed *polylectic;* they take pollen from whatever plants are available at the time. There are, of course, times when the foraging bees are seeking nectar and other times when pollen (or water) is sought, so the reason for flower visiting may vary from time to time and may not be too obvious. In the spring, apple trees are a good source of pollen for foraging bees but often a richer source of nectar might be dandelion flowers, growing as weeds in the orchard, and often honeybees can be seen to forsake the apple flowers and seek out flowering dandelions and other weeds. It is apparent that more research is needed to establish clearly which crop plants have a particular need for insect pollination, which insects normally visit them, and which insects might be suitably managed to ensure that the crop is pollinated to an adequate level.

At the present time, honeybees are being used extensively by farmers and growers to supplement natural pollination of some crops. But some seed merchants and some farmers are using blow flies (obtained as puparia from fish-bait rearers) to pollinate some flowers under glass, and some Cruciferae (cabbage, sprouts, etc.) when a local variety is propagated inside a netted corner of a field. But flies can only pollinate open-flowered plants for their mouthparts are not capable of probing.

The need for crop pollination has been appreciated for a long time in some countries; in the U.K. there has been a county-based pollination service available through local bee farmers for many years, and recently there has been publicity concerning the National Pollination Service offered by the U.K. Bee Keepers Association. Any farmer or grower needing pollination assistance can be given a local contact who will provide beehives at the appropriate time. The greatest need is generally in fruit orchards, and one hive per acre is regarded as sufficient; flower production is usually closely synchronized and in good weather the crop can be pollinated in 4–5 days, although in poor weather up to 14 days may be required. Hives are regularly used with field beans and clovers grown for seed, but 2–4 hives per acre are usually required; on these crops pollination results are often not good—honeybee tongues may be too short to reach the nectar at the base of the corolla tube. Long-tongued species of bumblebees are more effective. Honeybee hives are often used on oilseed rape in the U.K., but this crop flowers early and weather can be a problem in that sometimes the air temperature is too low to encourage bee foraging. Typically the bees do not often increase the yield, but their attendance causes the crop to mature uniformly and there is an increase in seed quality at harvest; the apiarists like to use the rape crops because the honey yield is high. Most professional apiarists (beekeepers) in the U.K. at the present time are making their basic living through the fees for the pollination service, and the money realized from honey and other products is often regarded as a bonus.

Apis mellifera occurs as a series of geographical subspecies and also as local races; the total number of intraspecific forms can be a topic of heated debate. Some forms are more efficient pollinators than others. There are several races in Africa, but the one usually referred to as the "African honeybee" is Apis mellifera scutellata; this race is regarded as a good pollinator and nectar gatherer for it starts foraging early in the morning and continues to dusk. The different races are preferred by some apiarists for different purposes. The Italian race is rather docile and is preferred for teaching displays as the workers are less likely to sting. Although assisted pollination usually involves honeybees, as mentioned, blow flies and other flies have been used on some crops for some time. One of the first striking cases was the cultivation of red clover for seed to grow forage in New Zealand; seed-set was poor for there was an absence of wild pollinators and honeybees are too short-tongued. Several species of long-tongued Bombus were imported and established and now it is reported that adequate levels of red clover pollination are achieved. Osmia bees have been used for some time in Japan for fruit pollination, and alfalfa growers in the U.S.A. have used several species of solitary bees to help pollination (this industry is said to be worth U.S. $100 million annually). In the U.S.A., Nomia melander (alkali bee) is now managed by alfalfa growers and in some areas Megachile rotundata (European alfalfa leaf cutter), imported from Europe, has been successfully established for alfalfa pollination. With the recent increase in protected cultivation (glasshouses and polythene tunnels) in Europe, and with the trend towards biological pest control of the pests of these crops, there has developed very recently an interest in trying to establish Bombus nests in glasshouses for the purpose of plant pollination. Several companies are engaged in commercial feasibility studies at the present time.

In the tropics, bees are still kept mainly for honey production (although tropical colonies of honeybee are low yielding; see page 66), but the importance of insect pollination of crops is becoming more widely appreciated, and international conferences on pollination in the tropics have been held in recent years. Each country in the world (excepting Greenland) has a branch of the Ministry/Department of Agriculture responsible for apiculture. The original object was to foster beekeeping for honey production, but now more and more concern is for adequate local crop pollination.

After the early catastrophes, when insecticide spraying of orchards and crops in flower resulted in large-scale bee destruction, the lesson has been learned (to some extent) in most countries. Now there is usually close collaboration between farmers and spray contractors and the local apiarists who are warned when widespread spraying is imminent so that hives can be closed and the bees kept inside until the period of greatest danger is over. Pesticide companies now all include honeybee-toxicity testing as part of the routine screening of all candidate chemicals for commercial pesticides. If an approved compound poses a danger to honeybees, this is carefully noted and advertised as part of the precautions needed to be taken. And there is a general trend to recommend that all (or most) crops should never be sprayed when actually in flower. Unfortunately, spray warnings are not heard by the wild bees, and so there is inevitably mortality among the other insect pollinators.

As mentioned in the next section, in the U.K. the International Bee Research Association is now devoting much of its attention to the problems of pollination, both in temperate regions and now in the tropics (see Crane & Walker, 1983), and is a source of bibliographical information. I.B.R.A. cosponsors the International Symposium on Pollination, with the International Commission for Bee Botany of the I.U.B.S.; the 1st was held in 1960 in Copenhagen and the 5th in 1983 in Versailles, France. The I.B.R.A. is the proposed world center for the advancement of apiculture in developing countries, and one of their projects is the preparation and publication of a Pollination Directory for crops grown in developing countries. In the U.S.A. the main center for apicultural information is the U.S. Department of Agriculture at Beltsville, Maryland. In the U.K. there are groups involved with beekeeping and plant pollination both at Rothamsted Experimental Station, Hertfordshire (Free, 1970, etc.), and in the School of Pure and Applied Biology, University of Wales, at Cardiff.

Useful sources of information include the I.B.R.A. publication by Crane & Walker (1983), Kevan & Baker (1983), and the *New Naturalist* volume, number 54 by Proctor & Yeo (1973).

APICULTURE

This is the rearing of honeybees (*Apis mellifera*) primarily for honey, wax and other products, but now ever increasingly as agents of pollination for use by farmers and growers. But the origin of apiculture was initially just for the honey. In the U.K. the main source of information is the International Bee Research Association, formerly at Gerrards Cross, Buckinghamshire, but now relocated in Bristol; started in 1949 it has progressively expanded and it now has a completely worldwide remit. I.B.R.A. publishes the journals *Bee World, Journal of Apicultural Research,* and *Apicultural Abstracts,* as well as a series of books, reports, pamphlets, bibliographies, etc., and it is the convener of the International Conferences on Apiculture in Tropical Climates (1976, 1980, 1984), as well as the conferences on pollination. In the U.S.A. there is the American *Bee Journal,* and there are other journals in other parts of the world.

The genus *Apis* is now thought to contain just four species. *Apis mellifera* is the main honeybee (Fig. 2.1) and it has a series of geographical subspecies (sometimes called races) regionally, and also strains or races locally, many showing a number of differences in their basic biology. There is disagreement among apiarists as to the precise status of most of these intraspecific forms. *Apis mellifera* is found wild in eastern Europe, Asia Minor and Africa. *Apis cerana* (eastern honeybee) is native to India and S.E. Asia; some cultivation is practiced in India; *A. c. indica* is the Indian honeybee. *Apis dorsata* is the sometimes dreaded giant or rock honeybee, which makes large open nests in the Himalayas; it is very aggressive when accidentally disturbed and people and pack animals have often been attacked on mountain paths, often with fatal results. It is thought there may be two subspecies. *Apis florea* (little honeybee) also exists as two subspecies according to recent information, native to India and parts of S.E. Asia.

The African honeybee of recent notoriety is *Apis mellifera scutellata* native to the eastern part of Africa south of the Sahara, although there are several other "races" to be found in Africa. This bee is responsible for unprovoked mass attacks on people and sometimes livestock, and in eastern Africa a couple of dozen people are stung to death each year. Some apiarists believe this race to be superior in various respects, and in the

Fig. 2.1 Honeybee workers (*Apis mellifera*) (Hymenoptera; Apidae); one with filled pollen baskets (corbiculae); South China.

1950s it was introduced into South America; inevitably some escaped from captivity and it is reported to be replacing the local honeybees quite rapidly. The local bees were imported European bees that do not adapt completely to tropical conditions and thus do not compete well against the African bees. The established feral African bees have spread from Brazil and are now starting to enter the southern states of the U.S.A., reportedly causing much concern, but since it is a race adapted to warm climates it is not expected to spread far into the U.S.A.

It seems that humans have always had a "sweet tooth" and in the early days of our evolution there were few sources of sugar. Sugar sorghums, wild sugarcane, and in North America the sugar maples, were really the only sources of sugar. But harvesting honey from wild bee colonies is a hazardous process, sometimes physically difficult, and often very painful, so probably thousands of years ago we learned that bee colonies could be enticed to nest in suitable containers, and apiculture was started.

The nest consists of a series of combs constructed of wax, containing hexagonal-shaped cells in rows. Some banks of cells are used for breeding purposes (brood combs) and typically contain eggs, larvae and pupae, and other combs are primarily for the storage of honey (a mixture of thickened plant nectars). In temperate regions most nectar is available during the spring and summer, when air temperatures during the day are suitable for insect foraging, and since the colony is permanent, food is required for the over-wintering period. Thus the economy of the colony is based on the storage of excess honey during the summer period for later consumption throughout the winter. In primitive apiculture either the whole honey store is removed (this is seldom) or else just part is taken, and sufficient store is left for the nucleus of the colony to survive. In the tropics, the dry season is usually the period of inclement weather—with a shortage of flowers and a shortage of water. But in some climates there is no major period of unfavorable weather and bee activity may continue all year. In good bee management, in temperate regions, the practice is to remove most (but not all) of the honey from the storage combs and then to provide the bees with sugar solution at regular intervals from late summer over the winter as their main source of food. In the U.K. typically 10,000–20,000 workers survive in the winter cluster in each hive.

In the tropics, as there is often no major dormant period, the bee colony economy centers upon the construction of combs, then the provisioning for a population buildup, and then swarm development. The parent colony may fragment into several swarms (occasionally a large number) and the old nest may be deserted. So, in the tropics there is often a high yield of wax per hive and, in some areas, wax production may be more important commercially than the honey. Average yields of honey in tropical bee colonies are low, being 6–8 kg per hive per year, although there may be much variation. Most of the colony energy goes into swarm production, and swarms usually take place every 12–13 weeks.

A striking feature of the bee colony is that a single large queen (fertile female) is responsible for all the egg production—all she does is lay eggs and at the height of the season she lays about 2000 eggs per day (up to 3000 are recorded). The work force is composed entirely of "sterile" females called workers; they have undeveloped gonads and could become reproductive in the event of the death of the queen. Within the colony there is a division of labor based on the age of the individuals. The youngest workers have duties of nursery attendance and feeding, seeing to the needs of the queen, nest cleaning, etc. Then the older workers have foraging duties and fly out to find sources of nectar and pollen, and sometimes water. Foraging is hazardous and exhausting and, at the height of the season, the life expectancy of workers is only a few weeks. At this time the colony may contain 50,000 workers. Early in the summer males are produced from unfertilized eggs, and young queens are also allowed to develop. Queens are females, produced from fertilized eggs, but the larvae are fed exclusively on a special diet—a substance known as "royal jelly" which is rich in protein, produced from the mandibular glands of the worker bees while they are young. When the young queen(s) is at the pupal stage it is usual for the old queen to be driven from the hive by the workers—she departs surrounded by about half the workforce of adult workers in a swarm to seek a new nest site and start another colony. If more than one young queen emerges then the extra one(s) may leave with further swarms containing drones, to fly away and start new colonies. In good bee management the apiarist is alert to these developments and he captures the emerging swarms and entices them to colonize new hives.

Beehives are composed of a series of square sections placed one on top of the other as the colony grows and as it requires more comb-space (Fig. 2.2). By the end of the summer the hives may consist of 4–5 comb sections. The top of the hive is the crown board and is covered with a metal sheet or some other waterproof material. The base has a landing platform resting on the stand, and the entrance is narrow so that bees can pass freely but too narrow to allow passage of predators or animals that would rob the hive. But field mice can enter, as can death's head hawk moths, and in the winter when the mice are most likely to attempt invasion it is customary to place a mouse-guard over the entrance. The first or lower chamber is the brood chamber and the more shallow section racks above this contain the honey storage combs. The modern beehive started in about 1850, and now in the U.K. there is a National Beehive (B.S. No. 1300; 1960), but other types of hive are favored by commercial apiarists and local variations of this basic pattern are used worldwide. The first (oldest) types of beehive in Europe were dome-shaped structures made of woven straw. In Africa, beehives are traditionally a hollow log or a cylinder of bark, positioned horizontally and often with a rain shelter over the top (Fig. 2.3).

Now that most commercial apiarists in the U.K. rent out their hives to farmers for crop pollination, hive transportation has become commonplace. One problem with apiculture is to ensure a continuous supply of nectar and pollen throughout the entire season for the bees. In the past most apiarists had fewer hives and they were stationary, so any one establishment could only support a small number of colonies otherwise there would not be enough nectar to go around. Now with the commercial producers the practice is to take the bee colonies to the sources of nectar, so an organized apiarist can have 100 hives or more. In the U.K. there is a succession of entomophilous flowers that are available as sources of nectar and pollen; when the weather is warm, enough workers will emerge and forage even in midwinter but, or course, not many flowers are available during this time. Most of the crop pollination is done early in the season; rape usually in April and apple and other fruits in May and early June. In the summer, the large-scale

Fig. 2.2   Honey farm in South China (Yuen Long Apiaries) with rows of beehives.

Fig. 2.3   Modern beehive in Ethiopia replacing the less accessible traditional forms; Alemaya, Ethiopia.

apiarists have to find other sources of flowers for the bees and many transport hives to the high moorlands to collect nectar from heather. Heather honey is said to have a special flavor that makes it a high-quality product. Rape honey is regarded as being less desirable. At the present time, in the U.K. it is recorded that there are 33,000 small-time beekeepers (averaging four hives each) and 900 bee-farmers with an average of about 100 hives each.

Bee management, so far as temperate regions are concerned, has several different aspects. The main objectives are as follows:

1. To produce a sustained yield of honey; that represents an economic return on the financial investment involved.

2. The survival of the bee colony, in healthy condition.

3. To suppress certain aspects of the bee life history; namely to reduce swarming and to maintain a large workforce, and to hold the worker population together.

With good management, swarming can be controlled so that it only occurs when the apiarist either needs new colonies for replacement purposes or for sale, or for business expansion. Generally, if a colony swarms, then the honey storage level will be low and the autumn yield will be low. It is reported that without careful management, after a colony swarms during the summer, then quite often over the winter period either the depleted parent colony or the swarm will actually die. High yield honey storage is basically an artificial situation (a product of bee management), for, in the wild, as soon as a good store has been accumulated it would be used for the purpose of swarming.

Apiculture has three basic products: honey, wax, and propolis, although "royal jelly" may be collected by some apiarists, and swarms are, of course, sold to establish new colonies. Honey was originally produced for local consumption but now several countries have areas of high production so that they are regular honey exporters. A major factor is having a large area with abundant suitable flowers (either wild or cultivated) for nectar production. Heather tracts and flowering trees are particularly important as nectar sources. The total value of honey production in the world cannot be realistically estimated, but in 1988 it was calculated that the value of honey and wax produced in the U.S.A. was about U.S. $100 million per year. The wax forms the comb in which the honey is held, and for every 6–7 kg (12–15 lb.) of honey produced there is about 1 kg of wax. The annual yield of honey from a bee colony (in excess of that left for the colony as winter food) varies tremendously; in Africa it may be as low as 6–8 kg (12–16 lb.) per hive; in the U.K. and southern U.S.A. 10–20 kg (20–40 lb.) and up to 100 kg (200 lb.) per hive in Australia from Eucalyptus trees.

Honey is a viscous sweet liquid made from plant nectar by removal of excess water, and by the inversion of most of the nectar sucrose into fructose and glucose. There are also small amounts of pollen, some minerals, vitamins, amino acids, plant oils, and many other substances in tiny amounts. The precise constitution varies according to the plants from which the nectar was collected and the environmental conditions (especially humidity); the total list of chemicals analyzed from honey is very extensive. Its main use is clearly for sweetening foods of a wide range. Honey also has slight antiseptic properties and it has been used for many different medicinal purposes for many years. Fermented honey makes a delicious pale yellow wine—this is called mead in the U.K. and earliest records go back to about AD 500; in Ethiopia its history is older and honey wine is called tej. Honey sold in many African countries comes complete with broken combs and bits of wax and propolis mixed together. But in the Western world most commercial honeys have been "refined" by careful sieving and cleaning. There have regrettably been some commercial "honeys" sold in England for years that are (were) allegedly not honey at all, but made entirely of ordinary sugars. In recent years, in Europe and North America, there has been a great public reversion to "natural" foods and as part of this trend there is interest in unrefined honey—some is now sold as raw honey with bits of wax and propolis included, and some is actually sold still intact in pieces of comb. These "contaminants" are now regarded as being highly beneficial to human diet; there is a great deal of accumulated evidence that this is so.

Wax, also called *"beeswax,"* is secreted as tiny scales or flakes from four pairs of abdominal glands on the ventral body surface. The scales are transferred to the mouth by the legs, where they are chewed by the mandibles before the wax is incorporated into the comb. It is estimated that bees eat 3.5–7.5 kg (7–16 lb.) of honey to produce 0.5 kg (1 lb.) of wax. Beeswax is a yellowish solid, insoluble in water but soluble in carbon tetrachloride, chloroform, and warm ether. It has traditionally had a number of uses, for some of which it has been replaced by modern synthetic waxes. For some purposes it is still superior, and there has been recent interest in the use of beeswax for wider uses. Most commercial beeswax in Europe and North America comes from *Apis mellifera* but wax from India and parts of Africa comes from *A. dorsata* and other species. At the current time the main uses for beeswax are for making religious candles, in polishes and waxes, some cosmetics, ointments, wax paper, lithographic inks, and a number of rather obscure products. Because of its inert nature it is used for filling spaces in bones in some human medical operations.

Propolis. Foraging bees often collect a resinous material from buds or bark of trees, particularly poplars, and it is carried in the pollen baskets on the hind legs of the workers. It is used as a form of cement to seal holes and cracks in the nest cover of the hive, and to join the combs to the roof. It is important that hives are draft-proof if the colonies are to survive the rigors of the temperate winter. This propolis is also used to embalm bodies in the hive (such as a dead mouse) that are too large to be evicted, to seal them and make them hygienic. Formerly, this substance was regarded as a nuisance by apiarists but in ancient times it was used medicinally and there is a revival of interest in propolis for its medicinal properties and it is now a very valuable byproduct of apiculture. It has been shown to possess antibiotic properties and there are claims that it has been used successfully to treat a wide range of human ailments. The earliest recorded writings about propolis and its medicinal value were by Pliny and Aristotle in about AD 70.

Studies on human longevity have revealed that in parts of eastern Europe around the Balkan States and western Asia, people often live to a great age (up to 120 years and thereabouts); this is an area of ancient and intensive apiculture and unrefined honey has long been a regular part of the local diet. The local strain of honeybee (Caucasian race) is a particularly effective nectar collector, and the local apiarists have long been aware of the medicinal properties of propolis and unrefined honey. There is now great interest in honey, pollen, and propolis in several European countries and in the U.S.S.R.; world demand for propolis is ever-increasing at the present time. Many proprietary formulations including propolis, pollen, royal jelly, etc., are being sold in most countries.

Pollen. Bees store pollen in certain cells on the brood combs as the source of protein for older worker (and drone) larvae. In recent studies it has been claimed that pollen has therapeutic properties, and it has been used to alleviate symptoms of a wide range of complaints and diseases (colds, influenza, asthma, hayfever, rheumatism, arthritis, prostate disorders, menstrual and menopause problems, virility problems, and others). It should be stressed though that the most frequently imbibed form of pollen has actually been either in unrefined honey or as a pollen/propolis mixture.

Pollen is basically proteinaceous and is composed of many different amino acids and vitamins; up to 20 different amino acids have been identified, many different enzymes, and the total list of chemicals analyzed is very extensive. Pollen may be collected from the storage combs or else a "pollen-trap" can be fitted to the hive entrance which causes the workers to lose most of their carried pollen.

Royal jelly is the substance secreted by the mandibular glands of young worker bees, fed to all young larvae for the first three days and to queen larvae for their entire larval life of 5½ days. The protein content is about 40–43%, and there are many vitamins, especially of the B-complex. The mandibular glands atrophy as the worker ages and the wax glands start to function. The jelly fed to developing queens contains about 35% hexose sugars, that fed to larvae destined to become workers contains only 10% sugar; the supplementation of the jelly with extra sugar is essential to the development of larvae into queens.

It has been claimed by various beauticians that a regular daily dose of honey bee royal jelly has pronounced rejuvenating effects, but there seems to be little direct evidence of this. It is, however, the basis of a few thriving businesses in Europe and other parts of the world.

In summary, it does appear that there are definite beneficial effects to having raw, unrefined honey (including bits of wax, propolis, pollen, etc.) as a regular item of diet. It is to be hoped that future research may cast light on the biological properties of the various components.

SERICULTURE

Silk is the secretion from modified salivary glands by pupating larvae (caterpillars) of the lepidopterous families Lasiocampidae, Bombycidae, and Saturniidae, used for the construction of the pupal cocoon. Many caterpillars produce silk, used for the aerial dispersal of tiny, first-instar larvae (Geometridae, etc.), for assisting larger larvae to roll leaves, making larval "tents," etc., and for construction of the pupal cocoon. In fact, the majority of families in the Order Lepidoptera include species who at some stage spin silken threads, for one purpose or another. But only in certain groups does the silk have properties that render it of possible commercial application. The cocoon is spun, or rather constructed, of a single long strand of silk. Silk is also produced from the posterior-positioned spinarettes of various Arachnida, namely the spiders (Araneida) and some mites (Acarina; Tetranychidae). Spider silk threads are very strong, and some have been used for various odd purposes by humans, including fishing lines made from *Nephila* webs in parts of coastal Africa and Madagascar.

Silk is first recorded as being economically important in China in 2600 BC, but the origination of sericulture is lost in the mists of antiquity. Raw silk was one of the first commodities traded between China and Europe and was of tremendous value. It is thought that smuggled eggs were taken to India in about 150 BC which started the Indian sericulture industry. Now sericulture is practiced in all parts of the world, and the main production areas are China, Japan, Korea, U.S.S.R., and India; in 1978, world production was reported to be 408,000 tons of cocoons giving 819,000 bales of raw silk. Present value of silk production is estimated as being more than $1000 million per year. Historically, the greatest sales of silk were from China to Japan to make silk clothing for the hot humid summer weather and, today, Japan is still the largest silk consumer for manufacture of kimonos, but now they have their own sericulture industry (dating from the third century AD), and they export the surplus silk.

The main species of moth used in commercial silk production is *Bombyx mori,* variously called the mulberry silkworm, oriental silkworm, of just the silkworm, but others have been used. A list of the species that have been used as a source of silk is shown below. Many of the other species are used in small numbers as silk sources in India, and the Chinese oak silkworm is reported to be widely cultivated in China. Other species have been tried as silk sources—for example, the unfortunate introduction of gypsy moth (*Lymantria dispar*) into North America was to test the species as a possible source of silk. But these are the species that are reported to have been used for silk production.

Bombyx mori occurs as several different races, which is not surprising considering the great age of silk-farming and the thousands of years that many populations have been isolated from each other. The four major races are known as the Chinese race, the Japanese, the European, and the Tropical race. A major factor separating the races is the type of lifecycle; the European race is univoltine, the Chinese and Japanese include both univoltine and bivoltine forms (races), and the Tropical race is multivoltine.

MOTH SPECIES RECORDED AS SOURCES OF SILK

BOMBYCOIDEA

Lasiocampidae (tent caterpillars; lappets, etc.).

Gonometra postica	southern Africa	*Acacia* spp.
Trabala vishnou (oriental lappet moth) (Fig. 12.85)	S.E. Asia	Polyphagous

Bombycidae (silkworm moths).

++*Bombyx mori* (silkworm; oriental silkmoth) (Fig. 2.4)	S.E. Asia, etc.	Mulberry
Bombyx spp. (6)	India	Mulberry

Saturniidae (emperor moths; giant silkmoths).

Actias selene (moon moth) (Fig. 12.87)	India to China	Many hosts
Antheraea assama (muga silkworm)	India	Polyphagous
Antheraea mylitta	India	Jujube tree
Antheraea paphia (tussor silkworm)	S.E. Asia	Polyphagous
Antheraea pernyi (Chinese oak silkworm)	China	Oaks
Antheraea yamamai (Japanese oak silkworm)	Japan	*Quercus* spp.
Attacus atlas (atlas moth) (Fig. 12.89)	Pantropical	Polyphagous
Eriogyna pyretorum (giant silkworm moth) (Figs. 2.5–2.7)	S.E. Asia	Camphor
Naudaurelia spp. (silkworms, etc.).	Africa	Many hosts
Samia cynthia (wild silkmoth; lesser atlas moth) (Figs. 2.8–2.9)	India to China	*Lantana, Michelia*
Samia ricini (evi silkworm)	India, S.E. Asia	? Castor

NOCTUOIDEA

Notodontidae (processionary caterpillars, etc.).

Anaphe infracta (African wild silkworm)	Uganda	*Bridelia*

++A very important species.

Each female moth will lay about 500 eggs. At 25°C (77°F) larval development takes some 20–25 days and pupation another 10 days. Hibernation takes place in the egg stage, which is very convenient for commercial production. The caterpillars of the different races vary considerably in color, growth rate, disease susceptibility, and other traits. The caterpillars are fed on fresh mulberry leaves (*Morus* spp., often *M. alba*) and they can be reared in dense numbers until they pupate. The small posterior "tail" is characteristic of larvae of the family Bombycidae. The silken cocoon (Fig. 2.4) is white, pink or yellowish in color, and is made of a single thread of silk between 500 and 1200 m in length. The thickness of the thread (and length) varies according to the race—the thicker fibers are generally shorter in length. In Japan it is now usual for much of the rearing to be done using a completely artificial (synthetic) diet—probably half the total silkworms reared in Japan are now on artificial diet. There is some variation in the thickness of each fiber—initially it is thinner, and it may be thinner at the end also. In Thailand the polyvoltine races produce an uneven fiber, but with a brilliant luster, which has to be hand-woven, and produces the very attractive but slightly uneven traditional Thai silk. The univoltine and bivoltine races produce a large cocoon with a longer, thinner and more uniform thread that can be mechanically unwound, so Chinese and Japanese silk is thinner, smoother, and of uniform consistency. Each commercial thread of silk is formed from two fibers as two cocoons are reeled simultaneously.

Silk produced from the cocoons of the other species of Bombycoidea differ in color, texture and quality, and invariably have to be hand-reeled, but the fabric can be attractive and with a few species it is a worthwhile local cottage industry in some parts of Asia and in Africa.

There have been long-established research institutes for sericulture in China, Japan, and also in Thailand. In 1980, the Japanese Society of Sericultural Science published a pamphlet titled "Sericulture in Japan" for participants of the XVI International Congress of Entomology. There was an informative article on silk production in Thailand in *New Scientist* (Morton, 1989). There are five or six silk farms in England, and others scattered throughout Europe and the U.S.A.

Fig. 2.4   Pupal cocoons of *Bombyx mori*—the source of commercial silk in China and Japan.

Fig. 2.5 Giant silkworm moth (*Eriogyna pyretorum*) (Lepidoptera; Saturniidae); wingspan 110 mm (4.4 in.); Hong Kong.

Fig. 2.6 Caterpillar (giant silkworm) of *Eriogyna pyretorum* feeding on foliage of camphor; Hong Kong.

Fig. 2.7 Two pupal cocoons of giant silkworm moth fastened to camphor foliage; Hong Kong.

Fig. 2.8 Adult of lesser atlas moth (*Samia cynthia*) (Lepioptera; Saturniidae); wingspan 110 mm (4.4 in.); Hong Kong.

Fig. 2.9   Caterpillar of lesser atlas moth eating leaf of *Michelia alba* in Hong Kong; body length 60 mm (2.4 in).

OTHER INSECT PRODUCTS

Shellac. A sticky resinous substance produced from the thick scale of the lac insect (*Laccifer lacca*—Lacciferidae; Coccoidea) in India. The hard brown scale is harvested for the resin, and the body of the female insect yields a bright red dye (lac dye). The scales encrust young twigs of *Acacia,* soapberry, and especially *Ficus religiosa* in India and parts of S.E. Asia. The encrusted twigs are collected twice a year and the scales scraped off— this crude substance is called "stick-lac"; the dye is extracted with hot water or hot sodium carbonate solution. The resin (minus dye) is now called "seed-lac," and it is melted and strained through canvas, and after it has cooled to a flaky texture it is the shellac of commerce.

Shellac has long been used for a wide range of purposes, including the making of electrical insulators, buttons, sealing wax, hair sprays, abrasives, cake glaze, and it is still in demand as a wood polish (French polish) for high-quality furniture. Some 17,000–90,000 insects are required to produce 0.5 kg (1 lb.) of shellac. Annual production of lac is more than 25 million kg (50 million lb.); most is exported from India to Europe and the U.S.A. The closely related *Tachardina* is called the forest lac insect; these species have a smaller scale and are not used as a source of lac; Fig. 2.10 shows another species of *Tachardina* on *Michelia alba*.

Cochineal. A red dye produced from the dried and powdered body of cactus mealybugs (*Dactylopius coccus*—the cochineal insect; Dactylopiidae; Coccoidea). This insect is native to Mexico and Peru. About 70,000 insects (dried and pulverized) weigh 0.5 kg (1 lb.). It was a major export of great value for Mexico and what started as a wild gathering became carefully ordered cultivation. In 1760 the annual export of cochineal, mostly to France, was worth French F 12 million; in 1850 the export from the Canary

Islands amounted to 65,000 kg (130,000 lb.). The insects feed on prickly pear cactus (*Opuntia*, especially *O. coccinellifera*). Although prickly pear has now spread with great vigor to all parts of the tropical world, the cochineal insect is less adaptable, but it is established as a commercial venture in the Canary Islands and in Honduras at the present time. The insects are swept off the cacti and collected, three times in the season, and they are killed by roasting in ovens or in hot water (Cloudsly-Thompson, 1986). Using different treatment methods, the dried insect bodies can yield a bright red dye or any shade to orange. Its commercial uses have largely been replaced by the synthetic aniline dyes but in recent years there has been a resurgence of interest, as cochineal is a natural product with minimal side effects to its usage, and it is now being used in cosmetics, homeopathic medicine and in foodstuff coloration. It is no longer being used as a cloth dye, however, for here it was superseded by the aniline dyes.

Cantharidin. This is an urticating substance (blistering) found in the bodies of various beetles in the family now called Meloidae (oil or blister beetles). Probably the best-known species is *Lytta versicator*, the "Spanish fly" of notoriety that is reputed to be an aphrodisiac, and whose careless use has led to many deaths. If ingested, cantharidin causes an inflammatory reaction in the ureters and the urethra but, in small calculated doses, the chemical has medicinal uses for the treatment of certain urinogenital disorders. For this purpose, there is an annual harvest of *Mylabris* beetles in India. *Mylabris* beetles are common and widespread throughout the tropics and should be handled with care; apparently there is a high concentration of cantharidin in the elytra (Fig. 10.67).

Human food. As already mentioned, it is clear that insects must have played a major role in providing food for early humans when at the hunter/gatherer stage of evolution but, since the advent of agriculture, their use has greatly diminished. At the

Fig. 2.10 Forest lac insect (*Tachardina* sp.) (Homoptera; Lacciferidae) encrusting twigs of *Michelia alba*; Hong Kong.

present time, in parts of the developed world some insects are eaten but purely out of gastronomic curiosity; these include honey-ants, chocolate-coated bees/termites, etc. But in parts of the tropics insects are still a dietary item of some importance. The cases most clearly recalled include the following. In Uganda the long-horned grasshopper (*Homorocoryphys nitidulus*) is atypical in that it feeds on Gramineae and periodically swarms in huge numbers; the swarming adults are collected and eaten locally, often raw but sometimes cooked—the local name is "Nsenene." *Macrotermes bellicosus* is very abundant in the vicinity of Lake Victoria and in most villages each nest mound is regarded as the property of one family; at swarming time the women put woven baskets over the entrance hole to trap the young alate forms as they emerge. In much of Africa and tropical Asia it is a common sight to see termite nests being dug open to collect the queen, which is sometimes eaten on the spot. Desert locust is apparently eaten in many parts of North Africa. Bombay locust (Fig. 7.15) is regularly eaten in Thailand and in many street cafés you can buy ten roasted locusts on a satay stick. Various beetle larvae are eaten in different countries, and the true witchety grub of the Australian Aborigines is the larva of *Xyleutes leucomochla* (Cossidae) that lives in a silk-lined tunnel in the soil where it feeds externally on the roots of *Acacia ligulata*. Because of the extensive fat body in the abdomen, insects can be squashed or rendered down to make cooking fat, and this is done sometimes using termites (swarming adults). With lake midges (Chironomidae), collected over Lake Victoria, swarming adults may be collected in vast numbers and be compressed into blocks of "midge butter"; this is also practiced on some of the other Rift Valley lakes in Africa. In Thailand Bombay locusts are collected and pulverized and the mess produced serves as cooking fat. In parts of Africa, honeybee larvae and pupae are eaten with the wax comb in which they live. Several interesting books have been written on the topic of insects as human food (Holt, 1988; Bodenheimer, 1951; Gilbert & Hamilton, 1990).

Insect farming/rearing. Insects have long been collected for museums, reference and teaching collections, as objects of interest or beauty, and for other purposes, but over recent years demand has greatly increased and now, to meet this need, many different insects are being reared either privately or commercially in considerable numbers and the majority are now being sold alive, rather than as dead specimens. It is not easy to categorize the different purposes for which insects are now reared, but an attempt needs to be made.

Pleasure is a somewhat vague category, but in England the company Worldwide Butterflies Ltd. has for many years offered Lepidoptera for sale, either as pinned/dead specimens or else alive as eggs, larvae or pupae for rearing, and they now offer many other insects for sale. These insects can be bought for home-rearing using locally available food plants. The attractiveness of flying butterflies has led to the establishment of many Butterfly Houses in zoos and leisure centers scattered around the U.K., and Butterfly Farms are being developed in many parts of the country.

Because of the attractiveness and rarity of the S.E. Asian birdwing butterflies (Fig. 2.11), some years ago they were declared to be protected species in Papua New Guinea and their capture made illegal. But an enterprising recent venture is the rearing in captivity of several spectacular species of *Ornithoptera* and others, in Papua New Guinea as a cottage industry, so that specimens can now be exported commercially without endangering the local wild populations. In England it is now possible to buy several different species of stick insects (Phasmidae) for home-rearing and as pets; several species are easy to feed as they take privet leaves or bramble, etc.; they are long-lived, easy to rear, and very popular with children.

Teaching and the previous category clearly overlap; most Butterfly Farms have regular visits from local schools. Because of the importance of the Insecta to human

Fig. 2.11 A selection of exotic tropical butterflies in a display case; six are spectacular birdwing butterflies from S.E. Asia (Lepidoptera; Papilionidae).

society there is clearly a need for specimens of many different types for teaching purposes at all levels from secondary schools to M.Sc. degree courses. Cockroach or locust specimens have traditionally been the basic insect type studied at school and at university/college; in countries such as England these cannot be collected locally, and they have to be either purchased from a biological supplier or locally reared.

Research. Different types of insects are used in a wide range of research work and some are bred especially for this purpose. Government research establishments and universities use a wide range of different insects for different purposes. All the Agrochemical Companies that manufacture insecticides rear their own insect pests for testing the candidate pesticides, as well as some of the beneficial species; many companies regularly rear more than 25 species of insects and mites.

Animal food. Many vertebrate animals kept in captivity include insects in their diet; some are fed entirely on insects others just partially, and the insect food is reared commercially, often with great financial success. In China and in parts of S.E. Asia, chironomid larvae are reared in small shallow concrete pans, usually adjacent to fish ponds; others are harvested from the shallow edges of fish ponds. After cleaning, the bright red larvae (Fig. 2.12) are sold as live food for tropical fish kept in household aquaria. Mealworms (larvae of *Tenebrio molitor*; Col.; Tenebrionidae) (Fig. 2.13), crickets, and locusts are reared as animal food in the U.K., many thousands being reared and sold each week.

Fish-bait. The most popular hobby in the British Isles is reputedly fishing—mostly freshwater coarse fishing. The most-used bait is blowfly maggots (Calliphoridae). The maggots are bred and reared commercially in dozens of small cottage industries throughout the British Isles, and a few large commercial companies. The final instar (third) larvae are used as fish bait. Farmers who wish to pollinate their own varieties of *Brassica* crops buy consignments of blowfly puparia from the fish-bait dealers and the emerging flies will pollinate the cabbage and sprouts plants very effectively.

Fig. 2.12 Bloodworms (*Chironomous* spp.) (Diptera, Chironomidae) collected and sold as food for aquarium fish in South China and Hong Kong.

Fig. 2.13 Yellow mealworms (*Tenebrio molitor*) (Coleoptera; Tenebrionidae) reared as food for cage birds and other small livestock; South China.

Bio-control projects. A number of insect predators and parasites are reared commercially for use in biocontrol programs, some in the open and others in commercial greenhouses and other forms of protected cultivation. Some phytophagous species are reared for weed control. But these aspects are looked at in more detail in the following section.

Crop Pests. Although crop pest insects do cause damage to the host plant they do not necessarily always cause a yield reduction. The final criterion when considering a crop pest is usually the yield at harvest, or more specifically, the monetary value of the yield. In recent years the quality of produce has often become paramount with the introduction of quality grading in the U.S.A. and the E.E.C. Size of "fruits" can be an important factor; with some crops, fruits that are too small (e.g. apples) are down-graded and thus worth less, whereas in other cases, such as potatoes, cauliflowers, etc., if too large they can be difficult to sell.

The several ways in which crop pests are regularly found to have a beneficial effect (albeit often slight) on final produce yield are as follows:

1. Compensatory yield (spacing/edge effect)—commercial crop plant spacing is often selected as a compromise between various (often conflicting) factors. If a plant stand is reduced due to pest attack the adjacent plants usually grow larger. Rows of cereal plants have often been destroyed and then the plants of the adjacent rows have been larger, produced more tillers and grown more strongly, so that, following a 5–10% reduction of plant stand, there have been many cases recorded where total grain yield was slightly increased overall.

2. Pruning effect—destruction of twigs/shoots/buds of trees or bushes by insect pests sometimes has a "pruning" effect, and the overall plant habit may be improved (physically sturdier and stronger). Reduction of flowers in some years may also be beneficial (see 4).

3. Uniformity of yield and simultaneous ripening—successful mechanical harvesting of large crops does necessitate to some extent simultaneous ripening of the crop; one of the problems of modern agriculture. One effect of pea midge in the U.K. is to destroy the growing tip and arrest plant growth. When this occurs after four trusses are set, the result is often four pairs of pods that ripen simultaneously; this is a very desirable quality in a crop of peas for freezing.

4. Improvement of "quality" or salability—some apple pests destroy flowers and reduce fruit set, and the result can be a small yield of fruit but larger apples which rate a higher grading and are thus of greater value. Also some potato pests do not reduce the number of tubers but they are smaller in size and more readily find a market at certain times of the year.

5. Biochemical changes—in some fruits it has been reported that slight pest damage has resulted in biochemical changes so that the fruit tastes sweeter (has a higher sugar content).

6. Change in agricultural practice—the classic case is in Georgia, U.S.A., where the cotton boll weevil became so devastating and uncontrollable that the local plantation crop was reluctantly changed from cotton to groundnut; it turned out that conditions were ideal for groundnut and many farmers became very prosperous. In gratitude, the farming community of one town has apparently erected a statue of cotton boll weevil!

NATURAL AND BIOLOGICAL CONTROL

Natural control is the insect population control (reduction) whereby population fluctuations are regulated by the actions of a complex of naturally occurring predators, parasites, and pathogens causing diseases. All these species belong to the same ecological community, share the same habitat and generally they have all evolved together over many millennia.

Under the artificial conditions created by widespread agriculture, often a pest population explosion is produced by the provision of unlimited food. It is often found, however, that the natural enemies increase in numbers correspondingly, and so natural mortality rates for the pest may remain high so that the final pest population may not become so large after all. There is, of course, a time-lag between the growth of the pest population and the growth of the predator and/or parasite population. Thus natural control of pests on short-lived annual crops such as most vegetables is seldom of great importance. Parasites are usually prey (host)-specific and some predators are prey-specific and so their population growth is completely dependent upon the host (pest) organism, and there will be a definite time-lag in their population growth. A few parasites and many predators are polyphagous and nonspecific in their choice of prey, however, and these enemies can have an immediate population-controlling effect on a pest.

For insect and mite pest populations in perennial orchards and on plantation crops, natural control is of tremendous importance because in these long-term crops the predacious and parasitic species have time to build up their populations to levels at which they suppress the pest population. With parasites, predators, and pathogens, it is important to remember that some are quite host-specific, probably restricted to a single genus of host insect (being monophagous). Whereas the oligophagous natural enemy may feed on members of the same family, the polyphagous forms will attack a wide range of host/prey insects; the crucial factor in host/prey selection may often be purely size rather than biochemistry. Parasites and pathogens tend to be more specific, and predators tend to be less specific in their choice of food; there are exceptions to both generalizations. Dietary limitations have pronounced results on the effectiveness of a natural enemy in controlling a pest population. Specific parasites will die out in the absence of the natural host, but polyphagous predators can turn to other sources of food and will thus survive.

Natural control is present in all pest ecosystems at all times (to varying degrees) so it is difficult to appreciate the extent of its population controlling effect. This is especially true since many insect pests have spectacular fecundity and, in crop situations, the only reason we are not inundated with pest insects is the high rate of natural mortality caused by the predators, parasites, and pathogens. In the tropical parts of the world, many researched lepidopterous pests causing extensive damage (such as cereal stem borers) have been shown to suffer a regular combined egg and first instar larvae mortality rate of 95% or thereabouts, and these are still pests of economic consequence! In the two life-tables shown in Davies (1988) it is clear that unprotected insect eggs show very high levels of mortality due to natural enemies—90% and more is typical.

Pest species are typically very aggressive biologically and physically robust, and are not easy to kill with insecticides, whereas predators and parasites are typically more delicate and more easily poisoned. The extent of natural control in some pest ecosystems has been displayed many times now—in the past, when the injudicious use of persistent and toxic pesticides (usually the organochlorine compounds: D.D.T., dieldrin, etc.) in orchards and plantations killed off most of the predators and parasites, there was a tremendous pest population resurgence with a corresponding spectacular increase in damage. Oil palms in parts of S.E. Asia were completely defoliated by bagworms and slug caterpillars (Conway, 1972). In Europe and North America, red spider mites became major orchard pests when the predatory mites on the apple foliage (and others) were killed by the D.D.T. used for codling moth suppression. Various Coccoidea also became serious orchard pests, especially mussel scale, when the organochlorine insecticides killed off the chalcid and braconid parasites with far greater ease than they killed the scales. Both apple and citrus orchards reached a point about 20 years ago where a scale or spider mite population resurgence could be guaranteed to follow a single foliar spray of D.D.T. or dieldrin.

The original concept of *integrated control* was developed in the 1960s when it became apparent that many chemical control programs were endangering the existing natural control already in effect locally. It was of even greater importance in California where the citrus industry had been saved from devastation by Coccoidea by several parasites and predators introduced from South China and Australia. Integrated control stressed that chemical application should be compatible with existing natural control and any applied biological control measures.

With the recently developed integrated pest management (I.P.M.) approach to pest control, it is of major concern to perpetuate existing levels of natural control, and to enhance these levels if possible. Generally, the importance of natural control of insect pests in all parts of the world in all situations just cannot be over emphasized.

Biological control of insect pests can be regarded very broadly when a diverse range of approaches can be included but, in the strict sense, it is limited to the deliberate introduction of predators, parasites or pathogens to deplete the pest population. The pathogens are viruses, fungi, and bacteria, and are beyond the scope of this chapter. Similarly, endoparasitic nematodes are excluded, and all the vertebrate and spider predators of insect pests.

Parasitic insects are mostly confined to a few groups.

Strepsiptera—attack some Homoptera and aculeate Hymenoptera; not of any real economic importance.

Diptera—Tachinidae (all are insect parasites—very important).
    —Phoridae, Pipunculidae, Bombylinidae, etc.; several families have some species that are parasites as larvae.

Hymenoptera—Braconidae (larval endoparasites of Homoptera, Lepidoptera, Coleoptera, and many other groups).
    —Ichneumonidae (also endoparasites of Homoptera, Lepidoptera, Coleoptera, and other groups).
    —Chalcidoidea (some 20 families involved) attack all groups of insects, and are very important in natural pest control.
    —a few other families, such as Bethylidae, Scelionidae, Proctotrupidae, etc.

Insect predators occur in almost all of the insect orders, but the major groups are those listed below:

Odonata—all families; adults aerial, nymphs in freshwater.

Dictyoptera—Mantodea; all families, both adults and nymphs.

Hemiptera—Heteroptera; many families. Anthocoridae, Reduviidae, aquatic bugs, etc., both adults and nymphs.

Neuroptera—Chrysopidae, Hemerobiidae (adults and larvae), and several other families.

Coleoptera—Carabidae, Staphylinidae, Coccinellidae (both adults and larvae); also some Histeridae, Dytiscidae, Hydrophilidae, Cleridae, Meloidae (larvae only), etc.

Diptera—some Cecidomyiidae (larvae), Asilidae (adults), Syrphidae (larvae); and Therevidae, Conopidae, etc.

Hymenoptera—Vespidae, Scoliidae, Formicidae (adults, and larvae in nest), and some other families.

Acarina—Phytoseidae and some other families; eat other mites and insect eggs usually.

Araneida—all families of spiders are now known to be very important predators of insects and insect pest populations.

In addition to the control of insect pests by other insects that are natural enemies, there have been some very spectacular cases of noxious weeds being controlled by certain phytophagous insects (see following page). Generalization concerning the groups of insects used in weed control is not very meaningful for it has mostly been a few species of the major phytophagous families in the Hemiptera, Coleoptera, Lepidoptera, Diptera, etc.

Most crop plants come originally from one small part of the world and, because of their desirable qualities, they are taken to other regions of suitable climate for extensive cultivation. Usually some of their major pests are taken with them from the country of origin to the new locations, often on other continents. In the new locations, the crop plant often flourishes, for various reasons; and, in these new locations, the insect pests that have traveled with the plants usually find a very reduced range of natural enemies to attack them. This absence of natural enemies permits the insect pest to increase its numbers and there is often great damage done to the crop plants. Typically in the country of origin, the pest insect was usually not too serious as its numbers were kept in check by the local natural enemy complex. If an exotic pest is causing a problem locally then a search in its native country might reveal a suitable predator or parasite that could be imported to control the local population outbreaks. Most of the early work on biocontrol was done by entomologists in California at the turn of the century when they went to South China to collect parasites and predators of scale insects that were threatening the newly established *Citrus* industry in the U.S.A.; at the University of California a unique Department of Biological Control was established, with great success.

The literature describing successful biocontrol projects against insect pests of crops is now quite extensive and there are several major books (DeBach, 1974; Wilson, 1960; Delucchi, 1976; Greathead, 1971; Huffaker, 1971; etc.) and a series of *Reports* and *Technical Bulletins* published by C.I.B.C. (of C.A.B.I.). There has of course been a very large number of failures, for a range of different reasons (see Hill, 1973; p. 89), but there are well-documented cases of some 200–300 insect and mite pests being spectacularly controlled by an imported natural enemy, with considerable financial savings; some of the most successful projects are summarized in Hill, 1973 (pp. 83, 85, 91) and Hill, 1987 (pp. 107, 108). In addition to a number of national and international biocontrol projects worldwide, biocontrol lends itself to use in the controlled environment of commercial greenhouses and for some years now pest control in commercial greenhouses in much of Europe and North America has been achieved with the use of a series of insect parasites and predators that are available commercially from special suppliers (summarized in Hill, 1987; p. 110, etc.).

Many noxious weed species are South American in origin and now established throughout Africa and Asia, and most are difficult to control. But a number of semi-xerophytic weeds or ruderals from the Mediterranean Basin have become established in dry areas of Australia in the livestock pastures. In Australia the plants have become rampant weeds and have attracted the attentions of C.S.I.R.O. A concerted biocontrol program was mounted by C.S.I.R.O. several decades ago and now several of the more noxious weeds are under a measure of control (Wilson, 1960, gives a history of earlier efforts). A classic case was the prickly pear cactus (*Opuntia* spp.) introduced into Australia in the 1800s; several species escaped from captivity and two species became established as weeds and by about 1900 it spread rapidly throughout Queensland and New South Wales until by about 1925 some 60 million acres were covered with the cactus and useless for agriculture (DeBach, 1974).

A number of different insects were collected from South America and brought back to Australia, but the outstanding success was the pyralid moth *Cactoblastis cactorum* imported in the 1920s—the larvae tunnel inside the stems of the cactus and they are host-specific. Vast acreages of *Opuntia* were effectively destroyed by the caterpillars. A similar project was conducted in California (Huffaker, 1971).

An interesting temperate example of weed biocontrol is the control of tansy ragwort (*Senecio jacobaea*), a biennial weed poisonous to cattle, by the monophagous defoliating caterpillar of the cinnabar moth (*Tyria jacobaeae*; Arctiidae). In the U.K. and Europe, the caterpillars occasionally defoliate areas of ragwort, sometimes quite spectacularly, but they have little effect on the long-term population spread of the weed. The weed has been widely dispersed throughout the New World, Australasia and South Africa; in parts of the U.S.A. and New Zealand the moth was introduced and a measure of control was achieved.

Some of the recent C.S.I.R.O. successes in exotic weed control in Australia were a multipronged attack on the weed plant by several different insects on different parts of the plant body simultaneously, and in some cases a specific pathogen was also used. The combined effect of the several natural enemies in concert has been sufficient to achieve a fair measure of population control.

In all pest situations it is important that the existing natural population control is not affected by any projected pest control measures; if at all possible, measures should be taken that enhance the existing natural control levels. But deliberate introduction of natural enemies into the agroecological community (i.e. biocontrol *sensu stricta*) is not often feasible. It is generally only possible if the pest is truly exotic (from another region/continent) and if there is a vacant niche in the local natural enemy spectrum, and then the candidate predator or parasite has to be imported from the pest's native home region. Most biocontrol projects are more likely to be successful in warm tropical regions where development continues all the time; in temperate regions with a long, cold winter and slow insect development, they are less likely to work, but winter moth has been successfully controlled in Canada by parasitic Tachinidae and Ichneumonidae.

When biocontrol of a pest population is possible then it does have several definite advantages over the use of chemical pesticides, namely:

1. A single introduction (of sufficient size), can result in long-term (permanent) control of the pest, and precise timing is not required.

2. No development of resistance by the pest.

3. Absence of toxic effects.

4. No poison residues or accumulation in food webs; no environmental pollution.

5. No destruction of natural enemies or pollinators.

Some of the more successful cases of pest biocontrol have as a frequent result a single introduction giving an annual saving of millions of dollars, which may continue indefinitely.

Fig. 2.14 shows parasitized aphids (mummies) on a stem of Chinese cabbage; Fig. 2.15 shows *Tachardina* scales with parasite emergence holes on a twig of lime, Penang; Fig. 2.16 is a parasitized *Ceroplastes* scale on a twig of *Annona* in Ethiopia.

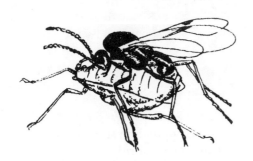

Nymph being parasitized

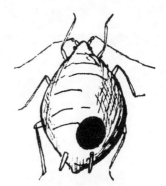

Mummified nymph
with parasite
emergence hole

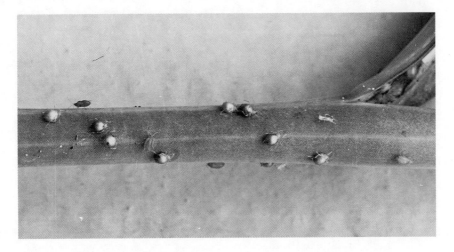

Fig. 2.14   Aphids parasitized by braconid wasps, and mummified aphid nymphs on stem of Chinese cabbage; Hong Kong.

Fig. 2.15 *Tachardina* scales on twig of lime tree, with parasite emergence holes; Penang, Malaysia.

Fig. 2.16 *Ceroplastes rusci* scale on twig of *Annona* with parasite emergence hole; Alemaya, Ethiopia.

3

Insect and Mite Pests

When considering an insect/host situation in regard to its being a pest problem, it is first necessary to be certain that the insect really is a pest. There has often been extensive activity and much expense over situations when the insect concerned was really not doing enough damage to be defined as a pest.

Pest. A pest is any insect (or organism) causing harm to man or his livestock, or damage to his crops, cultivated plants or possessions. *Damage* (or harm) is the key word, and it can be regarded very broadly to include even nuisance or disturbance—annoyance can result in a loss of mental concentration and hence work output. Thus, in Africa, face fly can at times be regarded as a pest in this respect, as can the nocturnal buzzing of a mosquito that can prevent sleep.

Pest species. This is a species whose members are often, or usually, of pest status, causing damage of economic consequence, and usually requiring population control measures. But the individuals may not necessarily be of pest status for they might be on hosts of no economic importance (wild plants) or they may be present only in very small numbers. Because an insect belongs to a pest species, it should not be assumed that it itself is a pest—evidence is required to prove it is a pest.

Economic pest. In agriculture, the important consequence of pest attack is damage (to the crop plants, etc.) leading to a loss of yield (both a loss of quality and quantity) and eventually a loss of money. When losses reach a certain level then the damage done warrants the insect being described as an economic pest (in that particular situation). The value of the produce is clearly a major factor in this assessment, but several authors have argued for a general guideline that an insect species reaches *pest status* when there is a 5–10% loss of yield. Related to this is *economic damage* which is the amount of damage done by a pest that will financially justify the cost of applying control measures (often an insecticide).

Economic injury level (E.I.L.). This is the lowest pest population that will cause economic damage. Thus, applied entomologists are concerned with pest population levels in order to try to prevent economic damage being done. The ultimate object of pest control programs is essentially the prevention of damage being done, although too often our efforts are directed at damage reduction because of unsuccessful anticipation.

Economic threshold. In the quest to prevent economic damage from being done, the object is to manipulate the pest population so as to keep it at a level where damage is insignificant. So the economic threshold is the definition of the population density of an increasing insect population at which control measures should be started to prevent the population from reaching the economic injury level. Thus a great deal of basic research into the ecology and life history of all the different insects that affect human society is needed in order to understand how best to manipulate their population development.

Clearly when an insect belongs to a pest species and is encountered in a situation where it may be harmful, it is best to be aware that it might well be causing damage but it should never be assumed that it is; positive evidence should be sought.

Pest populations. As already mentioned, an insect is only a pest (by the strict definition) at, or above, a certain population density when a particular level of damage is achieved. The basis for the great majority of pest control programs is just population reduction and not extermination. An interesting case has recently arisen in the U.K. where *Gryllotalpa gryllotalpa* (European mole cricket) has now become so scarce that it has been declared an "endangered species" and is afforded legal protection—a far cry from the situation with *Gryllotalpa africana* (African mole cricket) in Africa and southern Asia.

Pest damage. There are several ways in which damage to hosts is done, and sometimes for convenience they are grouped under two headings—direct and indirect damage. On a crop basis, direct damage is when the insect directly affects the plant—the most direct is when it attacks the part of the plant to be harvested (e.g. cotton bollworms; apple fruitworms) and the less direct is when other parts of the plant are attacked with a resulting general debilitation (e.g. a cotton stem borer or leafworm). Indirect effects include the insect acting as a disease vector, or producing copious honeydew (e.g. recent problems with whitefly honeydew making cotton lint sticky), or causing disturbance (the buzz of warble fly wings can cause cattle to stampede).

In matters of pest control, the first two most important criteria are the identity of the pest and an assessment of the damage being done to the host/crop.

Damage assessment. Accurate assessment of the damage is vital, for often a superficial inspection can be misleading. In the past much effort has been wasted over control efforts that were not really necessary and, in some cases, serious infestations were left unchecked because, at the time of examination, the damage potential was grossly underestimated. Each crop, or host, needs to be studied in detail in relation to the serious pests that attack it in each of the major regions of the world, so that control measures are only employed when really necessary. For any particular pest on a specific crop, it is usual to make a damage assessment (or infestation level) on a simple numerical scale for quick and easy reference, and to facilitate widespread use of the technique. The simplest scales are usually as follows:

Damage	Scale	Pest population (infestation) level
Very severe (VS)	4	Very common (VC)
Severe (S)	3	Common (C)
Mild (M)	2	Uncommon (U)
Very mild (VM)	1	Rare (R)
None	0	Absent

In some publications the numerical scale is reversed with 1 representing the highest level, but the consensus of opinion is that 1 is always the lightest level of damage. A scale of 1–4 is very simple to use but the categories are rather broad, and for some purposes may be too broad. Thus entomologists often use a scale of 1–6 for research purposes, and plant pathologists often use a scale of 1–9 as disease symptoms on plants are often more easily quantified. Research will often produce a specific figure for the economic injury level or the economic threshold for a particular pest on a crop, for example, two apple aphids per shoot, or five cereal aphids per flag-leaf. Thus the growers and farmers are advised to apply a spray of insecticide when these levels of infestation are observed, so long as weather conditions are fine.

New pests. In recent years there have been several dramatic upsurges of insect pests that have caused great concern. There are three particular ways in which a new pest is created, as shown by these recent developments.

Changing distribution is a problem that in parts of the world looms alarmingly—in the U.K. the fear of introduction of Colorado beetle is ever present; in California and Florida, in the U.S.A., various fruit flies annually threaten the success of the citrus and the fruit industry generally. The International Phytosanitary Agreement sponsored by F.A.O. is based on the need to restrict distributional spread of major pests. But a series of major pests have recently spread to new countries (indeed, new continents) and have successfully resisted eradication attempts and are now locally established and spreading; these include the following insects:

Cochliomyia hominivorax (screwworm fly) from the southern U.S.A. into North Africa. (Now [1992] thought to be eradicated.)
Macrosiphum albifrons (lupin aphid)—from the U.S.A. into southern England.
Mononychus tanajoa (cassava green mite)—from South America into East Africa.
Phenacoccus manihoti (cassava mealybug)—from South America into West Africa.
Prostephanus truncatus (larger grain borer)—from Central America into East Africa.

In 1988 several swarms of desert locust arrived in the West Indies, presumably on the high altitude airstreams from North Africa, but, so far as is known, the species did not establish locally.

New agricultural practices were alleged to be responsible for the rise of *Nilaparvata lugens* (brown planthopper of rice) from a minor pest to the most serious major pest of rice throughout S.E. Asia; these included the use of new varieties (high-yielding) of rice with tall lush foliage, and the use of nitrogenous fertilizers. Many of the new high-yielding crop varieties, and varieties more tolerant of growing conditions (now being grown in different seasons), are more susceptible to pest attack than the older varieties. Widespread cultivation of only a few varieties of a crop (e.g. maize in the U.S.A.) and monoclonal propagation and cultivation (e.g. oil palm in S.E. Asia) are both running great risks from the point of view of vulnerability to pest attack.

New physiological races (= biotypes) of pests occasionally develop locally and suddenly the insect becomes a more important pest, either because it causes a more serious damage or because it becomes resistant to pesticides. Brown planthopper of rice comes into this category also for it is creating new biotypes with alarming rapidity in different parts of its range in S.E. Asia, which makes control using insecticides very difficult. Since evolution is thought to be constantly in progress, bearing in mind the speed of development of many insect species, it would not be unexpected to find that some insects have virtually evolved into new species since the original description was made maybe 200 years ago.

AGRICULTURE

In this context, agriculture refers to the cultivation of plants for food and products, including ornamentals, but does not include forestry. In its very broadest sense (as in the title of this book), it can include the cultivation/rearing of biological organisms for human purpose; in its narrowest definition it refers only to the large scale cultivation of field crops (cf. horticulture). The bulk of this book relates to insects and mite pests of cultivated plants (including forestry) for in considering the world biota, by far the greatest proportion of the Animal Kingdom are phytophagous grazers.

When considering agricultural pests it is evident that some members of most of the 29 insect orders can be regarded thus, although some are very minor. The total list of families that are thus engaged is too long to contemplate inclusion here, so reference will be made only to the main groups (larger families) where the majority of species are phytophagous. The text in Part B of the book will reveal just how many groups of insects can be regarded as pests of cultivated plants.

The main groups of Insecta and Acarina that are of concern to the agricultural entomologist (*sensu stricta*) are as follows:

Order Thysanura (silverfish, etc.; primitive biting mouthparts).
    Family Lepismatidae—several species are domestic pests of foodstuffs.
Order Collembola (springtails; tiny, wingless; biting mouthparts).
    Several families—damage seedlings and soft-stemmed plants; often abundant.
Order Orthoptera
    Gryllidae (crickets)—soil dwellers; nocturnal; omnivorous.
    Gryllotalpidae (mole crickets)—fossorial; root-eaters.
    Tettigoniidae (long-horned grasshoppers)—arboreal; nocturnal defoliators.
    Pyrgomorphidae (stink grasshoppers)—aposematic diurnal defoliators.
    +Acrididae (locusts; grasshoppers)—grassland; diurnal; some gregarious.
Order Phasmida
    Phasmidae (stick insects)—arboreal; cryptic; defoliators; S.E. Asia mostly.
Order Isoptera (termites; polymorphic; social nest builders in tropics).
    +Termitidae (mound-building termites)—subterranean; large nest mounds usually.
Suborder Homoptera (plant bugs; all are sap suckers; adults and nymphs together).
    +Cicadellidae (leafhoppers)—on both Gramineae and bushes/trees; feeding causes "hopperburn"; virus vectors; some are migratory.
    +Delphacidae (planthoppers)—same as for Cicadellidae above.
    +Aleyrodidae (whiteflies)—on herbaceous hosts/bushes; produce honeydew; virus vectors.
    +Aphididae (aphids)—on foliage of all plants; honeydew; virus vectors.
    +Pseudococcidae (mealybugs)—white, waxy, wingless; mostly tropical; honeydew.
    +Coccidae (soft scales)—female reduced, scalelike; honeydew.
    +Diaspididae (armoured scales)—hard scale over female body; woody hosts mostly.
Suborder Heteroptera (all are sap suckers with toxic saliva, cause necrosis).
    +Miridae (capsid/mosquito bugs)—on all types of hosts; worldwide.
    Coreidae (plant bugs)—large brown bugs, or smaller thin ones.
    Pentatomidae (shield and stink bugs)—mostly large, brightly colored.
Order Thysanoptera (thrips; tiny; straplike wings; rasping mouthparts).
    Thripidae (thrips)—feeding causes scarification; all types of hosts.
    Phlaeothripidae (leaf-rolling thrips)—cause leaf-folding and leaf-rolling.

+Families marked thus are especially important.

Order Coleoptera (beetles; biting and chewing mouth-parts; diversity of habits).
    Scarabaeidae (cockchafers; white grubs)—larvae in soil eat roots/tubers.
    Coccinellidae (ladybird beetles)—Epilachninae phytophagous (others predatory).
    Meloidae (flower/pollen beetles)—adults eat flowers, some eat leaves.
    Cerambycidae (longhorn beetles)—larvae bore stems, and tree trunks.
    +Bruchidae (pulse beetles)—larvae eat seeds in legume pods.
    +Chrysomelidae (leaf beetles)—many different phytophagous groups here.
    Apionidae (pod weevils; sweet potato weevils)—larvae bore in seeds, tubers.
    +Curculionidae (weevils proper)—great diversity; adults may eat foliage, seeds; larvae bore in plant tissues, eat roots in soil, in seeds, etc.
Order Diptera (flies; larvae are the pest stage in crops—adults never).
    Cecidomyiidae (gall midges)—larvae gall all parts of all plants; worldwide.
    Agromyzidae (leaf miners)—larvae mine leaves, gall stems, etc.
    Anthomyiidae (root flies, etc.)—larvae in roots, stems, leaves, or in soil.
    Muscidae (shoot flies)—larvae bore cereal seedling shoots (dead-hearts).
    +Tephritidae (fruit flies)—larvae in ripening fruits of all types.
Order Lepidoptera (only larvae cause real damage; adults are fluid feeders).
    Gracillariidae (leaf miners)—tiny; larvae mine leaves of shrubs/trees.
    +Gelechiidae (stem borers, etc.)—larvae more in stems, shoots, fruits.
    Limacodidae (stinging/slug caterpillars)—larvae eat leaves of bushes/trees.
    +Tortricidae—diverse group; larvae bore fruits, pods, seeds; web leaves.
    +Pyralidae—large diverse group; larvae bore stems, roll leaves, bore fruits.
    Lycaenidae (blue butterflies)—larvae inside legume pods and fruits.
    Pieridae (white butterflies)—pests of Cruciferae, legumes, etc.
    Papilionidae (swallowtails)—larvae eat leaves of Rutaceae, etc.
    Hesperiidae (skippers)—larvae on leaves of Monocotyledonae; may roll leaves.
    Geometridae (loopers)—larvae eat leaves of trees and bushes; most temperate.
    Lasiocampidae (tent caterpillars)—hairy gregarious larvae eat leaves of trees.
    Sphingidae (hawk moths; hornworms)—large larvae eat foliage; solitary.
    Arctiidae (tiger moths)—hairy larvae eat leaves and fruits; aposematic.
    +Noctuidae (owlet moths)—very large, very diverse group; leaf eaters; stem/fruit borers; armyworms; cutworms; worldwide; very important.
    Lymantriidae (tussock moths)—hairy larvae eat leaves, may defoliate.
Order Hymenoptera
    Tenthredinidae (sawflies)—larvae eat leaves; mostly Holarctic; some gregarious.
    Formicidae (ants)—some pests with Homoptera; leaf-cutting ants in New World.
Order Acarina (Mites; tiny in size; mouthparts, stylets, piercing and sucking).
    Eriophyidae (gall/rust mites)—make foliage galls or bronze leaves; scarify.
    +Tetranychidae (spider mites)—scarify and web foliage; leaves wilt.

In Table 3.1 are listed some of the best-known plant diseases transmitted by insects and mites.

Table 3.1   Some Plant Diseases Transmitted by Insects and Mites

Vector	Disease	Parasite	Transmission	Distribution
Antestiopsis spp.	coffee berry rot	*Nematospora* spp.	Feeding punctures;	Africa
Dysdercus ssp.	cotton staining	*Nematospora* spp.	biological and mechanical	Pantropical
Calidea spp.	cotton staining	*Nematospora* spp.	transmission	Africa
Nezara viridula	cotton staining	*Nematospora* spp.		Pantropical
Scolytus spp.	Dutch elm disease	*Ceratostomella ulmi*	Mechanical; on insect body	Europe, N. America
Platygaster sp.	coffee leaf rust	*Hemileia vastratrix*	Spores on insect body	Africa
Frankliniella occidentalis	ear rot of corn	*Fusarium moniliforme*	Feeding	U.S.A.
Aphids and leafhoppers	fire blight	*Erwinia amylovora* (Bacteria)	Contamination	Europe, N. America
Aphis craccivora	groundnut rosette	Virus	Feeding	Pantropical
Myzus persicae (100 diseases)	cauliflower mosaic	Virus	Feeding	Europe, U.S.A.
	potato leaf roll	Virus	Feeding	Europe, U.S.A.
	sugar beet mosaic	Virus	Feeding	Europe, U.S.A.
	sugar beet yellows	Virus	Feeding	Europe, U.S.A.
Pentalonia nigronervosa	banana bunchy top	Virus	Feeding	Old World tropics
Toxoptera citricidus	tristeza disease	Virus	Feeding	Africa
Bemisia tabaci	cassava mosaic	Virus	Feeding	Pantropical
	cotton leaf curl	Virus	Feeding	Pantropical
	sweet potato B	Virus B	Feeding	Pantropical
Nephotettix spp.	rice dwarf	Virus	Feeding	India, S.E. Asia
	rice yellow dwarf	Virus	Feeding	to China
	transitory yellowing	Virus	Feeding	and Japan
Sogatella furcifera	rice yellows	Virus	Feeding	India through
	stunt disease	Virus	Feeding	to China
Laodelphax striatella	northern cereal mosaic	Virus	Feeding	Europe, Asia
	oat rosette	Virus	Feeding	Europe, Asia
	rice stripe	Virus	Feeding	Asia
Pseudococcus njalensis	swollen shoot of cocoa	Virus complex	Feeding	W. Africa
Thrips tabaci	tomato spotted wilt	Virus	Feeding (nymph)	Cosmopolitan
Cecidophyopsis ribis	currant reversion	Virus	Feeding	Europe
Eriophyes tulipae	wheat streak mosaic	Virus	Feeding	Cosmopolitan

FORESTRY

This refers to woody plants, usually trees, grown as a source of timber, or for wood pulp, for firewood, or for a wide range of different purposes. Trees grown for fuel wood typically receive no protection from insect pests, although they will often have to be protected from goat browsing while at the seedling and sapling stages. The usual practice in tree cultivation is to grow seedlings in a protected bed and to transfer them to narrow polyethylene bags/tubes (pots) as established seedlings. The seedlings are left exposed in the nursery, packed together, to grow and harden off. At this stage they may be given pesticide treatments if warranted; they might also have been given a pesticide dip at the time of potting. At tall seedling or small sapling stage the tiny tree is planted out, sometimes with a stake and a stem-protector, but sometimes naked, and typically it is then on its own. Planting-out takes place usually at the time when rains are expected. In some areas, aerial seeding is practiced but germination success and plant stand varies considerably; aerial seeding tends to be more successful as a means of cover establishment and erosion control rather than for timber production.

Seedlings are, of course, vulnerable to almost all of the pests that smaller crop plants suffer from, and protected seed beds may actually encourage field crickets which will destroy the tree seedlings. Seedlings in pots have passed the first and most vulnerable phase of growth but are still small enough to be easily killed or maimed. Seedlings and saplings are vulnerable to apical shoot destruction which can convert a standard tree into a bush by inducing secondary branching. But, once established, a tree can generally withstand considerable damage with minimal effect. Leaf loss is to be expected as all plants suffer natural grazing of the foliage being the basis of most terrestrial food webs. Sap sucking by bugs, and leaf eating (leaf loss) of up to 20% or more is usually tolerated quite easily, for most trees have a high leaf area index, and there will be compensatory growth. But trees grown for quality timber are vulnerable to trunk borers; although much research has now shown that most established trees are only liable to be attacked when stressed.

Natural temperate forests differ from the tropical in often having only a single dominant tree species over large tracts, either oak, beech, birch, or a conifer in more northern locations, and some insects have evolved to take advantage of this natural monoculture. These are really the only natural (that is, not human-induced) insect population explosions other than locusts and armyworms. From time to time vast areas of either coniferous forest or broad-leafed deciduous forest, both in Eurasia and North America, are defoliated by huge numbers of caterpillars or sawfly larvae; but the greatest population outbreaks tend to be in North America with insect species accidentally introduced many years ago from Europe and Asia.

The practice of forestry usually starts as exploitation of natural climax vegetation, and is then followed by planting of replacement trees after the felling of the "wild" trees. Eventually, most tracts end up being composed entirely of deliberately planted trees. Thus the original wild monoculture is replaced by another cultivated one of either the same species or else a different species. In many parts of the temperate northern zones, natural broad-leafed deciduous forest has been replaced by faster growing conifers, and similarly in many tropical areas *Eucalyptus, Acacia* and other legume trees, and other exotic fast-growing trees are being cultivated, rather than indigenous species. Such extensive artificial monocultures are vulnerable to various insect pests (and diseases), especially leaf-eating caterpillars and some bark beetles. In North America there have been many serious population outbreaks of gypsy moth, winter moth, pine sawflies, etc.;

in these outbreaks, vast tracts of forest have been completely defoliated by gregarious larvae and much damage was done. Pesticide control was seldom successful but various insect predators and parasites have been established and are exerting a measure of population control. In the modern world, the need for wood pulp is tremendous, and in the tropics there is a similar demand for fuel wood, so there is a tendency to replace the indigenous hardwood trees (which are slow-growing) by fast-growing, soft wood, exotic species that can be harvested sooner. In northern temperate regions, and others, this is often *Pinus* (or some other conifer) and in most of the tropics it is *Eucalyptus,* although a few tropical species of *Pinus* exist and some are used. The overall undesirability of this practice is really self-evident and it is slowly being recognized worldwide; there is now pressure in most countries to plant or replant with a mixture of local hardwood trees and fast-growing exotic species.

Tree seedlings in the nursery may be attacked by almost all or any of the groups listed for "Agriculture," as already mentioned, but some groups/species are of particular importance. Saplings are threatened by a few species—for example, in Ethiopia, cottony cushion scale attacked most species, *Acacia* were also susceptible to *Ceroplastes* scales, and young *E. globulus* were debilitated by eucalyptus psyllid populations on the shoots, most of which died. A number of species are of concern to large established trees, but very few cause any serious damage.

The groups of insects of some concern particular to forestry include:

Order Orthoptera (biting and chewing mouthparts).
  Gryllidae (crickets)—field crickets destroy seedlings, especially under covers.
  Tettigoniidae ("bush crickets")—nocturnal feeding destroys apical shoots.
Order Isoptera (termites; tropical only; social; large nests; biting mouthparts).
  Termitidae (bark-eating termites)—workers eat bark of trees under earth sheet
    on trunk; carton nests on trunks; some "seedling termites."
  Rhinotermitidae (wet-wood termites)—nest in moist tree stumps, etc.
  Kalotermitidae (dry-wood termites)—nest in dead dry branches and timbers.
Suborder Homoptera (plant bugs; all are sap suckers).
  Cicadidae (cicadas)—adults feed on branches; females split twigs ovipositing;
    nymphs in soil pierce roots for sap.
  Flattidae (moth bugs)—nymphs and adults gregarious on forest trees.
  Membracidae (treehoppers)—gregarious on twigs; seldom pests.
  +Diaspididae (armoured scales)—scales encrust twigs and branches.
Suborder Heteroptera (sap suckers, with toxic saliva; cause necrosis).
  Coreidae (twig wilters, etc.)—feeding adults kill woody shoots; bushes.
  Pentatomidae (shield/stink bugs)—some prefer woody hosts, bushes and trees.
Order Coleoptera (beetles; great diversity; biting mouthparts).
  Scarabaeidae (white grubs)—larvae in soil eat roots, may kill seedlings.
  +Anobiidae ("woodworm" beetles)—larvae tunnel dead wood and timbers.
  Bostrychidae (black borers)—adults bore branches to make breeding galleries.
  Buprestidae (flat-headed borers)—larvae tunnel trunks, branches, roots.
  +Cerambycidae (long-horn beetles)—larvae tunnel trunks, branches, roots.
  Curculionidae (pine sawers, etc.)—some larvae bore tree trunks.
  +Scolytidae (bark beetles)—adults bore under bark for breeding galleries; some
    species gregarious; some fungus vectors.

+Families marked thus are especially important.

Order Lepidoptera (moths; caterpillars with biting mouthparts; great diversity).
     Gracillariidae, etc. (leaf miners)—tiny larvae mine leaves, etc.
     Sesiidae (clearwing moths)—larvae tunnel tree trunks and roots.
     Cossidae (carpenter moths)—larvae tunnel branches and trunks of trees.
     Metarbelidae (wood borer moths)—larvae eat bark and tunnel trunk.
     Tortricidae (tortricids)—great diversity; larvae eat/roll/fold/web leaves.
   +Geometridae (loopers)—larvae are leaf eaters; may defoliate; some serious.
     Lasiocampidae (tent caterpillars)—hairy, gregarious larvae eat leaves.
     Saturniidae (emperor moths)—large caterpillars eat tree leaves.
     Notodontidae (processionary caterpillars, etc.)—gregarious larvae defoliate.
     Noctuidae (cutworms, etc.)—cutworm larvae destroy seedlings; some eat leaves.
     Lymantriidae (tussock moths)—many larvae eat leaves of forest trees.
Order Hymenoptera (sawflies, etc.; larvae are many-legged caterpillars).
     Siricidae (woodwasps)—larvae bore in trunks of conifers.
     Cimbicidae—larvae eat leaves of deciduous trees.
   +Diprionidae (conifer sawflies)—gregarious larvae eat needles on conifers.
     Tenthredinidae (sawflies)—larvae mine/skeletonize/eat leaves of many trees.
     Formicidae (ants)—leaf-cutting ants defoliate New World forest trees.
     Torymidae (seed chalcids)—*Megastigmus* larvae eat seeds of *Pinus*, etc.
Order Acarina (mites; minute in size; rasping/piercing mouthparts).
     Eriophyidae (gall/rust mites)—adults and nymphs scarify/bronze leaves; make erinea and galls on the leaves.
     Tetranychidae (spider mites)—scarify and web foliage; leaves wilt, may fall; most damaging to small plants.

LIVESTOCK (veterinary pests)

The production of livestock in most countries consists of the rearing of cattle, sheep, goats, ducks, and chickens for food, and oxen and horses as draft animals. Other species of animals are of local importance in some parts of the world, for example, quail and pigeons in China, camels, water buffalo, donkeys, turkeys, guinea fowl, and so on. Also, in some countries, there is now extensive game-ranching whereby antelopes, ostriches are reared in Africa, red deer, reindeer in Europe, and the latest species include crocodiles, iguanas, and birdwing butterflies in Papua New Guinea. The list of animal species being reared for commercial purposes worldwide is very extensive.

Aquaculture in the form of fish farming was traditionally the rearing of carp species in ponds, but is now extended to include trout, eels, freshwater crabs, crayfish, shrimps and frogs. Mariculture takes place in seashore ponds, and in trapped estuarine flood water (Tam baks in S.E. Asia) in large shallow ponds, and now in suspended cages and enclosures actually in the sea lochs and inlets (fjords, etc.). The ponds are used for some fish (mullet, etc.), shrimps, lobsters, and other Crustacea, and the marine enclosures for fish such as salmon, turbot, etc., in cooler temperate regions. Insects are not involved with cultivation of fish and Crustacea except as food. The inclusion of fish into paddyfield rice cultivation is important, partly as a source of animal protein for the farmers, and it is now clear that some of these fish can be very effectivè predators of many of the troublesome insect pests of rice.

Only the more general veterinary pest groups are listed below; they fall into two main categories—the direct pests that damage the host animal, usually by feeding on the blood or tissues, and the indirect pests that are vectors (or intermediate hosts) of parasites or pathogens causing diseases. Some of the most serious pests act in both capacities.

Order Mallophaga (biting lice; permanent ectoparasites; bite scabs and skin).
    Several families—mostly on birds such as ducks, chickens, pigeons, but a few on dogs and cattle, etc.
Order Siphunculata (sucking lice; mouthparts for piercing and sucking blood).
    Haematopinidae (cattle lice, etc.)—adults and nymphs ectoparasitic on cattle, horses, sheep, etc.
Order Heteroptera (animal bugs; blood suckers with toxic saliva).
    Cimicidae (chicken bugs, etc.)—two genera attack chickens in South America.
    Belostomatidae (giant water bugs)—large predacious bugs kill fish/frogs in ponds.
Order Siphonaptera (fleas; adults suck blood; larvae are detritivores).
    Many families—adults ectoparasitic on warm-blooded animals; suck blood; mostly minor pests.
Order Diptera (flies; adults are blood sucking, some disease vectors; larval habits diversified, but some practice myiasis).
    Culicidae (mosquitoes)—some species prefer livestock as hosts—adults only.
    +Tabanidae (horse flies)—adults suck blood; harass livestock; disease vectors.
    Simuliidae (black flies)—near water; feeding adults harass livestock, drain blood.
    +Muscidae (sweat flies/stable flies)—adults bloodsuckers, passive disease vectors; larvae saprophagous.
    Gasterophilidae (bot flies)—larvae intestinal parasites of Equidae, etc.
    +Oestridae (warble flies)—larval myiasis in cattle, sheep, horses, etc.
    Cuterebridae (rodent bots)—also attack cattle, horses in New World.
    +Calliphoridae (blow flies)—adults passive disease vectors; some larvae cause myiasis; screwworm obligatory larval parasite in cattle, goats, etc.
    +Glossinidae (tsetse)—trypanosomes cause "Nagana" in cattle in Africa.
    Hippoboscidae (louse flies)—pupiparous ectoparasites on mammals and birds.
Order Acarina (ticks and mites; adults and nymphs have piercing and sucking mouthparts; feed on blood; many are vectors of disease organisms).
    Dermanyssidae (red mites)—attack poultry and carry viruses.
    Argasidae (soft ticks)—most are pests of birds (ducks, chickens); serious.
    +Ixodidae (hard ticks)—very serious cattle and livestock pests; disease vectors.
    Demodicidae (hair follicle mites)—many species parasitic on Mammalia.
    Sarcoptidae (mange mites)—burrow in skin of mammals and birds.
    Psoroptidae (scab mites)—live in hair follicles and pierce skin.

+Families marked thus are especially important.

Table 3.2 lists some of the major animal diseases transmitted by insects and Acarina.

Table 3.2   Some Animal Diseases Transmitted by Insecta and Acarina

Vector	Disease	Parasite	Transmission	Distribution
Musca domestica	anthrax	Bacillus anthracis	Contamination	Worldwide
Culicoides spp.	—	Dipetalonema spp.	Bite	Africa
	blue tongue of sheep	Virus	Bite	S. Africa
Simulium ornatum, etc.	cattle filariasis	Onchocerca gutturosa, etc.	Bite	Europe, N. America
Simulium venustum	(ducks)	Leucocytozoon anatis	Bite	N. America
Glossina spp.	nagana	Trypanosoma spp.	Bite	Africa
Tabanidae	surra	Trypanosoma evansi	Bite	India, S.E. Asia
Tabanidae	—	Trypanosoma theileri	Bite	Cosmopolitan
Stomoxys calcitrans	mal de Caderas	Trypanosoma equinum	Bite	S. America
Ctenocephalides felis	dog tapeworm	Dipylidium caninum	Eating flea	Cosmopolitan
Dermanyssus gallinae	fowl spirochaetosis	Borrelia anserina	Bite	Australia
Argas persicus	duck tick paralysis	Anaplasma marginale	Bite	U.S.A.
	poultry piroplasm	Aegyptianella pullorum	Bite	U.S.A.
	fowl spirochaetosis	Borrelia anserina	Bite	Europe, Asia
Ixodes ricinus	pyaemia of lambs	Staphylococcus aureus	Bite	Britain
	redwater of cattle	Babesia bovis	Bite	Africa
Boöphilus annulatus	Texas cattle fever	Babesia bigemina	Bite	U.S.A.
Rhipicephalus spp.	East Coast fever (of cattle)	Theileria parva	Bite	E. and S. Africa
	redwater of cattle	Babesia bigemina	Bite	Africa
Amblyomma hebraeum	heartwater of cattle	Rickettsia ruminantium	Bite	Africa

HUMAN (medical pests)

A number of insects that feed mostly on animals and birds will also feed on humans, especially in the absence of the usual host. As with veterinary pests—some are direct pests, in that they attack the host and suck blood, whereas as indirect pests they transmit parasites or pathogens to the human host, acting either as vectors or also intermediate hosts.

The main groups concerned are as follows:

Order Siphunculata (sucking lice; suck blood; permanent ectoparasites).
  +Pediculidae (lice)—head, body and crab louse; also disease vectors.
Suborder Heteroptera (animal bugs; toxic saliva makes "bites" irritating).
      Reduviidae (assassin bugs)—domestic; nocturnal blood feeders; some vectors of Chaga's disease in South America.
  +Cimicidae (bed bugs)—*Cimex* are nocturnal blood feeders.
Order Diptera (flies; some adults bloodsucking, salivary enzymes; disease vectors).
  +Culicidae (mosquitoes)—feeding adults transmit diseases (malaria, etc.).
      Ceratopogonidae (biting midges)—*Culicoides,* etc., bloodsucking adults.
      Tabanidae (horse flies)—adults mechanically transmit some diseases.
      Simuliidae (black flies)—near water; *Simulium* and Onchocerciasis.
  +Glossinidae (tsetse flies)—*Glossina* and trypanosomiasis in Africa.
  +Muscidae (house flies, etc.)—adults and larvae may be pests; adults as disease vectors; larval myiasis.
      Cuteribidae (bot flies)—human bot fly, larval myiasis, in South America.
  +Calliphoridae (blow flies)—many adults domestic pests; disease vectors; some larval myiasis.
      Sarcophagidae (flesh flies)—some larval myiasis; adults synanthropic.
Order Siphonaptera (fleas; adults bloodsucking ectoparasites; larvae free-living detritivores).
      Pulicidae (human flea)—now rare worldwide; animal fleas mostly attack humans.
      Ceratophyllidae—cat flea and dog flea are widespread domestic pests.
      Leptopsyllidae (rodent fleas)—tropical rat flea is main plague vector.
      Tungidae (jigger/chigoe flea)—*Tunga penetrans* female embedded in human foot.
Order Hymenoptera (wasps, bees, ants, etc.—very diverse group).
      Vespidae (wasps)—various species regularly attack humans; stings are painful and can be injurious.
      Formicidae (ants)—as above.
      Apidae (honeybees)—African honeybee very aggressive and regularly attacks and kills humans in tropical Africa.
Order Acarina (ticks and mites; mouthparts adapted for piercing skin and sucking blood).
      Ixodidae (hard ticks)—many species will feed on humans, may transmit diseases.
      Trombiculidae (chiggers)—larvae are skin parasites in the tropics.
      Sarcoptidae (itch mites)—human itch mite burrows in human skin.
      Acaridae (domestic mites, etc.)—several species cause dermatitis.

+Groups marked thus are especially important.

Table 3.3 lists some of the more important Insecta and Acarina that are vectors of disease-causing organisms in people.

Table 3.3 Some Human Diseases Transmitted by Insecta and Acarina

Vector	Disease	Parasite	Transmission	Distribution
Triatoma spp.	chagas disease	*Trypanosoma cruzi*	Bite contamination by feces	S. America
Pediculus humanus	relapsing fever	*Borrelia recurrentis*	Skin contamination	Europe, Asia, Africa
	epidemic fever	*Rickettsia prowazeki*	Bite contamination	Eurasia
	trench fever	*Rickettsia quintana*	Bite contamination	Eurasia, Mexico
Xenopsylla cheopis	plague	*Pasturella pestis*	Bite	Pantropical
	murine typhus	*Rickettsia typhi*	Bite contamination	Pantropical
Anopheles spp.	malaria	*Plasmodium* spp.	Bite	Pantropical
Aëdes aegypti	yellow fever	Virus	Bite	Africa, S. America
	dengue	Virus	Bite	Tropics
Mosquitoes	encephalitis	Viruses	Bite	Africa, America
Mosquitoes (*Anopheles* and *Culex*)	Filariasis (Elephantiasis)	*Wuchereria bancrofti*	Bite and through skin	Pantropical
Phlebotomus spp.	leishmaniasis (Kala-azar, etc.)	*Leishmania* spp.	Bite, etc.	Africa, Asia, S. America
Glossina spp.	sleeping sickness	*Trypanosoma* spp.	Bite	Africa
Simulium spp.	onchocerciasis (River blindness)	*Onchocerca volvulus*	Invasion of bite	C. and S. America, Africa
Chrysops spp.	loasis	*Loa loa*	Bite	Africa
Chrysops, etc.	tularaemia	*Pasturella tularensis*	Bite	Holarctic
Cockroaches, *Musca domestica*	dysentery, amoebic	*Entamoeba histolytica*	Food contamination	Pantropical
Musca domestica and *M. sorbens*	dysentery, bacillary	*Bacillus* spp.	Food contamination	Pantropical
(also *Ophyra*, *Atherigona*, etc.)	conjunctivitis	Bacteria	Eye visits	Pantropical
	typhoid fever	*Salmonella typhi*	Food and water contamination	Pantropical
	cholera	*Vibrio cholerae*	Food contamination	Pantropical
	yaws	*Treponema pertenue*	Wound contamination	Pantropical
	trachoma	Virus	Contamination	Pantropical
Trombicula spp.	scrub typhus	*Rickettsia tsutsugamushi*	Bite	S.E. Asia
Ornithodorus spp.	African relapsing fever	*Borrelia* spp.	Bite (or contamination)	Africa
Amblyomma and *Dermacentor* spp.	Rocky mountain spotted fever	*Rickettsia*	Bite	U.S.A.

POST-HARVEST STORAGE (urban or domestic pests)

After harvest, grain is put into storage so that it can be used as food over the coming year (until the next harvest). Tubers such as potato, other root crops, bulbs (onions, etc.), dried pulses, dried vegetables, mushrooms, and a wide range of food materials are stored, many of them for up to a year. Flours, meals and other farinaceous products may also be stored, for variable lengths of time. On-farm stores are usually situated next to dwelling places or they may be inside, or below, the actual dwelling place. Often, certain types of food material are kept inside the house whereas the family store of food grain is often in a small structure adjacent. Bulk grain is often stored in regional stores or silos which may be large or very large in size. Dried fish and a wide range of dried/preserved meats are also stored in domestic premises as food for future use. Seeds for next season's planting also have to be stored very carefully to ensure their preservation.

Thus the insect pests of post-harvest storage (of agricultural products) are essentially the same as those that infest human dwellings and are broadly regarded as synanthropic. Some insect species have lived for so long in association with humans that they are seldom found in the truly wild state. A few of the truly domestic pests do not normally infest stored foodstuffs, and these include the clothes moths and some dermestid beetles whose larvae feed on wool and other forms of keratin.

In the broad sense, stored foodstuffs (products) include fresh fruit and vegetables that are taken into the domestic premises (homes, shops, restaurants, etc.) and kept for a while before being consumed. This is the approach taken in the book just published (Hill, 1990), but is not really appropriate here.

The main groups of insects and mites that are regarded as common and important domestic pests (pests of stored products) are as follows:

Order Thysanura (silverfish, etc.; primitive, wingless; biting mouthparts).
    Lepismatidae (silverfish)—eat paper and organic debris; worldwide.
Order Collembola (springtails).
    Several species are regularly found, but seldom are of importance.
Orthoptera (grasshoppers, etc.; large; biting mouthparts; stridulate at night).
    Gryllidae (crickets)—many species are urban and nocturnal.
Order Dictyoptera (cockroaches; omnivorous; polyphagous; cosmopolitan).
    +Blattidae—*Blatta, Periplaneta* major domestic biting pests.
    Epilampridae—*Blattella,* many species; smaller but very abundant.
Order Isoptera (termites; social nest-builders; pantropical; wood/plant feeders).
    Termitidae (mound-building termites)—sometimes invade stores and homes.
    +Rhinotermitidae (wet-wood termites)—attack damp timbers; small nests.
    Kalotermitidae (dry-wood termites)—small nest colony inside dry timbers.
Order Psocoptera (booklice; tiny, colorless; fungivorous with biting mouthparts).
    +Several families contain domestic pest species; can be very damaging.
Order Coleoptera (beetles; adults and larvae with biting mouthparts; diversity).
    +Dermestidae (larder/hide beetles)—adults and larvae eat leather, carpets, dried fish, meats; Khapra beetle is major grain pest.
    Anobiidae (wood beetles)—*Lasioderma* and *Stegobium* are major food pests.
    Ptinidae (spider beetles)—many species, worldwide; mostly minor pests.
    +Bostrychidae (black borers)—adults bore in wood, some in grain, to make breeding gallery; *Prostephanus* and *Rhizopertha* serious grain pests.

+Families marked thus are especially important.

Cleridae (checkered beetles)—*Necrobia* eat bacon, copra, oil seeds, etc.

Nitidulidae (sap beetles)—dried fruit beetles are serious pests.

Cucujidae (flat bark beetles)—*Cryptolestes* spp. attack grain, flours, etc.

Silvanidae (flat grain beetles)—*Oryzaephilus* spp. polyphagous, many foodstuffs.

+Tenebrionidae (flour beetles)—*Tribolium*, etc.; major pests worldwide.

+Bruchidae (seed beetles)—larvae eat seeds of legumes worldwide.

Apionidae (seed weevils, etc.)—larvae in seeds of legumes; *Cylas* in sweet potato tubers.

+Curculionidae (weevils proper)—*Sitophilus* (grain weevils) very important.

Order Diptera (flies; larval stages are the pests; adults can carry pathogens).

Drosophilidae (small fruit flies)—small maggots in ripe fruits.

Anthomyiidae (root flies, etc.)—some vegetable pests; some synanthropic.

+Muscidae (house flies)—adults and maggots spoil foods; synanthropic.

Calliphoridae (blow flies)—adults and maggots spoil foods; synanthropic.

Piophilidae (cheese skippers)—maggots in cheeses and bacon.

Tephritidae (fruit flies)—maggots in fruits of all types.

Order Lepidoptera (moths; only larvae are pests; biting mouthparts).

+Tineidae (clothes moths)—larvae eat keratin (wool); also grain moth.

Oecophoridae (house moths)—polyphagous domestic pests; very abundant.

Gelechiidae—potato tuber moth in stored tubers; Angoumois grain moth.

+Pyralidae—*Pyralis, Ephestia, Plodia, Corcyra* all major stored-products pests.

Order Hymenoptera

Formicidae (house ants)—several species are major domestic pests.

Order Acarina (mites: tiny in size; mouthparts modified for biting).

Acaridae (house mites)—many species are important domestic pests; flour mite, house mites, dust mite, cheese mites, etc.; sometimes split into several different families.

Part B

INSECT GROUPS
Biology, Life History;
Important Pest Species

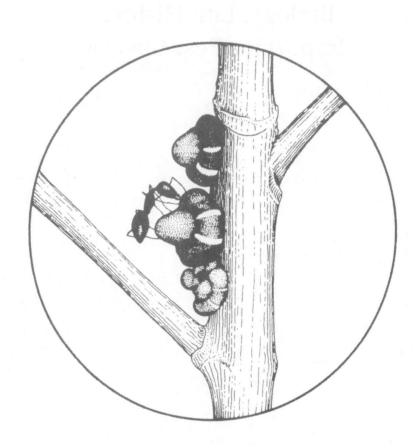

The object of this, the major section of the book, is to review briefly each Order of Insecta and Acarina with particular emphasis on basic biology, and life-history, and to discuss the importance of the main species as pests to humans and the practice of agriculture in its widest sense. Selected species of importance are illustrated either by a line drawing or a photograph, and in some cases damage is also illustrated. Beneficial species are also included.

Reference is made to the distribution maps produced by the C.A.B. International Institute of Entomology (I.I.E.), London (formerly the C.I.E.), where applicable, as sources of further detail.

As already mentioned, the basic classification used here follows that used by Richards & Davies (1977) since it is a very widely used textbook, and a simplified consistent classification is most suitable for general use.

The numbers included to the right of each Order represent the number of families and the approximate number of species belonging to each group.

Classification of the Insecta
(after Richards & Davies, 1977)

Class Insecta
Subclass (1) Apterygota

Orders	1.	Thysanura (bristletails)	5: 550
	2.	Diplura	6: 600
	3.	Protura	4: 200
	4.	Collembola (springtails)	11: 1500

Sublcass (2) Pterygota

Division A. Exopterygota (= Hemimetabola)

5.	Ephemeroptera (mayflies)	19: 2000
6.	Odonata (dragonflies and damselflies)	26: 5000
7.	Plecoptera (stoneflies)	14: 1700
8.	Grylloblattode	1: 16
9.	Orthoptera (crickets, grasshoppers, etc.)	17: 17,000
10.	Phasmida (stick and leaf insects)	2: 2500
11.	Dermaptera (earwigs)	8: 1200
12.	Embioptera (web-spinners)	8: 300+
13.	Dictyoptera (cockroaches and mantids)	9: 6000
14.	Isoptera (termites)	7: 1900
15.	Zoraptera	1: 22
16.	Psocoptera (barklice and booklice)	22: 2000
17.	Mallophaga (biting lice)	12: 2800
18.	Siphunculata (sucking lice)	6: 300
19.	Hemiptera (bugs)	38: 56,000
20.	Thysanoptera (thrips)	6: 5000

Division B. Endopterygota (= Holometabola)

21.	Neuroptera (lacewings, etc.)	19: 5000
22.	Coleoptera (beetles)	93: 330,000
23.	Strepsiptera	9: 370
24.	Mecoptera (scorpion flies, etc.)	9: 400
25.	Siphonaptera (fleas)	6: 1400

26.	Diptera (flies)	c. 130: 85,000
27.	Lepidoptera (moths and butterflies)	97: 120,000
28.	Trichoptera (caddis flies)	11: 5000
29.	Hymenoptera (sawflies, wasps, ants, bees, etc.)	60+: 100,000+

There is much disagreement internationally over the details of insect classification, and estimated numbers vary greatly. In some groups undoubtedly many more species remain to be described and named and final numbers of species will be much greater.

4

Subclass Apterygota

A primitively wingless group of insects with only a slight metamorphosis, and most adults bear vestiges of abdominal appendages (other than genitalia). Most molt a few times after reaching sexual maturity. They represent basically a diverse assemblage and possibly should not really be lumped together in one group. Most species are small or tiny and soil-dwelling, and as such are not particularly conspicuous but they are abundant in most soils and in litter. They are probably less abundant in the tropics than in temperate regions where they sometimes are recorded in vast numbers. As a group they are only of slight interest agriculturally.

ORDER THYSANURA (bristletails)   *5: 550*

Small (5–20 mm; 0.2–0.8 in.), active insects, found in leaf litter and domestic premises; a few species inhabit rocky seashores; seldom found in living vegetation. The body is elongate, tapering, white or brown in color, and usually bears a series of lateral styliform appendages, and terminally a pair of long cerci with a long median tail filament; antennae are long and many-segmented; the insects are agile and can move swiftly. Mouthparts are ectognathous and used for biting and chewing, and a few species are of some importance in domestic premises where foodstuffs, paper and books may be damaged; and they are usually to be found in food stores and warehouses. These include the ubiquitous silverfish (*Lepisma saccharina*) (Fig. 4.1) and several species of *Ctenolepisma*; in heated situations (bakeries, etc.) worldwide, the firebrat (*Thermobia domestica*) is common.

ORDER DIPLURA   *6: 600*

Small, slender, soil and litter insects, but a few species are quite long (up to 50 mm; 2 in.), mostly found in tropical regions, and not often seen. They are eyeless, have long antennae, and a pair of usually long terminal cerci. The mouthparts are biting but entognathous and inserted deeply into the head capsule. They feed mostly on fungi and detritus, but some species are herbivorous and others are predacious. They are of no concern to agriculture generally, but are clearly of some importance in the soil ecosystems.

Fig. 4.1 The ubiquitous domestic silverfish (*Lepisma saccharina*) (Thysanura; Lepismatidae).

ORDER PROTURA   *4: 200*

Minute (0.5–2.5 mm; 0.02–0.1 in.) whitish insects, without eyes or antennae, with piercing entognathous mouthparts. They are found throughout the world in moist soil, leaf litter, turf, and sometimes under tree bark; they feed on decaying vegetable matter.

ORDER COLLEMBOLA (springtails)   *11: 1500*

Small insects (1–5 mm; 0.04–0.2 in.), with entognathous biting mouthparts, shortish four-segmented antennae, and a forked springing organ under the abdomen with a ventral tube under the first abdominal segment, giving them, overall, a very characteristic appearance. Most are found in soil, leaf litter, some in the nests of ants and termites, some live on the surface of freshwater ponds, others are littoral or "marine", and a few are regularly found in domestic premises. They tend to be abundant only in moist situations. A couple of domestic species have been recorded living as ectoparasites in human public hair. Body coloration varies from white, through green and yellow, to gray and black; some are banded or patched, and a few have a metallic sheen. They feed on fungi, pollen grains, and dead or living plant material. The group is worldwide but most abundant in temperate soils, and some are arctic–alpine in distribution. Some nival species are recorded at altitudes of 6000 m in the Himalaya and are called "snowfleas".

The group is divided into two distinct suborders, as shown below.

SUBORDER ARTHROPLEONA

These have a more or less elongate body and segmentation is evident. Most species tend to be soil burrowers, although the freshwater and seashore species belong to this group. Some have the springing organ greatly reduced or even absent; the deepest soil burrowers show the greatest reduction. A number of species are recorded as agricultural pests; they damage roots in the soil, and the soft stems and early leaves of seedlings; populations in the U.K. may be so large that the soil surfaces in some fields are literally covered with springtails. A species of *Drepanocyrtus* was recorded to be damaging coffee seedlings in Tanzania during a period of drought. The main pest species generally vary from continent to continent; most are polyphagous and will attack many different soft-stemmed seedlings. *Onychiurus* is a cosmopolitan genus including a number of pest species. Typical appearance is shown by *Homidea* sp. (Fig. 4.2), a domestic species from Hong Kong feeding on food debris and fungi. Several different genera and species are to be found in households in different parts of the world. Several species are regularly recorded as pests in greenhouses in temperate regions.

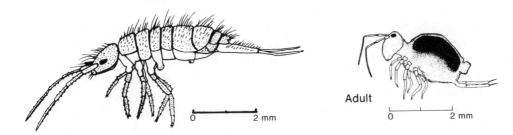

Fig. 4.2   A common domestic species of spring-tail (*Homidea* sp.); Hong Kong.

Fig. 4.3 Lucerne "flea" (*Sminthurus viridis*) (Collembola; Sminthuridae) from wheat field; Cambs., U.K.

SUBORDER SYMPHYPLEONA

The body is subglobular, with thorax and first four abdominal segments fused together. They are clearly not suited for soil burrowing, and most are to be found on the soil surface and in litter; only rarely do a few species live in the soil. A few species are crop pests, mostly in temperate regions, and on occasion, they are to be seen on the soil surface in vast numbers, when damage to seedlings can be very extensive—they usually eat holes in the epidermis of cotyledons and first leaves, and also in soft stems. The three most important genera recorded as crop pests are *Bourletiella*, *Sminthurides*, and *Sminthurus*; they are cosmopolitan but more abundant in cooler regions, although they are found around the Mediterranean, in South Africa, Australia, and China. The best-known species is probably *Sminthurus viridis* (Lucerne "flea") (Fig. 4.3) found throughout the world as far north as Alaska, Lapland, and Iceland, but equally abundant in the Mediterranean region; records are sparse in the New World (C.I.E. Map No. A. 63).

5

Subclass Pterygota:
EXOPTERYGOTA—Small Orders

These are the winged insects, although some are secondarily wingless, with a varied but definite metamorphosis; size is varied, but many are large. The group is subdivided into two divisions (Exopterygota and Endopterygota) on the basis of the type of metamorphosis and the manner in which development takes place.

DIVISION EXOPTERYGOTA (= HEMIMETABOLA)

Collectively regarded as the more primitive orders of true insects, with a simple metamorphosis and seldom a stage resembling a pupa. Wings develop externally; and as with all true insects the mouthparts are positioned externally (ectognathous). The immature stages have long been called nymphs, for they resemble adults but are smaller and lack wings and external genitalia. In many groups there is no ecological separation between adults and nymphs and they are found together in the same habitats; but in some orders (Odonata, Ephemeroptera, Plecoptera) the nymphs live in freshwater while the adults are aerial. The aquatic forms show a series of anatomical modifications to their life under water, especially the development of gills and respiratory structures.

The following orders are either of limited or little interest agriculturally and so are presented here together for convenience in a single chapter.

ORDER EPHEMEROPTERA (mayflies)   *19: 2000*

This group is of importance in freshwater habitats. The nymphs live submerged in ponds, lakes, streams, and rivers, and they have a series of seven pairs of lateral, platelike abdominal gills (tracheal gills filled with air) with which they obtain oxygen from the water by gaseous diffusion. Most nymphs feed on algae and plant debris but a few are carnivorous. Nymphal life typically lasts a year or more—development is slow and protracted through many instars. Adults are very short-lived (hence their name), delicate-bodied, and weak flying, with tiny antennae, long tail filaments, and either just a single pair of wings or else with tiny hind wings also. Adult life span ranges from a few hours to about a day, during which time they do not feed—in fact, their alimentary canal is nonfunctional. Both adults and nymphs play important roles in freshwater food chains, especially in the diet of many fish. Adults fly to lights at night and may be found in large numbers at some distance from the nearest water, and in the mornings may be found sitting on plants near the lights.

Nymphs are found in paddy fields (and are food for fish) and occasionally direct damage to rice plants has been reported—a species of *Cloeon* is said to tunnel in the

Fig. 5.1 Typical adult mayfly (Ephemeroptera)—a species of *Ameletus*; body length 10 mm (0.4 in.); South China.

stems of rice plants in Togo and Guinea. In the tropics, one or two species of mayflies act as phoretic hosts for larvae of certain species of *Simulium* (black flies). In East and Central Africa, *Pavilla adjusta* has nymphs which burrow into floating or submerged wood and timbers, but it appears that usually an aquatic wood-rotting fungus causes the initial damage permitting the mayfly nymphs to start burrowing operations. A typical adult mayfly (*Ameletus* sp.) is shown in Fig. 5.1.

ORDER ODONATA (dragonflies)   *26: 5000*

Fiercely predacious insects with large biting mouthparts, and two pairs of more or less equally-sized wings; eyes very large and prominent. Adults are strong fliers and have the thorax modified anatomically to allow all six legs to project anteriorly—they seize their insect prey in flight and hold the body to their mouth region, using all six legs. Nymphs are aquatic and found in all types of freshwater habitats, and some in brackish or saline waters. Development is lengthy and most species remain as nymphs for a year or more, depending to some extent on environmental conditions. The nymphs are fiercely predacious and eat aquatic invertebrates, tadpoles, and small fish. They possess a specially modified mouthpart structure called a "mask", developed from the labium, which can be suddenly extended for seizing the prey. The nymphs are major freshwater predators, and can be of some importance in paddyfields where they will eat semiaquatic defoliating caterpillars (*Nymphula* etc.) and some aquatic pests; in private and commercial fishponds the larger nymphs can cause economic losses by killing small fish.

The adults spend some time in the vicinity of water, especially for breeding purposes, but also spend much time flying in the shelter of tall trees, woodland edges, near tall buildings, where they hawk for their insect prey. But the greatest abundance and diversity of dragonflies is usually found by paddy fields. Most of the insects taken as prey over paddy fields are clearly rice pests, and so in such locations the adult dragonflies are important as part of the general insect predator complex.

The group is worldwide in distribution but most abundant in the Oriental and Neotropical regions; the majority of species are tropical or subtropical. The order is subdivided into two very distinct suborders, as described below.

SUBORDER ANISOPTERA (dragonflies)

These are the dragonflies proper, robust and stout-bodied, 3–10 cm (1.2–4 in.) in body length, very fast flying, agile and aggressive. Nymphs respire using an internal rectal gill basket, water being regularly taken into the rectum and then expelled—if expelled rapidly, the nymph achieves jet propulsion. A few species are strongly migratory, but this behavior is more typical of temperate and subtropical species. Fig. 5.2 shows a typical adult, and Fig. 5.3 a large nymph. There is generally much variation in body and wing coloration in the adults, and some diversity of body form in the nymphs according to the habitat where they live—stream species tend to be more flattened and with stronger legs.

Fig. 5.2   Typical adult dragonfly with clear wings; female *Anaciaeschna jaspidea* (Odonata; Aeschnidae); body length 60 mm (2.4 in.); Hong Kong.

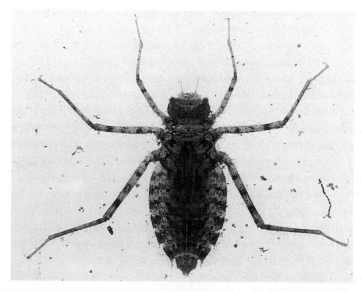

Fig. 5.3   Dragonfly nymph, almost fully grown, *Epophthalmia elegans* (Odonata; Macromiidae); Pokfulam Reservoir, Hong Kong; body length 37 mm (1.5 in.).

SUBORDER ZYGOPTERA (damselflies)

As the name suggests, these insects are of a more delicate build and slender in shape but neither adults nor nymphs are less ferocious! Adult eyes are large on a distinctly transverse head; wings are petiolate (narrow at the base) and at rest held vertically over the insect body (rather than laterally as in the dragonflies). The adults are generally more restricted to the vicinity of water, and are particularly abundant over paddy fields. Most adults are smaller than dragonflies, and some species are quite tiny and measure only 15 mm (0.6 in.) in body length; Fig. 5.4 shows a typical damselfly adult.

Nymphs are characterized by having three, terminal, platelike abdominal gills, which are gently wafted from side to side to increase ventilation. Basically, damselflies are equally as important as insect predators as dragonflies but they generally take smaller prey.

Fig. 5.4   Large damselfly (male *Mnais mneme*) (Odonata; Callopterygidae); body length 58 mm (2.3 in.), but shape is typical for the suborder; South China.

ORDER PLECOPTERA (stoneflies)   *14: 1700*

This is a small group with aquatic larvae that live in clean, unpolluted streams—very few inhabit standing waters. They are used sometimes as biological indicators for water pollution surveys with considerable success. Nymphs are mostly carnivorous and feed on other insect larvae but members of the genus *Nemoura* are reputed to be herbivorous and to feed on algae and diatoms. Gills are thoracic in location, at the bases of the legs, and tuft-like in appearance. Both adults and nymphs typically have long tapering antennae and long cerci. Adults have two pairs of large equally-sized wings. The group is of no importance agriculturally.

ORDER GRYLLOBLATTODEA   *1: 16*

A tiny group of obscure insects, found high in the mountains of North America, Japan, and Siberia. They are apterous and show many primitive features and are thus of phylogenetic interest.

ORDER PHASMIDA (stick and leaf insects)   *2: 2500*

A tropical group of phytophagous insects, mostly Oriental, remarkable for their cryptic resemblance to leaves or to twigs. They are moderate to large in size, apterous or winged, usually elongate and thin-bodied (a few are quite stout-bodied), colored green or brown. In addition to their cryptic appearance, they are nocturnal and during the day they sit quietly in foliage and are reluctant to move.

Phasmidae (stick insects)

The body is elongate, narrow usually, and twig- or stick-like in appearance. At rest in plant foliage they sit with the forelegs straight out in front. Fig. 5.5 shows a typical stick insect (*Baculum* sp.) from South China. Some species have a shorter, thicker body, and are sometimes called "twig insects" (Fig. 5.6) and they are usually found on leaf litter on the ground in S.E. Asia rather than arboreal in living foliage. The giant Australian species *Acrophylla titan* reaches a body length of 25 cm (10 in.) (excluding antennae). A few species are recorded as agricultural pests. In Fiji, and in many parts of the South Pacific, *Graeffea cocophaga* and *G. minor* have been reported defoliating coconut palms, and *G. crouani* causes similar damage in Polynesia and Melanesia. *Anchiale maculata* is a widespread defoliator in New Britain on cocoa and many other plants. Several different species regularly cause damage to eucalyptus trees in Australia.

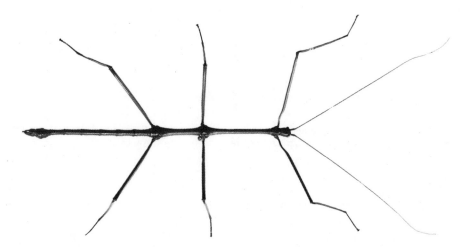

Fig. 5.5 A typical stick insect (*Baculum* sp.) (Phasmida; Phasmidae); body length 100 mm (4 in.); South China.

Fig. 5.6 Twig insect (*Datames* sp.) (Phasmida; Phasmidae); body length 50 mm (2 in.); South China.

Phyllidae (leaf insects)

These insects are quite elongate but also have lateral expansions of the legs and body skeleton to break up the body outline, and to make it flattened and leaflike. Most species are to be found in the Oriental region in tropical climates where they eat the leaves of trees and cause occasional defoliation, but mostly on forest trees rather than agricultural crops. The main genus, is clearly *Phyllium* which occurs throughout the Oriental region as well as in Papua New Guinea and in Queensland.

ORDER DERMAPTERA (earwigs)   *8: 1200*

A small, rather primitive group, widespread, commonly found but not particularly abundant anywhere. They are readily recognizable by their smallish size (6–20 mm; 0.24–0.8 in.) elongate antennae, forewings shortened into small leathery brown tegmina under which are held the large ear-shaped and complexly folded hind wings. They are nocturnal in habit and fly quite well. The cerci are very distinctively modified into a pair of unsegmented heavily sclerotized forceps. The forceps are reputed to be used some-times in copulation—the male forceps are semicircular in some species, while those of the female are straight. But some of the predacious species can apparently hold their prey in the forceps and, by tilting the abdomen, bring the prey to their mouth.

Some species of Dermaptera live inside deep caves (e.g. Batu caves of West Malaysia) in total darkness and feed on bat guano. Other groups are ectoparasites of bats in parts of S.E. Asia, and some live on the bodies of certain rodents in Africa.

Most of the Dermaptera appear to be omnivorous, but many of the Labiduridae are predacious—*Chelisoches morio* is an important predator of Cicadellidae, Aphididae and leaf-mining hispid larvae in many parts of the Pacific region. In Thailand, a local earwig is an important predator of the eggs of Asian corn borer on maize. Two well-known pest species are:

Euborelli stali—recorded damaging groundnuts in South India; feeding adults bite holes in the developing pods; also damages sorghum, brassicas, and onions in India; *Euborellia* adults wingless; *E.* spp. in produce stores.

Forficula auricularia (common earwig)—now widely distributed throughout the Palaearctic, North America and Australasia; the mouthparts are not strong but damage is done to some vegetables and fruit by biting, and to many orna-mental flowers (especially chrysanthemum).

ORDER EMBIOPTERA (web-spinners)   *8: 300+*

A small tropicopolitan group of fragile insects that are also found in some sub-tropical regions. Males have two pairs of equally-sized wings and fly at night to lights; they have quite large external genitalia, asymmetrical in shape. The females live gregariously on trees on the trunk bark or on rock surfaces, under silken webs in an extensive system of tunnels that is quite characteristic in appearance. They often emerge from the tunnel system at night to search for food. Both sexes and the nymphs collec-tively construct the silken tunnels.

Several "weed" species of *Oligotoma* and *Aposthonia* are common and widespread throughout the tropics and warm temperate regions. Several species of *Oligotoma* and others have been recorded damaging the roots of cultivated epiphytic orchids in Hawaii and parts of S.E. Asia.

ORDER DICTYOPTERA (cockroaches and mantids)　　9: 6000

This group was formerly lumped together with the Orthoptera and Phasmida, but is now regarded as being quite distinct, but comprising two separate homogeneous suborders. Antennae are filiform with numerous segments; mouthparts mandibulate; well developed legs (sometimes forelegs raptorial); forewings thickened into tegmina; cerci distinct and many segmented; eggs laid inside a protective oötheca stuck on to a solid substrate. Distribution in the main is tropical and subtropical, with a few species of cockroach in warm buildings in temperate regions.

SUBORDER BLATTERIA (cockroaches)　　　　　　　　　　　　　　4000 spp.

Cockroaches mostly have a depressed body with the head almost (or completely) covered by an anterior extension of the shield-like pronotum, the forelegs are unmodified. None are regarded as being crop pests, although some are said to be capable of damaging tender plants and young seedlings, both in the field in the tropics and in greenhouses in temperate regions; the pantropical *Pycnoscelis surinamensis* (Fig. 5.7) is the species most recorded doing this. Several species are quite important as domestic pests and damage stored foodstuffs and other stored products in urban habitats. A few species live on organic debris in caves (often on bat guano); some are intimately associated with people and their foodstuffs; others are completely woodland and forest dwellers in the tropics, including a few amphibious species; a few field/woodland species regularly invade human dwellings and tend to be more polydemic. One of the reasons for their importance as domestic pests is the adult longevity; many species can live for periods from 3–18 months, with *Periplaneta* being the longest lived at 12–18 months.

Formerly, it was usual to regard all Blattaria as belonging to a single family (Blattidae) but now in the latest edition of Imms (Richards & Davies, 1977) four different families are recognized. Some of the more important species are mentioned below:

Blaberidae

A large and diverse group, mostly being forest litter species, and including the now cosmopolitan Surinam cockroach (*Pycnoscelis surinamensis*) (Fig. 5.7) said to cause damage to some field crops in the tropics (tobacco, pineapple, etc.) and to greenhouse plants in Germany and North America; in warmer regions it sometimes invades buildings.

The spectacular *Blaberus* grows to 8 cm (3.2 in.) long in the tropical New World, found mostly in rain forest litter, but sometimes invading buildings. Some species are regularly associated with bananas and may be transported around the world in banana consignments.

Blattidae

A few species are cavernicolous but most of the important urban pest species belong to this family. The main pest species include the following:

> *Blatta orientalis* (oriental cockroach)—now thought to have originated in North Africa and to have spread throughout Europe and Asia from eastern Europe and also by trade in ships to both North and South America. Generally, this is

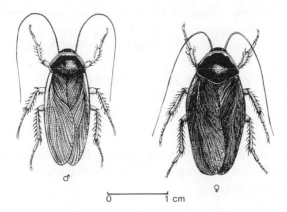

Fig. 5.7 Surinam or black litter cockroach (*Pycnoscelis surinamensis*) (Dictyoptera; Blaberidae).

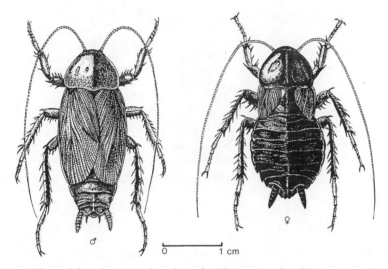

Fig. 5.8 Male and female oriental cockroach (*Blatta orientalis*) (Dictyoptera; Blattidae).

a more temperate species and does not occur in the humid tropics. The female is characterized by being almost entirely black and having vestigial wings (Fig. 5.8).

Leucophaea maderae (Madeira cockroach) (Fig. 5.9)—a large brown species, up to 6 cm (2.4 in.) long, widespread throughout Africa, North and South America, but apparently absent from S.E. Asia and Australasia generally. In the hot tropics, this species lives outdoors and may be associated with several crops, especially bananas.

Nauphoeta cinerea (lobster cockroach)—a tropical domestic species, now completely pantropical.

Periplaneta americana (American cockroach) (Fig. 5.10)—this is the single most important species of cockroach, thought to have originated in tropical Africa and now completely cosmopolitan, abundant in all warmer regions and established in heated buildings in cooler North America and northern Europe. Three other closely related species occur.

P. australasiae (Australian cockroach)—equally as cosmopolitan as the former species, but apparently preferring a slightly higher temperature; in the tropics often found in fields associated with crops; once recorded damaging greenhouse plants.

P. brunnea (brown cockroach)—a tropical species found in both Old and New Worlds, but not abundant.

P. fuliginosa (smokey-brown cockroach)—a subtropical species found in southern U.S.A.

Supella supellectilium (brown-banded cockroach)—another African species now worldwide in warmer regions. A domiciliary pest important locally in a number of regions.

Fig. 5.9 Madeira cockroach (*Leucophaea maderae*); body length 6 cm (2.4 in.); Uganda (Dictyoptera; Blattidae).

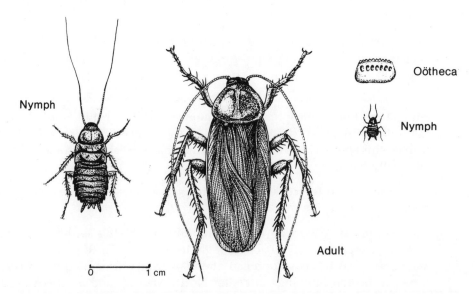

Fig. 5.10 American cockroach (*Periplaneta americana*); adult, nymphs, and oötheca (Dictyoptera; Blattidae).

Epilampridae

A very important pest species is *Blatella germanica* (German cockroach) (Fig. 5.11), a completely cosmopolitan domestic species. It is small in size, very adaptable, and is the most common cockroach on modern ships; is found as far north as Alaska in heated buildings. Adults live for some 3–4 months, but under optimal conditions the life cycle can be completed in 6 weeks. There are many species of *Blatella* both in Africa and in Asia, most of which are forest dwellers—the center of evolution for *Blatella* is now thought to have been in the region of N.E. Africa. The Blatellinae is a large group and contains some 1300 species.

One species of *Blatella (B. lituricollis)* is known as the Asian cockroach as it is thought to be native to S.E. Asia, and has recently become established in Florida (U.S.A.) where it is causing great concern locally; it breeds outside in leaf litter mostly, but flies readily and is attracted to lights at night; the nocturnal invasion of domestic premises in large tracts of Florida is a serious problem.

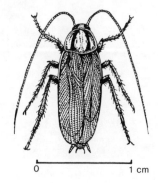

Fig. 5.11   German cockroach (*Blatella germanica*) (Dictyoptera; Epilampridae).

SUBORDER MANTODEA (mantids; praying mantises)   *2000 spp.*

An interesting group of totally predacious insects with raptorial forelegs with which insect prey are caught. The head is triangular, mobile, and with large eyes situated so as to give the insect good binocular vision. Formerly, all mantids were placed in a single large family (Mantidae) but now eight families are recognized. Fig. 5.12 shows a *Hierodula* (large green mantid) from Hong Kong and S.E. Asia. Mantids themselves are not very important as insect predators as their populations are usually scattered and rather sparse, but it should be remembered that they are part of the general natural insect predator complex and as such are of importance in the natural control of many insect pests; in agricultural areas their presence should never be discouraged.

There has been sporadic interest in the idea of using mantids as part of a biological control project. Certainly, the egg masses (oöthecae) could be easily collected and transported, as they are compact and quite conspicuous; Fig. 5.13 shows öthecae of *Hierodula* and *Tenodera*. The hatching nymphs need to be dispersed as soon as possible for if kept together they are cannibalistic; if a whole brood is kept together in a small cage almost invariably only one nymph survives. The cannibalistic tendency in the Mantidae is shown by the mating behavior; a pair of *Tenodera* were observed mating—the larger female was able to turn her head around and eat the head, prothorax and forelegs of the male, but they remained *in copula* for a further 18 hours (20 hours in all). Parasitism of mantid öthecae by *Podagrion* wasps (see page 527) is, however, quite common and could reduce populations significantly.

Fig. 5.12 Large green mantid; female (*Hierodula* sp.); Hong Kong (Dictyoptera; Mantidae).

Fig. 5.13 Mantid oöthecae—*Tenodera* sp., 35 mm (1.4 in.), and *Hierodula* sp., 25 mm (1 in.) (Dictyoptera; Mantidae); Hong Kong.

ORDER ISOPTERA (termites) 7: 1900

A truly tropical group of insects that are renowned for several different aspects of their biology. Their social life is complex and interesting, and associated with this is their extreme polymorphism and the very spectacular nests constructed in the warmer parts of the tropics. In addition, they have the ability, one way or another, of breaking down cellulose into food sugars, and so they are able to destroy wooden structures and trees with alarming ease. According to Fletcher (1974), on a worldwide basis termites are responsible for U.S. $500 million worth of damage to growing crops and timber structures and trees per annum, and so they are major crop pests throughout the tropical regions of the world.

Young adults are winged, with two pairs of equal-sized wings, but with fracture lines at the bases so that after the brief mating flight the wings are shed prior to the adults starting their subterranean life. Their flight is weak and most of the mating flights are held on moist, still evenings and only last for an hour or less. After the mating flight the adults fall to the ground, and on a wet surface they may be trapped by surface tension effects, otherwise they leave a pile of discarded wings (especially under lights at night). The paired, young, mated adults find a suitable location, dig a tunnel in the earth (or whatever) and start the nest and the new colony. Most (but not all) nests are underground and eventually reach a large size with a population of many thousands of individuals. The original female becomes the queen termite and she may reach a prodigious size as her abdomen swells with eggs and fat body. The queen lives, usually with the attendant male (king), in a hollow royal chamber deep in the nest (termitarium) where it is well protected (Fig. 5.14). She functions solely as an egg-layer, but is also probably the source of some social hormones or pheromones that may control the development of the different castes within the colony.

Fig. 5.14 Royal chamber of *Macrotermes bellicosus* opened to show huge queen and attendant male; Uganda (Isoptera; Termitidae).

The main castes are wingless and sterile, and those with smaller-sized heads with small and pale bodies are the workers. They are responsible for attending to most of the domestic chores within the nest as well as to most of the food gathering. Those with a large brown head and large mandibles are soldiers and have a defensive function in the nest and an offensive function when accompanying the foraging workers. Some soldiers have a frontal gland on the face from which formic acid can be squirted, and in extreme cases there are nasute soldiers with the face extended into a snout-like projection for acid squirting. For identification purposes, the head of the soldier caste is mostly used. Some of the workers do have a reproductive capability which is available in case the queen dies or meets with an accident.

Termite damage to growing plants is somewhat problematical. Healthy, vigorously growing plants are seldom attacked by most species of termites, but if the plant is water stressed, sickly, or physically damaged it is much more likely to be attacked. And, of course, in the drier tropics during the dry season virtually all of the plants are water-stressed. Many species are collectively known as "bark termites" and they construct an earthen sheet over the bark of trees, sometimes to a height of 3–4 m (10–13 ft.), under which they gnaw off the bark and outer layers of sapwood (Fig. 5.15) and they may pene-trate deeper into the living tissues. Some species regularly gain entry into living trees (palms, tea, etc.) through the dead ends of pruned branches (tea) or leaf petioles (palms), from which they may then invade the living tissues. Other species tend to be general gatherers of leaf material, seeds, etc., from the soil surface which is taken back to the nest as a source of food material.

Fig. 5.15   Bark termite damage to coconut palm trunk; Penang, Malaysia.

Although many termites are pests, it should be remembered that the subterranean species do perform valuable functions in respect to soil ventilation and drainage; some of the vegetable material deposited underground ends up as soil humus.

Some of the more important pests include the following:

Kalotermitidae (dry-wood termites)   *250 spp.*

These are small termites, of limited agricultural importance, but very destructive to interior (dry) timbers and woodwork. These termites have an intestinal microfauna of mutualistic (symbiotic) flagellate protozoa (especially in their large rectal pouch) which are apparently responsible for cellulose degradation, so these termites are effectively "eating" dry wood and they are able to fully utilize metabolic water. A typical member of this group is *Cryptotermes brevis* (Fig. 5.16), a tropicopolitan species from South China. In most parts of the world these dry-wood termites are not too abundant, but in Trinidad they are reported to be very serious domestic pests.

Cryptotermes brevis (dry-wood termite)—S.E. Asia up to South China.

Cryptotermes domesticus (Pacific dry-wood termite)—parts of Asia and the Pacific region.

Cryptotermes dudleyi—established in Central America, parts of S.E. Asia and parts of Africa.

Kalotermes jepsoni—in the Mediterranean attack fruit trees and grapevines.

Kalotermes spp.—mostly pests of dry timber and internal woodwork; some species can attack living tea bushes in Sri Lanka and hollow out the stems.

Neotermes spp. (stem termites)—are pests of cocoa in West Africa, S.E. Asia and Central and South America where they also attack coconut palms and various trees, and they attack tea bushes in Sri Lanka; they basically inhabit dead branches or twigs and then they invade the living tissues.

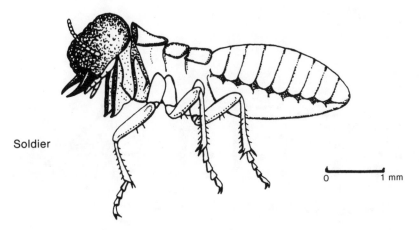

Soldier

0   1 mm

Fig. 5.16 Dry-wood termite (*Cryptotermes brevis*) (Isoptera; Kalotermitidae); Hong Kong.

Hodotermitidae (harvester termites)

A small group of subterranean termites characterized by their collecting pieces of grass and other plant material from the soil surface. The cut grass and leaf material is stored underground in the nest system and is apparently eaten by the termites. Precisely how the cellulose material is utilized by these termites is still not clear. The group is more or less confined to arid areas of Africa, the Middle East and parts of India.

Hodotermes mossambicus (harvester termite) (Fig. 5.17)—has soldiers with distinctively stout mandibles and a dark coloration; this species occurs in arid areas of Africa from Ethiopia to Cape Province; their damage is usually confined to grassland where they apparently eat all species of grasses, and they have been recorded damaging cotton crops. Other species of *Hodotermes* also occur in eastern Africa.

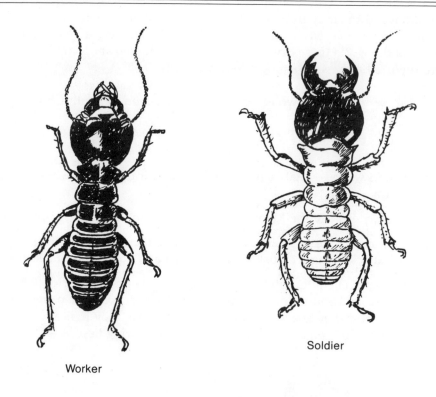

Worker

Soldier

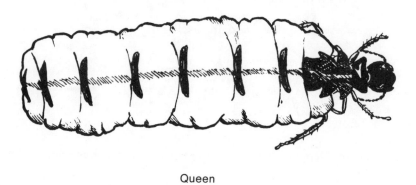

Queen

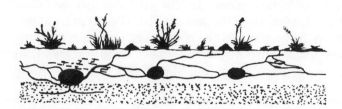

Part of underground nest system

Fig. 5.17 Harvester termite (*Hodotermes mossambicus*) from eastern Africa (Isoptera; Hodotermitidae).

Rhinotermitidae (wet-wood termites, etc.)

This is the first group (in this arrangement) where the frontal gland is apparent and in some soldier castes it is a well-developed defensive organ. They are all small, wood-eating termites and subterranean in habits, and they all have the mutualistic intestinal microorganisms that permit digestion of cellulose material. There are several distinct subfamilies with associated ecological differences. A few of the more notable pest species include:

Coptotermes—a truly tropical genus with 45 species, best represented in S.E. Asia and Australasia. They are basically forest species that live in moist stumps of dead trees, but have adapted to both urban and agricultural conditions. They eat structural timbers and woodwork that is damp, as well as dead trees, and they also take wood from living trees and also destroy herbaceous plants— they are quite polyphagous in diet so long as the material is plant cellulose.

Coptotermes formosanus (Chinese wet-wood termite)—is one of the best-known species, from southern China, now spread by Man into Hawaii, Japan, U.S.A. and South Africa; illustrated in Fig. 5.18. In urban areas of Hong Kong (Kowloon) in 1979, this species caused a massive electricity failure by eating through the thick insulation (3 mm; 0.12 in.) surrounding underground power cables thereby permitting entry of water to the copper cores and causing a short circuit. They had apparently eaten through the outer rubberized plastic insulation to get at the hessian wrapper around the copper cores and, in eating the hessian, they had also eaten the plastic sheath around the copper cores. Crops reported as damaged by this species include rice, sugarcane, groundnut, fruit trees, and various other food crops as well as many forest trees.

Other species of *Coptotermes* are damaging to rubber and coconut palms throughout S.E. Asia.

Reticulitermes—restricted to the northern temperate region, and is found in southern Europe, North Africa, parts of Asia and North America.

Within this family are a number of different small genera from different parts of the tropical world but few are of any particular importance.

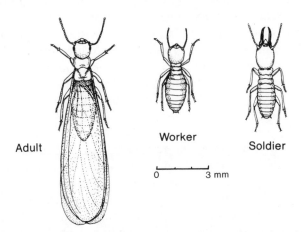

Adult   Worker   Soldier

0   3 mm

Fig. 5.18 Chinese wet-wood termite (*Coptotermes formosanus*); adult, worker and soldier; South China (Isoptera; Rhinotermitidae).

Termitidae (mound, subterranean, or bark-eating termites, etc.)   *1300 spp.*

A very large family found throughout the tropics and subtropics; most make a subterranean nest, usually with a large mound, but a few tropical species make arboreal nests. The Macrotermitinae are characterized by having fungal combs inside the nests. The chopped up plant material is formed into the honeycomb-like structures in special chambers within the nest, and these "gardens" are inoculated with a special fungus called *Termitomyces* (other fungi may also be involved). The fungal mycelium produces small round white bodies (swollen hyphal ends) called brometia which are eaten by the termites. This large family shows considerable variation in habits, and the precise mode of feeding/nutrition is not too clear. Some species appear to have some microorganisms in the intestine that may be capable of breaking down cellulose, but they also eat a certain amount of decaying vegetable matter which is already infested with fungi. It is possible that many Termitidae actually feed to a considerable extent on foraged fungi (other than brometia) rather than entirely on cellulose. Experiments have shown that some species of termites apparently secrete cellulases themselves in the midgut.

The pest species forage for plant material in several ways—roots may be eaten underground, seedlings may be entirely destroyed; the most usual damage to trees is for the tree trunk or plant stem to be covered by a sheet of soil and chewed bark fragments, underneath which the workers chew away the bark and the sapwood (Fig. 5.15). An important form of agricultural damage is the destruction of the underground parts of fence posts and wooden stakes; posts of 10 cm (4 in.) diameter may be entirely eaten away in a matter of a few months; if the fence post is a small tree trunk covered in bark then the termites may eat under the bark quite a height above the ground. The workers have pale delicate bodies with a thin integument and are strongly negatively phototaxic and if exposed would be in serious danger of desiccation, so they usually only forage above ground under the shelter of covered pathways. One species in Malaysia (*Macrotermes carbonarius*) is unusual in that workers forage along open pathways during daylight—from a distance they look like foraging ants.

Some of the more important pest species include:

Macrotermes spp. (bark-eating termites; mound termites)—about ten species occur throughout the Old World tropics (not Australasia). They damage a wide range of field crops as well as trees. Some of the largest termite species belong to this genus. Nest mounds may be large and very spectacular; in Africa the mound of *M. bellicosus* may be up to 2.4 m (6–7 ft.) in height (Fig. 5.19) and that of *M. subhyalinus* a similar height; but in Malaya *M. carbonarius* makes a mound in the forest not more than 1 m (40 in.) tall and in South China *M. barneyi* has the entire nest underground with no surface mound at all. It seems likely that the size of the mound is correlated to local temperature conditions, and the winter in South China is quite cold. Generally, these termites have hyaline wings (Fig. 5.20). Fig. 5.21 shows excavation of the nest of *M. bellicosus* in Uganda.

Microcerotermes spp. (live wood-eating termites)—most abundant in Africa, but also occurring in India and S.E. Asia and a few in South America. Some species destroy crop seedlings, others can apparently invade living wood and live plant tissues (generally with ease) and do not require to enter the plant body through dead tissues. Some make nests underground, a few make carton-nests on the side of trees, and some live in small communities inside hollowed branches.

Fig. 5.19 Mound of *Macrotermes subhyalinus;* 1.5 m (5 ft.) high; central Ethiopia (Isoptera; Termitidae).

Fig. 5.20 Adult *Macrotermes barneyi;* wingspan 50 mm (2 in.); South China (Isoptera; Termitidae).

Fig. 5.21 Nest mound of *Macrotermes bellicosus* being excavated; Kampala, Uganda (Isoptera; Termitidae).

Microtermes (seedling termites)—a small genus of small termites with small nests; but are very abundant in parts of Africa and S.E. Asia, being the commonest ground termites in some areas and often destroy crop seedlings.

Nasutitermes—a large genus with many species throughout the tropical world; the soldier caste has the frontal gland extended into an anterior "snout" and is very distinctive in appearance. This is one of the groups that typically construct a carton-nest on the side of a tree trunk or palm; they eat away the sap wood and inner bark underneath the older more lignified outer bark layers, and also make covered pathways down to the ground and up to the crown of the palm. Many species typically have a main nest established in the ground before they establish secondary aerial nests. The carton-nest material is mostly worker excreta, consisting of lignin, cellulose, and other vegetable residue which dries to form a hard dark brown carton material; in certain species it may be mixed with soil particles.

Odontotermes—an Old World genus, not present in Australasia, and Harris (1971) lists a total of 23 species as pests of crops throughout the Old World. They are generally smaller in size than *Macrotermes,* with little or no mound-building, and colonies extend rather more into cooler regions, such as South China, South Africa; in many species the wings are dark.

Odontotermes badius (crater termite) (Fig. 5.22)—a species with a very distinct nest community with a series of crater-like entrances and shafts over an area of several square meters. It is one of the fungus-growing species, found throughout most of tropical Africa. Usually not a serious crop pest, except in very dry times or dry areas when crop seedlings may be taken, but a nuisance and disfigurement when the nest is in the middle of a road (Fig. 5.23) or in a lawn or flower garden.

Pseudacanthotermes militaris (sugarcane termite)—there are actually two closely related species widely distributed throughout tropical Africa (south of the Sahara). Although sugarcane is the usual crop most damaged, this species is also recorded damaging tea bushes and young tung trees in Malawi. Most damage is done to the newly planted pieces of cane (setts) rather than to established plants (Fig. 5.24).

Harris (1971) lists 59 species of termites recorded damaging sugarcane worldwide but, generally, serious damage is sporadic and usually quite local.

Termite nests are typically inhabited by a series of commensals and inquilines, collectively referred to as "termitophiles", including other species of termites, many different insects, snakes, lizards, and the occasional small mammal and bird species.

Houses in Malaya are often built on stilts with a 60 cm (2 ft.) concrete base to each wooden post to combat termites.

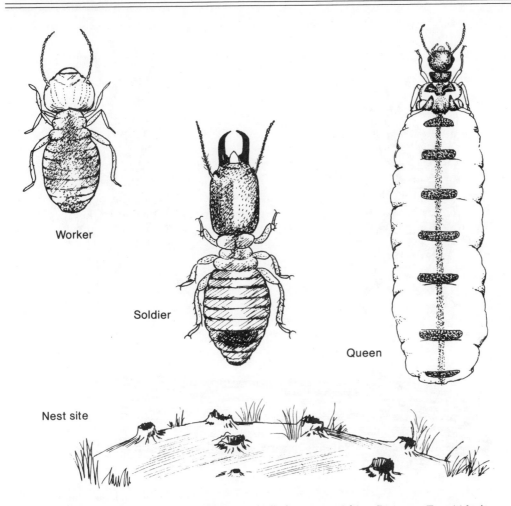

Fig. 5.22 Crater termite (*Odontotermes badius*); eastern Africa (Isoptera; Termitidae).

Fig. 5.23 Nest entrance holes of crater termite (*Odontotermes badius*); Rift Valley, Ethiopia (Isoptera; Termitidae).

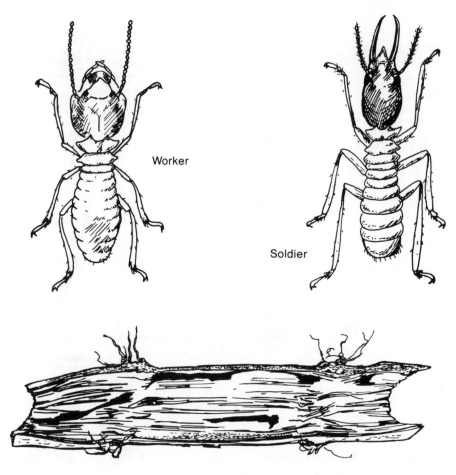

Fig. 5.24 Sugarcane termite (*Pseudacanthotermes militaris*); Kenya (Isoptera; Termitidae).

ORDER ZORAPTERA 1: 22

Minute, winged or apterous, primitive insects, mostly tropical in distribution, found under dead tree bark and in leaf litter. None of the few species is of any importance agriculturally.

ORDER PSOCOPTERA (barklice and booklice) 22: 2000

Small to minute insects, soft-bodied, with long filamentous antennae, and a rather bulging head which gives them a characteristic appearance. They may be winged, micropterous or apterous; the wings are delicate, sometimes maculate or dark colored. Mouthparts are rather weak but of the biting type; most species feed on fungi, algae or lichens but some eat general organic debris. They are mostly found in tree foliage on twigs and branches, on rocks, and weathered fencing; a few inhabit nests of birds and mammals. Many species live gregariously and can be seen gathered in clusters of 50 or more on tree trunks or rock surfaces. Some species make extensive sheets of silken webbing over tree trunks or rock surfaces under which they live gregariously; such silk sheeting differs from that made by Embioptera in that there are no discrete tunnels and the sheet is usually found on a smooth surface rather than on the rough surface favored by embiids.

There is a recent awareness of the importance of psocids as domestic pests; in the past they have usually been ignored since they are so small and inconspicuous. In Europe, with the spread of central heating in private dwellings and other domestic buildings, there has been a concomitant spread of various stored-products pests. Similarly, since the widespread adoption of fitted carpets in houses, other pests have been encouraged. In museums, it has long been known that Psocoptera can be very damaging; their tiny size enables them to gain access to pinned insect collections and, once in the specimen drawers, they eat out the dried body contents of pinned insects which eventually just disintegrate. In libraries they mainly feed on fungi that grow on damp book covers. In private dwellings they tend to be psychological pests in that the occupants are alarmed at the presence of the insects and also at their common name of booklice. The louse connotation frequently causes concern.

Several species, especially of *Liposcelis* (Fig. 5.25), are the well-known booklice of domestic situations, tiny in size, apterous, flattened in shape, often pale in color. They

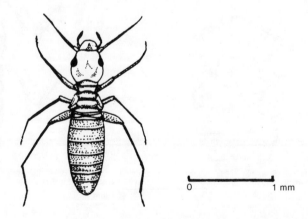

0   1 mm

Fig. 5.25 Typical wingless domestic psocid (*Liposcelis* sp.) (Psocoptera; Liposcelidae).

feed on tiny scraps of food, flours, meals, and almost any type of dried plant and animal material, although many species seem to prefer fungi. *L. bostrychophilus* is a common household species, thought to originate in Africa but now virtually cosmopolitan in warmer situations. This species is parthenogenetic, only females being known; the adult lives for up to six months, laying some 100–150 eggs of relatively enormous size. This is a tropical species and egg laying ceases at temperatures below about 20°C (68°F) optimum conditions are a temperature of 25–30°C (77–86°F) and 75% RH; in experiments it was found that about 80% of nymphs reached maturity under these conditions. Species of *Trogium* and *Lepinotus* are regularly found in stored produce in temperate regions.

Certain species are regularly found on some woody crop plants (tea, cinchona, rubber, etc.) but they do not damage the plant at all—presumably they are feeding on lichens or fungi growing on the bark.

ORDER MALLOPHAGA (biting lice)   *12: 2800*

These ectoparasites are sometimes called "bird lice" but this is not strictly accurate as some are found on mammals. But they all have biting mouthparts and feed on fragments of skin, feathers, blood scabs, skin debris; will also take blood if the opportunity arises, such as from scratches or wounds. All species are wingless and the body shape depressed. Claws are well developed, and most harm to the host is done by skin irritation.

Some species show interesting ecological preferences in that they are only found on certain parts of the host body, such as wings, head, flanks, etc. One bird host may have several different lice species all living on different parts of the body. They are all permanent, obligate ectoparasites; eggs are laid on the host and stuck to hairs or feathers. Some

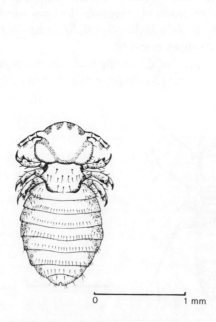

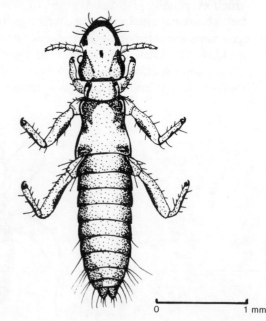

Fig. 5.26 Dog-biting louse (*Trichodectes canis*); South China (Mallophaga; Trichodectidae).

Fig. 5.27 Typical bird louse (*Lipurus* sp.) (Mallophaga; Lipuridae).

dispersal of lice takes place in the nest, by bodily contact of adults and from adults to the young birds as they fledge.

One group of biting lice is confined to elephants and warthogs as hosts—suborder Rhychophthirina, family Haematomyzidae. The suborder Ischnocera contains four families all found on mammals, and one on birds. Species of agricultural interest include *Bovicola bovis,* a worldwide ectoparasite of cattle. *Trichodectes canis* (dog-biting louse) (Fig. 5.26) is commonly found on dogs, and *Felicola* is worldwide on cats. Other species occur on some rodents, some monkeys, but mostly on other Carnivora. On domestic pigeons (reared for food, for racing, and ornamental birds) the elongate and slender *Columbicola columbae* is often very abundant.

The suborder Amblycera includes three families that infest birds, and three that are found on marsupials, rodents, and dogs. A typical bird louse is shown in Fig. 5.27—*Lipurus* sp. The common chicken louse is *Menopon gallinae,* and several species occur on ducks.

ORDER SIPHUNCULATA (sucking lice)   6: 300

These are permanent obligatory ectoparasites confined to the Mammalia as hosts. Mouthparts are of a piercing and sucking nature and used to extract blood from the host. The species showing striking host-specificity (as also do most Mallophaga) but the group is not abundant and infested individuals are often few in number. One group (family) is confined to seals and other Pinnipedia; others occur on rodents, insectivores, monkeys, and other mammals.

Species of agricultural importance include the following:

Haematopinus asini (horse louse)—on horses and zebras.

H. erysternus (cattle louse)—widespread on domestic cattle.

H. suis (hog louse)—on pigs worldwide.

H. tuberculatus (buffalo louse)—on buffalo in Asia.

Linognathus spp.—on sheep, goats, and cattle.

Fig. 5.28 shows a species of *Haematopinus* taken from a cow in South China.

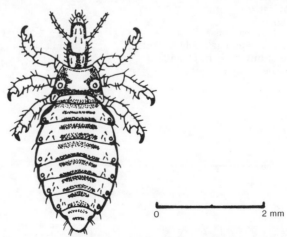

0   2 mm

Fig. 5.28 Typical animal louse (*Haematopinus* sp.); South China (Siphunculata; Haematopinidae).

Other species of importance are the human louse ectoparasites which are now quite cosmopolitan and abundant, especially in the warmer regions of the world. The human louse (*Pediculus humanus*) (Fig. 5.29) occurs as two distinct varieties; var. *corporis* is the human body louse to be found on the body between the clothing and skin; var. *capitis* is the human head louse, which lives only in the fine hair on the head. Eggs are laid firmly fixed to hairs (as illustrated), and are colloquially known as "nits". Body louse transmits various diseases under suitable conditions of overcrowding and poor personal hygiene; trench fever and epidemic fever have been responsible for several major historical epidemics when many thousands of people died. With general improvement in levels of domestic and personal hygiene, the human body louse has gradually become more scarce in many parts of the world. But in recent years, the human head louse has become alarmingly abundant in Europe and North America, presumably due to changing fashion habits, especially men and boys with long hair.

The human crab louse (*Pthirus pubis*) (Fig. 5.30) with its stouter legs and larger claws is adapted for living in only the coarser body hair of humans, such as the hairs of the pubic region, armpits, eyebrows, and sometimes even the eyelashes. The bites of this species are particularly irritating and infestations are difficult to eradicate without careful insecticidal treatment.

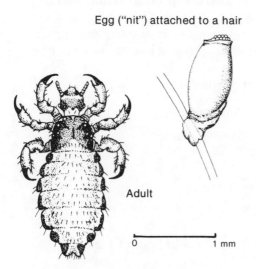

Egg ("nit") attached to a hair

Adult

0   1 mm

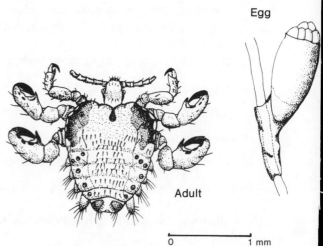

Egg

Adult

0   1 mm

Fig. 5.29 Human louse (*Pediculus humanus*); adult and egg (called a "nit") attached to a human hair; Uganda (Siphunculata; Pediculidae).

Fig. 5.30 Human crab louse (*Pthirus pubis*); adult and egg; Uganda (Siphunculata; Pediculidae).

6

Subclass Pterygota:
ENDOPTERYGOTA—Small Orders

These are the more advanced insects and all have a very pronounced metamorphosis which involved a quiescent pupal stage. The complete metamorphosis permits great ecological diversity and divergence so that the immature stages (termed larvae) are often widely separated from the adults in quite different habitats—this divergence has great effect on the importance of various groups of these insects as pests, including agricultural, medical, and veterinary pests.

ORDER NEUROPTERA (lacewings, etc.)   *19: 5000*

A somewhat heterogenous group of predacious insects, best represented in the tropical parts of the world. They are small to quite large in size, soft-bodied and with the antennae usually elongate, two pairs of similar wings, typically held roof-like over the abdomen at rest, and the wings with many extra veins. The group is subdivided into two distinct suborders.

SUBORDER MEGALOPTERA (alderflies and snakeflies)

Sialoidea (**alderflies**)—adults are large insects, generally regarded as primitive, with aquatic larvae to be found in both ponds and streams (according to the species concerned). Adults are usually found close to water. The larvae have abdominal gills for underwater respiration; they are fiercely predacious and feed on insect larvae, worms, and other small invertebrates.

Rhaphidoidea (**snakeflies**)—a terrestrial group of very specialized insects to be found in forested regions, especially in the warmer parts of the world (excluding Australia). The adults have an elongated prothorax giving them their common name. The predacious larvae are often found under tree bark, particularly on conifers, where they prey on soft-bodied insects.

These two groups are mostly of ecological importance, and generally of no agricultural interest except that they do contribute to the general predator complex responsible for the natural control of many insect populations.

133

SUBORDER PLANIPENNIA (lacewings, ant-lions, etc.)

The majority of Neuroptera belong to this group; all adults have wings with many extra secondary veins. The larvae have large mandibles modified for seizing, piercing, and sucking—prey are grasped in the long mandibles, and penetrated by the row of long "teeth" and then the body fluids are sucked out along either grooves or hollows in the "teeth". Of the 16 families listed by Richards & Davies (1977), four are of some importance agriculturally and require special mention.

Coniopterygidae   *240 spp.*

Small, fragile-bodied insects, similar to aphids in appearance and with a body coating of fine, waxy powder. As a group they are of some importance in that they usually occur with aphid infestations, and both adults and larvae prey on aphids, scales, and phytophagous mites. They are most abundant in the warmer parts of the world.

Hemerobiidae (brown lacewings)   *800 spp.*

Adults are very similar to the more abundant green lacewings but are slightly smaller and brown in color, and nocturnal in habits. Their larvae are generally more slender and fusiform in shape without lateral body tubercules, and the more delicate mandibles are devoid of teeth along the inner margins. It has been recorded in the U.K. that a single adult *Hemerobius* could eat between 13,000 and 15,000 aphids (*Adelges*), and a larva could eat 3000 during its short lifetime, so the agricultural importance of these predators is obvious.

Chrysopidae (green lacewings)

A conspicuous group of small fierce predators with a distinctive green coloration and the diurnal adults have bright golden eyes (Fig. 6.1). Adults have biting mouthparts with which they eat their insect prey. Larvae are fusiform in shape or broader, and with many body tubercules and spines.

Fig. 6.1 Green lacewing adult (*Chrysopa* sp.) (Neuroptera; Chrysopidae); wingspan 28 mm (1.1 in.); Hong Kong.

Sometimes the larvae carry the empty body shells of their devoured prey impaled upon their spines, presumably as a form of camouflage. They are active predators and frequently to be seen at the site of aphid infestations and also those of Coccoidea. The elongate mandibles are toothed along the inner margin; the teeth are grooved along their ventral surface and fit together with a similarly grooved extension of the maxilla forming a series of functional suctorial grooves. The eggs of *Chrysopa* are distinctive in being pedunculate and attached to vegetation.

As predators of crop pests they are extremely important. Their prey usually consists of aphids, scales, psyllids, cicadellids, thrips, and some mites. A single *Chrysopa* larva is reported to eat 300–400 aphids during its larval life. *Chrysopa* spp. are now amongst the insects reared commercially for sale as biological control agents in both Europe and North America.

Myrmeleontidae (ant-lions) *1200 spp.*

These are more of academic interest than of direct agricultural importance, but they are insect predators quite abundant in most parts of the tropics. The adults resemble dragonflies but have short, knobbed antennae and slender legs. The larvae live individually in pits in sandy soil where they lurk buried in the sand but with the mandibles just protruding. Since the most abundant insects walking on the ground in these dry sandy areas are ants they usually form the major part of the diet of *Myrmeleon* larvae.

The other members of the Neuroptera are individually of little significance but, collectively, they do form part of the natural predator complex which is responsible for the natural population control of many insects.

ORDER STREPSIPTERA 9: 370

A very small group of small endoparasitic insects, with free-living males that have functional hind wings; the females remain inside the host insect. The larvae are endoparasitic, mostly inside the bodies of Homoptera (Auchenorrhyncha), Hymenoptera (Sphecoidea, Vespoidea, Apoidea), and some Diptera. Pupation takes place within the body of the host, and the adult female remains inside the old puparium which protrudes slightly from the host body. Most species are known from the Holarctic region, but some are tropical.

ORDER MECOPTERA (scorpion flies, etc.) 9: 400

Another small group of woodland insects with larvae living in the soil and leaf litter. Adults have two pairs of wings, often maculate, and the face extended into a beaklike projection. Both adults and larvae may be carnivorous but their numbers are generally too low for their roles as natural predators to be of any importance.

ORDER SIPHONAPTERA (fleas) 6: 1400

Small, apterous, ectoparasitic insects of characteristic appearance with body laterally compressed and large hind legs used for jumping, although walking is the usual mode of locomotion. Mouthparts are adapted for piercing skin and sucking blood from the host animal.

Adults are obligatory ectoparasites of birds and mammals and are both widespread and abundant. They are negatively phototactic and respond positively to warmth.

The larvae are free-living in the nests and homes of the hosts, feeding on organic debris, dried blood, etc. Pupation takes place in these sites, inside a silken cocoon. Young adults remain inside the pupal cocoon until their emergence is triggered off by vibrations caused by the host. The whole life cycle is thus geared to availability of hosts using the nest (or habitat), to ensure survival of the flea population. Under warm conditions, the life cycle can be completed in 3–4 weeks, but, in the absence of suitable hosts, and warmth, both eggs and pupae may remain quiescent for many weeks or months.

The human flea (*Pulex irritans*) (Fig. 6.2) is now quite scarce in many parts of the world where it was formerly abundant. In most places the most abundant domestic flea is the cat flea (*Ctenocephalides felis*) (Fig. 6.3) which normally lives on both domesticated cats and dogs, and regularly feeds on humans. This flea is of importance domestically as the vector of the small dog tapeworm (*Dipylidium caninum*). The most important flea species is probably the tropical rat flea (*Xenopsylla cheopis*) (Fig. 6.4) which is the major vector of plague (*Pasturella pestis*) throughout the world. The epidemic of plague that swept through Europe in the Middle Ages was known as the Black Death. Plague is still endemic in many parts of the tropics and still represents a major environmental hazard throughout the warmer parts of the world.

An interesting human parasite is the Chigoe flea (*Tunga penetrans*)—the young female flea burrows into the skin usually at the sides of the toenails, and she swells to a spherical size of 3–4 mm (0.12–0.16 in.) by the enlargement of the abdomen; a strange life-style for a flea but most irritating to the human host.

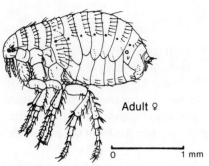

Fig. 6.2 Human flea, adult (*Pulex irritans*) (Siphonaptera; Pulicidae); Hong Kong.

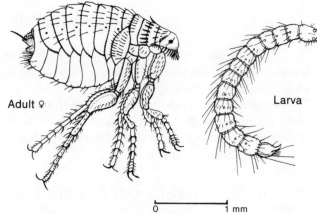

Fig. 6.3 Cat flea, adult female and larva (*Ctenocephalides felis*) (Siphonaptera; Ceratophyllidae); Makerere, Uganda.

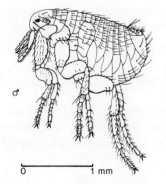

Fig. 6.4 Tropical rat flea (*Xenopsylla cheopis*), adult male (vector of plague) (Siphonaptera; Leptopsyllidae); South China.

ORDER TRICHOPTERA (caddisflies)   *11: 5000*

Adults are small to medium-sized insects (2–40mm; 0.8–1.6 in.), of a delicate moth-like appearance, with long tapering antennae. The two pairs of wings are somewhat hairy and may be scaled. Adults are weak fliers and do not survive long, although some have been recorded living for up to four weeks. The larvae are aquatic and live in both streams and ponds; they resemble caterpillars slightly and they construct a case in which the body is sheltered. Most cases are transportable and made of pieces of vegetation or sand grains; but in torrents some species construct nets of silk in which they live, stuck on to the rocks; the net traps food particles. Most larvae appear to be omnivorous or detritivorous, but some are quite carnivorous.

The group is of great ecological interest and important in limnological studies, but relatively unimportant agriculturally. However, a few species of Trichoptera cause damage to paddy rice, and to other aquatic plants. In some cases the damage consists of cutting pieces of leaf from rice seedlings to build larval cases, and some omnivorous species actually eat pieces of root or submerged stem from seedlings. Occasionally, serious damage has been reported, sometimes following a migration of larvae from ditches or drainage channels into the rice fields. The species recorded damaging rice are as follows:

Limnephilus spp.—in China, Korea, Japan, and East Siberia.

Oecetis nigropunctata (black-dotted caddisfly)—China and Japan.

Setodes argentata (rice caddisfly)—Japan.

Triaenodes bicolor—northern Italy.

An unidentified species was recorded damaging rice in Australia (N.S.W.).

7

ORDER ORTHOPTERA
crickets, grasshoppers, etc.
17: 17,000

In some textbooks this order still includes cockroaches, mantids, stick insects, etc., but this view is not generally held nowadays.

They are medium-sized or large insects mostly; winged, brachypterous, or in some cases apterous; mouthparts are strong and of the biting type; prothorax is large, and hind legs usually enlarged for jumping. The forewings are thickened and form tegmina (for protection); hind wings are large and used for flying, and often have a bright distinctive color patch basally (often red or yellow); many species have specialized stridulatory and auditory organs; metamorphosis is slight. There are many important pest species.

SUBORDER ENSIFERA

These insects all have elongated tapering antennae, as long as the body, and the female often has an elongate ovipositor.

TETTIGONOIDEA

There are two other small families of no agricultural importance.

Tettigoniidae (long-horned grasshoppers; bush crickets; katydids)   *5000 spp.*

A large group, predominantly tropical, with species divided into 19 subfamilies. Some are apterous, but in winged forms the male has specialized areas on the tegmina bases (cubito-anal regions) for stridulation. Winged forms are predominantly green or brown and live arboreally in tall herbage or bushes, and the males can be heard chirping (rather like baby chickens) in bushes in the early evening and also at night; a few females also stridulate. Most species are nocturnal in habits (as distinct from the diurnal Acrididae) and are cryptic in behavior during the day. Some species are carnivorous, and some are omnivorous, but the great majority are herbivorous and feed on foliage.

One interesting subfamily is the Pseudophyllinae—these are leaf-grasshoppers which simulate leaves as they sit quietly in the plant foliage, with their expanded tegmina looking very much like either a brown or a green leaf.

A few species oviposit in the soil but the great majority oviposit in the foliage where they live; the prominent swordlike ovipositor is used to cut a slit in a twig (or sometimes the edge of a leaf) and the eggs are laid in a row in the slit. The twig is often killed by the process of oviposition, and favored tree hosts may have a dozen or more dead twigs producing conspicuous patches of dead brown leaves over an otherwise green canopy. Most of the trees used are wild forest species (e.g. *Ficus* spp., etc.); sometimes the female may strip bark from twigs as well as eating leaves.

Since they are basically forest species, not very many are recorded as crop pests, but a few species of note are listed below:

Anabrus simplex (Mormon "cricket")—polyphagous, including Gramineae; widespread in the U.S.A., and in Canada.

Caedicia spp.—damage *Citrus* fruits in Australia.

Chloracrus prasina—sometimes oviposits in twigs of cocoa in Indonesia.

Conocephalus spp.—a large and widespread genus in Asia; many species are omnivorous and they feed on eggs and young nymphs of rice stem borers and rice bugs (*Leptocorisa* spp.) as well as eating rice leaves, in parts of S.E. Asia.

Decticoides spp. (Bush "crickets")—polyphagous on cereals and other plants; Ethiopia.

Elimaea spp.—several species found eating foliage of *Hibiscus,* tea, sugarcane, tobacco and soybean in parts of S.E. Asia (Fig 7.1).

Holochlora pygmaea—recorded damaging tea in Java, by eating leaves and splitting twigs.

Homorocoryphus spp. (edible grasshoppers)—several species swarm at regular intervals in eastern Africa; they are somewhat atypical in that they feed mostly on Gramineae (grasses and cereal crops).

Fig. 7.1 Long-horned grasshopper (*Elimaea punctifera*); adult female, resting on *Hibiscus* leaf; body length 25 mm (1 in.); Hong Kong (Orthoptera; Tettigoniidae).

Mecopoda elongata (brown leaf grasshopper) (Fig. 7.2)—this species lays eggs in the soil but eats plant foliage; recorded damaging beans and sugarcane in parts of S.E. Asia.

Phlugis mantispa—an interesting pest that attacks unripe bananas in Surinam, and it oviposits into the peel of the fruits causing disfiguration and the fruit is rendered unsalable.

Sexava spp.—several species occur throughout S.E. Asia and they defoliate palms and eat banana leaves; eggs are laid in the soil under the trees.

In most tropical countries there are recorded a few species of Tettigoniidae that damage cultivated plants, but most are local species of very restricted distribution, and most are only minor pests.

Fig. 7.2   Brown leaf grasshopper (*Mecopoda elongata*); female; body length 60 mm (2.4 in.); Hong Kong (Orthoptera; Tettigoniidae).

GRYLLOIDEA

Grylloidae (crickets)   *2300 spp.*

A large group of nocturnal insects that live underground or in leaf litter on the ground. Most are brown or blackish in color, with long antennae and long cerci, stout hind legs (but only jump a little), the female usually has a long straight ovipositor, and they stridulate by friction between the tegmina. Eggs are usually laid underground, often in a nest. The Oecanthinae are termed tree crickets, usually greenish brown in color, and some of these lay eggs in the pith of living twigs; the life cycle is totally arboreal. Many species of crickets are detritivorous and quite omnivorous, but some are mostly herbivorous and can be pests of some importance. Some larger species live in subterranean tunnels with a nest and they cut off herbaceous seedlings at ground level at night; the seedlings are left on the surface for a day in order to induce wilting, then the following night the limp and flexible seedlings are dragged down into the nest where they are eaten by both adults and the young nymphs. Adults are quite long-lived and 2–4 months is usual, according to the species.

Some of the more important pest species include the following:

Acheta domesticus (house cricket) (Fig. 7.3)—cosmopolitan throughout the warmer parts of the world; a domestic species in buildings. Several other species of cricket are to be found in houses and food stores.

Brachytrupes membranaceus (tobacco cricket) (Fig. 7.4)—polyphagous; Africa south of the Sahara.

Brachytrupes portentosus (large brown cricket) (Fig. 7.5)—polyphagous; tropical Asia, Indonesia, Papua New Guinea.

In Ethiopia, the species *Oecanthus pellucens* is the tobacco leaf cricket and is quite damaging to tobacco plants, eating holes in the leaf lamina. In Canada and the U.S.A. several species of *Oecanthus* (tree crickets) are minor pests of orchard trees and bush fruits, both by foliage eating and oviposition.

Several species are known collectively as "field crickets" throughout the world, and they have been variously referred to the genera *Acheta, Gryllus* and *Teleogryllus,* and several synonyms are known. Some of the species most frequently recorded are listed below:

Acheta assimilis—found throughout the U.S.A. and South America.

A. bimaculata (two-spotted cricket)—in Africa, southern Europe, and parts of Asia.

Teleogryllus mitratus (= Acheta testaceus) (oriental field cricket) (Fig. 7.6—a widespread species in S.E. Asia and South China.

Teleogryllus spp. (field crickets)—several species recorded throughout S.E. Asia, Indonesia and Australia.

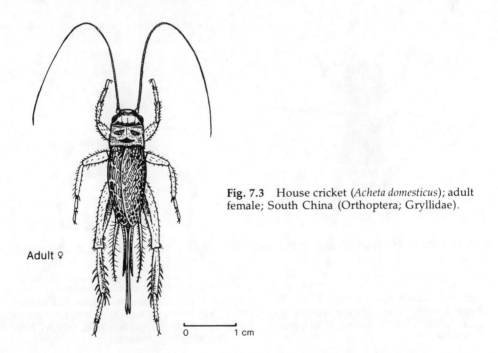

Adult ♀

Fig. 7.3   House cricket (*Acheta domesticus*); adult female; South China (Orthoptera; Gryllidae).

0   1 cm

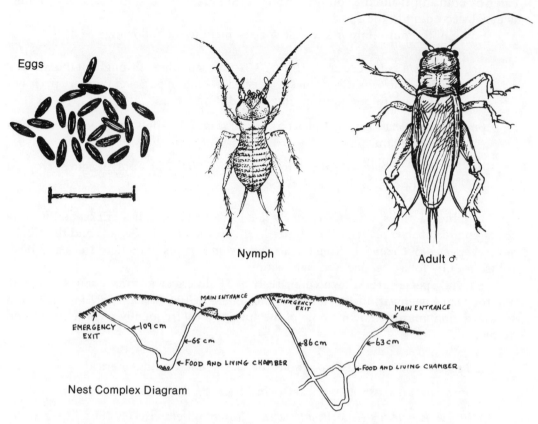

Eggs

Nymph

Adult ♂

Nest Complex Diagram

Fig. 7.4 Tobacco cricket (*Brachytrupes membranaceus*) from East Africa (Orthoptera; Gryllidae).

Fig. 7.5 Large brown cricket (= big-headed cricket) (*Brachytrupes portentosus*); adult male; South China (Orthoptera; Gryllidae).

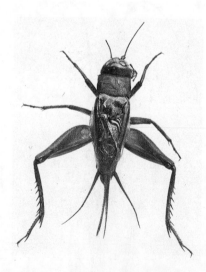

Fig. 7.6 A common Asian field cricket (*Teleogryllus mitratus*) (formerly called *Acheta testaceus*); adult female; South China (Orthoptera; Gryllidae).

Gryllotalpidae (mole crickets) *50 spp.*

A small group of insects of striking appearance with their anatomical modifications for a fossorial life, in particular the large shovel-like forelegs with which they dig. Despite the small size of the folded wings they do fly quite well, and may come to lights at night. Mole crickets generally lead a subterranean life and feed on the roots of growing plants as well as tubers and other underground stems and storage structures. They are omnivorous in diet and do eat some other insects, but overall they are quite damaging crop pests. Damage tends to be sporadic and seldom very serious, but they are widespread and abundant throughout the warmer parts of the world. Most of the pest species are tropical and belong to the genus *Gryllotalpa* but the American genus *Scapteriscus* is widely distributed in the U.S.A. The main pest species are as follows:

Gryllotalpa africana (African mole cricket) (Fig. 7.7)—cosmopolitan throughout the warmer parts of the Old World (C.I.E. Map No. A. 293); but recently specimens from the Far East have been separated off as *G. orientalis* although the two species appear to be virtually identical in appearance.

G. gryllotalpa (European mole cricket)—found throughout the warmer parts of Europe, Asia, North Africa and also in parts of the U.S.A.; in the U.K. now rare and protected.

G. hexadactyla (American mole cricket)—the U.S.A. and Central America.

G. hirsuta (large mole cricket)—to date only known from parts of Indonesia.

Scapteriscus vicinus—a species endemic to the West Indies and Central America.

Scapteriscus spp.—mostly found in North America, but reports also from Indonesia.

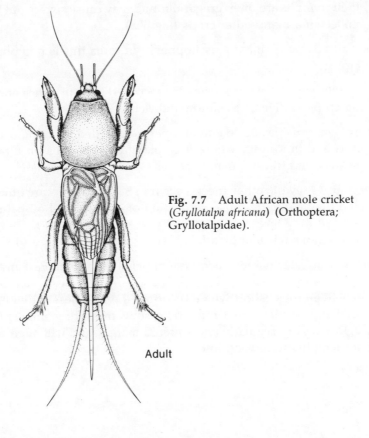

Fig. 7.7 Adult African mole cricket (*Gryllotalpa africana*) (Orthoptera; Gryllotalpidae).

Adult

SUBORDER CAELIFERA

ACRIDOIDEA

These are the short-horned grasshoppers and the locusts, together with a few less well-known families. Antennae are short (shorter than body length), and the ovipositor (when present) is short and stout with reduced inner valves.

Pyrgomorphidae   *400 spp.*

Most of these species are brightly colored (aposematic coloration), some with a pointed head, and most have stink glands which give them a characteristic and repulsive odor, presumably to deter predators. Some species are known as stink grasshoppers. Ecologically they appear to be restricted to arid areas in Africa, southern Europe, and desert areas of Asia. A few species are pests of economic importance in that they damage the crops grown in the dry areas, mostly cotton, sorghum, and the millets, but the genus *Zonocerus* in Africa would appear to be more widely distributed ecologically, and it attacks a wider range of plants. They are diurnal in habits. Some of the more important species include:

Attractomorpha spp.—recorded from India to South China; quite polyphagous.

Chrotogonus senegalensis (frog grasshopper)—in northern Africa damaging to cotton.

Chrotogonus spp. (surface grasshoppers)—Africa, India; polyphagous.

Phymateus spp. (bush locusts)—several species in eastern Africa are abundant; their nymphs are often gregarious and may cause severe but local damage to millets and many other crops (Fig. 7.8).

Poekilocerus pictus (painted grasshopper)—in India this is polyphagous on many crops.

Pyrgomorpha spp.—several species are recorded both in Africa and India, feeding on crops as diverse as cotton and rice.

Valanga nigricornis (Javanese grasshopper)—recorded from India, throughout S.E. Asia and Indonesia, where it is totally polyphagous on Gramineae, herbs, bushes, and trees of many species.

Valanga spp.—several other species occur in S.E. Asia; they are quite polyphagous, and they regularly cause defoliation of coconut and oil palms.

Zonocerus elegans (elegant grasshopper)—from South, S.E., and Central Africa.

Zonocerus variegatus (variegated grasshopper)—from both East and West Africa.— these species are quite polyphagous; they occur as both alate and brachypterous (short-winged) forms; the flightless individuals climb very well and can defoliate *Citrus* and cashew trees three meters (10 ft.) tall (Fig. 7.9). They are regularly encountered mating, and will often remain *in copula* for lengthy periods of time.

Fig. 7.8 Adult bush locust (*Phymateus viridipes*); body length 45 mm (1.8 in.); Ethiopia (Orthoptera; Pyrgomorphidae).

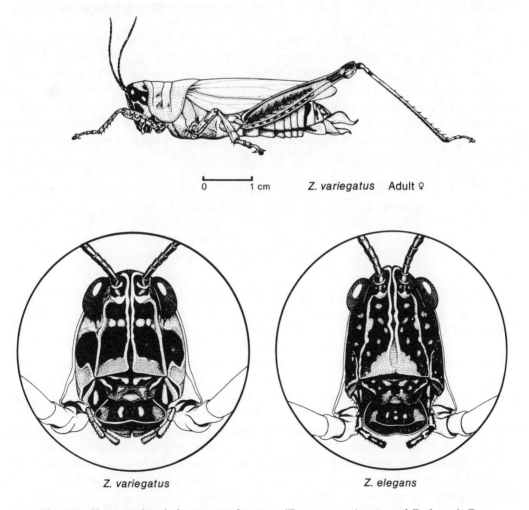

0   1 cm   *Z. variegatus*   Adult ♀

Z. variegatus   *Z. elegans*

Fig. 7.9 Variegated and elegant grasshoppers (*Zonocerus variegatus* and *Z. elegans*); East Africa (Orthoptera; Pyrgomorphidae).

Acrididae (short-horned grasshoppers and locusts)   *9000 spp.*

A large family found worldwide but predominantly in warmer regions, and as many as 500 species can be regarded as agricultural pests. The group is subdivided into 19 different subfamilies. In temperate regions, damage is mostly confined to pastures and grasslands; many species are of importance on the prairies of North America and on the grasslands of Australia and Asia. In the tropics, some species damage broadleafed crops in addition to their widespread damage to Gramineae and palms. Most males practice stridulation, and several different methods of sound production are used. Oviposition takes place in soil, usually a light sandy soil being preferred, and the entrance to the hole is sealed as part of the frothy egg-pod (ootheca) which protects the egg mass (30–100 eggs usually). In many situations the egg-pods are parasitized by tiny parasitic Hymenoptera, and they are also preyed upon by larvae of Meloidae (blister beetles). This natural population control is often of great importance in suppressing grasshopper/locust outbreaks. There are, in fact, many different insects that feed on grasshopper eggs in the soil, including a number of different beetles and also fly larvae from several different families.

Most grasshoppers have a vivid patch of color at the base of the hindwings, which they flash as they take flight, and this can help in the field recognition of some species—the usual colors being red, purple, and yellow, in different shades. It is assumed that this color serves a disruption function in startling predators.

In the great grassland habitats of the world (see Fig. 1.2, p. 23) there are complexes of Acrididae that are the natural insect grazers; each major grassland area having a different insect fauna so far as genera and species are concerned. In the grasslands of the U.S.A. and Canada (prairies and Prairie Provinces), much of which is now used for cereal production, there has long been a complex of pest species responsible for much damage to both grass and cereal crops, and some other crops. The five main species causing crop damage in North America are probably those listed below, but the total list of damaging Acrididae is probably 40–50 species.

> *Camnula pellucida* (clear-winged grasshopper)
> *Melanopus bivittatus* (two-striped grasshopper)
> *Melanopus devastator* (devastating grasshopper)
> *Melanopus femurrubrum* (red-legged grasshopper)
> *Melanopus sanguinipes* (migratory grasshopper)

It is difficult to generalize for there are considerable differences from state to state and from province to province, and there are annual and long-term fluctuations in population size.

Many species of Acrididae have great fecundity and vast powers of population growth, and they regularly show population irruptions, usually associated with migratory dispersal; this happens regularly in Europe, Asia, Africa, Australia, and the New World. Some species have developed this behavior to a point where they have distinct solitary and gregarious phases, with different coloration and different habits; these species often have the largest and most damaging swarms, and they are usually called "locusts" to emphasize their distinctiveness from the other grasshoppers.

Locust swarms threaten almost one-third of the world's land surface, and, from time immemorial, have caused devastation in the warmer parts of the world. Because of the importance of locusts as crop pests and the need to understand the reasons for swarming, in 1929 the Anti-Locust Research Centre was established in London. Later, the parent organization became redesignated as the Centre for Overseas Pest Research

(C.O.P.R.), after the original brief had been broadened to include other pests of international importance. Finally, in 1982, it was amalgamated with the Tropical Products Institute (T.P.I.) and reduced in size, as part of the Tropical Development and Research Institute (T.D.R.I.), for the moment at the same address in London. T.D.R.I. is now part of the N.R.I. (Natural Resources Institute) and relocated at Chatham in Kent. In the meantime it was realized that the locust problems were essentially international and that each of the three main locust species required separate study, and so the Desert Locust Control Organization (D.L.C.O.) was established in 1960, and later in 1971 the International Red Locust Control Organization for Central and Southern Africa (I.R.L.C.O.—C.S.A.) was formed. Now the main international center for coordination of locust study is at the headquarters of the Food and Agriculture Organization of the United Nations (F.A.O.) in Rome. The most important locust species are dealt with in the *Locust Handbook* (C.O.P.R., 1966) but an exhaustive treatment is given in the recent book, *The Locust and Grasshopper Agricultural Manual* (C.O.P.R., 1982) where several hundred pest species are considered.

The main locust species breed in temporary habitats, often at the edges of deserts, and usually after the rains when there is lush grassy growth for the nymphs to feed on. As the nymphs grow, the vegetation gradually dries (and is also eaten up) and the habitat shrinks as the surface water diminishes, forcing the locust nymphs to crowd together. The crowding may result in the formation of the "gregarious phase" with its brighter and more distinctive coloration, and greater activity and the social interaction which causes the swarm to form. Swarming hoppers march across the countryside resembling a vast living carpet of insects, and after metamorphosis is completed the adults fly. Aerial swarms may contain many millions of individuals, and estimates of up to 1000 million have been made. The flying swarm can literally block out the sunlight when overhead. Fortunately, swarms only develop occasionally when ecological conditions are precisely suitable. However, an individual plague may actually last for 10 years or more as it slowly sweeps across the country (and continent), stopping to breed periodically, and eventually traversing 1000 miles or more.

Since the establishment of the international control organizations there had not been a serious outbreak or plague since 1948, but in 1986/88 the program foundered when confronted with developing swarms. The swarms that do develop are usually controlled in their infancy due to careful monitoring and surveillance, followed by ground and aerial spraying of the incipient swarms with insecticides. However, it was widely reported in the press that the 1986 weather conditions were suitable for swarm development in Africa, and, following a breakdown of surveillance which coincided with administrative difficulties in the respective control organizations, swarms of both desert locust and red locust developed and were threatening to cause serious agricultural losses in many parts of Africa. Simultaneously the brown locust of South Africa was also swarming. The outcome of this unhappy situation was a major outbreak of desert locust.

Solitary phase locusts are found throughout their geographical range but, since they are only found in small numbers and in scattered populations, they are scarcely noticeable; they merely form part of the local grasshopper complex. The three most important species of locust are as follows:

> *Locusta migratoria* (migratory locust) (Fig. 7.10)—prefers a diet of Gramineae (grasses and cereal crops) but will eat other plants during the dry season. Widely distributed Old World species; now recognized as three distinct geographical subspecies. *L. m. migratoria* is Palaearctic (in the warmer parts) in distribution, mostly in central Asia, and is called the Asiatic migratory locust.

L. m. migratorioides (African migratory locust) is found in tropical Africa and parts of western Asia; the main outbreak area lies in the Niger Valley of West Africa. The most recent spectacular outbreak started in 1928 near Timbuktu, and, by 1932 it had reached the Red Sea. By the following year (1933) the swarms had traversed eastern Africa and finally reached Cape Province after which the swarm collapsed. The last major outbreak was in 1949, but it was successfully contained. It has been suggested that there are a further two subspecies about which relatively little is known: *L. m. capito,* the Madagascar locust, and *L. m. rossica,* the central Russian locust.

Locusta migratoria manilensis (Oriental migratory locust)—occurs in the Philippines and on Borneo, and swarms have occasionally reached Malaya; swarms have also been recorded in China. This subspecies has been identified in Australia, but there swarms are quite uncommon, and the solitary phase is reputed to occur in New Zealand.

Nomadacris septemfasciata (red locust) (Fig. 7.11)—this locust has two main outbreak areas: one in the Rukwa rift valley of Tanzania, and the other in the Mweru marshes of Zambia and the Chilwa plains in Malawi. The area at risk from this locust is eastern, central and southern Africa. Basically a polyphagous pest with a preference for Gramineae, it is similar to many other Acrididae. There is a single annual generation with breeding coinciding with the rainy period usually. Breeding of the solitary phase is of regular occurrence throughout Africa south of the Sahara. The last major outbreak was in 1930 and by 1934 it had covered almost all of the southern half of Africa; the plague did not finally die out until 1944.

Fig. 7.10 Migratory locust (*Locusta migratoria*); adult male, solitary phase, African subspecies (Orthoptera; Acrididae). Photo: Glenda Colquhoun, Centre for Overseas Pest Research.

Fig. 7.13   Sahelian tree locust (*Anacridium melanorhodon*) in *Citrus* bush; body length 55 mm (2.2 in.); Alemaya, Ethiopia (Orthoptera; Acrididae).

Fig. 7.14   Marbled grasshopper (*Gastrimargus marmoratus*); adult female; body length 40 mm (1.6 in.); South China (Orthoptera; Acrididae).

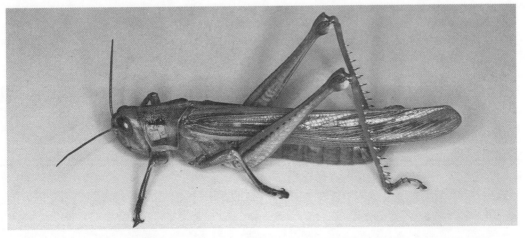

Fig. 7.15   Bombay locust (*Patanga succincta*); body length 60–80 mm (2.4–3.2 in.); South China (Orthoptera; Acrididae).

Oedaleus spp.—recorded widely from West and East Africa, India, S.E. Asia, to China and Japan; mostly on Gramineae.

Orthacris spp. (wingless grasshoppers)—India; on cereals, Gramineae, legumes.

Oxya spp. (small rice grasshoppers)—from North Australia, S.E. Asia, India, to China and Japan; on rice and grasses (C.I.E. Map No. A. 295); now also recorded from Africa (C.I.E. Map No. A. 439).

Patanga succincta (Bombay locust)—India, S.E. Asia, Indonesia, China, Japan; on Gramineae, some herbaceous crops and *Citrus* (Fig. 7.15); in China it does not behave gregariously.

Phaulacridium vittatum (wingless grasshopper)—Australia; polyphagous on many crops.

Schistocerca americana (American grasshopper)—U.S.A.; on Gramineae.

Schistocerca cancellata (South American locust)—South America; polyphagous.

Schistocerca flavofasciata—found in the West Indies and Central America.

Schistocerca spp. (locusts/grasshoppers)—throughout the U.S.A. and Canada; on Gramineae and forage legumes mostly.

Tetrigidae (grouse locusts) *1000 spp.*

A distinctive group mostly to be found in the tropics; they are small insects with the pronotum extended posteriorly over the abdomen. They are of interest ecologically in that they are mostly riparian in habits (living on streambanks) and some are semiaquatic. Only one or two species are recorded as pests; *Pantelia horrenda* occasionally damages cocoa in West Africa, and *Tetrix japonica* on several crops in Japan.

8

ORDER HEMIPTERA (= Rhynchota)
bugs
38: 56,000

A very large group, and of great importance agriculturally as many species are pests of crops. The group is characterized by having suctorial mouthparts adapted for piercing and sucking juices from plant or animal tissues. The elongate, thin labium is grooved dorsally and inside lie the paired piercing stylets (modified mandibles and maxillae). Metamorphosis is gradual and the nymphs usually share the same habitat as the adults.

There are two quite distinct groups within this order that according to some authorities should be regarded as separate orders in their own right.

An important new book on Hemiptera has recently been published (Dolling, 1991).

SUBORDER HOMOPTERA (plant bugs)

A very heterogeneous group but showing some characteristics in common, as follows: head more or less deflexed; wings usually held sloping (roof-like) over the sides of the body, the forewings of uniform consistency; wingless forms are frequent; base of rostrum extending posteriorly between anterior coxae. The saliva of most Homoptera is not usually seriously phytotoxic, but some pests do cause foliar destruction or distortion by their feeding and occasionally death, apparently due to enzymes in the saliva. With other pests there may be no symptoms of direct damage at all apart from leaf-curling. Many species are agricultural pests. Tiny groups (families) and those of no agricultural importance are not mentioned in the following text; for full details of all families, Richards & Davies (1977) should be consulted.

SERIES AUCHENORRHYNCHA

Antennae are variable, usually short with a terminal arista; rostrum plainly arising from the back of the head; they are active forms that fly and often jump readily.

FULGOROIDEA   *20: 9000*

Tettigometridae   *120 spp.*

An obscure little group, rather like jassids but with fulgoroid characteristics; they occur mostly in Europe and Africa. The only recorded pest species is *Hilda patruelis* (groundnut hopper) (Fig. 8.1)—of tropical Africa which damages various legumes and other crops occasionally, including *Citrus.*

Cixiidae   *1100 spp.*

A large cosmopolitan group but little known generally; *Oliarus* is recorded feeding on the roots of rice, sorghum, and grasses in East Africa and Australia.

Derbidae   *750 spp.*

These are delicate little insects with long mottled wings, often found sitting on the underside of leaves. The nymphs are thought to be fungal feeders in rotten wood. The adults suck sap from the phloem system, and they are at times quite abundant on a wide range of tropical plants. They are frequently found in the foliage of palms (coconut and oil palm especially) and bananas, but they seem to do little damage.

Diostrombus spp.—found on banana and sugarcane in East Africa.

Proutista moesta—a common species on oil palm in Malaysia.

P. tesselata—occurs on cocoa in West Africa, and cereal crops in India.

Meenoplidae   *50+ spp.*

A small tropical group; *Nisia atrovenosa* is reported infesting rice crops in India, and *Kermesia* can be seen on coffee bushes in East Africa.

Delphacidae (planthoppers)   *1300 spp.*

An important group, characterized by the presence of a movable serrulate spur on the hind tibiae; worldwide in distribution, small in size, they jump readily when disturbed and fly to lights at night. Several species are important cereal pests and do direct damage as sap-suckers as well as transmitting virus diseases. Eggs are laid directly into the plant tissues which offers them a measure of protection. Some species have been widely distributed by humans, particularly on sugarcane setts.

In the field of tropical agricultural entomology, one species is of particular importance. *Nilaparvata lugens* is the brown rice planthopper (Fig. 8.2), an important monophagous crop pest confined solely to species of *Oryza* (C.I.E. Map No. A. 199). Up until about 1970 this was a minor pest of rice in parts of S.E. Asia, only rarely causing any serious local damage; but during the so-called "green revolution", new high-yielding varieties of rice were introduced with lush foliage after the increased fertilizer application which was required to ensure the high yield, and in this lush vegetation the brown rice planthopper flourished. At the same time it started to show resistance to some of the usual rice insecticides. For these reasons, and some others, brown planthopper of rice has dramatically become, in all probability, the single most important pest of rice from

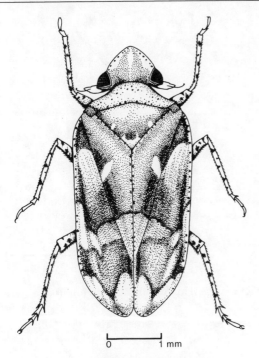

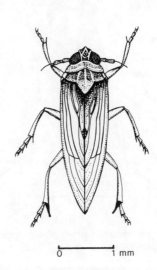

Fig. 8.2 Brown rice planthopper (*Nilaparvata lugens*) (Homoptera; Delphacidae).

Fig. 8.1 Groundnut hopper (*Hilda patruelis*) (Homoptera; Tettigometridae).

India throughout S.E. Asia to China and Japan. At the present time it is developing resistant biotypes to most of the insecticides being used on rice crops, and is a cause for considerable concern.

Many planthoppers have saliva somewhat toxic to the host plant and heavy bug infestations cause "hopperburn" when the leaves become brown and dry, and in addition most species transmit virus diseases into the crop. Many are important migratory pests and several regularly invade southern Japan in the summer from China and S.E. Asia.

Other delphacid crop pests of importance to agriculture include:

Laodelphax striatella (small brown planthopper) (Fig. 8.3)—a pest of all cereal crops and sugarcane, and some grasses throughout the Palaearctic Region and also parts of S.E. Asia and South China (C.I.E. Map No. A. 201).

Paurohita fuscovenosa—the bamboo planthopper of the Orient.

Peregrinus maidis (corn planthopper)—a tropicopolitan pest of maize, sugarcane and sorghum (C.I.E. Map No. A. 317).

Perkinsiella saccharicida (sugarcane planthopper)—very destructive to sugarcane in both Australia (Queensland) and Hawaii (C.I.E. Map No. A. 150). There are at least 22 species of *Perkinsiella,* several of which are reported from sugarcane, found widespread throughout the tropics (Fig. 8.4).

Sogatella furcifera (white-backed planthopper)—a pest of rice and some grasses; from India, S.E. Asia, Australia, China, Korea, and Japan (C.I.E. Map No. A. 200) (Fig. 8.5).

Sogatodes oryzicola (rice delphacid)—on rice in southern U.S.A.

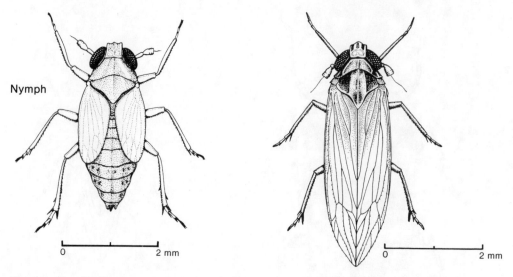

Nymph

Fig. 8.3 Small brown planthopper (*Laodelphax striatella*); nymph and adult (Homoptera; Delphacidae).

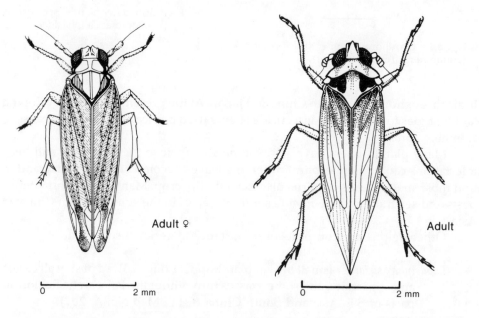

Adult ♀

Adult

Fig. 8.4 Sugarcane planthopper (*Perkinsiella saccharicida*); adult female (Homoptera; Delphacidae).

Fig. 8.5 White-backed planthopper of rice (*Sogatella furcifera*) (Homoptera; Delphacidae).

Fulgoridae (lantern bugs)   *600 spp.*

A tropical group of large and often strikingly colored bugs with many species having a pronounced anterior proboscis. Not of importance on crops except for some damage done to litchi and longan trees by *Pryops candelaria* (litchi lantern bug) in southern China (Figs 8.6, 8.7). The bugs typically rest on branches and trunks underneath the foliage canopy; when disturbed they fly to another tree showing the bright yellow hindwings vividly. *Pyrops tenebrosus* is recorded from *Citrus* in West Africa. The purpose of the "snout" is not known; it was once thought to be luminescent, hence the common name of lantern bug (fly). *Pyrops* is of interest in that it is parasitized by a small caterpillar. The adult moth is small and brown and is called *Epipyrops* (p. 439)—it has no mouthparts and so does not feed. Eggs are laid on the tree branches and the young active caterpillar attaches itself to the wings of the bug and feeds on blood from a vein. As the caterpillar grows, it moves to the upper surface of the abdomen where it bites a hole in the integument with its long thin mandibles and scoops up blood and tissues with its mouthparts. Only a small proportion of a *Pyrops* population are parasitized and any one bug only has a single *Epipyrops* larvae attacking it. The fully grown caterpillar is covered with a white silky material over its body, and it leaves the body of the bug to settle on a branch where it pupates in a white fluffy pupal case.

Fig. 8.6 Litchi lantern bug (*Pyrops candelaria*) (Homoptera; Fulgoridae); South China.

Fig. 8.7 Litchi lantern bug adult resting on trunk of litchi tree; South China.

Dictyopharidae   *500 spp.*

Small bugs, often with an anterior head process of a snoutlike appearance. A few species are found on crop plants but damage is usually slight. *Dictyophara* is often to be seen on *Citrus* foliage in S.E. Asia and South China, and *Retiala viridis* is recorded from coffee bushes.

Flatidae (moth bugs)   *950 spp.*

A group of tropical bugs, often brightly colored and beautiful in appearance. Both nymphs and adults often rest gregariously on the foliage, and in some species adults occur in two color forms. When resting gregariously one species in Africa (*Ityraea gregorii*) sits with the two color forms separated so as to resemble a large brightly colored flower. Nymphs usually produce copious quantities of wax in the form of long filaments so that infested branches and fruits are usually very distinctive in appearance. Several species are of some importance to crops grown at the edge of tropical rain forest. In S.E. Asia *Colobesthes falcata* (Fig. 8.8) and *Lawana candida* are found on many species of Sterculiaceae, including cultivated cocoa, kapok, and also on coffee. *Pulastya discolorata* is recorded on *Citrus* in South China. In East Africa *Cryptoflata* is recorded from coffee, and in Papua New Guinea *Colgar* infests avocado; but most species of Flatidae are to be found on trees in the tropical rain forests.

Ricaniidae (Ricaniid planthoppers)   *350 spp.*

Another distinctive tropical group with a moth-like appearance owing to their large forewings held in a somewhat lateral position. An interesting group, widely distributed throughout the tropics but most abundant in Africa and S.E. Asia. Not of much importance agriculturally but regularly found in small numbers associated with cultivated plants. The nymphs produce considerable quantities of waxy filaments carried on the posterior part of the body, as shown in the brown species of *Ricania* from grasses in Hong Kong (Fig. 8.9). The black and white *Ricania speculum* is found in S.E. Asia from Malaya to South China, and is a regular minor pest on oil palm and *Citrus*. *R. cervina* is a greenish species recorded from cocoa in West Africa; and another green species (*Ricania* sp.) is regularly found in small numbers on *Citrus* in South China. Also on cocoa in West Africa are species of *Ricanopsis*. *Scolypopa australis* is a minor pest of passion vine in Australia and is now introduced into New Zealand where damage to plants has been observed. In Papua New Guinea, *Euricania villica* and *Tarundia glaucesenus* are pests of avocado.

Lophopidae   *120 spp.*

Another group from the Old World tropics, generally obscure and little known, but one pest species in India is of some importance as a pest of sugarcane and several cereal crops—this is *Pyrilla perpusilla* (Indian sugarcane leafhopper) (Fig. 8.10), and C.I.E. Map No. A. 151. This has recently become a pest of rice in parts of S.E. Asia.

Fig. 8.8 Asian cocoa moth bug (*Colobesthes falcata*) (Homoptera; Flatidae) on cocoa pods in Malaya.

Fig. 8.9 *Ricania* sp. (Homoptera; Ricaniidae); adults and nymphs on *Miscanthus* grass; body length 8 mm (0.32 in.); Hong Kong.

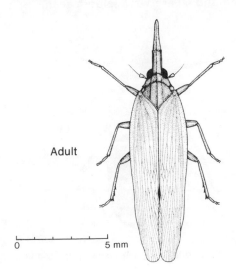

Adult

0   5 mm

Fig. 8.10 Indian sugarcane leaf-hopper (*Pyrilla perpusilla*) (Homoptera; Lophopidae).

CERCOPOIDEA   *4 families*

Cercopidae (froghoppers and root-spittlebugs)   *1400 spp.*

What was formerly regarded as a single large family is now subdivided into four separate families. This one is cosmopolitan but best represented in the tropics. The nymphs tend to be shallowly subterranean rather than aerial and usually found on the roots of the host plant.

The adults are stout little bugs, often brightly colored, which jump readily and fly with ease. All species have a characteristic group of stout terminal setae on the hind tibiae surrounding the tarsal base. Many adults are quite polyphagous and will feed on a wide range of host plants, including trees, bushes, herbs and grasses; the nymphs of a few species tend to be polyphagous but most are quite host specific. The largest group are probably specific to Gramineae as hosts. Although mostly found on the roots of the host plant, some nymphs do settle on the leaves; eggs are generally laid on the soil or in surface roots. The most usual infestation sites of *Tomaspis/Aenolamia* are at soil level on the sugarcane stool and root bases; in heavy infestations a joint spittle mass 20–40 cm (8–16 in.) in diameter over the surface of the stool may be formed.

Crop damage is done mostly by the adults owing to the presence of toxic elements in the saliva—necrotic leaf lesions are referred to as "froghopper blight". Diapause may occur in the egg stage. The only crop regularly seriously damaged is sugarcane in the West Indies, South and Central America, where 4 spp. of *Tomaspis,* 14 of *Aeneolamia,* and 4 of *Delassor* (all formerly regarded as *Tomaspis* spp.) are abundant. *Tomaspis* is a red and black colored insect of distinctive appearance (Fig. 8.11). The adults of *Aeneolamina* and *Tomaspis* rest inside the leaf funnel and their feeding on the young leaves causes reddish streaks (froghopper blight) which turn brown; in heavy infestations the leaves wither and die. Large, uncontrolled populations can reduce the sucrose content of the cane by 30–70%. They are the major (key) pests of sugarcane in Trinidad and parts of South America. In the West Indies and South America the commonest pest species is *Aeneolamia varia* which occurs as a distinct series of subspecies in different localities—*A. varia saccharina* is the race in Trinidad. There are usually four generations per year in Trinidad; if unchecked they can build up very large populations.

In Europe and in the U.K. the commonest species to be encountered is the small red and black froghopper (*Cercopis vulnerata*). It is seen resting on the foliage of most types of low herbage in gardens and other areas of cultivation, and in many wild habitats; the nymphs are reported to live underground and so are very seldom seen.

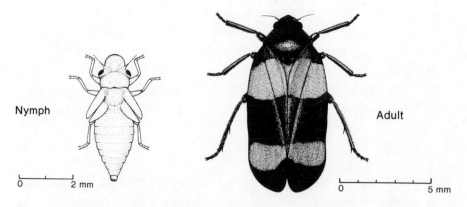

Nymph

0   2 mm

Adult

0   5 mm

Fig. 8.11 Sugarcane froghopper (and root-spittlebug) (*Tomaspis* sp.); Trinidad.

Aphrophoridae (froghoppers and spittlebugs) *800 spp.*

These are the insects commonly seen as spittlebugs or "cuckoo-spit insects", for the nymphs in their white froth are aerial on plant foliage and very conspicuous. However, nymphs of a few genera in this family are subterranean like the Cercopidae. Eggs are usually laid embedded into the host plant tissues, and they are usually capable of a phase of diapause when conditions are either too cold or too hot, so that hatching may require 2–40 weeks. The nymphal froth is produced from the intestine as a clear liquid and air is blown through a film of the liquid by a special body modification at the end of the abdomen. The froth is supposed to protect the developing nymphs from both predators and the danger of desiccation. About half the species are Palaearctic in distribution, and the others mostly in the Old World tropics. Some of the more notable pest species include:

Aphrophora spp.—several pest species in Japan and Europe recorded on trees and grapevine, etc.

Cosmocarta abdominalis—the common red and black froghopper of South China (Fig. 8.12). Adults polyphagous on trees and shrubs, but the nymphs are specific on *Melastoma* bushes.

Locris spp.—African, and red/orange and black in color; recorded from cereal crops and other Gramineae, and from *Citrus*.

Philaenus spumarius (common meadow spittlebug)—in Europe and North America the spittle mass and brown adults are seen on a vast range of host plants.

Ptyelus—a widespread genus, recorded breeding on *Cerbera manghas* trees in South China, castor, bananas, etc., in Uganda, and yams, sorghum and other plants in West Africa. Adults are often yellow in color and about 10 mm (0.4 in.) in length.

Poophilus costatus—widespread in tropical Africa on sorghum, maize, other cereal crops, and many grasses.

Sepullia murrayi—recorded feeding on *Citrus* and cotton in large numbers in the Sudan.

Fig. 8.12 Common spittlebug of Hong Kong and South China (*Cosmocarta abdominalis*) (Homoptera; Aphrophoridae); larval froth masses on stem of *Melastoma candidum* bush.

Machaerotidae (tubicolous spittlebugs)   *100 spp.*

This group is again being regarded by British entomologists as being a subfamily of the Cercopidae. A small group from the Old World tropics. The adult is characterized by having a long curved posterior spine on the scutellum. Nymphs are found inside a tube attached to the host plant and they live immersed in spittle within the tube. They are of little importance agriculturally, but may be found quite often; Fig. 8.13 shows the typical tubes of *Machaerota coronata* in Hong Kong.

Fig. 8.13   Tubicolous spittlebugs (*Machaerota coronata*) in their nymphal tubes on stems of *Helicteres*; Hong Kong.

CICADOIDEA

Cicadidae (cicadas)   *4000 spp.*

A large, spectacular group of tropical insects where the males produce a shrill sound of great intensity. In many parts of the tropics the daily chorus of cicada calling is quite deafening. The male insect has special sound-producing organs at the base of the abdomen ventrally, covered externally by a triangular flap of skeleton. In any one area there is usually a succession of song activity, seasonally according to the species, and each cicada species has a different song. With practice it is possible to identify the species concerned by identifying the song. The best-known species is probably the American periodical cicada (*Magicicada septemdecim*); the nymphs spend 17 years developing underground in the north or 13 years in the southern states of the U.S.A. The adults of each generation emerge with surprising synchronization in vast numbers. Most of the tropical species are little studied despite their abundance and local dominance. Generally, the female lays her eggs inserted into twigs that have been split by her ovipositor. In parts of the U.S.A. considerable damage is done to forest trees by ovipositing cicadas—the foliage distal to the oviposition site often dies and withers. The eggs hatch into tiny nymphs which fall to the ground and burrow into the soil using their stout, modified forelegs for digging. Nymphal life is spent underground; they pierce plant roots as a source of sap. Development is a lengthy process, as noted for the American periodical cidada, and the larger species take several years to develop in the tropics. When the nymphs are fully grown they emerge from the soil at night and crawl up either a tree trunk or other vegetation, where metamorphosis is completed; the adult emerges from the nymphal exuvium which is left firmly attached to the vegetation (Fig. 8.14). The newly emerged adults often congregate on the branches of certain trees where they feed and start to call. Cicadas are unusual in the Homoptera in that they feed from the xylem vessels of the plant, and they have a special alimentary mechanism for short-circuiting excess water straight to the hind gut (via a filter chamber) for prompt excretion.

The larger species of cicada are invariably forest species, most abundant in the canopy of the tropical rain forests of Africa, Asia, and South America, but there are some small species known as "grass cicadas" that feed on the stems of large grasses and sugarcane. A typical large brown cicada is *Cryptotympana* (Fig. 8.15) found feeding on a wide range of forest trees throughout S.E. Asia, China, and Japan, including camphor, tung, *Dalbergia* and others, and in Japan is a minor pest of *Citrus.* A speckled brown cicada (*Platypleura kaempferi*) is a serious pest of loquat trees in southern Japan where up to 30% of the trees in orchards suffered dieback due to fungal infestation of the cicada feeding punctures; Fig. 8.16 shows the closely related Chinese brown speckled cicada (*Platypleura hilpa*); other species occur in eastern Africa. *Mogannia* is one of the grass cicadas to be found on *Miscanthus,* wild cane, and other large grasses in S.E. Asia, and is a serious pest of sugarcane in Taiwan and Okinawa (Fig. 8.17). In Australia, the species *Melampsalta puer* and *Parnkalla muelleri* are pests of sugarcane and various grasses, adults damaging the foliage and nymphs the roots.

Fig. 8.14 Nymphal exuvium of large brown cicada (*Cryptotympana mimica*) (Homoptera; Cicadidae) clinging to a blade of *Miscanthus* grass; Hong Kong.

Fig. 8.15 Large brown cicada (*Cryptotympana mimica*) from S.E. Asia and China; body length 33 mm (1.32 in.) (Homoptera; Cicadidea).

Fig. 8.16 Chinese brown speckled cicada (*Platy-pleura hilpa*); body length 20 mm (0.8 in.); South China (Homoptera; Cicadidae).

Fig. 8.17 Grass/sugarcane cicada (*Mogannia* sp.); body length 15 mm (0.6 in.); South China (Homoptera; Cicadidae).

CICADELLOIDEA (= Jassoidea) (leafhoppers)

The taxonomic constitution of this superfamily is somewhat in dispute at present.

MEMBRACIDAE (treehoppers)   *2500 spp.*

These small, often brown bugs are characterized by having spines and other anatomical projections on the thorax; in some cases the insect has a totally bizarre appearance. *Tricentrus* is a common genus with an anterior pair of "horns" projecting laterally from the pronotum and a long posterior spine over the abdomen. Most species prefer trees and bushes as host plants, and the gregarious infestations are invariably attended by ants. Tree legumes are often favored hosts, but a number of herbaceous plants (including legumes) are recorded as hosts. In some species there is maternal solicitude in that adults guard their offspring. Eggs are laid in groups in two parallel slits cut into twigs of trees or shrubs and the nymphs usually remain together on the twigs, often attended by the adults as well as the usual ants. In some species the eggs are laid in twigs of trees or bushes but the nymphs feed and develop on legumes and Gramineae of the ground layer underneath the trees.

Gargara—a widespread Palaearctic genus with several pest species in southern Japan, and Europe, and there are several pest species recorded on cocoa in West Africa.

Tricentrus—occurs as many different species, and some are minor pests of a range of cultivated plants. *T. bicolor* is common in India and S.E. Asia and quite polyphagous—recorded from various forage legumes (alfalfa, etc.), other legumes and cereal crops (sorghum, wheat, sugarcane, etc.); another Indian species is *T. congestus*.

Leptocentrus and *Oxyrachis* are both genera with many species, and some are minor pests of several crops and ornamental trees in West and East Africa, the Near East, India, and S.E. Asia.

The list of minor pest species recorded throughout the tropics is quite long—on crops such as cocoa, coffee, legumes, cereals, and sugarcane—but they are placed in a large number of small and little-known genera which makes generalization difficult.

Cicadellidae (= Jassidae) (leafhoppers)   *8500 spp.*

A very large and very important group; usually a dominant group ecologically in almost all parts of the world, and with many serious crop pest species. In size, they range from tiny (2–3 mm; 0.08–0.12 in.) *Cicadulina* to a few tropical species up to 14 mm (0.56

in.) in body length (*Bothrogonia*), but most are 5–8 mm (0.2–0.3 in.) in body length. One morphological character of note is the double line of largish spines along the outer edge of the hind tibiae. Some species are confined entirely to the Gramineae as hosts, others to woody dicotyledons such as fruit trees; most species are fairly host-specific (oligophagous) but a few important pest species appear to be polyphagous. Most Cicadellidae are phloem feeders but the Typhlocybinae (*Empoasca, Erythroneura, Typhlocyba, Zygina*) feed in the mesophyll tissues of the leaves. The female ovipositor is used to pierce the plant tissues and eggs are inserted underneath the epidermis. Nymphs and adults are typically found together on the underneath of leaves, and when disturbed they run rapidly sideways over the leaf edge on to the other side. Adults fly readily and come to lights at night; in many parts of the tropics, household light globes are filled with the corpses of Cicadellidae (and Delphacidae). Sometimes in the mornings the leafhoppers may be seen sitting on vegetation under the lights (Fig. 8.18).

The grass-infesting species usually have periodical population irruptions when they disperse in large numbers (often with Delphacidae); and several species are important migratory pests in parts of S.E. Asia and North America when they spread annually from the tropical and subtropical regions up into the warm temperate areas. As with the Delphacidae, many of the Cicadellidae are very important vectors of virus diseases, although in the large genus *Empoasca* apparently only *E. papayae* is recorded as a virus vector. Some species, when present in large numbers, cause a browning of the leaves called "hopperburn" and it is thought that the saliva is somewhat toxic. In small numbers most species produce few symptoms, except possibly slight leaf-curling at the edges. An interesting example of plant resistance to jassid attack is shown by cotton in relation to *Empoasca* spp.—it was discovered that certain "hairy" strains of cotton (from Cambodia) were more or less unacceptable to *Empoasca* bugs. The "hairy" quality was easily incorporated into the main commercial varieties of cotton being grown at the time, and ever since then the importance of *Empoasca* on cotton crops has drastically declined.

In temperate regions it is usual for some Cicadellidae to hibernate over winter as adults (predominantly females) and they may be found under the leaves and in the foliage of both shrubs and evergreens; on warm days during the winter they may be active. Species of *Empoasca* and *Zygina* hibernate as adults in Europe.

In Richards & Davies (1977) this group is regarded as a single large family, but it contains a number of distinct groups (subfamilies) which some authorities regard as separate families, but there is no universal agreement as to the precise limits of some

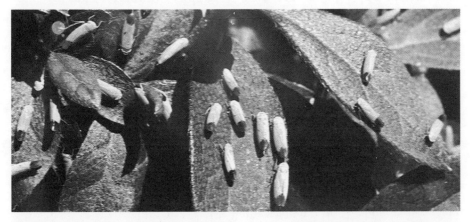

Fig. 8.18   Green rice leafhoppers (*Nephotettix* spp.) resting on plant foliage under verandah lights; body length 4–5 mm (0.16–0.2 in.); Hong Kong (Homoptera; Cicadellidae).

groups. And the literature is bedeviled by endless name changes and reshufflings as well-known species are moved from genus to genus. It now seems that some apparent species are probably a complex of sibling species or biological races. So the following list of genera and species should be viewed circumspectly. Some of the more important and widespread pest species include:

Bothrogonia spp. (large brown leafhoppers)—conspicuous large brown leafhoppers, 14 mm (0.56 in.) in length on several different hosts in S.E. Asia (Fig. 8.19).

Cicadella spp. (green/white leafhoppers)—on rice, potato, etc.; southern Europe, southern U.S.A., and Japan. (Fig. 8.20).

Cicadulina spectra (white rice leafhopper)—of India, S.E. Asia, China, and Japan.

Cicadulina spp. (maize leafhoppers)—on maize, other cereals, sugarcane; vectors of streak disease; tropical Africa (Fig. 8.21).

Circulifer spp. (beet leafhoppers)—on beet, spinach; Europe, Middle East, South Africa and western U.S.A.

Coelidia sp. (citrus leafhopper)—distinctive yellow and black coloration; on *Citrus*; S.E. Asia and South China.

Dalbulus maidis (corn leafhopper)—on maize, etc.; U.S.A., Central and South America.

Edwardsiana spp. (rose/fruit tree leafhoppers) (formerly grouped with *Typhlocyba*)—on rose, many temperate fruit trees, bushes, ornamentals (mostly woody plants); Europe, Asia, Australasia, U.S.A., South America (C.I.E. Map No. A. 432).

Empoasca spp. (green leafhoppers, etc.)—the genus is cosmopolitan with several important pest species and many minor ones; polyphagous but each species is fairly host specific, except for the polyphagous *E. lybica* (cotton jassid) (C.I.E. Maps Nos. A. 28, 223, 250, 326) (Fig. 8.22). Only *E. papaya* is recorded to transmit viruses.

Erythroneura spp. (grape/fruit tree leafhoppers, etc.)—on grapevine, cassava, sunflower, rice, and fruit trees; Africa, S.E. Asia, China, Japan, Canada, and U.S.A.

Euscelis spp. (apple leafhopper, etc.)—on apple, strawberry; Europe.

Graminella spp. (cereal/grass leafhoppers)—on cereals and grasses; U.S.A.

Idioscopus spp. (mangohoppers)—confined to *Mangifera* on which they destroy the flowers; India and S.E. Asia.

Macrosteles fascifrons (six-spotted leafhopper)—on vegetables, clovers, cereals, etc.; Canada and U.S.A.; regular migrant from U.S.A. into Canada; virus vector.

Nephotettix spp. (many) (green rice leafhoppers)—on rice, cereals, and grasses; India, S.E. Asia, China, Japan (migratory) (C.I.E. Maps Nos. A. 286, 287) (Fig. 8.23).

Orosius argentatus—on legumes and Solanaceae; S.E. Asia and Australasia.

Poecilocarda mitrata—on bananas, millets; East Africa.

Recilia dorsalis (zigzag-winged rice leafhopper)—on rice, wheat, barley, sugarcane; throughout South and S.E. Asia (Fig. 8.24).

Thaia oryzivora—on rice, sorghum, wheat, sugarcane, and grasses; India, S.E. Asia, Philippines, China.

Typhlocyba spp. (fruit tree leafhoppers, etc.)—a cosmopolitan genus, formerly with many species, now only a few; most polyphagous on fruit trees and bushes in temperate regions.

Zygina pallidifrons (glasshouse leafhopper)—on rose, cucurbits, tomato, etc.; in glasshouses throughout Europe.

Zygina spp. (blue leafhopper, etc.)—on fruit trees, etc.; much of Europe and Asia.

Fig. 8.19 Large brown leafhopper (*Bothrogonia* sp.) (Homoptera; Cicadellidae), one of the largest leafhoppers known, body length 14 mm (0.56 in.) South China.

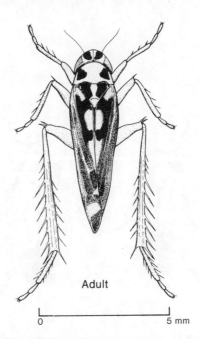

Adult

Fig. 8.20 Green/white leafhopper of potato (*Cicadella aurata*) (Homoptera; Cicadellidae).

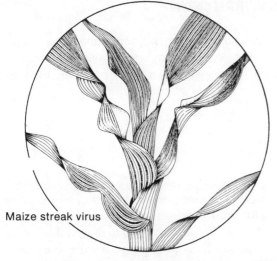

Maize streak virus

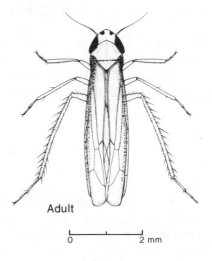

Adult

Fig. 8.21 Maize leafhopper (*Cicadulina mbila*) from tropical Africa (Kenya) (Homoptera; Cicadellidae).

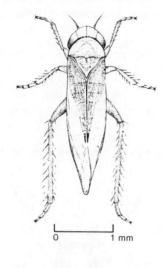

Fig. 8.22   Green leafhopper (*Empoasca* sp.) (Homoptera; Cicadellidae); Skegness, U.K.

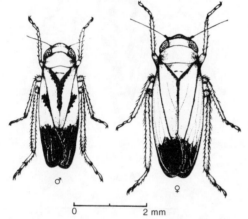

Fig. 8.23   Green rice leafhopper (*Nephotettix apicalis*); South China (Homoptera; Cicadellidae).

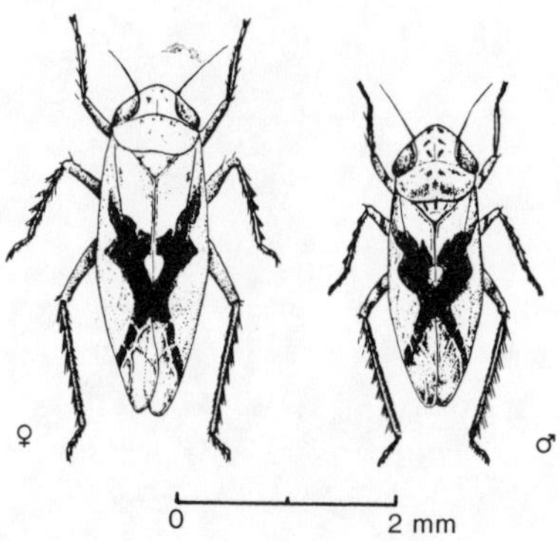

Fig. 8.24   Zigzag-winged rice leafhopper (*Recilia dorsalis*); South China (Homoptera; Cicadellidae).

SERIES STERNORRHYNCHA

Antennae usually well developed, no terminal arista, sometimes atrophied; rostrum appearing to arise from between the fore-coxae; many species with females and young incapable of locomotion. These are the more specialized Homoptera and the four large groups contain many crop pests.

PSYLLOIDEA

Psyllidae (jumping plant lice)   *1300 spp.*

These insects resemble tiny cicadas, but are the same order of size as aphids. They have two pairs of well-developed wings, with the forewings more sclerotized. Adults jump and fly if disturbed but are weak fliers. Some authorities prefer to regard this group as comprising six separate families, but for general teaching purposes the simplified approach used by Richards & Davies (1977) in regarding the group as a single family is preferable.

Eggs are laid stuck on to the plant foliage, or between folded leaves in buds—the conspicuous ovipositor of the female is apparently not used for piercing host plant tissues. The nymphs (of which there are five instars) are typically flattened and depressed with conspicuous lateral wing buds, and are often termed "suckers" by some agriculturalists. In the Triozinae many nymphs live in leaf depressions seen as pits ventraly and round lumps dorsally (Fig. 8.25); the nymphs are morphologically reduced and immobile. Other nymphs live gregariously, on shoots or stems, and their bodies are covered with fine whitish wax; and a few species produce long waxy filaments and their infestations are very conspicuous, e.g. *Macrohomotoma* and *Mesohomotoma* (Fig. 8.27). Some nymphs make galls on the host plant; *Livia* spp. infest sedges and rushes and make swollen shoot galls, but are not of agricultural importance. Some Triozinae make bright red leaf-roll galls on the edges of leaves (Fig. 8.34). In a few cases spectacular galls are formed on the leaves of the host, as shown by *Pauropsylla udei* on *Ficus chlorocarpa* (Fig. 8.30); these globular leaf-galls split open on maturity to allow the adult insect to emerge; occasionally the entire upper leaf surface is covered with galls (Fig. 8.31).

Some of the Australian psyllids (Spondyliaspinae) are renowned for their production of an elaborate test (scale) or "lerp" under which the sessile nymph lives; they are pests of eucalyptus trees.

Damage to crop plants is usually not too serious and ranges from leaf-curling and wilting due to loss of sap, to leaf galls of various types, and leaf-pitting which is not serious on older leaves but very young leaves can be killed. With a few species, however, the infested shoot may be killed, and on some crop plants (cocoa, etc.) the inflorescence may be destroyed. In Malawi, psyllid damage to cotton plants is called "psyllose"—the foliage turns a dark reddish color, leaves, young flowers, and young bolls may be shed, and the plant growth is restricted. Sooty molds are often found growing on the honeydew excreted by the bugs, which can be quite copious.

The group shows a very interesting diversity of habits in its relations with the host plants and entomologists will encounter psyllids with some regularity though most are not important pests, even though they may be quite conspicuous. They are most abundant in warmer regions.

Host specificity is usually marked; in many cases the hosts are members of one genus of plant, but in some cases the host appears to be a single species. An interesting diversity of types of host infestation and host reaction is seen both in the Malvaceae and the Moraceae (especially *Ficus*). The two largest genera are clearly *Psylla* which is mostly found on woody host plants (trees and shrubs), and *Trioza* whose species are divided between woody hosts and herbaceous plants, and some species appear to be polyphagous.

Fig. 8.25 Leaf of *Citrus* showing pits made by the nymphs of *Trioza eritreae* (Homoptera; Psyllidae); Alemaya, Ethiopia.

Some of the more notable psyllids to be encountered in both temperate and tropical regions include:

Ctenarytaina eucalypti (eucalyptus psyllid)—infests and may kill shoots of *E. globulus* (only) in Ethiopia (Fig. 8.26).

Diaphorina citri (citrus psylla)—on *Citrus* and other Rutaceae; India, S.E. Asia, South America (C.I.E. Map No. A. 335).

Euphyllura olivina (olive psyllid)—only on olive; Mediterranean Region, Ethiopia.

Heteropsylla cubana (leucaena psyllid)—from Central America, it is now a serious pest on *Leucaena* trees in India and Sri Lanka, and more recently in Africa.

Homotoma ficus (and spp.) (fig psyllid)—on *Ficus* and other Moraceae; southern Europe, Mediterranean Region, and Africa.

Livia spp.—on *Carex* and *Juncus* in damp pastures, making shoot galls; Palaearctic.

Macrohomotoma striata (fig-shoot psyllid)—in shoots of *Ficus microcarpa*; S.E. Asia and South China (Fig 8.27).

Megatrioza vitiensis (syzygium psyllid)—nymphs pit leaves of *Syzygium* spp.; S.E. Asia to South China (Fig. 8.28, 8.29).

Mesohomotoma hibisci (hibiscus psyllid)—on *Hibiscus* and some other Malvaceae; East Africa through S.E. Asia to South China.

Mesohomotoma tessmanni (cocoa psyllid)—on cocoa, other Sterculiaceae, and *Hibiscus*; West and Central Africa.

Paratrioza cockerelli (potato psyllid)—on potato and tomato; U.S.A.

Paurocephala gossypii (cotton psyllid)—only on cotton; Malawi, S.E. Africa, Congo and Sudan.

Pauropsylla depressa (fig leaf psyllid)—on *Ficus carica* and other *Ficus*; India.

Pauropsylla udei (fig leaf-gall psyllid)—leaf galls on *Ficus variegata* var. *chlorocarpa*; S.E. Asia to South China (Fig. 8.30, 8.31).

Psylla fatsiae (Shefflera shoot psyllid)—on *Shefflera octophylla* (a popular ornamental shrub); S.E. Asia and South China.

Psylla mali (apple sucker)—only on apple; Europe, Asia, North America (C.I.E. Map No. A. 154); other temperate species on pear and *Prunus* (Fig. 8.32).

Psylla pruni—on *Prunus*; throughout Europe, U.K., and parts of West Asia.

Psylla pyri (pear psyllid)—restricted to pear as host; Europe (not U.K.) and parts of the U.S.S.R. (C.I.E. Map No. A. 155).

Psylla pyricola (pear sucker)—only found on pear; Europe, parts of Asia, Near East, North America, and Uruguay (C.I.E. Map No. A. 156) (Fig. 8.33).

Psylla spp.—many species mostly on woody hosts, mostly temperate.

Psyllopsis fraxini (ash psyllid)—red leaf-rolls on ash; Europe.

Trioza alacris (bay leaf-roll psyllid)—on bay and avocado; Mediterranean Region and Europe (Fig. 8.34).

Fig. 8.26 Eucalyptus psyllid (*Ctenarytaina eucalypti*) (Homoptera; Psyllidae) infesting shoot of *E. globulus;* Alemaya, Ethiopia.

Fig. 8.27 Fig-shoot psyllid (*Macrohomotoma striata*) (Homoptera; Psyllidae) nymphs in shoot of *Ficus microcarpa;* inset of adult resting on leaf; South China.

Fig. 8.28 Leaf of *Syzygium* showing nymphal pits caused by *Megatrioza vitiensis* (Homoptera; Psyllidae); Hong Kong.

Fig. 8.29 Close-up of *Syzygium* leaf showing nymphs of *Megatrioza vitiensis in situ* in their pits; Hong Kong.

Fig. 8.30 Fig leaf-gall psyllid (*Pauropsylla udei*) infestation on *Ficus variegata* var. *chlorocarpa;* Hong Kong.

Fig. 8.31 Leaves of *Ficus variegata* completely covered with galls of *Pauropsylla udei*—one nymph per globular gall; Hong Kong.

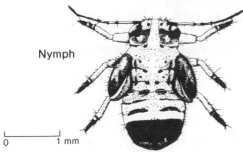

Nymph

0   1 mm

Fig. 8.32 Apple sucker nymph (*Psylla mali*) (Homoptera; Psyllidae).

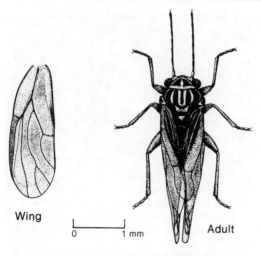

Wing

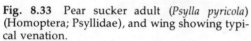

Adult

Fig. 8.33 Pear sucker adult (*Psylla pyricola*) (Homoptera; Psyllidae), and wing showing typical venation.

Fig. 8.34 Bay leaf-roll psyllid (*Trioza alacris*) (Homoptera; Psyllidae); leaf galls on bay; Alford, U.K.

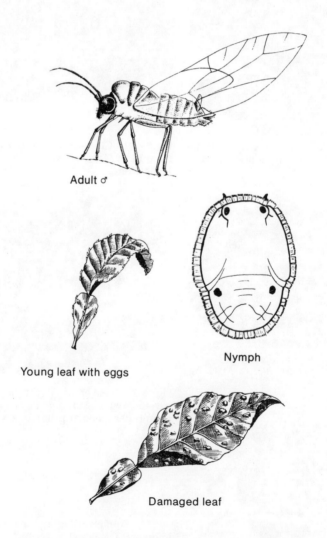

Adult ♂

Young leaf with eggs

Nymph

Damaged leaf

Fig. 8.35 Citrus psyllid (*Trioza eritreae*) (Homoptera; Psyllidae).

Trioza apicalis (carrot psyllid)—on carrot, parsnip, potato; Europe.

Trioza brassicae (brassica psyllid)—on brassicas; Russia.

Trioza camphora (camphor psyllid)—leaf pits on camphor; China and Japan.

Trioza diospyri (persimmon psylla)—on persimmon; U.S.A.

Trioza eritreae (citrus psyllid)—on *Citrus* and other Rutaceae; tropical Africa (C.I.E. Map No. A. 234) (Figs 8.25 and 8.35).

Trioza nigricornis—a polyphagous species found on many different plants; Europe, Mediterranean Region, West Asia.

Trioza tremblayi from onions and other hosts in Italy.

Trioza trigonica (carrot psyllid)—on carrot; Europe and Mediterranean.

Spondyliaspinae—a large group of more than 200 species found in Australia, and mostly on eucalyptus and a few other ornamental and forest trees.

Some of the many species recorded from *Ficus* may also attack other Moraceae, some of which are valuable timber trees—*Chlorophora* (Mvule) in Africa, etc. Many wild species of *Ficus* are important shade and ornamental trees, as well as being primary and secondary forest species throughout the tropics.

ALEYRODOIDEA

Aleyrodidae (whiteflies)   *1156+ spp.*

Tiny winged bugs, wings usually whitish but sometimes mottled dark, with scalelike nymphs, either whitish or black with a peripheral fringe of white wax. Their taxonomy relies heavily on the morphology of the scale, either nymphal or pupal. Natural infestations are often heavy with dense populations on the leaves (either upper or lower surface). The group is essentially tropical but with some well-known temperate species, and some pests are serious in greenhouses in temperate regions. Parthenogenesis is recorded in some species. The egg is typical in that it is fixed to the leaf surface by a short pedicel. The nymphs have initial mobility but after the first molt the antennae and legs atrophy and the nymph becomes sessile. Honeydew is excreted in large quantities and sooty molds and foraging ants usually accompany whitefly infestations. The final nymphal instar has developed, on its oval exoskeleton, a number of pores and waxy filaments (which are used for taxonomic purposes) and the mouthparts are permanently fixed into the plant host tissues. At the end of the developmental period the pharate adult may be seen through the translucent integument of the "pupa".

Most whitefly infestations are found on the underneath of the leaves (as shown in the illustrations) but in a few cases they are found only on the upper surface (see Fig. 8.37)—these nymphs are more scalelike and very firmly attached to the leaf surface. Most species infest dicotyledenous plants, but some infest ferns, and some feed on bamboos and other monocots—very few are found on grasses.

Damage to crop plants is done by sap removal—in a heavy infestation the thousands of bugs remove considerable quantities of sap. This can cause leaf curl and stunting of the plant, as shown by *Pealius fici* on *Ficus microcarpa*. Sooty mold development on the foliage may be serious, as it is messy and it interferes with photosynthesis. But a recent development that is now quite serious in many parts of Africa is contamination of the cotton lint by honeydew from *Bemisia tabaci*—the resulting "sticky" cotton is difficult to process, and results in economic losses. Many Aleyrodidae are vectors of viruses causing plant diseases and in this respect their damage to crops can be very serious. *Bemisia* is the vector for cotton and tobacco leaf curl, cassava mosaic, and many other diseases. The temperate *Trialurodes vaporariorium* is now very successfully controlled in European greenhouses using an aphelinid parasite (*Encarsia formosa*); a nice example of classical biological control. Natural levels of parasitism by Hymenoptera are usually high; there may be very large population fluctuations from year to year in any one location.

Some of the more important and widespread Aleyrodidae recorded as pests of cultivated plants in the warmer parts of the world include:

Acaudaleyrodes citri—on citrus, guava, pomegranate, legumes, etc.; Mediterranean, Africa, Pakistan, India.

Aleurocanthus spiniferus—polyphagous on citrus, etc.; East Africa, India, S.E. Asia, China, and Japan.

Aleurocanthus woglumi (citrus blackfly)—totaly polyphagous on citrus, coffee, mango, etc.; pantropical (C.I.E. Map No. A. 91) (Fig. 8.36).

Aleurodicus destructor (coconut whitefly)—on coconut and pineapple; S.E. Asia and Australasia.

Aleurodicus dispersus (American coconut whitefly)—on coconut, mango, banana, coffee, citrus, etc.; U.S.A., Central and South America.

Aleurolobus barodensis (sugarcane whitefly)—on sugarcane and some other Gramineae; India and S.E. Asia.

Aleurolobus marlatti (bauhinia blackfly)—on *Bauhinia, Citrus, Ficus, Morus*; S.E. Asia, China, and Japan (Fig. 8.37).

Aleurolobus olivinus (olive whitefly)—only recorded from olive; Mediterranean Region, Ethiopia.

Aleurothrixus floccosus—polyphagous on cashew, mango, coffee, citrus, eggplant, etc.; pantropical.

Aleurotuberculatus spp.—on guava, etc.; Africa, India, S.E. Asia, China.

Aleyrodes lonicera (strawberry whitefly)—polyphagous, mostly on Rosaceae and Labiatae; Europe and Asia (Fig. 8.38).

Aleyrodes proletella (brassica whitefly)—polyphagous, but mostly on temperate Cruciferae and Compositae; Europe, U.S.S.R., parts of Africa, New Zealand, and Brazil.

Aleyrodes spp.—on many different host plants (especially citrus, coconut, bamboo, *Ipomoea, Ficus*, Betel-pepper, etc.); the genus is quite cosmopolitan.

Bemisia tabaci (tobacco/cotton whitefly)—totally polyphagous on many crops; pantropical (C.I.E. Map No. A. 284) (Fig. 8.39).

Dialeurodes citri (citrus whitefly)—polyphagous on citrus and many other crops; pantropical (C.I.E. Map No. A. 111) (Fig. 8.40).

Dialeurodes citrifolii—recorded on *Citrus, Ficus, Gardenia*; India, Japan, U.S.A., Central and South America.

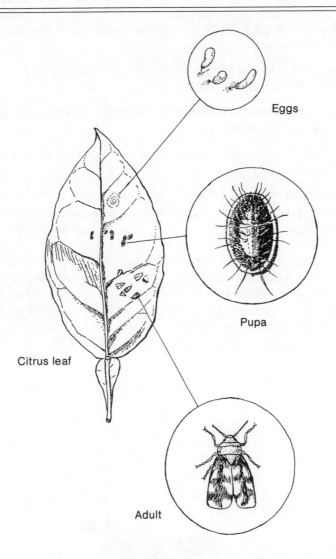

Eggs

Pupa

Citrus leaf

Adult

Fig. 8.36 Citrus blackfly (*Aleurocanthus woglumi*); all stages (Homoptera; Aleyrodidae).

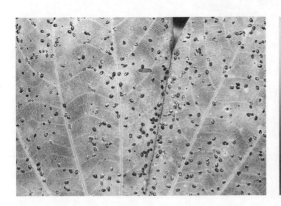

Fig. 8.37 Close-up of bauhinia blackfly (*Aleurolobus marlatti*) nymphs on *Bauhinia* leaf; Hong Kong (Homoptera; Aleyrodidae).

Fig. 8.38 *Oxalis corymbosa* leaf heavily infested with strawberry whitefly (*Aleyrodes lonicera*); Hong Kong (Homoptera; Aleyrodidae).

Pealius fici and spp. (fig blackflies)—on many species of *Ficus, Morus, Artocarpus* and other Moraceae; cosmopolitan.

Siphoninus phillyreae—on olive, pomegranate, peach, apple, pear, etc.; Palaearctic, North Africa, India, Pakistan.

Tetraleurodes mori (American mulberry whitefly)—on mulberry, citrus, etc.; U.S.A., Central and South America.

Trialeurodes floridensis—recorded from many different host plants; U.S.A., Central and South America.

Trialeurodes vaporariorium (European glasshouse whitefly)—totally polyphagous; cosmopolitan; native to highlands of Ethiopia, in greenhouses in Europe and North America (Fig. 8.41).

Some crops are attacked by large numbers of Aleyrodidae, different, both locally and on a worldwide basis. Mound & Halsey (1978) recorded 81 species of Aleyrodidae on *Ficus* spp., 65 species on *Citrus,* 44 on *Psidium,* 36 on *Quercus,* 34 on *Cocos,* 20 on *Rosa,* 18 on *Bambusa* spp. worldwide. The family appears to be equally well represented on woody and on herbaceous hosts; clearly more abundant in the tropics, but quite plentiful in temperate regions.

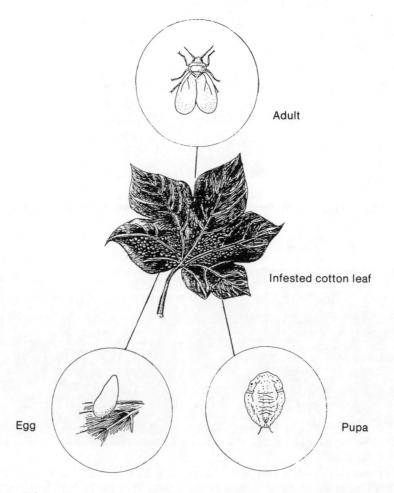

Fig. 8.39 Tobacco/cotton whitefly (*Bemisia tabaci*); all stages (Homoptera; Aleyrodidae).

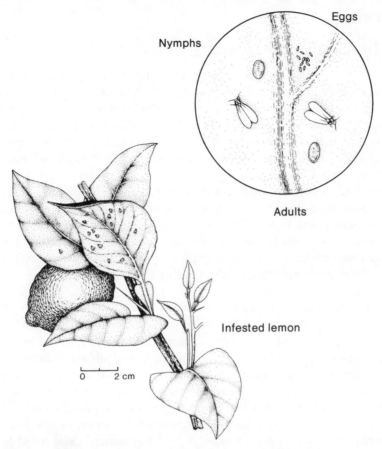

Nymphs Eggs

Adults

Infested lemon

0   2 cm

Fig. 8.40 Citrus whitefly (*Dialeurodes citri*) (Homoptera; Aleyrodidae).

Fig. 8.41 European glasshouse whitefly (*Trialeurodes vaporariorium*); infestation on underside of tobacco leaf; Eastern Highlands, Ethiopia (Homoptera; Aleyrodidae).

APHIDOIDEA

Aphididae (aphids; greenfly, etc.)   *3500 spp.*

One of the most abundant groups of Homoptera and in the northern temperate zone one of the dominant insect groups, despite their small individual size. What aphids lack in size they make up for in sheer numbers! In the tropics this particular niche is largely occupied by the Aleyrodidae and some Coccoidea. In the really hot tropics aphids are uncommon or even rare; but, in the less extreme tropics, in the highlands, and in the cool season, and in the subtropics, a number of species are quite abundant and are serious crop pests.

Aphids are small, soft-bodied insects with two pairs of wings (hind pair smaller), but apterous adults are abundant. The body bears a pair of terminal siphunculi which are thought to secrete wax and various pheromones. Honeydew is excreted from the anus. The size and shape of the siphunculi are useful taxonomic characters. Some aphids produce large amounts of wax on the nymphal body and are often called "mealy aphids", but most of the waxy aphids are in the family Pemphigidae and are collectively called "woolly aphids".

The group is equally well represented on woody plants and on herbaceous plants, and a number of species are confined to cereals, grasses, and Gramineae. Some aphids are quite host-specific and feed on only one genus of host plant, but, at the other extreme are polyphagous species recorded from dozens of different families. Actually, the majority of the non-economic species of aphids are monophagous. And some of the "polyphagous" cosmopolitan species appear to consist of geographical races that do show some degree of host specificity. The oligophagous species tend to be specific to one family of host plants. Some of the more important crops may be attacked by a dozen different aphid species (maize, for example, is attacked by 25 species) but the average figure recorded for most widespread crops is some 5–10 species of aphids per plant species. Some of the monophagous aphids are confined to one genus of host plant but some are restricted to a single plant species (like some Psyllidae), which is not common with phytophagous insects. What is confusing to the agricultural entomologist is that many of the important pest species of aphids are polyphagous, whereas the norm for Aphididae is host specificity and the great majority of 3500 species are monophagous. Generally, each genus of aphid tends to be associated with a family of plants and each aphid species with a genus or a species of plant. The host-alternating temperate aphids are found on primary and secondary host plants that are botanically quite separate.

Aphid life cycles usually consist of parthenogenetic generations that exploit rapidly growing herbaceous plants in the summer, and a sexual generation that results in diapausing eggs that overwinter on such "permanent" sites as twigs on a woody (primary) host. But, in some cases, the sexual phase had been either reduced or lost. Host-alternation (*heteroecy*) occurs as a regular seasonal migration between two host plants (usually unrelated)—the primary host is used for sexual reproduction and is often a woody shrub or tree. The secondary host is usually herbaceous and used only by the parthenogenetic morphs. *Monoecious* aphids stay on the same host plant year-round.

Cyclical parthenogenesis in the Aphidoidea enables the two basic reproductive functions to be efficiently separated. The repeated parthenogenesis builds up a large population very quickly, especially when viviparity is also involved, and even more so when most offspring are apterous, and then the sexual reproduction with its usual gene recombination will produce new genotypes to maintain basic variation. The short-lived herbaceous plants (such as most agricultural field crops) are thus exploited to a maximum by the parthenogenetic forms. This tremendous increase in insect biomass leads to the annual summer migration when alate female forms are produced and the population disperses to new hosts elsewhere.

There is so much variation in aphid life cycles that it is difficult to generalize, but some attempt has to be made, if only to introduce some of the specialized terminology. On the primary host in the spring the eggs hatch into *fundatrices,* usually large, apterous, viviparous, parthenogenetic females which utilize the flush spring growth of the primary host to start the population buildup, which will later lead to the spring migration of alates to the secondary hosts. The first offspring are called *fundatrigeniae.* After several spring generations there are winged parthenogenetic viviparous females produced (called *migrantes*) which develop initially on the primary host and later fly off to seek the secondary hosts. On the secondary hosts the offspring are called either *alienicole* or *virginoparae,* and there are usually several to many generations on the herbaceous host. Both winged and apterous forms are produced (in varying proportions). At the end of the summer, the final generation on the secondary hosts are the *sexuparae* or *gynoparae*—these are parthenogenetic, viviparous, mostly alate, females that migrate back to the primary hosts. This is usually termed the autumn migration. On the primary hosts they give rise to the *sexuales* which are males and females that will reproduce sexually—the females are oviparous, usually apterous, large, and robust. Males may be either winged or apterous. All these different stages show morphological differences, and some are quite distinct, leading to pronounced polymorphism; the *apterae* and *alatae* are in the major forms, but, from species to species, there is considerable variation in detail.

As already mentioned, in the tropics there is generally no alternation of hosts, and males are very rarely found. *Anholocycly* is the loss of the sexual phase, as best seen in the tropics, or in greenhouses, and typically the aphid then stays on the secondary host all the time.

This basic life cycle is apparently linked to both seasonal photoperiod and aspects of the diet, and crowding can be a decisive effect. Crowding often (but not always) leads to development of *alate virginoparae* on the secondary host, which, of course, can then disperse.

The fecundity of aphids is quite prodigious, especially with continuous viviparous parthenogenesis. The *fundatrix* may produce 50–70 offspring (which, of course, are all parthenogenetic viviparous females); the many generations of *viviparae* may produce some 20–60 young each. In many species, the *sexuparae* tend to produce only a small number (maybe 10) of *sexuales* which in turn lay only a small number of eggs (5–10) each. But, with a total of 9–12 generations per year, it is said that a single female aphid could give rise to 600,000 million offspring in one year. But, of course, natural predation levels are very high, as is the usual level of parasitism, and adverse weather conditions deplete populations further. Many of the migrating alatae also fail to find a suitable host plant and thus die. It is easy to see why, as a group, the Aphididae ranks as one of the most successful in the Insecta, and certainly one of the most damaging to human interests.

Some of the truly cosmopolitan and polyphagous "species" such as *Myzus persicae* and *Aphis fabae* occur as distinct physiological races with different host preferences. Most aphids are phloem-feeders, but some can also use the xylem system. Honeydew excretion is general, but some species tend to produce little and others copious quantities. With most aphid infestations there will be sooty molds on the excreted sugar, and ants are invariably in attendance. The saliva of a few species appears to be quite toxic and young shoots can be killed by heavy infestations, but the usual damage symptoms are wilting and leaf-curl following sap removal, with some plant stunting (Fig. 8.46). The most important damage done to agricultural crops is probably done indirectly by their acting as virus vectors (see page 90). One of the most serious virus vectors is *Myzus per-sicae*, which is recorded as able to transmit more than 100 different virus diseases in plants belonging to some 30 different families.

Aphid populations are preyed upon very heavily by many different insect predators, especially Coccinellidae, Syrphidae, lacewings, and by many parasitic Hymenoptera, and natural control is a very important aspect of pest infestations of growing crops. Because of their small size and soft bodies, aphid infestations are subject to many environmental pressures—dry weather reduces the population size as does a heavy rainstorm. Parasitized aphids have a characteristic brown globular appearance and are known as "mummies" (Fig. 2.14)—later there will be a small round emergence hole through which the parasite departed (see Fig. 2.16).

The literature concerning aphid pests, owing to the extensive synonymy, is very confusing; in some widespread and polyphagous species there has been a tendency to give a new name for each locality and for each major host plant! Generally, most of the taxonomic characters are somewhat esoteric and the average field entomologist is guided mostly by the more obvious characters, such as wing venation, body size and coloration, head shape, size and shape of siphunculi, host plant, and location/effect on host plant. The recent publication by Blackman & Eastop (1984) helps very con-siderably in identification of species by the nonspecialist.

In the single departure from the classification of Insecta used by Richards & Davies (1977) here are separated off the Pemphigidae (woolly aphids) from the Aphididae, as the group is biologically so distinct.

Some of the more important aphid species are listed below, but it has to be a small selection; it should be borne in mind that some temperate aphid species are common at high altitudes in the tropics, and conversely some tropical species may be found in temperate greenhouses.

> *Acyrthosiphon pisum* (pea aphid)—on pea and other legumes; cosmopolitan, but more temperate (C.I.E. Map No. A. 23).

> *Aphis craccivora* (groundnut aphid)—polyphagous on many crops, prefers legumes; completely cosmopolitan (C.I.E. Map No. A. 99) (Fig. 8.42).

> *Aphis fabae* (black bean aphid)—polyphagous; cosmopolitan, but scarce in the tropics (C.I.E. Map No. A. 174) (Fig. 8.43).

> *Aphis gossypii* (cotton/melon aphid)—polyphagous on many crops; cosmopolitan (C.I.E. Map No. A. 18) (Fig. 8.44).

> *Aphis nasturtii* (buckthorn–potato aphid)—on potato, some Cruciferae; Europe, Asia and North America.

> *Aphis nerii* (oleander aphid)—on oleander and some other ornamentals (distinc-tive yellow body with black siphunculi); pantropical.

> *Aphis pomi* (green apple aphid)—on apple, pear, and other Rosaceae; Europe, West Asia, North America (C.I.E. Map No. A. 87).

> *Aphis spiraecola* (spiraea/green citrus aphid)—on many fruit trees, bushes, etc.; cosmopolitan (C.I.E. Map No. A. 256) (also *A. citricola* in some publications).

> *Aulacorthum solani* (potato aphid)— on potato and many other crops (poly-phagous); almost worldwide (C.I.E. Map No. A. 86) (Fig. 8.45).

> *Brachycaudus helichrysi* (leaf-curling plum aphid)—on *Prunus* and Compositae; Palaearctic origin but now worldwide.

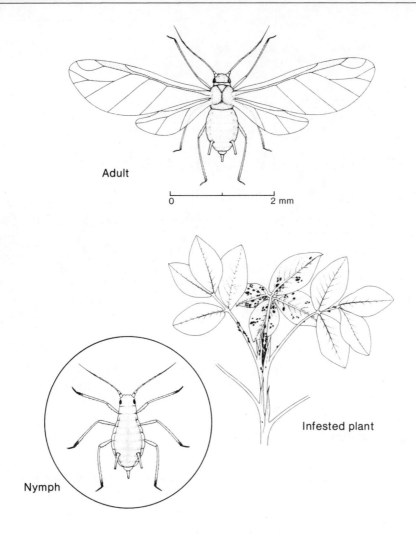

Fig. 8.42 Groundnut aphid (*Aphis craccivora*) (Homoptera; Aphididae).

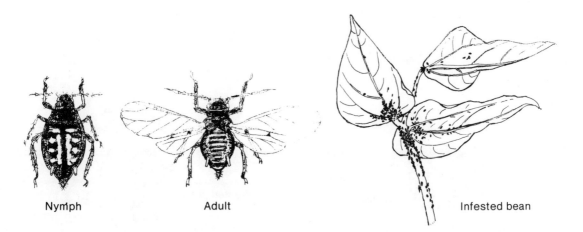

Fig. 8.43 Black bean aphid (*Aphis fabae*) (Homoptera; Aphididae).

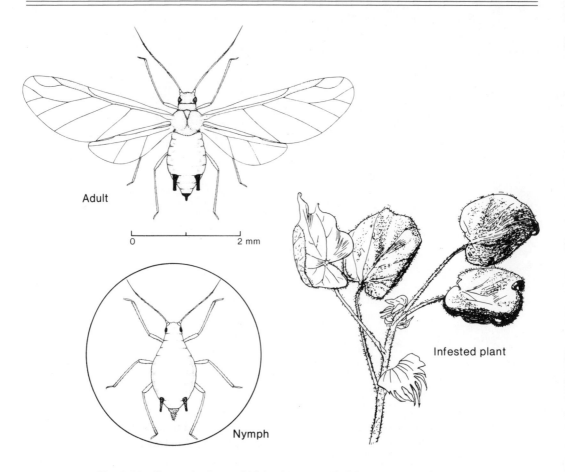

Adult

0   2 mm

Nymph

Infested plant

Fig. 8.44   Cotton/melon aphid (*Aphis gossypii*) (Homoptera; Aphididae).

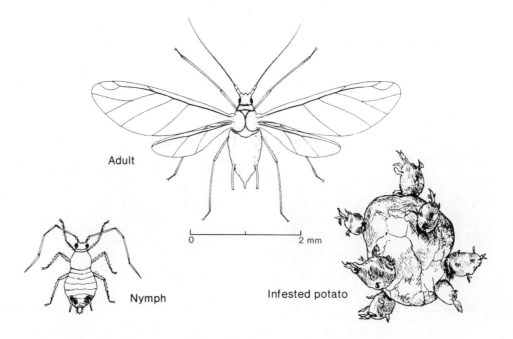

Adult

0   2 mm

Nymph

Infested potato

Fig. 8.45   Potato aphid (*Aulacorthum solani*) (Homoptera; Aphididae).

Brevicoryne brassicae (cabbage aphid)—on Cruciferae; cosmopolitan (at high altitudes in the tropics) (C.I.E. Map No. A. 37) (Figs. 8.46, 8.47).

Cavariella aegopodii (willow–carrot aphid)—on willows and carrot, etc.; widespread in temperate regions. About 30 species of *Cavariella,* most in Eurasia and North America on *Salix* and Umbelliferae; several species are pests.

Cavariella konoi (celery aphid)—on celery, etc; North America, now Holarctic.

Chromaphis juglandicola (walnut aphid)—specific to walnut; Europe, Asia, U.S.A.

Cinara spp. (black pine aphids) on *Pinus, Larix,* and other conifers; quite cosmopolitan.

Cryptomyzus ribis (red currant blister aphid)—on red currant and *Stachys;* Europe, Japan, and North America. Other species on black currant.

Diuraphis noxia (Russian wheat aphid)—on barley and wheat mostly; Palaearctic but now also in Africa and Argentina; becoming a very serious pest.

Drepanosiphum platanoides (sycamore aphid)—on sycamore only (monoecious); Europe.

Dysaphis plantaginea (rosy apple aphid)—on apple (and pear) and *Plantago;* Europe, Asia, North America. Other species also on apple, pear, etc.

Greenidea ficicola (fig aphid)—on *Ficus;* S.E. Asia and China (Fig. 8.48).

Hyalopterus pruni (mealy plum aphid)—on *Prunus;* Europe, Asia.

Hystoneura setariae (rusty plum aphid)—on *Prunus* and many Gramineae and some palms; North America, now Australia, India, Australia (C.I.E. Map No. A. 255).

Lipaphis erysimi (turnip aphid)—polyphagous, but serious on Cruciferae in S.E. Asia; cosmopolitan (C.I.E. Map No. A. 203) (Fig. 8.49).

Macrosiphum albifrons (lupine aphid)—on lupines, which may be killed; U.S.A. but now established in the U.K. (1984) (Fig. 8.50).

Macrosiphum euphorbiae (potato aphid)—on *Rosa,* potato, chili, etc., polyphagous; North America, now worldwide, except India (C.I.E. Map No. A. 44).

Macrosiphum rosae (rose aphid)—on *Rosa* and *Dipsacus;* worldwide except in East Asia.

Fig. 8.46 Cabbage aphid (*Brevicoryne brassicae*); infested cabbage showing leaf-curl symptoms; Ethiopia (Homoptera; Aphididae).

Fig. 8.47 Cabbage aphid (*Brevicoryne brassicae*) in typical colony on upper leaf surface of cabbage plant; Ethiopia (Homoptera; Aphididae).

Fig. 8.48 Fig aphid (*Greenidea ficicola*) (Homoptera; Aphididae) with ants, on *Ficus microcarpa*; Hong Kong.

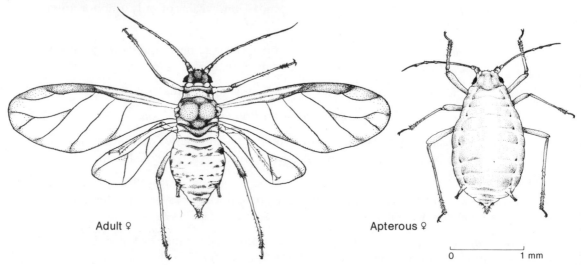

Adult ♀   Apterous ♀

0   1 mm

Fig. 8.49 Turnip aphid (*Lipaphis erysimi*); alate and apterous females; South China (Homoptera; Aphididae) (ex Lee, H.Y.).

Fig. 8.50 Lupine aphid (*Macrosiphum albifrons*) (Homoptera; Aphididae) infestation of lupine flower spike, which later died; Skegness, U.K.

Megoura viciae (vetch aphid)—on vetches, *Lathyrus,* field bean, pea, etc.; Europe, West Asia, Ethiopia.

Melanaphis sacchari (sorghum/sugarcane aphid)—on sorghum, sugarcane, and some cereals; pantropical (C.I.E. Map No. A. 420).

Metopolophium dirhodum (rose–grain aphid)—on *Rosa* and many cereals and grasses, and potato; cosmopolitan.

Metopolophium festucae (grass aphid)—only on grasses (rarely cereals); Europe.

Myzus ascalonicus (shallot aphid)—on onions and many vegetable and field crops; Europe, U.S.A., Canada (C.I.E. Map No. A. 113).

Myzus cerasi (cherry blackfly/aphid)—primary host is cherry, secondary hosts bedstraws, etc.; Europe, parts of Asia, Australasia, North America.

Myzus ornatus (violet aphid)—totally polyphagous on many families; cosmopolitan (C.I.E. Map No. A. 264).

Myzus persicae (peach–potato aphid; green peach aphid)—primary host peach, but secondary hosts in 40 families, totally polyphagous; probably of Asian origin but now worldwide (C.I.E. Map No. A. 45) (Fig. 8.51).

Nasonovia ribis-nigri (currant–lettuce aphid)—overwinters on currants, then on lettuce, chickory, and some weeds; Europe, North and now South America.

Pentalonia nigronervosa (banana aphid)—on banana and some other plants; pan-tropical and in greenhouses (C.I.E. Map No. A. 242) (Fig. 8.52).

Phorodon humuli (damson-hop aphid)—on *Prunus* and then hop; Europe, Central Asia, Ethiopia, and now North America.

Phyllaphis fagi (beech woolly aphid)—very abundant on beech foliage in the U.K. and parts of Europe; with copious honeydew.

Rhopalosiphoninus latysiphon (bulb and potato aphid)—on potatoes in storage, glasshouse bulb crops, and grasses and wheat; worldwide.

Rhopalosiphum insertum (apple–grass aphid)—apple and other Rosaceae, on to oats and grasses; Europe.

Rhopalosiphum maidis (maize/corn leaf aphid)—on maize, sorghum, barley, other cereals and grasses; cosmopolitan in warmer regions (C.I.E. Map No. A. 67) (Figs. 8.53, 8.54).

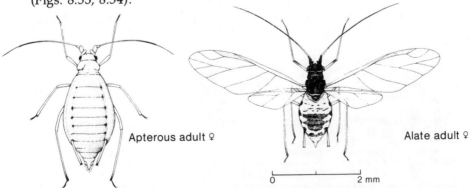

Apterous adult ♀   Alate adult ♀

0   2 mm

Fig. 8.51 Peach–potato aphid (*Myzus persicae*); alate and apterous females (Homoptera; Aphididae).

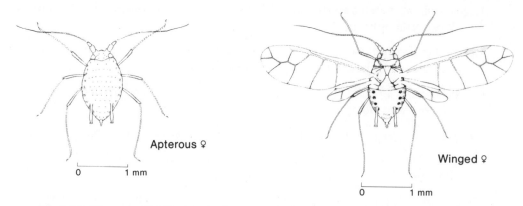

Fig. 8.52 Banana aphid (*Pentalonia nigronervosa*); alate and apterous females (Homoptera; Aphididae).

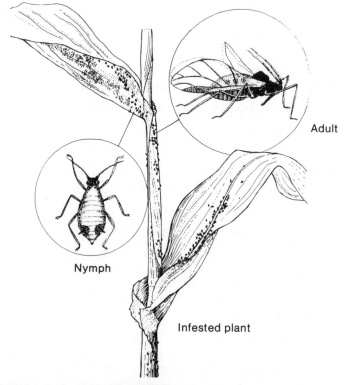

Fig. 8.53 Maize/corn leaf aphid (*Rhopalosiphum maidis*) infestation on maize leaf, attended by ants; Ethiopia (Homoptera; Aphididae).

Fig. 8.54 Maize leaf infested with *Rhopalosiphum maidis* (Homoptera; Aphididae); Dire Dawa, Ethiopia.

Rhopalosiphum padi (bird cherry–oat aphid)—on *Prunus padus,* then on cereals and grasses, rice in S.E. Asia; virtually worldwide (C.I.E. Map No. A. 288).

Rhopalosiphum rufiabdominalis (rice root aphid)—primary host is *Prunus,* and secondary are various Gramineae, potato, tomato; serious on roots of upland rice in S.E. Asia; pantropical (C.I.E. Map No. A. 289).

Schizaphis graminum (wheat aphid)—monoecious on Gramineae, anholocyclic unless winters are very cold; worldwide probably, but the validity of some records is questioned (C.I.E. Map No. A. 173) (Fig. 8.55).

Sipha flava (yellow sugarcane aphid)—monoecious, holocyclic on Graminae; North, Central, and South America (sugarcane in the West Indies).

Sitobion avenae (grain aphid)—only on cereals and grasses; worldwide but most abundant in temperate regions (C.I.E. Map No. A. 204).

Sitobion fragariae (blackberry–cereal aphid)—on *Rubus,* and then on Gramineae; Europe, Near East, South Africa, Australasia, western North America.

Therioaphis maculata (spotted alfalfa aphid)—on alfalfa and some other forage legumes; Europe, Middle East, India, and now U.S.A. and Australia (C.I.E. Map No. A. 126) (Fig. 8.56).

Therioaphis trifolii (yellow clover aphid)—on clovers and forage legumes; Europe, Middle East, India, Japan, and to the U.S.A. and Canada.

Toxoptera aurantii (black citrus aphid)—completely polyphagous, recorded from more than 120 hosts; cosmopolitan throughout the warmer parts of the world (C.I.E. Map No. A. 131) (Fig. 8.57).

Toxoptera citricidus (brown tropical citrus aphid)—on *Citrus* and other Rutaceae; tropical Africa, S.E. Asia, Australia and South America (C.I.E. Map No. A. 132).

Most of the aphid species selected above are major pests, and the majority are serious virus vectors spreading a vast range of plant virus diseases. Also the majority are honeydew producers, which in the wild are associated with ant colonies and the plant foliage often infested with sooty molds on the upper surfaces.

Fig. 8.55 Wheat aphid (*Schizaphis graminum*) on flag leaf of wheat; Skegness, U.K. (Homoptera; Aphididae).

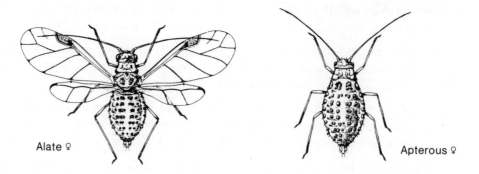

Alate ♀   Apterous ♀

Fig. 8.56   Spotted alfalfa aphid (*Therioaphis maculata*) (Homoptera; Aphididae).

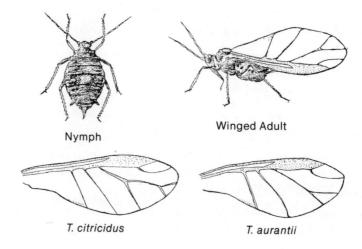

Nymph   Winged Adult

T. citricidus   *T. aurantii*

Fig. 8.57   Black citrus aphids (*Toxoptera* spp.) (Homoptera; Aphididae).

Pemphigidae (woolly aphids)

These are aphids characterized by having no long, protruding siphunculi—the openings are merely small lateral slits, the nymphs produce large quantities of waxy filaments (hence their name of "woolly aphids") and some species produce spectacular galls on woody hosts (see Fig. 8.59). Most species produce copious quantities of honeydew, some (bamboo woolly aphid) to such an extend that the ground underneath is wet with the constant dripping. Some infestations are extremely dense and the entire plant body may be covered with insects, and some young shoots may be killed.

A selection of species of note include:

Astegopteryx nipae and spp. (palm woolly aphids)—on the common species of Palmae; S.E. Asia.

Cerataphis lataniae (Asian palm aphids)—on palms; S.E. Asia.

Cerataphis palmae (palm aphid)—on coconut, oil palm, other Palmae; pantropical.

Cerataphis variabilis (coconut woolly aphid)—on coconut, oil palm, other Palmae, and mango; East Africa and India.

Ceratovacuna lanigera (sugarcane woolly aphid)—on sugarcane, wild cane, and some other large Gramineae; S.E. Asia, China, Japan (Fig. 8.58).

Chaitoregma sp. (privet woolly aphid)—makes large twig galls on Chinese privet; South China (Fig. 8.59).

Eriosoma lanigerum (apple woolly aphid)—on apple and many closely related Rosaceae; almost completely cosmopolitan, but absent from the hottest parts of the tropics; (C.I.E. Map No. A. 17) (Fig. 8.60). *E. pyricola* is a similar species in the U.S.A. on pears, etc.; other species of *Eriosoma* are temperate (Holarctic) on oaks and elm trees.

Forda spp.—make distinctive leaf galls on several species of *Pistacia* in the Near East; on roots of Gramineae; Europe, Asia, North America.

Pemphigus spp.—Palaearctic species that overwinter as eggs on trees; in the spring they make leaf-petiole galls (mostly on poplars; Fig. 8.61) and in the summer live on the roots of sugarbeet, lettuce and *Brassica* crops, and some weeds.

Pseudoregma bambusicola (bamboo woolly aphid)—on some larger species of bamboo; S.E. Asia and China (Figs 8.62, 8.63).

Fig. 8.58   Close-up of sugarcane woolly aphid (*Ceratovacuna lanigera*) on *Miscanthus* leaves; Hong Kong (Homoptera; Pemphigidae).

Fig. 8.59   Twig galls on Chinese privet made by privet woolly aphid (*Chaitoregma* sp.) (Homoptera; Pemphigidae); Botanical Gardens, Hong Kong.

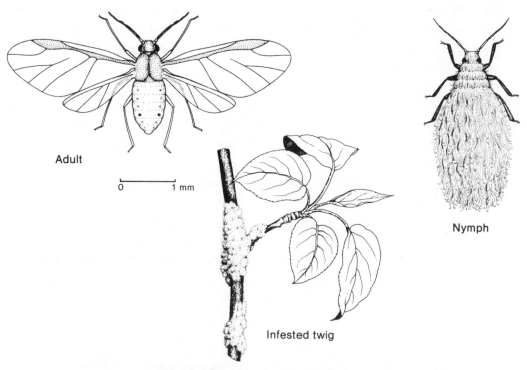

Adult

Infested twig

Nymph

Fig. 8.60 Apple woolly aphid (*Eriosoma lanigerum*) (Homoptera; Pemphigidae).

Fig. 8.61 Lettuce root aphid (*Pemphigus bursarius*) in poplar leaf petiole galls (Homoptera; Pemphigidae); Cambridge, U.K.

Fig. 8.62 Bamboo woolly aphid (*Pseudoregma bambusicola*) (Homoptera; Pemphigidae) encrusting a stem of bamboo; Hong Kong.

Fig. 8.63 Bamboo shoot covered with bamboo woolly aphids (*Pseudoregma bambusicola*)—the shoot later died; Hong Kong.

Phylloxeridae (Phylloxeras)

A small group of bugs found on deciduous trees and vines with a very complicated life cycle; most are temperate in distribution. The polymorphism associated with the alternation of generations/hosts is extreme. The leaf galls on grapevine caused by the grape phylloxera are well known. This was a very serious pest on grapevine and at times has jeopardized the entire European wine industry. On European vines the life cycle is confined to the roots, but on American vines there is alternation between roots and leaves. American vines have roots which tolerate the galls by special compensatory growth, whereas the roots of European vines die and rot killing the vine. Since the widespread practice of grafting European vines on to American phylloxera-resisting rootstocks this pest has become relatively unimportant.

The few important species include:

Aphanostigma pyri (pear phylloxera)—only recorded on pear where it infests the tree bark and destroys opening buds; Israel, Italy, and the Crimea.

Phylloxera spp. (oak leaf phylloxera)—on leaves of oaks in Europe, Asia and South Africa.

Viteus vitifoliae (grape phylloxera)—makes leaf-galls on grapevine; Mediterranean, South Africa, China, Japan, Australia, New Zealand (C.I.E. Map No. A. 339).

Adelgidae (= Chermesidae)

A small group confined to conifers in temperate regions. The two-year life cycle requires both *Picea* (spruces) as the primary host, and the secondary host can be *Pinus, Larix,* or *Abies;* but a few species live only on the secondary host. The two common genera are *Adelges* and *Pineus.* Some of the small waxy nymphs usually produce galls on the spruce, but others remain infesting the needles. The life cycle is generally very complicated.

COCCOIDEA

A large group with apterous, degenerate, larviform females and nymphs, scalelike or covered with a waxy exudate. Males (when they occur) have a single pair of wings and vestigial mouthparts. In many older textbooks and reports this superfamily is regarded as a single family—the Coccidae—which causes great confusion since the main families are really quite distinct.

Because the females of this group are largely immobile, provision for dispersal is made with the first instar nymphs; these are tiny and have well-developed legs and sense organs, and are very active. This is essentially a dispersive stage, and usually they do not feed. Dispersal is effected both directly by active crawling, and indirectly by air currents (wind), on the clothing of field workers, and the pelts of animals. Some Diaspididae have been recorded to be carried 1000 m (1100 yd.) downwind in sugarcane fields, on quite gentle winds.

There is, especially in the older literature, great confusion over names. There have been many name changes in recent years, much synonymy, and a few reshuffles as to which family certain species belong, and there have been many misidentifications made and published.

The group is of great importance as crop pests, particularly in the tropics and the warmer parts of the world, on all types of crop plants, but fewer on cereals and grasses. With these insects being sessile, and many small in size, there has, in general, been extensive transportation around the world on fruits, produce, and planting stock.

Margarodidae (fluted scales; ground "pearls", etc.)

The females usually have well-developed legs and antennae and remain mobile, and they have simple eyes. The male is winged, quite large, and has compound eyes. Nymphs and females are covered with a white waxy exudation, and the adult female carries a white ovisac attached posteriorly to the body. Underneath the white waxy coating the body is usually yellow, orange, or red in color. *Margarodes* are said to live underground and the female body is covered with a hard pearl-like waxy scale—these are called "ground pearls" and in some parts of the world are collected and strung together in a necklace.

Coccinellid predators play an important role in suppressing populations of these bugs, and some very successful biological control projects have been conducted.

In some species (especially *Icerya purchasi*) the young nymphs settle initially on the leaves where presumably it is easier to feed, but older instars return to the twigs (and sometimes, even quite thick branches) where they are more likely to survive. Heavy infestations of *Icerya* have been observed to kill shoots and small branches of both trees and bushes; they can be very serious pests on tree seedlings.

The group is essentially tropical, with some species found in warm subtropical areas. Some of the more important pest species include:

Drosicha mangiferae (mango giant "mealybug")—on mango, citrus, and many other plants; India and S.E. Asia.

Drosicha spp. (giant "mealybugs")—on mango, fig, persimmon, and other trees; India and S.E. Asia.

Icerya aegyptica (Egyptian fluted scale)—totally polyphagous on trees, bushes, some palms, etc.; widespread throughout the Old World Tropics (C.I.E. Map No. A. 221) (Fig. 8.64).

Icerya purchasi (cottony cushion scale)—totally polyphagous—most important on citrus, mango, guava, jacaranda, etc.; cosmopolitan throughout the warmer parts of the world (endemic to Australia), extending into southern Europe (C.I.E. Map No. A. 51) (Figs 8.65, 8.66).

Icerya seychellarum (Seychelles fluted scale)—polyphagous on many crops, especially citrus, mango, guava, jackfruit, and palms; recorded from scattered localities in East Africa, Indian Ocean Islands, India, S.E. Asia to Japan (C.I.E. Map No. A. 52).

Icerya spp.—several species are recorded from citrus, mango, coffee, guava, and many forest trees and ornamentals, throughout Africa, India and S.E. Asia.

Matsucoccus spp. (pine scales)—on *Pinus* spp.; Europe, Israel, Asia to Japan.

Monophlebus spp.—minor pests on cashew, citrus, pear, etc.; India.

Steatococcus assamensis—on *Croton* and other ornamentals; South China (Fig. 8.67).

Walkeriana spp.—minor pests on citrus, guava, pomegranate, etc.; India.

Fig. 8.64 Egyptian fluted scale (*Icerya aegyptica*) on leaf of *Litsea rotundifolia;* Hong Kong (Homoptera; Margarodidae).

Fig. 8.65 Cottony cushion scale (*Icerya purchasi*) on *Cassia surattensis;* Hong Kong (Homoptera; Margarodidae).

Fig. 8.66 Cottony cushion scale (*Icerya purchasi*) on 4 cm (1.6 in.) thick stem of *Jacaranda* tree; Ethiopia.

Fig. 8.67 *Steatococcus assamensis* (Homoptera; Margarodidae) on *Croton* leaf; Hong Kong.

Orthezidae

A very small group, native to the Palaearctic Region and America; females have normal legs and antennae and move on the host plant quite readily. The only species of agricultural importance is:

Orthezia insignis (jacaranda/lantana bug)—a minor pest on coffee, Solanaceae, citrus, roses, *Jacaranda* trees, and many other ornamentals; in some locations used to control *Lantana* with some success; probably pantropical, but few records to date from Asia, and none from Australasia (found in greenhouses in Europe and U.S.A.) (C.I.E. Map No. A. 73) (Fig. 8.68). Other species recorded from sugarcane in South America.

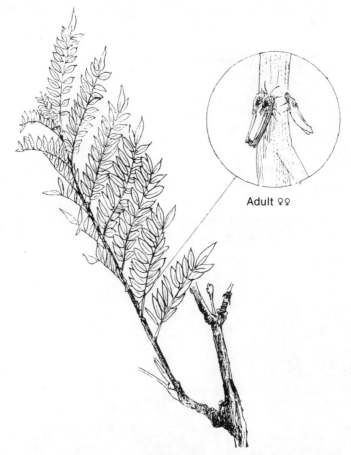

Adult ♀♀

Fig. 8.68   Jacaranda/lantana bug (*Orthezia insignis*); Kenya (Homoptera; Orthezidae).

Lacciferidae (lac insects)

The females have their body covered by a thick resinous scale—the body is highly degenerate. After the first active instar, both nymphs and females are totally immobile and they are firmly stuck on to twigs of their woody hosts. Reproduction is mostly by parthenogenesis, but males are produced in India. The group is almost entirely confined to the tropics and subtropics.

The more important species are as follows:

Kerria javanus (Java lac insect)—on durian, *Ficus,* and macaranga trees; Indonesia.

Laccifera lacca (Indian lac insect)—recorded from many different trees in India.

It is possible that *Kerria* and *Laccifera* are synonymous but both names are widely used in current literature. These are the true lac insects, and they congregate densely on the woody twigs and branches of the host trees. The lac-resin glands are situated all over the body and their secretion forms a thick communal encrustation covering the insect bodies and the twig to a depth of 6–12 mm (0.24–0.48 in.). Four other species of *Laccifera* are recorded from India. The natural lac encrustation is stripped off the trees and melted into stick-lac and from this the commercial shellac is produced, which is still in demand for French-polishing quality furniture. Cultures of *L. lacca* have been taken to China and Java where commercial production is underway—but the main production area is still India. In this context the chalcid parasites which are normally regarded as highly beneficial biological (natural) control agents are regarded as being pests!

Tachardina aurantica (forest lac insect)—found throughout S.E. Asia on a wide range of trees; the scales typically are less crowded on the twigs, the scale is thinner, and there is no communal crust formed (Fig. 8.69); they do not yield lac. In point of fact, this species is regarded as a nuisance in commercial lac production areas as it encourages a population buildup of chalcid parasites. Infestations are usually attended by ants.

Tachardina spp.—are recorded from the ornamental *Michalia figo* in Hong Kong (see Fig. 2.10), and others from coffee, guava, *Annona,* in S.E. Asia and East Africa.

Fig. 8.69 Forest lac insect (*Tachardina aurantica*) on twig of lime; parasite emergence holes evident; Penang, Malaysia (Homoptera; Lacciferidae).

Pseudococcidae (mealybugs)

The mealybugs are essentially a tropical group with a few species adapted to live in greenhouses in temperate Europe and U.S.A. Adult females are quite mobile and may move about the host plant freely. They are usually found on the aerial parts of the plant, especially in leaf-axils, round fruit stalks, etc., where there is a little physical shelter, but, sometimes, extensive colonies inhabit the roots of plants, particularly shrubs. Honeydew is usually excreted in large quantities, and infestations are invariably attended by ants and associated with sooty molds on the lower foliage. The relationship between mealybugs and ants may be very complex, with some species apparently protecting the bugs while feeding on the honeydew, and other ant species preying on the mealybugs and sometimes also on their attendant ants. These relationships are so complex and variable that it is not feasible to make generalizations.

Most mealybugs are white externally owing to a covering of wax, but the body underneath is usually reddish or orange in color. The surface wax is often extended into long filaments anteriorly and posteriorly, and several species are known as "long-tailed mealybugs".

A few species are particularly damaging as crop pests in that their saliva is toxic, and their feeding damages young shoots causing a "bunchy-top" condition, and the apical shoots may be killed. *Maconellicoccos hirsutus* does this to *Hibiscus* and many other trees and shrubs, and the cassava mealybugs (*Phenacoccus* spp.) are causing much concern in tropical Africa by killing the shoots of cassava plants.

Many species of agricultural importance have been variously placed in the genera *Planococcus* and *Pseudococcus*; some are now removed into new genera, and different authorities have different views as to the constitution of these large genera, so the literature is a mass of confusing names and references.

Some of the more important tropical mealybug pests include the following:

Antonina spp. (grass/bamboo/sugarcane mealybugs)—on stems of Gramineae, underground and above; pantropical (C.I.E. Map No. A. 216).

Brevennia rehi (rice mealybug)—on paddy rice and Gramineae; India, Pakistan, S.E. Asia (C.I.E. Map No. A. 401).

Centrococcus insolitus (eggplant mealybug)—on eggplant; India.

Dysmicoccus boninensis (gray sugarcane mealybug)—on sugarcane and grasses; Egypt, Mauritius, Taiwan, Japan, New Guinea, Australia, Pacific Islands, U.S.A., Central and South America (C.I.E. Map No. A. 116).

Dysmicoccus brevipes (pineapple mealybug)—on pineapple, sugarcane, rice, palms, coffee, etc.; pantropical (C.I.E. Map No. A. 50) (Figs 8.70, 8.71).

Ferrisia virgata (striped mealybug)—quite polyphagous on plantation, field, and vegetable crops (C.I.E. Map No. A. 219) (Fig. 8.72).

Geococcus coffeae (coffee root mealybug)—on roots of coffee, cocoa, citrus, etc.; pantropical (C.I.E. Map No. A. 285).

Geococcus oryzae (rice root mealybug)—on rice; Japan and Korea.

Geococcus spp. (root mealybugs)—on a wide range of host plants (citrus, etc.); pantropical and including Japan.

Maconellicoccus hirsutus (hibiscus mealybug)—polyphagous on *Hibiscus*, cotton, legumes, many tropical and subtropical fruit, shade and forest trees; a serious pest capable of killing shoots with saliva toxins, the most debilitating mealybug known; N.E. and East Africa, S.E. Asia, China, Papua New Guinea (C.I.E. Map No. A. 100) (Fig. 8.73).

Nipaecoccus nipae—polyphagous on sweet potato, potato, coconut, avocado, guava, etc.; North and South Africa, India, Vietnam, Hawaii, U.S.A., Central and South America (C.I.E. Map No. A. 220).

Nipaecoccus viridis (yellow mealybug)—on citrus, cotton, mango; pantropical (C.I.E. Map No. A. 446).

Phenacoccus aceris (apple mealybug)—on apple; Europe, U.S.A., and Canada.

Phenacoccus hereni (cassava mealybug)—on cassava; Central and South America.

Fig. 8.70 Pineapple mealybug (*Dysmicoccus brevipes*) (Homoptera; Pseudococcidae).

Fig. 8.71 Pineapple mealybug (*Dysmicoccus brevipes*) infesting a coconut; Seychelles, Mahé (Homoptera; Pseudococcidae).

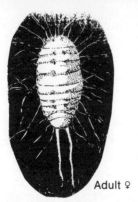

Adult ♀

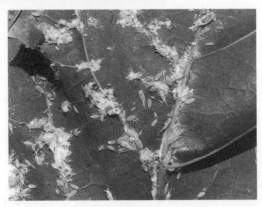

Fig. 8.72 Striped mealybug (*Ferrisia virgata*) (Homoptera; Pseudococcidae).

Fig. 8.73 Hibiscus mealybug (*Maconellicoccus hirsutus*) infestation of *Celtis* killing young shoots; Hong Kong (Homoptera; Pseudococcidae).

Phenacoccus manihoti (cassava mealybug)—a pest of recent importance causing serious damage to cassava since its accidental introduction into Africa; losses of up to 80% are recorded; South America, now West and Central Africa (C.I.E. Map No. A. 466).

Planococcoides njalensis (cocoa mealybug)—on cocoa, coffee, cola, and *Ceiba;* important as the vector of cocoa swollen shoot virus; West and Central Africa (C.I.E. Map No. A. 332).

Planococcus citri (citrus/root mealybug)—on coffee, cocoa, citrus (polyphagous); pantropical (C.I.E. Map No. A. 43) (Fig. 8.74).

Planococcus kenyae (Kenya mealybug)—polyphagous on coffee and many other coffee crops; East, Central and West Africa (C.I.E. Map No. A. 384) (Fig. 8.75).

Planococcus lilacinus (cocoa mealybug)—on cocoa, guava, citrus, coffee, and other trees; East Africa, India, S.E. Asia, China, Japan (C.I.E. Map No. A. 101).

Pseudococcus citriculus (long-tailed citrus mealybug)—on *Citrus, Ficus, Hibiscus,* orchids, and other plants; South China (Fig. 8.76).

Pseudococcus comstocki (Comstock's mealybug)—on apple, citrus, coffee, pear, litchi, *Morus,* etc.; West and East Asia, including China and Japan, U.S.A.

Pseudococcus longispinus (= *adonidum*) (long-tailed mealybug)—polyphagous on citrus, coffee, cocoa, sugarcane, coconut, and other palms, etc.; cosmopolitan (C.I.E. Map No. A. 93) (Fig. 8.77).

Pseudococcus maritimus (grape mealybug)—on grapevine, apricot, citrus, peach, tea, potato, walnut, etc.; Iran, Egypt, Cape Province, Canary Islands, Sri Lanka, New Zealand, U.S.A., Central and South America (a subtropical species) (C.I.E. Map No. A. 404).

Rastrococcus invadens (mango mealybug)—polyphagous; from S.E. Asia but now invaded West Africa.

Riparia spp. (root/ground mealybugs)—widespread in Asia, including Japan, and North America.

Saccharicoccus sacchari (pink sugarcane mealybug)—on sugarcane, rice, sorghum, and various grasses; pantropical up to Spain and to Japan (C.I.E. Map No. A. 102) (Fig. 8.78).

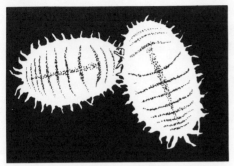

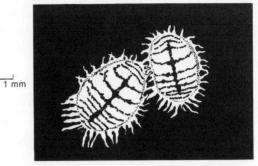

Fig. 8.74 Citrus/root mealybug (*Plano-coccus citri*) (Homoptera; Pseudococcidae).

Fig. 8.75 Kenya/coffee mealybug (*Plano-coccus kenyae*) (Homoptera; Pseudococci-dae).

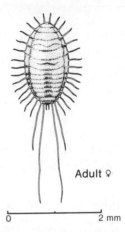

Adult ♀

Fig. 8.76 Long-tailed citrus mealybug (*Pseudococcus citriculus*) on kumquat; South China (Homoptera; Pseudococcidae).

Fig. 8.77 Long-tailed mealybug (*Pseudo-coccus longispinus*) (Homoptera; Pseudo-coccidae).

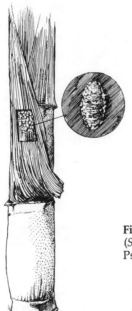

Fig. 8.78 Pink sugarcane mealybug (*Saccharicoccus sacchari*) (Homoptera; Pseudococcidae).

Eriococcidae

A small group formerly included in the Pseudococcidae, of little direct agricultural importance, most occurring on temperate forest trees.

Cryptococcus fagi (felted beech scale)—only on *Fagus;* a recent pest in the U.K.; Europe, Middle East, eastern U.S.A.

Eriococcus coccineus (cactus scale)—on cacti; Japan, Australia, U.S.A.

Eriococcus coriaceus (gum tree scale)—an Australian species confined to *Eucalyptus.*

Eriococcus spp.—on various trees, woody ornamental shrubs, cactus, bamboo, and grasses; Japan, Australia, and the U.S.A.

Kermes spp.—on oaks; Palaearctic.

Kuwanina parva (cherry bark scale)—on trunk and branches of cherry, etc.; Japan.

Micrococcus silvestri—in the Middle East these insects live on the roots of barley, oats, and other cereals, inside nests of the ant genus *Tapinoma* in a state of mutalism (fully grown females are globular and measure 10 mm (0.4 in.) in diameter.

Phoenicoccus marlatti (red date scale)—on date palms; Israel, Middle East, North Africa, U.S.A (Arizona, Texas, California). Regarded by Williams (1969) as probably belonging to the Eriococcidae rather than the Diaspididae.

Dactylopiidae (cactus mealybugs)   *9 spp.*

This tiny family was established for a single genus with nine species, restricted to Cactaceae for hosts, and native to Mexico and parts of South America. Adult females are dark red in color with a pigment used as a source of dye (cochineal).

Dactylopius coccus (cochineal insect)—this insect is the source of the red dye cochineal which used to be the principal source of red coloring for foodstuffs and cosmetics, before the recent development of synthetic dyes. Still produced commercially in Mexico, Honduras, and the Canary Islands, on *Opuntia* species.

Dactylopius opuntiae (prickly pear mealybug)—imported from Mexico into Australia and South Africa as a biological control agent to control prickly pear, with considerable success as it has saliva particularly toxic to the cactus.

Coccidae (= Lecaniidae) (soft scales)

The females are characterized by having a dorsal "scale" over the body which is actually part of the body integument. The scale may be naked (e.g. *Coccus*) or may be covered extensively with wax (e.g. *Ceroplastes, Gascardia*). Body segmentation is obscure, and the development of legs and antennae variable. An extremely important group throughout the tropics and the warmer parts of the world, with a few species in temperate regions and in glasshouses.

Younger instars are usually mobile and some quite large scales apparently have the power to move should their location prove untenable. Mostly, they infest leaves and some fruits and stems of dicotyledenous plants, and are seldom recorded on Gramineae. Female scales are usually found on the underside of leaves, either alongside the larger veins, or sometimes around the leaf periphery. A few species apparently prefer either the upper surface of the leaf or twigs as their site for settlement.

Honeydew excretion varies extensively from copious to very little, but most species are found attended by ants. With some species the ant association is apparently important; in South China *Crematogaster* ants were seen to have constructed a series of small arboreal subnests on branch forks in a tree, and inside each was a small colony of large female *Saissetia* scales (possibly *S. formicarius*) completely enclosed.

Some of the more important pest species to be found on cultivated plants in the warmer parts of the world include:

Coccus alpinus (alpine soft green scale)—on coffee, citrus, guava, and other plants, at altitudes above 1300 m (4300 ft.) in eastern Africa.

Coccus elongatus (elongate green scale)—mostly on ornamentals, but also on coffee, *Annona,* and *Acacia*; Europe, eastern Africa, China, and Japan.

Coccus hesperidum (soft brownish scale)—totally polyphagous on many crop plants (citrus, coffee, tea, papaya, banana, ferns, date palm, bay, etc.); cosmopolitan (C.I.E. Map No. A. 92) (Fig. 8.79).

Coccus mangiferae (mango soft scale)—a serious pest on mango, jackfruit, guava, avocado; less serious on citrus, persimmon, and *Ficus*; pantropical.

Coccus pseudomagnoliarum (citricola scale)—only recorded from *Citrus*; eastern Europe, China, Japan, Australia, U.S.A. and South America (C.I.E. Map No. A. 428).

Coccus viridis (soft green scale)—totally polyphagous on citrus, coffee, guava, etc.; pantropical (C.I.E. Map No. A. 305) (Fig. 8.80).

Ceroplastes floridensis (Florida wax scale)—on citrus, guava, fig, mango, tea, etc.; pantropical (C.I.E. Map No. A. 440).

Ceroplastes japonicus (sinensis) (Chinese wax scale)—on citrus, fig, grapevine, bay, pear; cosmopolitan but records scattered (C.I.E. Map No. A. 412) (Fig. 8.81).

Ceroplastes rubens (pink waxy scale)—polyphagous on citrus, coffee, tea, mango, *Ficus,* and many other trees and shrubs; East Africa, India, S.E. Asia, Pacific Islands, Australia, China, Japan, and Hawaii (C.I.E. Map No. A. 118) (Figs. 8.82, 8.83).

Ceroplastes rusci (fig wax scale)—polyphagous on *Ficus, Citrus, Camellia,* olive, pistachio, etc.; Mediterranean, Africa, West Iranian, South America.

Ceroplastes spp. (waxy scales)—several other species recorded as minor pests of citrus and many other crops and plants; throughout the tropics (Fig. 8.84).

Chloropulvinaria psidii (guava mealy scale)—on guava, coffee, citrus, tea, mango, and many other shrubs and trees; pantropical (C.I.E. Map No. A. 59).

Eulecanium tiliae (hazelnut scale)—on almond, apricot, apple, pear, plum, peach, cherry, loquat; India, Europe.

Gascardia brevicauda (white waxy scale)—on coffee and citrus; tropical Africa.

Gascardia destructor (white waxy scale)—polyphagous on citrus, coffee, guava, persimmon, other trees and shrubs; East, South and West Africa, Papua New Guinea, Australia, New Zealand, Florida and Mexico (C.I.E. Map No. A. 117) (Fig. 8.85).

Fig. 8.79 Soft brownish scale (*Coccus hesperidum*) on guava fruit; Ethiopia (Homoptera; Coccidae).

Fig. 8.80 Soft green scale (*Coccus viridus*) on hibiscus stem; Seychelles (Homoptera; Coccidae).

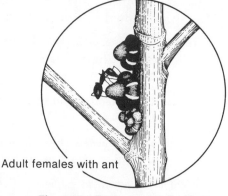

Adult females with ant

Fig. 8.81 Chinese wax scales, or *Ceroplastes japonicus (sinensis)* (Homoptera; Coccidae) on leaf of bay; Rome, Italy.

Fig. 8.82 Pink waxy scale (*Ceroplastes rubens*) on *Ficus* twig; Uganda (Homoptera; Coccidae).

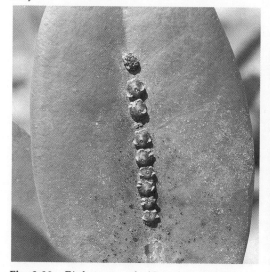

Fig. 8.83 Pink waxy scale (*Ceroplastes rubens*) on leaf of mangrove (*Avicennia marina*); South China (Homoptera; Coccidae).

Fig. 8.84 Orange waxy scale (*Ceroplastes* sp.); diameter 15 mm (0.6 in.); on twigs of *Acacia albida*; Ethiopia (Homoptera; Coccidae).

Eggs

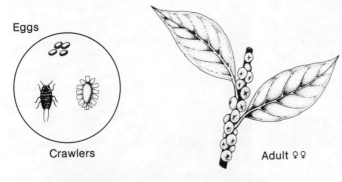

Crawlers

Adult ♀♀

Fig. 8.85 White waxy scale (*Gascardia destructor*) (Homoptera; Coccidae).

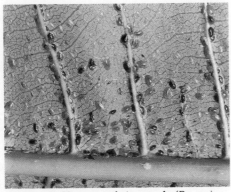

Fig. 8.86 Close-up of nigra scale (*Parasaissetia nigra*) under leaf of frangipani; Hong Kong (Homoptera; Coccidae).

Adult ♀

0   2 mm

Fig. 8.87 Plum scale (*Parthenolecanium corni*) (Homoptera; Coccidae).

Gascardia spp. (white waxy scales)—several species recorded on many different hosts, both crop plants and wild hosts.

Parasaissetia nigra (nigra scale)—polyphagous on citrus, rubber, kolanut, etc.; pantropical (Fig. 8.86).

Parthenolecanium corni (plum scale)—on many fruit trees, grapevine, walnut; Europe, North Africa, West Asia, eastern U.S.S.R. (C.I.E. Map No. A. 394) (Fig. 8.87).

Parthenolecanium persicae (peach scale)—on peach, grapevine, apple, guava, lemon, quince; Europe, North Africa, Middle East, Australia, New Zealand, U.S.A., and South America.

Pulvinaria spp.(woolly/cottony scales)—on citrus, mango, grapevine, mulberry, jujube, etc.; India, China, Japan, Europe, U.S.A., and Canada.

Saissetia coffeae (helmet scale)—polyphagous on coffee, tea, citrus, mango, fig, avocado, rubber and many other crops and plants; cosmopolitan (C.I.E. Map No. A. 318) (Figs 8.88, 8.89).

Saissetia oleae (black/olive scale)—totally polyphagous; cosmopolitan in warmer regions (C.I.E. Map No. A. 24) (Fig. 8.90).

Sphaerolecanium prunastri (spherical plum scale)—on fruit trees (Rosaceae); Old World tropics.

Vinsonia stellifera (star scale)—on citrus, mango, coconut, etc.; India, Seychelles, Indonesia, S.E. Asia, West Indies (Fig. 8.91).

Fig. 8.88 Helmet scale (*Saissetia coffeae*) totally infesting foliage of *Dipladenia boliviensis* vine; Hong Kong (Homoptera; Coccidae).

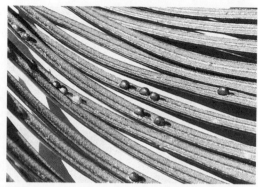

Fig. 8.89 Helmet scale (*Saissetia coffeae*) on leaf of sago palm; Hong Kong (Homoptera; Coccidae).

Fig. 8.90 Black/olive scale (*Saissetia oleae*) on twig of *Ficus microcarpa;* Hong Kong (Homoptera; Coccidae).

Fig. 8.91 Star scale (*Vinsonia stellifera*) on leaflet of coconut palm; Seychelles (Homoptera; Coccidae).

Asterolecaniidae (pit scales)

A small group with variable form, legs and antennae greatly reduced or absent. Most of the species (females) live in pits on the twigs and small branches, and some are crop pests, including the following:

Asterodiaspis minus (oak pit scale)—on the twigs of oaks and olive; Holarctic.

Asterodiaspis variolosa (oak scale)—found in Europe, Mediterranean, South Africa, Australia, U.S.A., and South America (C.I.E. Map No. A. 453).

Asterolecanium coffeae (coffee star scale)—on coffee, loquat and jacaranda trees; Central Africa (Fig. 8.92).

Asterolecanium pustulans (oleander pit scale)—polyphagous on apple, cocoa, coffee, cotton, *Ficus,* grapevine, peach, and oleander; pantropical.

Stictococcidae

This small family was erected solely for the African genus *Stictococcus* of which the female has a flattened circular body impregnated with wax; there are several pest species.

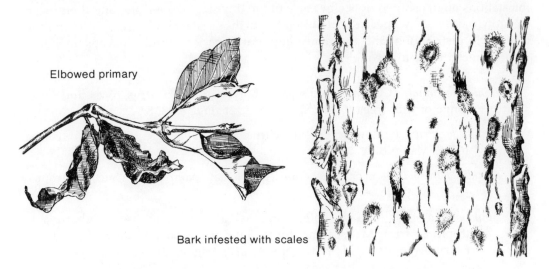

Elbowed primary

Bark infested with scales

Fig. 8.92 Coffee star scale (*Asterolecanium coffeae*) (Homoptera; Astrolecaniidae).

Stictococcus formicarius—occurs in myrmecodomatia on coffee and other trees; Ethiopia, Uganda and West Africa.

Stictococcus sjostedti—a polyphagous and abundant species.

Stictococcus spp.—eight species are recorded infesting a wide range of hosts, including cocoa, coffee, *Ficus, Annona, Hibiscus, Ceiba, Cola,* banana, *Morus,* jackfruit, pigeon pea, guava, etc., in different parts of tropical Africa. They are usually attended by *Oecophylla longinoda* ants which construct special shelters for them, but occasionally the ants are *Crematogaster* species.

Diaspididae (hard or armored scales)

A large family of great economic importance agriculturally; the adult female is characterized by having vestigial legs and antennae, and is covered by a hard scale not attached to the insect body, and usually fixed firmly to the plant surface. First instar nymphs (crawlers) are very active and disperse readily, being so small they are carried on air currents, by crawling from tree to tree, by falling off onto the ground, on the bodies of birds, animals, other insects and field workers. Experiments in East Africa showed that crawlers of sugarcane scale were carried as far as 1000 m (1100 yd.) downwind in cane fields by quite light air currents (Greathead, 1971). Most infestations are on twigs, but many are found on leaves (often on the upper surface) and also fruits; these latter species often have a shorter life-cycle correlated with the more temporary nature of their infestation site, but species on fruits tend to be dispersed more widely. Once the nymph is settled it inserts its proboscis into the plant tissues and it so remains, fixed to the plant surface for the rest of its life.

Adult females are either oviparous or viviparous, and, as procreation proceeds, the eggs or nymphs are laid under the protection of the female "scale", and her body shrinks progressively until she is dead and the entire space under the scale is occupied by the eggs or the nymphs of the next generation.

Many important pest species have been widely distributed throughout the warmer parts of the world by female scales attached to fruits and planting stock—in the past, light

infestations often escaped notice because of the tiny size of the scales and their inconspicuous nature. A few species are temperate, but the great majority are tropical and subtropical in distribution, and the many important species found on crops and ornamentals include the following:

Aonidiella aurantii (California red scale)—polyphagous on citrus, roses, and a wide
range of fruit trees (C.I.E. Map No. A. 2) (Figs 8.93, 8.94).

Aonidiella citrina (yellow scale)—on citrus, tea, *Prunus,* etc.; almost pantropical
(C.I.E. Map No. A. 349).

Aonidiella orientalis—on citrus, coconut, guava, date palm, papaya, etc.; pantropical
(C.I.E. Map No. A. 386).

Aonidomytilus albus (cassava scale)—on cassava, *Solanum,* and some other plants;
tropical Africa, Madagascar, India, Taiwan, Florida, Mexico, West Indies,
Brazil, Argentina (C.I.E. Map No. A. 217) (Fig. 8.95).

Aspidiella hartii (yam scale)—on yam, ginger, turmeric, etc.; West Africa, India,
Pacific Islands, Central America and the West Indies (C.I.E. Map No. A. 217).

Aspidiotus destructor (coconut/transparent scale)—on coconut, other Palmae,
mango, banana, avocado, cocoa, citrus, ginger, papaya, guava, rubber,
sugarcane, yam, jackfruit, etc.; pantropical (C.I.E. Map No. A. 218) (Fig. 8.96).

Aspidiotus hederae (= *nerii*) (oleander/white/ivy scale)—polyphagous on olive,
apple, mango, tung, Palmae, citrus, oleander, etc.; Mediterranean, Africa,
Australasia, U.S.A., Central and South America (C.I.E. Map No. A. 268).

Aulacaspis tegalensis (sugarcane scale)—on sugarcane; East Africa, Madagascar,
Mauritius, Seychelles, Malaysia, Java, Philippines, Taiwan (C.I.E. Map No. A.
187) (Fig. 8.97).

Chionaspis spp. (snow scales)—mostly on willows and various forest and ornamental trees, both deciduous and evergreen, and on some fruit trees in North
America; cosmopolitan.

Chrysomphalus aonidum (purple/Florida red scale)—on citrus, coconut, date palm,
mango, cinnamon, and many other plants; cosmopolitan (C.I.E. Map No. A. 4)
(Fig. 8.98).

Chrysomphalus dictyospermi (Spanish red scale)—on citrus, palms, deciduous fruit
trees, and various shrubs; cosmopolitan (C.I.E. Map No. A. 3).

Fiorinia japonica (conifer scale)—on conifers and bamboo palm; China and Japan
(Fig. 8.99).

Hemiberlesia lataniae (latania scale)—polyphagous on almond, avocado, banana,
coconut, *Acacia,* mango, rose, tung, etc.; pantropical (C.I.E. Map No. A. 360).

Ischnaspis longirostris (black line scale)—on coffee, coconut, citrus, banana, oil palm,
mango, *Annona* and other plants; pantropical (C.I.E. Map No. A. 235) (Fig.
8.100).

Lepidosaphes beckii (citrus mussel/purple scale)—on *Citrus* and other Rutaceae,
Croton, and other shrubs; cosmopolitan (C.I.E. Map No. A. 49) (Fig. 8.101).

Lepidosaphes gloverii (long scale)—mostly on *Citrus;* pantropical (C.I.E. Map No. A.
146).

Fig. 8.93 California red scale (*Aonidiella aurantii*) on orange; South China (Homoptera; Diaspididae).

Fig. 8.94 California red scale (*Aonidiella aurantii*) on oranges; Dire Dawa, Ethiopia (Homoptera; Diaspididae).

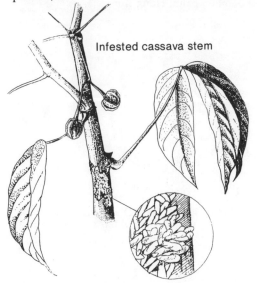

Infested cassava stem

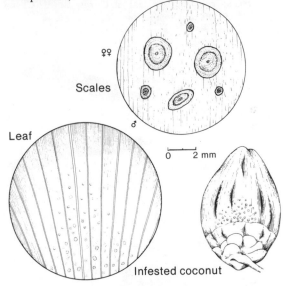

♀♀ Scales

♂

Leaf

0   2 mm

Infested coconut

Fig. 8.95 Cassava scale (*Aonidomytilus albus*) (Homoptera; Diaspididae).

Fig. 8.96 Coconut/transparent scale (*Aspidiotus destructor*) (Homoptera; Diaspididae).

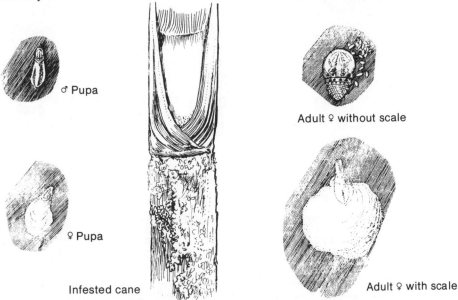

♂ Pupa

♀ Pupa

Infested cane

Adult ♀ without scale

Adult ♀ with scale

Fig. 8.97 Sugarcane scale (*Aulacaspis tegalensis*) (Homoptera; Diaspididae).

Fig. 8.98 Purple/Florida red scale (*Chrysomphalus aonidum*) on leaf of *Citrus*; South China (Homoptera; Diaspididae).

Fig. 8.99 Conifer scale (*Fiorinia japonica*) on leaf of bamboo palm; Hong Kong.

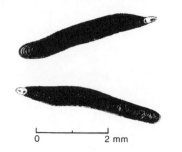

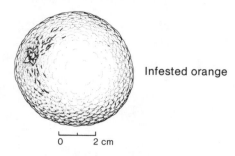

Infested orange

Fig. 8.100 Black line scale (*Ischnaspis longirostris*) (Homoptera; Diaspididae).

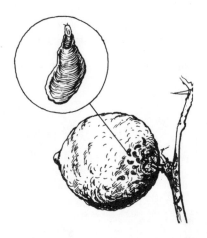

Fig. 8.101 Citrus mussel/purple scale (*Lepidosaphes beckii*) (Homoptera; Diaspididae).

Lepidosaphes ulmi (oystershell scale)—on deciduous fruit (apple, pear, plum, etc.) and other trees, bush fruits, and ornamentals; cosmopolitan in temperate regions (C.I.E. Map No. A. 85) (Fig. 8.102).

Parlatoria blanchardii (date palm scale)—on date palm and other Palmae; Africa, Near and Middle East, and Brazil (C.I.E. Map No. A. 148).

Parlatoria oleae (olive scale)—on olive, apple, pear, stonefruits, and many trees and shrubs; Mediterranean, India, North and South America (C.I.E. Map No. A. 147).

Parlatoria pergandi (chaff scale)—only recorded on Citrus; pantropical (C.I.E. Map No. A. 185).

Parlatoria zizyphus (black parlatoria)—monophagous on *Citrus;* pantropical, but most records from Old World (C.I.E. Map No. A. 186) (Fig. 8.103).

Phenacaspis cockerelli (mango/oleander scale)—on mango, coconut, oil palm, oleander, etc.; East and South Africa, Madagascar, Seychelles, Australia, South China, Japan, Hawaii (Fig. 8.104).

Pinnaspis aspidistrae (fern scale)—on palms, banana, citrus, mango, ferns, etc.; pantropical (C.I.E. Map No. A. 369).

Pinnaspis buxi (box scale)—on coconut, other Palmae, *Buxus, Pandanus;* scattered records worldwide (C.I.E. Map No. A. 233).

Pseudaonidia trilobitiformis (trilobite scale)—on avocado, cocoa, citrus, coffee, coconut, mango, passionvine, etc.; pantropical (but not Mediterranean) (C.I.E. Map No. A. 418).

Pseudaulacaspis pentagona (white peach scale)—on mulberry, peach, apricot, and other fruit trees (polyphagous); cosmopolitan (C.I.E. Map No. A. 58).

Quadraspidiotus perniciosus (San José scale)—totally polyphagous on all deciduous fruit (apple, pear, peach, plum, currants, etc.), yam, and a very wide range of trees and shrubs (more than 700 host species recorded); probably the single most damaging pest of deciduous (fruit) trees; cosmopolitan (C.I.E. Map No. A. 7) (Fig. 8.105).

Unaspis citri (citrus snow scale)—only recorded from *Citrus;* pantropical, but not the Mediterranean (C.I.E. Map No. A. 149).

Unaspis euonymi (euonymus scale)—on *Euonymus, Prunus, Hibiscus,* and other ornamentals; southern Europe, Mediterranean, China, Japan, U.S.A., and Argentina (C.I.E. Map No. A. 269).

The numbers of scale insect pests in most warmer countries are quite large; in Japan 101 species of Diaspididae are recorded as agricultural pests, including 21 species of *Lepidosaphes.* The total number of Coccoidea recorded from *Citrus* species worldwide is about 110 species, the majority of which are Diaspididae.

Fig. 8.102 Oystershell scale (*Lepidosaphes ulmi*) (Homoptera; Diaspididae) on twig of plum; Skegness, U.K.

Fig. 8.103 Black parlatoria (*Parlatoria zizyphus*) on leaf of *Ficus tinctoria;* South China (Homoptera; Diaspididae).

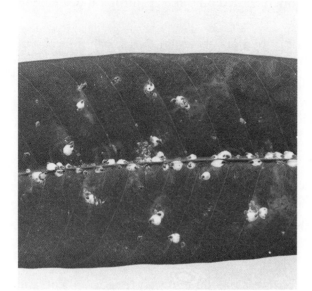

Fig. 8.104 Mango/oleander scale (*Phenacaspis cockerelli*) on leaf of mango; South China (Homoptera; Diaspididae).

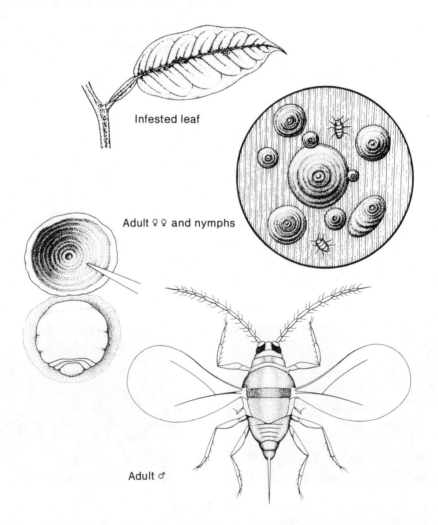

Infested leaf

Adult ♀ ♀ and nymphs

Adult ♂

Fig. 8.105 San Jose scale (*Quadraspidiotus perniciosus*) (Homoptera; Diaspididae).

SUBORDER HETEROPTERA (animal and plant bugs)

These bugs all have the head porrect (extended forward), wings folded flat over the abdomen, forewings usually sclerotized basally but membranous distally; rostrum base not touching anterior coxae; pronotum usually large. In most groups within the suborder the classification is still somewhat in dispute, and various different versions have been published. As previously stated, the somewhat simplified system used by Richards & Davies (1977) is followed. As agricultural pests these insects are not too important, although there are a number of species that are widespread pests and some that are quite serious pests. Most Heteroptera have toxic enzymes in their saliva, so the animal bugs have a very painful "bite", and the plant bugs usually produce necrotic spots at the feeding sites, so a few bugs can cause a great deal of damage to a plant.

SERIES GEOCORISAE

Mostly terrestrial, but a few littoral or semiaquatic, but body without ventral hydrofuge hairs; antennae longer than head; legs not modified for swimming.

SECTION CIMICOMORPHA

Antennae four-segmented; hemelytra usually with costal fracture and cuneus; eggs usually operculate and implanted into substrate by female ovipositor—four superfamilies.

TINGOIDEA

Body surface with lacelike reticulate sculpturing; antennae with segments 1 and 2 short, third longest; rostrum four-segmented; pronotum usually extended posteriorly to cover scutellum and clavus; two families, second one very small.

Tingidae (lace bugs) *800 spp.*

Many species Mediterranean, and many in Ethiopia; prothorax usually with extensively laminate outgrowths or else spiny. All are plant feeders, with eggs inserted into plant tissues. Nymphs generally without the sculptural ornamentation. Many different genera are recorded damaging plants of economic interest. Most species are only minor pests but occasionally serious damage is recorded. Some of the regularly encountered species include:

Compseuta spp.—on Malvaceae (especially *Hibiscus*); tropical Africa.

Corythucha gossypii (cotton lace bug)—a pest of cotton; U.S.A. and West Indies.

Corythucha spp.—on eggplant; West Indies.

Corythucha spp. (8+)—on chrysanthemum and various ornamental trees; U.S.A.

Diconocoris hewetti (pepper blossom lace bug)—on *Piper* spp.; Indonesia.

Elasmognathus greeni—on pepper; Sri Lanka, Indonesia.

Gargaphia solani (eggplant lace bug)—on eggplant; southern U.S.A.

Habrochila spp. (coffee lace bugs)—on coffee (mostly *arabica*); East, Central and West Africa (Figs. 8.106, 8.107).

Monanthis spp.—on peppermint and various ornamentals; East and West Africa.

Monosteira spp.—on jujube; India.

Stephanitis pyri (pear lace bug)—on pear and apple; Europe, Near East to Central Asia.

Stephanitis typica (banana lace bug)—on banana, coconut, cardamon, manila hemp, *Alpinia,* and other plants (C.I.E. Map No. A. 308).

Stephanitis spp.—on pear, camphor, *Azalea,* etc.; Japan and U.S.A.

Teleonemia australis (olive lace bug)—on olive; N.E. and South Africa.

Teleonemia scrupulosa (lantana lace bug)—used extensively as a biological control agent to destroy *Lantana;* pantropical.

Uhlerites spp.—on walnut and chestnut; Japan.

Urentius spp. (eggplant lace bug, etc.)—on eggplant, *Hibiscus,* jujube, etc.; Africa and India.

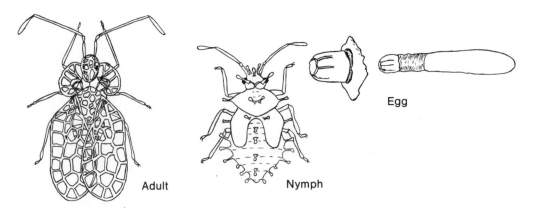

Fig. 8.106   Coffee lace bug (*Habrochila ghesquierei*) (Heteroptera; Tingidae).

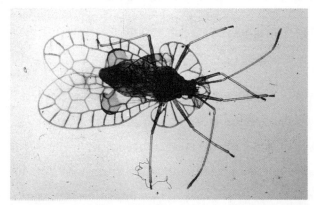

Fig. 8.107   A typical lace bug (possible *Habrochila* sp.) (Heteroptera; Tingidae); Hong Kong.

REDUVIOIDEA

A group of fiercely predacious bugs with an elongate head; antennae often geniculate (elbowed) after first segment; a short, stout, curved rostrum of three segments; ocelli usually present; hemelytra strongly veined but no cuneus.

Phymatidae (ambush bugs) *100 spp.*

A tropical group characterized by having the last (fourth) antennal segment enlarged and the forelegs raptorial with chelate tibiae; mid and hind tarsi two-segmented. Most species hide in flowers where they wait in ambush for other insects that are attracted to flowers. They mostly feed on the adults of other insect groups, but also take some larvae, especially Tenthredinidae.

Reduviidae (assassin bugs) *3000 spp.*

A very large family of predacious bugs, either bloodsucking from other insects or animals of a wide range of groups. The head is characteristically constricted behind the eyes so that a distinct "neck" is formed. The antennae may have extra segments (intercalary) up to a total of 40; tarsi are usually three-segmented. The group shows a tremendous diversity of form—far greater than any other insect family. Members of most animal families show sufficient morphological characters in common to have a recognizable form, but the Reduviidae is an exception. There is a total of 29 subfamilies according to Imms. Most species are tropical in distribution with a small number of temperate species, and the group is found throughout the world.

They are largely unspecialized in regard to their prey and feed on other insects, slugs and snails, and on vertebrates. On mollusks they often feed gregariously, and a large *Achetina* can be seen surrounded by a dozen or more reduviids each with its proboscis sunk into the flesh of the dying snail. They would not be expected to exert a controlling effect on the population of any one insect pest species, but the group must be important as part of the whole natural control complex preying on phytophagous insects and mollusks. One or two species are often associated with a particular pest, for example, *Phonoctonus* spp. and *Harpactor* spp. resemble cotton stainers (*Dysdercus* spp.) and are reputed to prey on them in Africa and India.

Many members of this group are quite large and agile and are terrestrial in habits, hunting on the ground and in low herbage, bushes, and sometimes trees. If handled, these insects "bite", and the enzymes pumped into the wound with the saliva produce a nasty reaction and the "bite" is very painful. When feeding on other insects the bug sucks out blood and juices from the body of the prey and presumably the enzymes digest some of the body contents, which can then be sucked up by the bug. Some species are diurnal and often brightly colored, and others (especially the domestic species) are nocturnal and drab in coloration.

The Triatominae are predominantly New World in distribution, but a few are Asian, and they are bloodsuckers feeding mostly on mammals or birds. In S.E. Asia *Triatoma rubrofasciata* (Fig. 8.108) is known as the domestic assassin bug; it inhabits buildings in rural villages or old dirty buildings in towns and cities. Analysis of stomach contents in Hong Kong showed that it feeds mainly on the blood of urban rats, but some 25% also contained human blood. The "bite" of this species is very painful and the site usually becomes inflamed and itching; in some hypersensitive individuals there is an anaphylactic reaction resulting in a severe illness lasting for several days. In Central and South America *Triatoma* and *Rhodnius* are the vectors for *Trypanosoma cruzi*, a blood

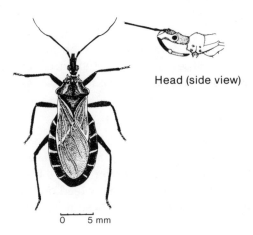

Head (side view)

0   5 mm

Fig. 8.108 Domestic assassin bug (*Triatoma rubrofasciata*) (Heteroptera; Reduviidae); Hong Kong.

parasite causing human trypanosomiasis (Chaga's Disease) which is frequently fatal. At present it is thought that 24 million people are infected. In the U.S.A. the Triatominae are referred to as "cone-noses".

Eggs are laid in groups of 10–40, each female laying several groups. Nymphal development is often slow, and nymphs are able to fast for 1–2 weeks without ill-effects. The nymphs sometimes remain together gregariously for a while and may collectively share large prey (e.g. snails). The life cycle from egg to adult is sometimes completed in two months but usually takes longer. Most adults feed regularly and live for 3–5 months.

Reduvius personatus is the European domestic assassin bug and feeds on *Cimex*, other insects, rats, and humans, and is now well established in the U.S.A. The genus *Sycanus* is characterized by having a narrow elongate head. *S. croceovittatus* from Hong Kong is dark brown with a yellow band across the base of the elytra and is similar to the black *S. annulicornis* of Indonesia, recorded feeding on many crop pest caterpillars and *Helopeltis* bugs. A few other important species commonly recorded preying on crop pests and other insects are as follows:

Acanthaspis siva—preys on honeybees in India.

Alleocranum spp.—dark-bodied predators of stored-products insects throughout the warmer parts of the world.

Arilus cristatus (wheel bug)—preys on soft-bodied insects in fruit trees in North America.

Coranus spp.—small black bugs (*c*. 9 mm/0.36 in. long) found in rice stores in Indonesia.

Platymeris laevicollis—preys on *Oryctes* adults in India.

Rhinocoris spp.—prey on caterpillars of *Heliothis*, *Spodoptera*, and others, also on aphids, cotton stainers, and epilachna beetles; in Africa, India, and Indonesia. Adult bugs are recorded consuming 1–2 half-grown *Spodoptera* larvae per day.

Scipinia spp.—narrow, spiny bugs (10 mm/0.4 in. long), found in S.E. Asia in rice fields where they feed on caterpillars, coccinellid larvae and *Leptocorisa* bugs.

Enicocephalidae   *50 spp.*

A small tropical group but widely occurring; rostrum four-segmented; head constricted basally; hemelytra entirely membranous; some species appear in swarms like midges.

CIMICOIDEA

Antennae not geniculate; second segment longest; pronotum trapezoidal; hemelytra with costal fracture and cuneus in winged forms.

Nabidae (damsel bugs)   *300 spp.*

A small group, mostly tropical but worldwide in distribution, mostly small in size with thin antennae (4–5 segments); rostrum four-segmented; cuneus absent; forelegs somewhat raptorial and tarsi three-segmented. They are mostly found in herbage where they hunt small phytophagous insects. Eggs are laid inserted into plant tissues (stems). *Nabis* and *Reduviolus* are well-known genera.

Cimicidae (bedbugs)

A very small but clearly defined group of bloodsucking ectoparasites of birds and mammals. Oval in body shape; very short hemelytra (almost completely wingless); rostrum lying in a ventral groove; ocelli absent, and tarsi three-segmented. The bedbugs which plague humans living under primitive or dirty conditions belong to the genus *Cimex*; there are two species, *Cimex lectularis* (common bed bug) (Fig. 8.109) with a rounded body shape, common throughout Europe and North America and now almost cosmopolitan; *C. hemipterus* (tropical bed bug) (Fig. 8.110) has a more elliptical body shape and is found mostly in Africa and southern Asia. The bugs are nocturnal and hide in crevices in the beds during daytime and at night they emerge to feed on the blood of the sleepers. Eggs are laid in crevices on the wooden frames of the beds. Other members of *Cimex* parasitize bats and birds. *Oeiacus hirundinis* lives in the nests of house and sand martin in Europe. *Haematosiphon inodorus* is a parasite on poultry in North and Central America, as is *Ornithocoris*.

0   2 mm

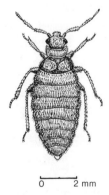

0   2 mm

Fig. 8.109 Common bed bug (*Cimex lectularis*) (Heteroptera; Cimicidae).

Fig. 8.110 Tropical bed bug (*Cimex hemipterus*) (Heteroptera; Cimicidae).

Polyctenidae (bat bugs)   *18 spp.*

A very small group of bugs, eyeless, flightless, and permanent ectoparasites deep in the fur of tropical bats. They possess one or more ctenidia (combs of short flat spines), and viviparity is practiced with young born at an advanced stage.

Anthocoridae (flower bugs)   *300 spp.*

A small group of predacious bugs, flattened and oval in shape; ocelli present; rostrum and tarsi three-segmented. Some authorities place them within the Cimicidae. They are worldwide and are found in vegetation; the genus *Anthocoris* is widespread and very common, on flowers, under bark, and in leaf litter. Eggs are embedded in plant tissues in most species. Some are myrmocophiles, some live in birds nests, some in human habitations (*Lyctocoris*). *Orius* spp. are sometimes important caterpillar predators, and also attack aphids, *Oxycarenus,* and thrips. *Xylocoris* is found in rice stores in S.E. Asia, and is presumed to prey on stored products pests.

Miridae (= Capsidae) (capsid bugs; plant bugs)   *6000 spp.*

A very large group, worldwide in distribution. Medium to small bugs, often delicate in appearance, characterized by a four-segmented rostrum; no ocelli; cuneus usually present, and tarsi three-segmented. These insects are very important agriculturally—the vast majority are phytophagous, and many are crop pests, but a few are predacious or else opportunistically omnivorous. Some of the more important predacious species include:

Blepharidopterus angulatus—preys on *Metatetranychus ulmi* in Europe, Asia and North America.

Cyrtorhinus lividipennis—on rice planthoppers in S.E. Asia.

Letaba bedfordi (sometimes placed in family Isometopidae)—on red citrus scale (*Aonidiella aurantii*) in South Africa.

Psallus spp.—prey on thrips on legumes in India and elsewhere.

As with most other Heteroptera these capsids have toxic enzymes in their saliva and when they feed they inject saliva into the plant tissues and the result is a small necrotic spot around the feeding puncture. Some species feed almost continuously— *Helopeltis* bugs make 20–150 feeding punctures during a 24-hour period, feeding a little from each site. *Sahlbergella* makes 24–36 punctures per day. The result is very extensive spotting of young leaves or fruits and a great deal of damage inflicted by a single bug (Fig. 8.112); older leaves show a characteristic tattering.

Eggs are laid singly, embedded in the plant tissues with only the apical cap showing (see Fig. 8.113); the usual site for oviposition being the stalk or main veins of young leaves. Some of the nymphs and adults (especially *Helopeltis*) are characterized by having a long, knobbed, hairlike projection upward from the thorax. There are five nymphal instars, and development usually takes from 2 to 4 weeks in the tropics. Adults usually live from 1–4 months, and both adults and nymphs are equally damaging as crop pests. The feeding bugs may kill flower buds and young fruits, which are shed. Older leaves and fruits remain on the plant but with necrotic spotting.

One or two genera have enlarged hind legs and they jump like flea beetles; the commonest genera are *Alticus* and *Halticellus*. Some of the more important crop pests in the warmer parts of the world include the following:

Adelphocoris apicalis—on cotton and legumes; East Africa.

Adelphocoris lineolatus (alfalfa plant bug)—on alfalfa; Japan and U.S.A.

Adelphocoris spp.—on various hosts in Japan and the U.S.A.

Alticus tibialis (legume jumping capsid)—Africa, S.E. Asia, Pacific Islands.

Alticus spp. (garden fleahoppers)—on various hosts; U.S.A.

Calocoris angustatus (sorghum earhead bug)—on sorghum; India.

Calocoris fulvomaculatus (hop capsid)—on hops, apple, and other fruits; Europe.

Calocoris norvegicus (potato capsid)—polyphagous; Europe, North America (Fig. 8.111).

Creontiades elongatus—polyphagous; East Africa.

Creontiades pallidus (shedder bug)—on sorghum, cotton, legumes, etc.; India and Africa.

Cyrtopeltis modestus (tomato bug)—on tomato; U.S.A.

Cyrtopeltis tenuis (tobacco/tomato capsid)—on tomato, tobacco, etc.; Africa, S.E. Asia, Japan.

Deraecoris spp.—on cotton, coffee, cashew, etc.; tropical Africa.

Distaniella theobroma (cocoa capsid)—on cocoa, etc.; Africa.

Halticellus insularis (oriental garden fleahopper)—Japan.

Helopeltis anacardii (cashew helopeltis)—on cashew and sweet potato; East Africa.

Helopeltis antonii—polyphagous; India.

Helopeltis bergrothi—polyphagous; Africa (Fig. 8.112).

Helopeltis bradyi—on tea and cinchona; Malaysia.

Helopeltis clavifer—polyphagous; S.E. Asia, Papua New Guinea.

Helopeltis schoutedeni (cotton helopeltis)—polyphagous; tropical Africa (C.I.E. Map No. A. 297) (Fig. 8.113).

Helopeltis theivora (tea helopeltis)—on tea; India.

Helopeltis theobromae (cocoa helopeltis)—on cocoa; Malaysia.

Helopeltis spp. (mosquito bugs)—on cocoa, coffee, tea, etc.; tropical Africa.

Hyalopeplus clavatus (tea capsid)—on tea; Indonesia.

Hyalopeplus spp.—polyphagous; tropical Africa.

Lamprocapsidea coffeae (coffee capsid)—on coffee; Africa.

Lygocoris pabulinus (common green capsid)—polyphagous; Europe (Fig. 8.114).

Lygocoris spp. (7–8)—on many different hosts; Europe, Asia to Japan, U.S.A.

Lygus apicalis—polyphagous; Africa.

Lygus disponsi (Japanese tarnished plant bug)—on many hosts; Japan.

Lygus elisus (pale legume bug)—on legumes; U.S.A. and Canada (C.I.E. Map No. A. 38).

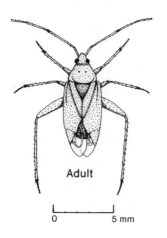

Fig. 8.111 Potato capsid (*Calocoris norvegicus*) (Heteroptera; Miridae).

Fig. 8.112 Tea leaf showing *Helopeltis bergrothi* feeding damage (Heteroptera; Miridae); Malawi.

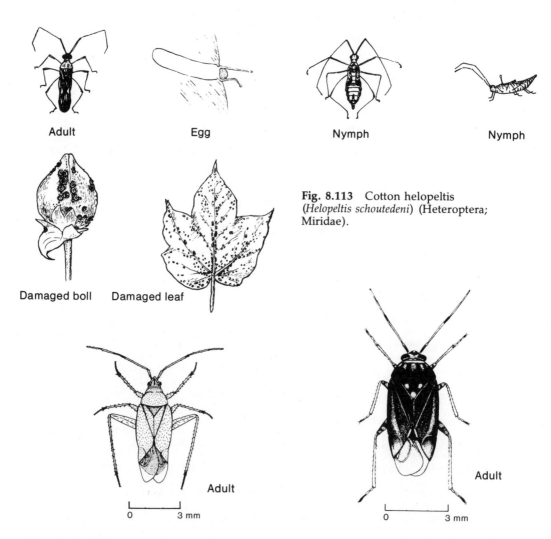

Fig. 8.113 Cotton helopeltis (*Helopeltis schoutedeni*) (Heteroptera; Miridae).

Fig. 8.114 Common green capsid (*Lygocoris pabulinus*) (Heteroptera; Miridae).

Fig. 8.115 Tarnished plant bug (*Lygus rugulipennis*) (Heteroptera; Miridae).

Lygus lineolaris (tarnished plant bug)—polyphagous; Mexico, U.S.A. and Canada.

Lygus pratensis (tarnished plant bug)—polyphagous; Europe, Asia (C.I.E. Map No. A. 39).

Lygus rugulipennis (tarnished plant bug)—polyphagous; Europe (Fig. 8.115).

Lygus solani (potato capsid)—on potato, etc.; Indonesia.

Lygus spp. (tarnished plant bugs)—on many different hosts, including coffee, cotton, beans, etc.; Africa, Asia, U.S.A. and Canada.

Mertila malayensis (orchid bug)—on various orchids; S.E. Asia.

Orthops campestris (stack bug)—on cultivated Umbelliferae; Europe, North America (Fig. 8.116).

Pachypeltis humeralis—on cinchona, etc.; India.

Pachypeltis vittiscutis—polyphagous; Indonesia.

Phytocoris spp.—on coffee; Africa.

Plesiocoris rugicollis (apple capsid)—on apple; Europe (Fig. 8.117).

Ragmus importunitas (sunn hemp capsid)—on sunn hemp; India, S.E. Asia.

Sahlbergella singularis (cocoa capsid)—on cocoa; Africa (C.I.E. Map No. A. 22) (Fig. 8.118).

Sahlbergella spp.—on cocoa; tropical Africa.

Stenotus rubrovittatus (sorghum plant bug)—on sorghum; Japan.

Stenotus spp.—on millets; Africa.

Taylorilygus vosseleri (cotton lygus)—polyphagous; Africa (Fig. 8.119).

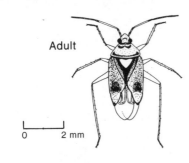

Adult

0   2 mm

Fig. 8.116 Stack bug (*Orthops campestris*) (Heteroptera; Miridae).

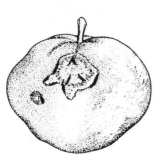

Capsid damage to apple

Adult bugs in apple shoot

Fig. 8.117 Apple capsid (*Plesiocoris rugicollis*) (Heteroptera; Miridae).

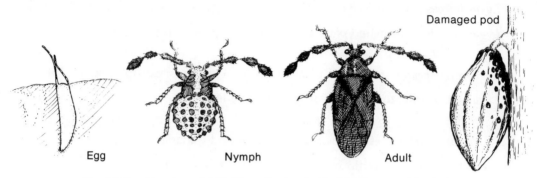

Fig. 8.118   Cocoa capsid (*Sahlbergella singularis*) (Heteroptera; Miridae).

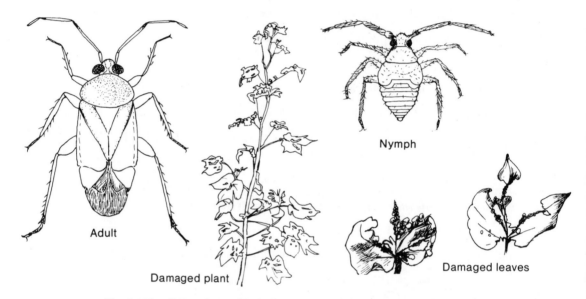

Fig. 8.119   Cotton lygus (*Taylorilygus vosseleri*) (Heteroptera; Miridae).

Isometopidae   *60 spp.*

A tiny family found throughout the world; very similar to Miridae but with ocelli, and they have antennae like Lygaeidae, and some species have enormous eyes. They live in herbage, and some species can jump. *Letaba bedfordi* is an important predator of California red scale on *Citrus* in South Africa.

SECTION PENTATOMORPHA

Antennae 4–5 segmented; hemelytra without costal fracture or cuneus; eggs not operculate and not implanted into substrate; it contains six superfamilies.

SALDOIDEA

Head broad and with prominent eyes; ocelli close together; labrum short and broad; both antennae and rostrum are long and four-segmented; pronotum narrowed anteriorly.

Saldidae (shore bugs) *150 spp.*

A small group, mostly Holarctic, and of great interest ecologically, but of no interest agriculturally. They live on salt marshes or in wet areas near to streams, and are predacious on other insects. They have large eyes, fly and jump well; ocelli almost always present, but not pedunculate; tarsi three-segmented.

ARADOIDEA

Very flattened insects living under tree bark (subcorticolous) or in fungi; no ocelli; tarsi two-segmented; hemelytra when present always narrower than abdomen.

Aradidae *400 spp.*

A worldwide group, apparently mycetophagous, little studied, characterized by head bearing two antennal tubercules.

LYGAEOIDEA

Small, oval-shaped insects (2–8 mm/0.08–0.32 in. long), strongly sclerotized; antennae and rostrum four-segmented; ocelli present in winged forms.

Lygaeidae (lygaeid bugs; seed bugs) *2000 spp.*

Small, dark or brightly colored bugs; tarsi three-segmented; antennae inserted well down the sides of the head. Most are phytophagous but a few are predatory. The group is worldwide. *Geocoris* are predators, and in the U.S.A. are called "big-eyed bugs"; *G. punctipes* preys on the cotton mite (*Tetranychus cinnabarinus*) in the U.S.A. and Mexico; *G. tricolor* preys on gray weevil (*Myllocerus* spp.), mealybugs, aphids, and other bugs in India.

Some of the more important crop pests are as follows:

Aphanus sordidus—attacks stored groundnuts; West Africa.

Aphanus spp.—on coffee, cotton, etc.; Africa.

Blissus diptopterus—on wheat; South Africa.

Blissus insularis (southern chinch bug)—on cereals; U.S.A.

Blissus leucopterus (chinch bug)—on cereals in the U.S.A. and West Indies (C.I.E. Map No. A. 333) (Fig. 8.120).

Blissus occiduus (western chinch bug)—on cereals; U.S.A. and Canada.

Blissus pallipes (wheat bug)—on wheat; China.

Lygaeus spp.—on cotton, sorghum, etc.; Africa, India.

Nysius binotatus—on fruit and vegetable crops; South Africa.

Nysius caledoniae (Caledonia seed bug)—on many crops; U.S.A.

Nysius clevelandensis (grey cluster bug)—polyphagous; Australia.

Nysius cymoides—on *Brassica* spp.; Mediterranean region.

Nysius ericae (false chinch bug)–polyphagous; U.S.A. and Canada.

Nysius inconspicuous—on sesame; India.

Nysius plebejus—polyphagous; Japan.

Nysius raphanus (false chinch bug)—U.S.A.

Nysius spp. (seed bugs)—on a wide range of crops; Mediterranean, India, Australasia, East Africa, Canada, and U.S.A. (Fig. 8.121).

Oncopeltis fasciatus (milkweed bug)—U.S.A.

Oxycarenus hyalipennis (cotton seed bug)—on cotton and other Malvaceae; throughout southern Europe, Africa, India, and in South America (C.I.E. Map No. A. 433) (Fig. 8.122).

Oxycarenus spp. (4+) (cotton seed bugs)—on cotton, etc.; Africa and Australasia.

Spilostethus spp.—on many different fruits; India.

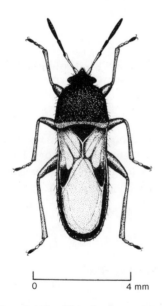

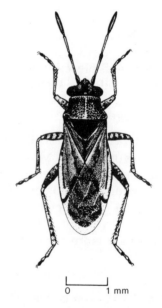

Fig. 8.120 Chinch bug (*Blissus leucopterus*) (Heteroptera; Lygaeidae).

Fig. 8.121 Seed bug (*Nysius* sp.) (Heteroptera; Lygaeidae).

Berytidae (= Neididae) (stilt bugs) *200 spp.*

Delicate, elongate insects, resembling Lygaeidae but with geniculate antennae; and long slender legs, with apically clavate femora. This small group is widely distributed and is found in herbage and meadows.

Colobathristidae *70 spp.*

Very closely related to the Berytidae, but the insect body has a basally constricted

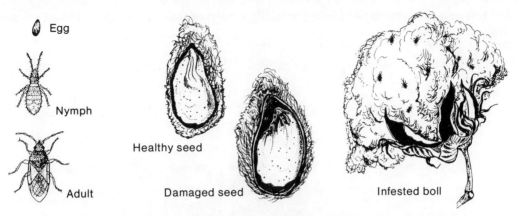

Fig. 8.122 Cotton seed bug (*Oxycarenus hyalipennis*) (Heteroptera; Lygaeidae).

abdomen. They occur mainly in South America and the Indomalayan region. *Phenacantha saccharicida* has been recorded damaging sugarcane in Java—there were population outbreaks in 1926–29 and 1940.

Piesmatidae (lace bugs)

A small group, of small insects that resemble Tingidae with which they were formerly grouped. The body is elongate oval; ocelli usually present; rostrum of four segments; mandibular extensions resemble small horns in front of head. Forewings and pronotum are characteristically reticulo-punctate; scutellum visible; tarsi of two segments, and pulvilli present. The group is mainly Palaearctic, phytophagous in habits, and feeding on wild and cultivated Chenopodiaceae.

Piesma cinerium (beet lace bug in the U.S.A.)—a minor pest of beet in the U.S.A., and alfalfa in Canada.

Piesma quadratum (beet lace bug)—similarly a pest of beets and spinach; Europe and parts of Asia; transmits a virus in sugar beet.

PYRRHOCOROIDEA

Medium-sized, usually brightly colored; eyes conspicuous; ocelli absent; rostrum and antennae slender and four-segmented.

Pyrrhocoridae (cotton stainers; red bugs)

Quite a small group but well known because of the very conspicuous and abundant cotton stainers (*Dysdercus* spp.). Brightly colored, usually red and black, these insects are abundant in cotton growing regions. The cotton stainers are important pests of cotton—the genus is completely pantropical (tropicopolitan). The adults are large bugs (*c.* 15 mm/0.6 in. long) that drop off the plant when approached, but nymphs on opened bolls usually remain on the lint (see Fig. 8.125). The common name is derived from their habit of piercing unripe cotton bolls and then contaminating with the fungus *Nematospora* which stains the lint.

Most members of this group are basically phytophagous, sucking sap from seeds, but will also prey on other insects especially when seeds of the usual host plant are not available. Important predators are:

Antilochus coquebertii—predator on *Dysdercus* (eating four stainers per day); S.E. Asia.

Dindymus rubiginosus (predacious red bug)—preys on caterpillars, etc. (Fig. 8.123); S.E. Asia.

Fig. 8.123   A predacious red bug (*Dindymus rubiginosus*) killing a small shield bug; South China.

Each female lays a large number of eggs (800–1000) in the soil litter. First instar nymphs do not feed but need moisture, and they stay near the egg shells in the litter. Second instar nymphs feed mainly on fallen ripe seeds of the host plant but later nymphs climb the plants to seek the seeds, and they usually feed in the opened bolls. Adults can feed on the unripe (closed) bolls because their stylets are long enough to reach the seeds inside. Apparently they do feed on other parts of the plant also in order to obtain sufficient water. Stainers need to be polyphagous since they primarily feed on seeds (and occasional insects). Adults fly freely and have been recorded dispersing up to 15 km (10 miles); they generally live for 2–3 months.

Dysdercus are rated as serious pests of cotton in Africa and India, but only as minor pests in the New World. At least ten species of *Dysdercus* are recorded from cotton in Africa, and more than ten species in the New World.

Some of the more important pest species are listed below:

Dindymus versicolor—polyphagous on fruit and vegetables; Australia.

Dysdercus cingulatus (oriental cotton stainer)—this and the other species feed mostly on Malvaceae, but also on sorghum, *Ceiba,* and a number of other plants; India, S.E. Asia, Australia (C.I.E. Map No. A. 265).

Dysdercus fasciatus—in Africa and Madagascar (C.I.E. Map No. A. 266).

Dysdercus nigrofasciatus—in Africa (Figs. 8.124, 8.125).

Dysdercus sidae (Australian cotton stainer)—Australasia (C.I.E. Map No. A. 267).

Dysdercus superstitiosus (African cotton stainer)—Africa (Fig. 8.126).

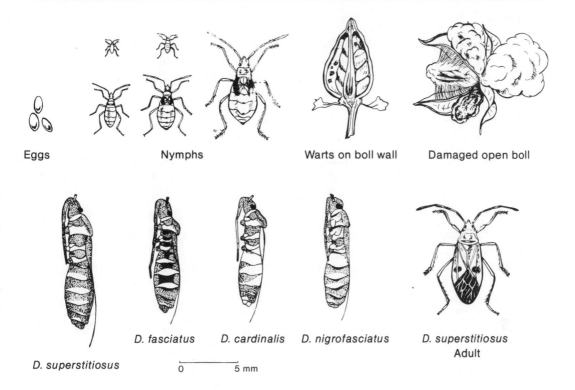

Eggs   Nymphs   Warts on boll wall   Damaged open boll

D. superstitiosus

D. fasciatus   *D. cardinalis*   *D. nigrofasciatus*   *D. superstitiosus*
Adult

0   5 mm

Fig. 8.124 Cotton stainers (*Dysdercus* spp.) (Heteroptera; Pyrrhocoridae).

Fig. 8.125 Cotton stainer nymphs (*Dysdercus* sp.) on opened cotton boll; Uganda.

Fig. 8.126 African cotton stainer (*Dysdercus superstitiosus*) a mating pair of adults on leaf of *Althaea*; Alemaya, Ethiopia (Heteroptera; Pyrrhocoridae).

Largidae

A small group, very closely related to Pyrrhocoridae but they differ in having the female with the seventh abdominal sternum divided; a few large (5 cm/2 in. or more) species are known; India, S.E. Asia, Australia and Papua New Guinea.

COREOIDEA

Mostly large, elongate insects, with strong legs; antennae four-segmented; in Imms (1977), five families are included but three are small and of little concern agriculturally.

Coreidae (brown bugs, etc.)   *2000 spp.*

Most are brownish in color, 10–30 mm (0.4–1.2 in.) in length, but a few are brightly colored; they are strong fliers, all phytophagous, and capable of emitting a pungent odor. Body shape is more or less oval (as distinct from elongate); head is much narrower than pronotum. They are abundant throughout the tropics. They all have very toxic saliva and their feeding causes conspicuous tissue necrosis in the host plants. This takes three main forms; firstly, the seed feeders shrivel and may kill the seeds on which they feed (e.g. *Clavigralla* and legume seeds); secondly, bugs such as *Mictis* and *Anoplocnemis* feed on young shoots of some crops and ornamentals, and the shoot distal to the feeding site wilts and dies, and these bugs are often called "twig wilters"; thirdly, bugs feeding on thick stems (*Notobitus* on bamboo) or large fruits (*Pseudotheraptus* on coconut) cause necrotic areas that elongate as the stem/fruit grows into streaks of brown tissue. In many cases the feeding lesions become infested by fungi and rotting may become more extensive; often the distinction between direct damage and secondary fungal invasion is difficult to assess.

Eggs are laid on the plant foliage either singly or in short rows. Nymphal development takes generally 1–2 months; there are five nymphal instars as is usual with Heteroptera, and the nymphs (and occasionally adults) may remain together gregariously on the plant foliage. There may be one to several generations per year, depending in part upon local climate. Adults are quite long-lived and many species live for 1–3 months, but they generally only feed once or twice per day. *Leptoglossus* and some other genera have large foliate expansions of the hind tibiae in the males, whose function is unknown. Some others have swollen and curved hind femora. On some crops, egg parasites (Hym., Chalcidoidea) are very important and 60–90% of the eggs may be destroyed.

Some of the more important coreid pests of crops include:

Acanthocoris spp. (squash bugs)—on Cucurbitaceae; S.E. Asia, China, Japan.

Amblypelta cocophaga (coconut bug)—on coconut and other crops; Pacific Region.

Amblypelta spp.—on banana, macadamia, etc.; Australia.

Anasa tristis (squash bug)—on Cucurbitaceae; U.S.A. and Canada.

Anoplocnemis spp. (giant twig wilters)—polyphagous on citrus, fig, mango, okra, cowpea, etc.; Africa, S.E. Asia.

Clavigralla spp. (= *Acanthomyia*) (spiny brown bugs)—on legumes; tropical Africa (C.I.E. Map No. A 445) (Fig. 8.127).

Dasynus piperis (pepper bug)—on pepper fruits; Malaysia, Indonesia.

Dasynus spp.—on coconut, cashew, citrus, banana, etc.; Africa and India.

Leptoglossus australis (= *membranaceus*) (leaf-footed plant bug)—polyphagous; found throughout the Old World tropics (C.I.E. Map No. A. 243) (Fig. 8.128).

Leptoglossus zonatus (leaf-footed plant bug)—on *Citrus;* Central and South America.

Leptoglossus spp. (leaf-footed plant bugs)—on *Pinus* and various ornamentals; U.S.A.

Leptocoris spp.—on coffee, cashew, etc.; tropical Africa.

Mictis spp. (giant twig wilters)—on fruit trees and ornamentals; Australia, S.E. Asia, China (Fig. 8.129).

Notobitus meleagris (bamboo bug)—on some species of bamboo; South China.

Paradasynus spp. (cashew bug, etc.)—on cashew, etc.; India, S.E. Asia, China, Japan.

Pseudotheraptus spp. (coconut bugs)—on coconut, cashew, cassava, etc.; East, Central and West Africa (Fig. 8.130).

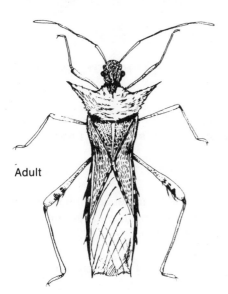

Fig. 8.127 Spiny brown bug (*Clavigralla* sp.); Tanzania (Heteroptera; Coreidae).

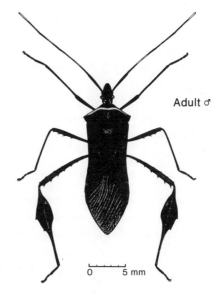

Fig. 8.128 Leaf-footed plant bug (*Leptoglossus australis*) (Heteroptera; Coreidae).

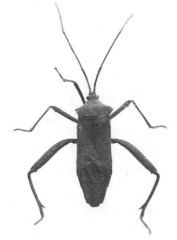

Fig. 8.129 Giant twig wilter (*Mictus* sp.); female; body length 20 mm (0.8 in.); Hong Kong (Heteroptera; Coreidae).

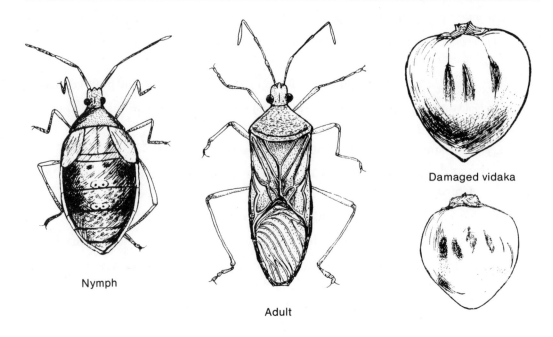

Nymph

Adult

Damaged vidaka

Fig. 8.130 Coconut bug (*Pseudotheraptus wayi*) from the Kenya coast (Heteroptera; Coreidae).

Alydidae

Formerly, these bugs constituted a subfamily of the Coreidae, from which they are now separated because of their slender elongate body form, pronotum and head equally broad, and fourth antennal segment curved and longer than third. There are species found in all regions of the world, but only four genera are recorded as crop pests. They are very similar to Coreidae in their biology and their effects on the crop plants.

Cletus spp. (cletus bugs)—on many different hosts (e.g. rice, citrus, cowpea, sweet potato, other legumes, etc.); Africa, Asia up to Japan.

Leptocorisa acuta (Asian rice bug)—on rice, cereals, and grasses (regularly cause rice crop losses of up to 50%); India, S.E. Asia, Australasia, China (C.I.E. Map No. A. 225) (Figs. 8.131, 8.132).

Leptocorisa chinensis (Chinese rice bug)—on Gramineae; Malaysia, Indonesia, China, Japan.

Leptocorisa pallida—on cocoa; tropical Africa.

Leptocorisa spp. (rice bugs)—on Gramineae; Africa and southern Asia.

Riptortus spp. (bean pod bugs)—on the pods of many cultivated legumes, etc. (crop losses can be very high); Africa, Asia.

Stenocoris southwoodi (African rice bug)—on rice and other cereals; tropical Africa (Fig. 8.133).

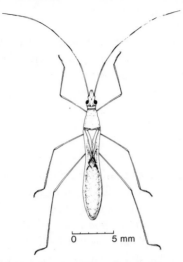

Fig. 8.131 Asian rice bug (*Leptocorisa acuta*) (Heteroptera; Alydidae).

Fig. 8.132 Asian rice bug (*Leptocorisa acuta*) on leaf of rice; body length 15 mm (0.6 in.); South China (Heteroptera; Alydidae).

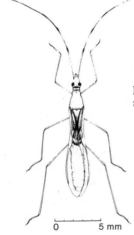

Fig. 8.133 African rice bug (*Stenocoris southwoodi*) (Heteroptera; Alydidae).

PENTATOMOIDEA

Stout-bodied insects, strongly sclerotized, antennae usually with five segments; pronotum usually six-sided; mesoscutellum very large (reaching to either middle of abdomen or covering it entirely); legs with two or three tarsal segments. The higher classification of the superfamily is difficult and disputed—in Imms (Richards & Davies, 1977) eight families are recognized.

Plataspididae (= Coptosomatidae) (dwarf shield bugs; helmet bugs) *500 spp.*

Rounded shiny insects with a very convex shape, and scutellum covering almost entire abdomen; hemelytra long and folded beneath scutellum at rest; tarsi two-segmented. Found throughout the Old World Tropics. Most are small in size—*Coptosoma* is 2–5 mm (0.08–0.2 in.), and *Brachyplatys* is 5–7 mm (0.2–0.3 in.).

> *Brachyplatys testudanigra*—occurs on various legumes, rice, maize, eggplant, etc.; East and West Africa.

Brachyplatys spp.—on a wide range of crops; India.

Coptosoma—contains a number of species that can be regarded as minor pests on a wide range of fruit and vegetable crops (on the flowers and young shoots) in tropical Africa, and Asia up to Japan (and parts of Central Asia), and also in the U.S.A.

Plataspis spp.—recorded from coffee, cocoa; East Africa.

Cydnidae (burrowing stink bugs)

Quite a large group and found worldwide; they are small dark bugs adapted for digging, and found under stones, in leaf litter, or in the soil around plant roots. The legs are flattened and shovel-like; body length varies from about 4–9 mm (0.16–0.36 in.). The nymphs resemble root-aphids but have larger legs. *Geotomus pygmaeus* is found throughout Asia and North America, and *Cydnus* spp. are widespread; the adults fly to lights at night, and they breed on the roots of grasses and various herbaceous weeds. Several species of *Stibaropus* are pests attacking the roots of tobacco, sorghum, bulrush millet, and palms in India, and another species attacks sugarcane roots in Java.

Scutelleridae (shield bugs)

A widely distributed but largely tropical group, quite large in size (most 8–20 mm/0.3–0.8 in. long); scutellum large and covering entire abdomen; tarsi three-segmented. Some species are bright metallic and very colorful; several species of *Calliphara* occur in S.E. Asia, China, and Australia, and they are dark red underneath with a brilliant green upper surface (Fig. 8.134)—they are conspicuous but not often found on plants of economic importance, but are occasionally seen on citrus, grapevine, and some local fruits in India. *Cantao ocellatus* is a brightly colored orange bug with green and dark spots, and slight sexual dimorphism found breeding gregariously in the foliage of *Mallotus paniculatus* (an important shade tree) in Hong Kong (Fig. 8.135); this species is renowned for the maternal care practiced by the female, in that she guards her egg mass (20–40 eggs), and broods over the young bugs, even to the extent of driving away parasitic wasps.

In cooler regions, the adults may overwinter in hibernation; in the U.S.S.R., adults of *Eurygaster* migrate from the wheat fields up into hibernation sites on mountainsides, often 10–20 km (6–12 miles) away; in the spring, the adults complete the migration by returning to the wheat fields. In cooler regions, there are usually only 1–2 generations annually, but in the hotter tropics there are often several generations; adults are long-lived (1–6 months), often fly to lights at night, and produce pungent odors from the stink glands.

Some of the more important agricultural pests include:

Calidea spp. (blue bugs)—several species on a wide range of host plants (cotton especially); throughout tropical Africa, and Arabia (Fig. 8.136).

Catacanthus nigripes (tea shield bug) on *Camellia* spp.; China.

Chrysocoris grandis (large white shield bug)—on bamboo and litchi; South China (Fig. 8.137).

Chrysocoris javanus—on *Ricinus*, etc.; Indonesia.

Chrysocoris stollii (litchi shield bug)—India, S.E. Asia.

Eurygaster austriaca (wheat shield bug)—on wheat and other cereals; southern Europe, Near and Middle East, U.S.S.R. (C.I.E. Map. No. A. 361).

Eurygaster integriceps—on wheat and cereals; Near East, Middle East, Pakistan, and U.S.S.R. (C.I.E. Map No. A. 40) (Fig. 8.138).

Eurygaster koreana—on cereals; Japan, Korea.

Poecilocoris dives (tea fruit bug)—on tea; Java.

Poecilocoris latus (camellia shield bug)—on *Camellia* spp.; South China (Fig. 8.139).

Scotinophara coarctata (black paddy bug)—on rice; India, S.E. Asia. (Fig. 8.140).

Scotinophara lurida (Japanese black rice bug)—on rice; from India through S.E. Asia to China and Japan.

Tectocoris diophthalmus (harlequin bug)—on cotton, citrus, etc.; Australia, Papua New Guinea, Fiji.

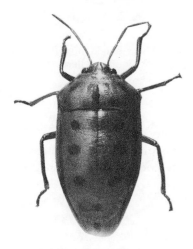

Fig. 8.134 Blue shield bug (*Calliphara nobilis*) (Heteroptera; Scutelleridae); body length 14 mm (0.56 in.); South China.

Fig. 8.135 *Cantao ocellatus* (Heteroptera; Scutelleridae) from *Mallotus* foliage; body length 17 mm (0.7 in.); Hong Kong.

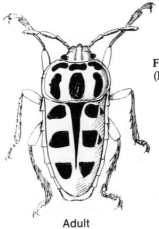

Adult

Fig. 8.136 Blue bug (*Calidea* sp.) (Heteroptera; Scutelleridae).

Fig. 8.137 Large white shield bug (*Chrysocoris grandis*) (Heteroptera; Scutelleridae); body length 21 mm (0.84 in.); South China.

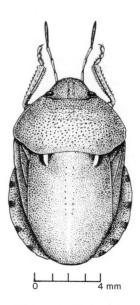

Fig. 8.138 Wheat shield bug (*Eurygaster integriceps*) (Heteroptera; Scutelleridae).

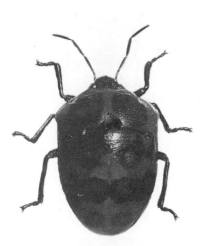

Fig. 8.139 Shield bug (*Poecilocoris* sp.) from tea bushes; Tai Mo Shan, South China; body length 16 mm (0.64 in.) (Heteroptera; Scutelleridae).

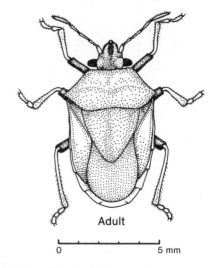

Fig. 8.140 Black paddy bug (*Scotinophara coarctata*) (Heteroptera; Scutelleridae).

Pentatomidae (stink bugs)   *2500 spp.*

Following the classification in Imms (1977) to avoid contention, this large family is worldwide, but most abundant in Neotropical, Ethiopian, and Indo-Malayan regions. The scutellum usually only covers half the abdomen; antennae are five (rarely four) segmented; tarsi two or three segmented; hemelytra well developed. Some species are pests but there are also some useful predators, and a very small number appear to be omnivorous. Considering the abundance of the group, the number of pest species is relatively low.

Most of the recorded life histories are rather similar. Eggs are barrel shaped and laid in groups (10–50) stuck to the plant foliage—each female usually lays 100–300 eggs. The litchi stink bug appears to lay precisely 14 eggs per batch! The newly hatched nymphs usually stay by the egg mass for the first instar (Fig. 8.141). The five instars generally take 25–50 days (mostly about 30) to complete and most species have 1–3 generations per year, but there are differences due to climate and to the species concerned and also to the host plant. In regions with a winter or with a pronounced dry season adults often hibernate. All species have "stink" glands and can produce obnoxious odors; in *Tessaratoma* the nymphs have four pairs of dorsal abdominal odoriferous glands which atrophy in the adult and are replaced by ventral thoracic glands from which the disturbed bugs can eject a fine stream of stinking yellow liquid for a distance of 50 cm (20 in.) or more, and it stains both skin and clothing and is extremely irritating to the eyes. Both sexes of *Tessaratoma* are able to stridulate, using a "file" on the abdominal dorsum and a row of teeth (comb) on the undersurface of each wing base.

Fig. 8.141   Stink bug; first instar nymphs newly emerged from egg mass; South China (Heteroptera; Pentatomidae).

Feeding damage to plants is partly direct, resulting from the toxic saliva, and often indirect in that the feeding punctures are used as invasion sites by fungi; but several important pest species regularly transmit the fungus *Nematospora* through infected mouthparts during the feeding process.

Following the subfamily classification used in Imms (1977) some of the more important pest species are listed below:

PHYLLOCEPHALINAE (Body elongate, head and pronotum prolonged.)

Megarrhampus spp. (grass stink bugs)—on grasses, cereals, and sugarcane; S.E. Asia to China (Figs. 8.142, 8.143).

ASOPINAE (Predatory stink bugs)

Andrallus spp.—prey on caterpillars on herbaceous plants; India, and S.E. Asia.

Fig. 8.142 Miscanthus stink bug (*Megarrhampus hastatus*); length 21 mm (0.84 in.); Hong Kong (Heteroptera; Pentatomidae; Phyllocephalinae).

Fig. 8.143 Miscanthus stink bug (*Megarrhampus hastatus*) lying along a *Miscanthus* leaf blade; Hong Kong (Heteroptera; Pentatomidae; Phyllocephalinae).

Cantheconidea spp.—prey on caterpillars (especially Limacododae) on shrubs and trees; India and S.E. Asia.

Cazira chiroptera—prey on *Epilachna* larvae (which they closely resemble); India and S.E. Asia.

TESSARATOMINAE (Nymphs with flattened oval-shaped body)

Musgraveia spp. (orange bugs)—on citrus; Australia.

Tessaratoma javanica (litchi stink bug)—on litchi and longan; India and S.E. Asia.

Tessaratoma papillosa (litchi stink bug)—on litchi and longan; S.E. Asia up to China (Figs. 8.144–8.146).

Tessaratoma quadrata—on litchi and apple; India.

Fig. 8.144 Litchi stink bug (*Tessaratoma papillosa*); adult in litchi foliage; Penang, Malaysia (Heteroptera; Pentatomidae; Tessaratominae).

Fig. 8.145 Litchi stink bug (*Tessaratoma papillosa*); nymph showing characteristic body shape; body length 19 mm (0.79 in.); South China (Heteroptera; Pentatomidae; Tessaratominae).

Fig. 8.146 Litchi stink bug (*Tessaratoma papillosa*); nymph with white waxy body surface, in foliage of longan tree; Penang, Malaysia (Heteroptera; Pentatomidae; Tessaratominae).

PENTATOMINAE

Aelia spp. (wheat bugs)—on wheat and cereals; Middle East, S.E. Europe.

Agonoscelis spp. (cluster bugs)—polyphagous on legumes, sorghum, citrus, sunflower, Solanaceae; Africa, S.E. Asia.

Antestia spp. (coffee bugs)—on arabica coffee; tropical Africa.

Antestiopsis spp. (antestia bugs)—on arabica coffee; tropical Africa, S.E. Asia (C.I.E. Map Nos. A. 381, 382) (Fig. 8.147).

Bagrada spp. (harlequin bugs)—on Cruciferae, etc.; Africa, India, S.E. Asia (Fig. 8.148).

Biprorulus bibax (spined citrus bug)—on citrus; Australia.

Chlorochroa ligata ("Conchuela")—polyphagous; U.S.A.

Diploxys fallax (rice stink bug)—on rice; Swaziland, Madagascar (Fig. 8.149).

Erthesino fullo (tallow stink bug)—on tallow tree; China (Fig. 8.150).

Eurydema ornata (cabbage bug)—polyphagous on cabbage, potato, cereals, cotton, etc.; Africa.

Eurydema pulchrum (cabbage bug)—on *Brassica* spp.; S.E. Asia, Australasia.

Murgantia histrionica (harlequin bug)—on *Brassica* spp.; U.S.A.

Nezara immaculata (cereal green stink bug)—on maize, wheat; Ethiopia.

Nezara viridula (green stink bug)—polyphagous; cosmopolitan (C.I.E. Map No. A. 27) (Fig. 8.151).

Oebalas poecilus (rice stink bug)—on rice, etc.; Indonesia.

Oebalas pugnax (rice stink bug)—on rice, maize, sorghum, and grasses; southern U.S.A., Cuba (Fig. 8.152).

Oebalas spp. (rice stink bugs)—on rice and other Gramineae; Central and South America.

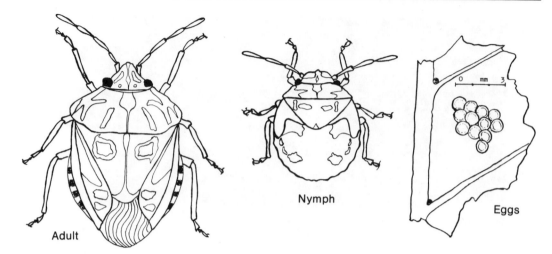

Adult   **Nymph**   **Eggs**

Fig. 8.147 Antestia bug (*Antestiopsis* sp.); Kenya (Heteroptera; Pentatomidae; Pentatominae).

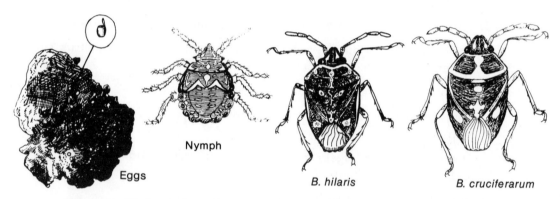

Eggs   **Nymph**   **B. hilaris**   **B. cruciferarum**

Fig. 8.148 Harlequin bugs (*Bagrada* spp.); ex Kenya (Heteroptera; Pentatomidae).

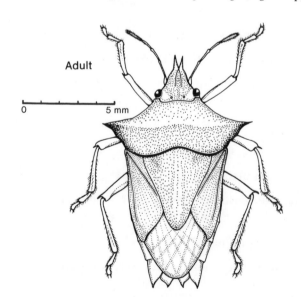

Fig. 8.149 Rice stink bug (*Diploxys fallax*) (Heteroptera; Pentatomidae).

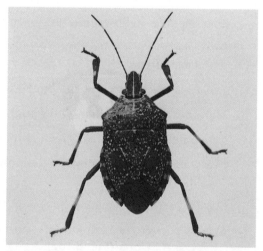

Fig. 8.150 Tallow stink bug (*Erthesino fullo*) (Heteroptera; Pentatomidae); body length 20 mm (0.8 in.); South China.

Pentatoma rufipes (forest bug)—on various fruit trees and bushes—a woodland species; Europe (including the U.K.).

Piezodorus spp.—on many different hosts, including legumes, cotton, millets, etc.; Europe, eastern Africa, India, West Indies.

Pygomenida variepennis—on rice and maize; S.E. Asia.

Rhynchocoris humeralis (citrus stink bug)—on *Citrus* spp.; India, S.E. Asia, China (Fig. 8.153).

Rhynchocoris longirostris (citrus stink bug)—on citrus; Philippines (Fig. 8.153).

Tolumnia spp.—on coffee, Solanaceae, etc.; S.E. Asia.

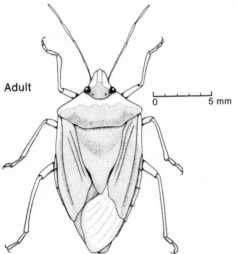

Fig. 8.151 Green stink bug (*Nezara viridula*) (Heteroptera; Pentatomidae).

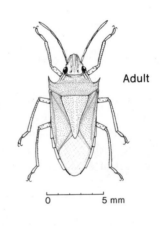

Fig. 8.152 Rice stink bug (*Oebalus pugnax*) (Heteroptera; Pentatomidae).

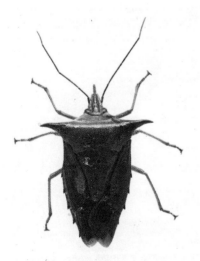

Fig. 8.153 Citrus stink bugs, *Rhynchocoris longirostris* from the Philippines, and *R. humeralis* from China; body length about 22 mm (0.9 in.) (Heteroptera; Pentatomidae).

SERIES AMPHIBICORISAE

These are the aquatic bugs that live on the surface of water bodies; mostly ponds, paddy fields, irrigation ditches, and the edges of slow-flowing streams. They have long antennae, and the ventral body surface with a hydrofuge pubescence. The group is of some importance to rice growers in that these insects are predacious and feed on insects fallen on to the water surface—thus they eat many species of rice pests in paddy fields. But they also feed on dead insects found on the water surface. Five families are recognized here.

GERROIDEA

Hydrometridae (water measurers)   *70 spp.*

The majority of species are tropical; they are delicate with an elongate body and head; antennae with four or five segments; rostrum three-segmented. Legs are thin and stilt-like, tarsi with three segments and apical claws. The main genus is *Hydrometra*; it stalks slowly over the surface of the standing water, seeking small insects for food.

Veliidae (water crickets)   *250 spp.*

A worldwide group, but predominant in oriental and neotropical regions. *Velia* occurs mostly at the edge of streams, but another genus swims in fast streams. The middle legs have tarsi with a terminal fan of bristles, which they use in the manner of oars. One group is marine and found on the Indian and Pacific Oceans. They are all very small in size, mostly about 2 mm (0.08 in.) long.

A group of even smaller species forms a separate family, the Mesoveliidae.

Gerridae (pond-skaters)   *200 spp.*

These are usually much larger insects, with a body length of 15 mm (0.6 in.) or so, and legs many times longer. The middle and hind legs are longest and fringed with bristles, with which the insects skate or run over the water surface with rapidity. They sometimes congregate in numbers; this is often seen in sheltered locations on streams. On paddy fields they may be numerous. Adults fly at nights on occasions. The common genera include *Gerris* in Europe, and *Ptilomera* in S.E. Asia (Fig. 8.154). *Halobates* and others are marine insects and wingless, and found on the open ocean or in mangroves.

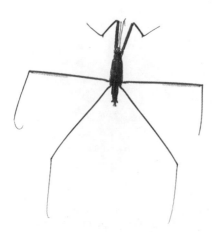

Fig. 8.154 Adult pond-skater (*Ptilomera* sp.) (Heteroptera; Gerridae); body length 15 mm (0.6 in.); South China.

SERIES HYDROCORISAE (= Cryptocerata)

Also aquatic bugs, rarely semiaquatic, mostly to be found in ponds, fish ponds, paddy fields, but some prefer streams and deeper water; they live in the water and swim using their legs, usually coming to the water surface to breathe. Most are carnivorous, and some will take insects on the water surface. The larger species will kill frogs and fish and in fish ponds can be quite serious pests. The antennae are very short and usually concealed under the head; most adults are winged and fly freely at times at night, and may come to lights. There are three superfamilies and a total of nine families.

NOTONECTOIDEA

Truly aquatic forms, with elongate rostrum, three- or four-segmented; antennae concealed; fore tarsi normal (not spatulate); no ocelli.

Naucoridae (saucer bugs) *150 spp.*

Smallish, oval, flattened bugs, with hind legs fringed for swimming and forelegs strongly raptorial; antennae are four-segmented and simple; rostrum stout and three-segmented. The group is mostly tropical and to be found in both ponds and streams, where they prefer aquatic vegetation. The oriental genus *Cheirochela* could be beneficial in paddy fields, but could equally be a pest in fish ponds. However, *Cheirochela* and *Heleocoris* are quite small (8–10 mm; 0.32–0.4 in. long) and so would only kill fish fry.

Belostomatidae (giant water bugs) *100+ spp.*

The common name for these bugs in Australia is "fish-killers", which is not surprising for *Lethocerus* occurs throughout the tropics as a series of species, up to 8–10 cm (3.2–4 in.) (Fig. 8.155). They are fiercely predacious and kill small fish, frogs, and other insects—the rostrum is short and curved with a sharp tip, and the bite of small species is very painful, so that of *Lethocerus* should be terrible. Adults fly strongly at night on occasions, and are attracted to lights. There are small species in the group, and *Sphaerodema* is only about 10–12 mm (0.4–0.48 in.), and males have the strange habit of carrying the eggs stuck onto the wingcases (Fig. 8.156).

Nepidae (water "scorpions") *150 spp.*

The appearance of these insects is quite characteristic in that there is a long, posterior respiratory tube, the legs are long and thin and used for walking in the aquatic vegetation, but the forelegs are prehensile and raptorial. In Europe *Nepa* is the common genus, in S.E. Asia *Laccotrephes* is very similar but larger, and in Africa some species have a large, oval-shaped body. *Ranatra* differs from these in having a stick-like body. They prefer standing water with plenty of aquatic vegetation—they typically sit in the vegetation with the respiratory siphon touching the water surface. Fig. 8.157 shows an adult *Laccotrephes* sp. from a fish pond in South China.

Notonectidae (backswimmers)

A distinctive group in that they swim on their backs using the long oar-like hind legs; the body is sharply convex or keeled dorsally, and it bears bristles. Air is carried trapped beneath the wings, for respiration when submerged; it is renewed by repeated visits to the water surface. The larger species measure 10–12 mm (0.4–0.48 in.); they swim fast, and fly readily. They should be handled with care for the "bite" is painful; they feed on tadpoles, small fish, and other insects. In Europe there are many species of *Notonecta*, but the common species in South China is the very similar *Enithares* (Fig. 8.158).

Fig. 8.156 Adult small water bug (*Sphaerodema* sp.), male with egg mass stuck to its back; body length 12 mm (0.5 in.) (Heteroptera; Belostomatidae); South China.

Fig. 8.155 Adult giant water bug (*Lethocerus indicus*) (Heteroptera; Belostomatidae); body length 90 mm (3.6 in.) South China.

Fig. 8.158 Backswimmer adult (*Enithares* sp.) (Heteroptera; Notonectidae); body length 10 mm (0.4 in.); South China.

Fig. 8.157 Typical water "scorpion" (*Laccotrephes* sp.) (Heteroptera; Nepidae); body length 32 mm (1.3 in.); South China.

CORIXOIDEA

Actively swimming bugs, using the middle and hind legs as oars, but the body is flattened and they swim upright; fore-tarsus usually spatulate. They are unusual in that they are phytophagous and feed on diatoms and tiny algae, piercing the cells with their short proboscis (rostrum one- or two-segmented).

Corixidae (water boatmen)   *200 spp.*

In paddy fields, ponds, etc., these bugs can be very numerous; they fly freely and soon invade new bodies of water, but they are entirely phytophagous and do no damage. They are eaten by many species of fish, and by predacious insects in the water.

9

ORDER THYSANOPTERA
thrips
6: 5000

These are small to minute insects 1–8 mm (0.04–0.32 in.) long; mostly 1–4 mm (0.04–0.16 in.), slender bodied; wings very narrow with long, fringing setae; antennae short (6–10 segments); mouthparts asymmetrical and piercing; tarsi one or two segmented, each with a protrusible vesicle; no cerci; metamorphosis through inactive pupal-like stage. Adults are mostly black, brown, or yellow in color, but nymphs are more often red, orange, or yellow.

Most feed on plants by piercing the surface tissues with their mouthparts and sucking up the exuding sap. The epidermal cell rupturing usually occurs on young leaves and flowers and results in a distortion of tissues as growth takes place, and thus there are a number of important pest species. Damage to leaves is usually epidermal scarification causing withering, browning, and occasionally death of the leaves. A few species are important plant pollinators; others are predacious on small insect pests and thus quite beneficial to agriculture. Many species are mycetophagous and to be found in leaf litter, in grass tussocks, or on dead wood, or else in the surface layers of the soil.

Reproduction may be either sexual or parthenogenetic; in some species males seem to be very rare. Eggs are laid either in groups on the foliage or leaf litter (Tubulifera) or inserted into the plant tissues using the sawlike ovipositor (Terebrantia). The nymphs are often a different color from the adults—commonly the adults are blackish and the nymphs red or orange. In some species, the nymphs carry a drop of excrement at the tip of the upturned abdomen. Development is usually quite rapid and the nymphs feed continuously—in some species the nymphs are gregarious and found in groups. Metamorphosis, called "pupation", usually takes place in the soil—entire field infestations can apparently disappear overnight as the large nymphs descend into the soil to pupate. *Heliothrips haemorrhoidalis* is unusual in that pupation takes place on the foliage of the host plant.

Many adults look remarkably similar in appearance. Field identification is often very difficult, and to some extent reliance on host plant data may give an indication as to identity. Adults may have a period of active dispersal by flight. In Europe the cereal and grass thrips (*Limothrips* spp.) swarm in mass flights in the summer during hot dry sunny weather in enormous numbers, and they are sometimes called "thunder-flies". Some species are additionally important as crop pests in that they act as vectors of some virus diseases, and a few may carry fungal spores.

SUBORDER TEREBRANTIA

Female abdomen with conical apex (rounded in male) and ovipositor sawlike; wings with microtrichia and at least one longitudinal vein reaching the apex.

AEOLOTHRIPOIDEA

Ovipositor curved dorsally; wings relatively broad; antennae nine-segmented.

Aeolothripidae (predacious thrips)

The genera *Aeolothrips* and *Aleurodothrips* are predacious on scale insects (Coccoidea) and other small insects on plants and in grass tussocks.

THRIPOIDEA

Ovipositor curved downwards; wings more or less narrow and pointed; antennae with 6–10 segments, the last 1–3 forming a thin style; three families are placed here. A major taxonomic guide to the Thysanoptera has just been published by C.A.B. International Institute of Entomology (Palmer, Mound & du Heaume, 1989).

Thripidae (thrips) *c. 2000 spp.*

A large family including more than 160 genera, characterized by having antennae 6–9 segmented; completely worldwide in distribution, and containing most of the important pest species. They are all phytophagous, sap-feeding forms. Some of the more important pest species to be found on cultivated plants are listed below:

Anaphothrips obscurus (grain thrips)—on cereals and grasses; Japan, U.S.A., Canada.

Anaphothrips orchidaceus (orchid thrips)—on orchids; tropical Asia to Japan.

Anaphothrips spp.—50 species worldwide; some on cereals and sugarcane.

Aptinothrips spp. (cereal and grass thrips)—on cereals and grasses; Europe, U.S.A., and Canada.

Astrothrips spp. (12)—on bananas, and leaves of cotton, castor; Old World tropics.

Baliothrips biformis (rice thrips)—on rice; India, S.E. Asia to Japan (C.I.E. Map No. A. 215) (Fig. 9.1).

Baliothrips minutus (sugarcane thrips)—on sugarcane, etc.; U.S.A.

Caliothrips striatoptera (black maize thrips)—on maize, etc.; S.E. Asia.

Caliothrips spp.—on maize, cotton, pulses, flax, onions, grasses, etc.; 18 species known; both Old World and New World in warmer regions.

Chaetanaphothrips orchidii (orchid/banana rust thrips)—polyphagous and pan-tropical; pest of banana in Central America, and orchids in temperate greenhouses.

Chaetanaphothrips signipennis (banana thrips)—cause "rust" on banana fruits; pantropical.

Chaetanaphothrips spp. (8)—cause "rust" on bananas, leaf distortion of tea, etc.; pantropical.

Diarthrothrips coffeae (coffee thrips)—on coffee; East and Central Africa (Fig. 9.2).

Dichromothrips spp. (14) (orchid thrips)— on orchids; S.E. Asia, Old World tropics.

Drepanothrips reuteri (grapevine thrips)—on grapevines in southern Europe and California, and oak trees in England.

Frankliniella fusca (tobacco thrips)—on tobacco, cotton, tomato, and cucumber in greenhouses (a virus vector); U.S.A. and Canada.

Frankliniella intonsa (flower thrips)—on many flowers/crops; Palaearctic.

Frankliniella occidentalis (western flower thrips)—on citrus, cotton, many vegetables and many flowers (many greenhouse crops) (virus vector); U.S.A., and now established in the U.K. in greenhouses; a very serious pest.

Frankliniella schulzei (cotton bud thrips)—on cotton, and many other crops; tropical Africa, but cosmopolitan; vector of viruses and fungal spores.

Frankliniella tritici (eastern flower thrips)—on alfalfa, oats, beans, asparagus, etc.; Canada, and U.S.A.; virus vector and very serious pest.

Frankliniella spp. (flower thrips)—many different species worldwide, on many crops and cultivated plants (Fig. 9.3); 100 species known, mostly New World.

Heliothrips haemorrhoidalis (black tea thrips, etc.)—polyphagous on many tropical crops, and greenhouse crops in temperate regions; cosmopolitan (C.I.E. Map No. A. 135) (Fig. 9.4).

Hercinothrips bicinctus (banana/smilax thrips)—on banana and other plants (also greenhouses); Africa, southern Europe, Australia, Hawaii, North, Central and South America (Fig. 9.5).

Hercinothrips femoralis (banded greenhouse/sugarbeet thrips)—polyphagous on many field crops in warmer regions, and on ornamentals in greenhouses; worldwide, but records are scattered (C.I.E. Map No. A. 402).

Kakothrips robustus (pea thrips)—on pods of pea and beans; Europe.

Limothrips cerealium (grain thrips)—on wheat, maize, rye, oats, barley, grasses, and also *Citrus;* cosmopolitan (C.I.E. Map No. A. 245).

Limothrips denticornis (barley thrips)—on cereals; Canada.

Limothrips spp. (cereal and grass thrips)—on cereals, grasses, and some other crops; the genus is quite cosmopolitan.

Odontothrips spp. (flower thrips)—in legume flowers; Europe.

Parthenothrips dracaenae (palm thrips)—on Palmae in the tropics, and indoor monocots in temperate regions; worldwide.

Retithrips syriacus (castor thrips)—on castor, etc.; tropical Africa.

Rhipiphorothrips cruentatus (grapevine thrips)—on grapes, cashew, almond, rose, castor, etc. (polyphagous); Oriental region.

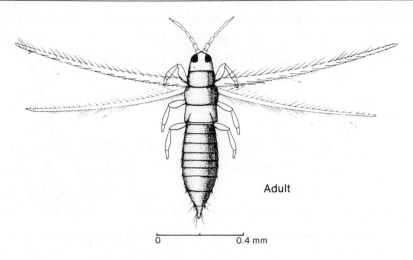

Fig. 9.1   Rice thrips (*Baliothrips biformis*) (Thysanoptera; Thripidae).

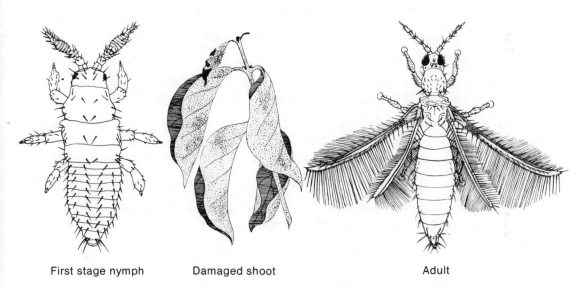

Fig. 9.2   Coffee thrips (*Diarthrothrips coffeae*) (Thysanoptera; Thripidae).

Scirtothrips aurantii (citrus/tea thrips)—polyphagous; tropical Africa (C.I.E. Map No. A. 137) (Fig. 9.6).

Scirtothrips citri (citrus thrips)—on *Citrus;* southern U.S.A.

Scirtothrips dorsalis (chilli thrips)—polyphagous; India, S.E. Asia, Austalasia (C.I.E. Map No. A. 475).

Scirtothrips mangiferae (mango thrips)—on mango; India.

Scirtothrips manihoti (cassava thrips)—on cassava; South America.

Scirtothrips spp.—50+ species are known worldwide, several pests on cereals and sugarcane; several crop pests in India.

Selenothrips rubrocinctus (red-banded thrips)—on mango, cocoa, cashew, and other crops; pantropical (C.I.E. Map No. A. 136) (Fig. 9.7).

Sericothrips adolfifriderici—on cashew and legumes; tropical Africa.

Taeniothrips (now *Megalurothrips sjostedi*) (bean flower thrips)—in flowers of beans and other legumes, coffee, etc.; tropical Africa.

Taeniothrips inconsequens (pear thrips)—in flowers of fruit trees in Europe and U.S.A.; now pest of sugar maple in eastern North America.

Thrips angusticeps (cabbage thrips)—on brassicas, beet, pea, apple, and pear flowers; Europe.

Thrips australis (gum tree thrips)—in flowers of *Eucalyptus,* etc.; now pantropical.

Thrips atratus (carnation thrips)—in carnations, etc.; Europe.

Thrips flavus—damages flowers of apple, citrus, alfalfa, and leaves of cotton and mustard; Europe, Africa, Asia, Pacific, North America.

Thrips hawaiiensis (flower thrips)—polyphagous on many crops, very common and variable (pollinator of oil palm); Oriental and Pacific regions.

Thrips imaginis (plague thrips of Australia)—on apple flowers, but polyphagous; Australasia and the Pacific region.

Thrips linearis (flax thrips)—on flax; Europe (not U.K.), U.S.S.R.

Thrips major (rose thrips)—in flowers of Rosaceae, etc.; U.K. and most of Europe.

Thrips nigropilosus (pyrethrum thrips)—on pyrethrum, various Compositae, and greenhouse crops in temperate Europe; Africa, Europe, Hawaii, Fiji.

Thrips palmi (palm thrips)—damaging to cucurbits, eggplant, beans, cotton, potato, tobacco, other legumes and flowers; spread from the Pacific to China, S.E. Asia and India (C.I.E. Map No. A. 480).

Thrips parvispinus (tobacco thrips)—basically polyphagous; S.E. Asia.

Thrips simplex (gladiolus thrips)—flowers of gladiolus, lilies, pea, beans, etc.; Europe, N.E. Africa, Asia, North America.

Thrips tabaci (onion thrips)—onions, cotton, tomato, lettuce, tobacco, and many other crops; widespread, abundant, and polyphagous; also virus vector; cosmopolitan (C.I.E. Map No. A. 20).

Thrips spp.—several hundred species known, on a wide range of host plants; worldwide but most abundant in temperate regions.

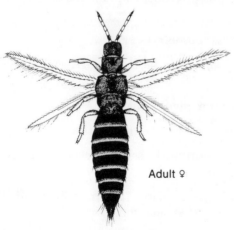

Adult ♀

Fig. 9.3 Flower thrips (*Frankliniella* sp.) (Thysanoptera; Thripidae).

0   1 mm

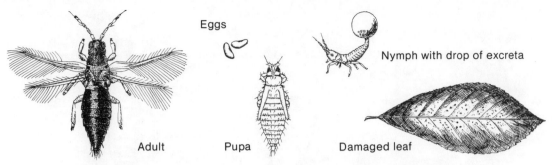

Fig. 9.4 Black tea thrips (*Heliothrips haemorrhoidalis*) (Thysanoptera; Thripidae).

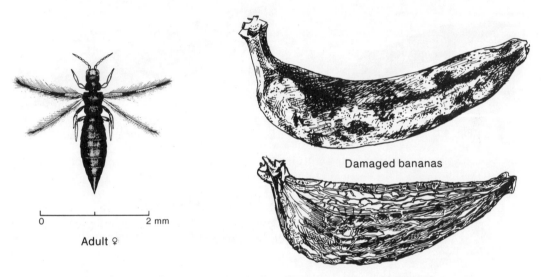

Fig. 9.5 Banana/smilax thrips (*Hercinothrips bicinctus*) (Thysanoptera; Thripidae).

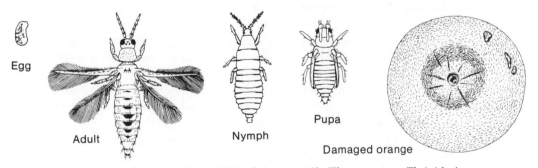

Fig. 9.6 Citrus/tea thrips (*Scirtothrips aurantii*) (Thysanoptera; Thripidae).

MEROTHRIPOIDEA

Ovipositor reduced or absent; antennae 8–9 segmented.

Merothripidae (fungus thrips)   *23 spp.*

Nearly all the species are placed in *Merothrips;* they are widely distributed, found in leaf litter and on dead wood; many species with apterous morphs.

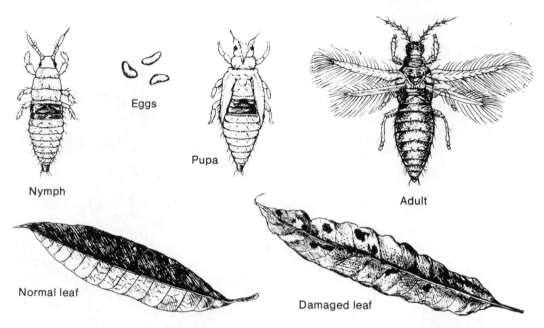

Eggs

Pupa

Nymph

Adult

Normal leaf

Damaged leaf

Fig. 9.7   Red-banded thrips (*Selenothrips rubrocinctus*) (Thysanoptera; Thripidae).

SUBORDER TUBULIFERA

Ovipositor absent, but tenth abdominal segment tubular; wings without veins or microtrichia.

Phlaeothripidae (leaf-rolling thrips, etc.)   *c. 2000 spp.*

This family is actually larger than the Thripidae, with some 200 genera, showing a diversity of habits. Most are phytophagous, and some cause leaf rolling or shoot deformation, and some are agricultural pests. *Cryptothrips, Phlaeothrips,* and others live on the bark of trees; several genera inhabit plant galls; *Megathrips* and others feed on fungal spores under bark, or on dead leaves in leaf litter, and *Urothrips* are found in soil or leaf litter. Several genera (see page 250) are predacious and feed on small insects and mites, and can be part of the natural control complex of predators. Some of the tropical genera have species that are relatively very large; *Gigantothrips* up to 5 mm (0.2 in.), and *Phasmothrips* up to 12 mm (0.48 in.) in length.

Eggs are laid on the surface of the foliage, often in groups on young leaves, and the leaf grows over in a folded or distorted manner sometimes producing an ill-defined gall; within the shelter of the distorted leaf the nymphs develop. In many species there is parental attendance and the egg cluster is guarded by either the female alone or by both sexes; the young nymphs are also guarded. *Gigantothrips elegans* on *Ficus* leaves causes no distortion—the egg cluster is usually laid on the underside of a mature leaf and the large female (4–5 mm/0.16–0.2 in.) stays in attendance (Fig. 9.8).

Some of the more interesting species of Phlaeothripidae recorded on plants of agricultural importance are listed below:

Gigantothrips elegans (giant fig thrips)—under leaves of *Ficus;* tropical Asia, and also
    Africa. There are 20 species known—mostly on *Ficus* (Fig. 9.8).

Gynaikothrips ficorum (banyan leaf-rolling thrips)—folds leaves of *Ficus retusa* in California and the Pacific, and *F. microcarpa* in tropical Asia (Fig. 9.9).

Gynaikothrips kuwani (Cuban laurel thrips)—causes extreme leaf distortion on several ornamental shrubs in tropical Asia and the U.S.A. (Fig. 9.10); there are 80 species known in warmer temperate regions of the world.

Haplothrips articulosus (nug flower thrips)—on *Guizotia,* pigeon pea, etc.; Ethiopia.

Haplothrips froggatti—on grapevine; Australia.

Haplothrips tritici (wheat leaf-rolling thrips)—on wheat; Europe.

Haplothrips spp.—300 species are known; some are recorded from a wide range of crops and host plants, including mango, rice, clovers, Compositae, etc.; the genus is found worldwide.

Hoplandothrips marshalli (coffee leaf-rolling thrips)—only recorded on coffee; East Africa (Fig. 9.11). All the other species (70) are fungus-feeders.

Liothrips oleae (olive leaf thrips)—roll leaves of olive; Mediterranean Region and eastern Africa.

Liothrips spp. (200) (leaf-roll thrips)—many trees and shrubs, grapevine, etc., are attacked (fig, oaks, hickory, *Piper,* willow, etc.); worldwide.

Fig. 9.8 Giant fig thrips (*Gigantothrips elegans*) (Thysanoptera; Phlaeothripidae); adult female guarding egg cluster underneath a leaf of *Ficus microcarpa;* Hong Kong.

Fig. 9.9 Banyan leaf-rolling thrips (*Gynaikothrips ficorum*) on *Ficus microcarpa;* South China (Thysanoptera; Phlaeothripidae).

Fig. 9.10   Cuban laurel thrips (*Gynaikothrips kuwani*) causing extreme leaf deformation on *Aporusa chinensis;* South China (Thysanoptera; Phlaeothripidae).

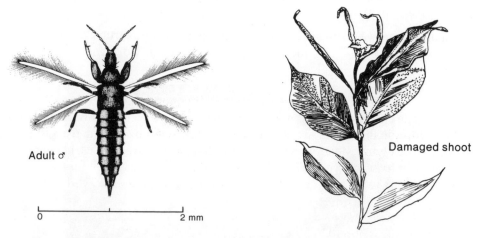

Fig. 9.11   Coffee leaf-rolling thrips (*Hoplandothrips marshalli*) (Thysanoptera; Phlaeothripidae).

Because of the number of predacious species involved, and their possible importance, they are listed separately below:

Aleurodothrips fasciatus (whitefly thrips)—found on *Citrus* where they prey on Diaspididae.

Haplothrips spp.—a few of the many species in this genus are recorded feeding on small insects.

Karnyothrips spp. (15)—predators of scale insects; worldwide.

Leptothrips spp. (20) (black hunter, etc.)—predatory on mites; New World.

Podothrips spp. (20)—some are predatory on Diaspididae; pantropical.

In the C.A.B.I. *Guide to Thysanoptera of Importance to Man* (Palmer, Mound & du Heaume, 1989) a number of other genera are included together with both taxonomic detail and basic biological information.

10

ORDER COLEOPTERA
beetles
95: 330,000

This is the largest order of insects, and is also the largest group in the animal king-dom, with many new species being described annually. In size, they range from minute (0.5 mm/0.02 in. body length) to large (150 mm/6 in. long); they are characterized by having the forewings, which are not used in flight, modified into thick, hard, horny elytra, which, at rest, meet down the midline of the body. The elytra are protective and cover the large folded membranous hindwings, although some species are apterous. Mouthparts are the biting and chewing type and basically unspecialized. The prothorax is usually large and mobile. The mesothorax lies under the elytra and is much reduced, and the abdominal tergites are often only a little sclerotized and usually protected dorsally by the elytra. Metamorphosis is complete; larvae are very varied in form, from campodeiform to eruciform and in some groups apodous; usually with well-developed mandibles. Sometimes the larvae share the same habitat as the adults, and both may be crop pests, but usually there is great ecological diversity between larvae and adults. There is so much diversity shown by the 330,000 species known that it is only possible to generalize broadly about their habits, etc., at family level. In general there is con-siderable similarity between the adults of each family, and usually a striking similarity between the larvae.

Many species are agricultural pests, sometimes it is the larva, sometimes the adult, sometimes both together, sometimes both but in different ways. Very few species cause any damage medically or veterinarily. Damage to cultivated plants is invariably direct, by the removal of pieces of tissue by biting with the mandibles, but a few species are vectors of fungi. The great majority of beetles are associated with the ground, either in the soil and leaf litter, or feeding on decaying plant and animal material there, or with fungi. But a large number of species are phytophagous, and all parts of the plant body are utilized by them as food. Some groups are entirely aquatic. Some species are so successful living terrestrially that they can dwell in hot deserts. Because of their cryptic, and often noc-turnal, habits, beetles are not seen as frequently as one might expect.

In this account only the families of agricultural importance, in the widest sense, are mentioned; for further details about the Coleoptera, Imms' textbook (Richards & Davies, 1977) and the tome by Crowson (1981) should be consulted.

SUBORDER ARCHOSTEMATA

SUBORDER MYXOPHAGA

These groups contain a few small families with various primitive characteristics and are somewhat specialized, and of no particular importance agriculturally.

SUBORDER ADEPHAGA

A large group of predacious beetles, somewhat specialized for an active predatory life, but in other respects (particularly larval structure) to be regarded as primitive. A few species have become secondarily phytophagous. The group contains one superfamily and eight families.

CARABOIDEA

Carabidae (ground beetles)   *25,000 spp.*

A very large group, worldwide in distribution, in temperate regions they are all ground beetles, to be found in leaf litter, under stones and logs, under bark, and in the soil. In the tropics many species are arboreal and good fliers. Most are black in color but the tiger beetles (Cicindelinae) are brightly colored and often metallic. Some of the ground beetles (*s.s.*) are wingless, with the elytra fused together, and with legs adapted for digging. Both larvae and adults are fiercely predacious on other insects and small invertebrates and they are an important group of natural predators in virtually all agricultural situations, eating eggs, larvae, pupae, and adults of many crop pests, depending in part on the relative sizes of the insects concerned. A typical ground beetle is shown in Fig. 10.1. The species that have become secondarily vegetarian feed mostly on seeds and are regarded as minor pests on several crops in the U.S.A. and Europe. *Stenolophus* and *Clivina* are seedcorn beetles and eat sown cereal grains; in the U.K. and Europe *Harpalus rufipes* is the strawberry seed beetle (it removes the seeds from the surface of ripe fruits) and several species of *Pterostichus* (= *Feronia*) are strawberry fruit beetles and they usually eat parts of ripe fruits. *Clivina rugithorax* in New Zealand is recorded preventing germination of maize seeds and also damaging strawberry fruits. *Harpalus natalensis* damages young beetroot in South Africa. In Europe, *Zabrus tenebroides* larvae regularly damage young cereals.

Eggs (simple, white, and ovoid) are laid in the soil, and larval development is usually slow; typical larvae are elongate and tapering, with ten abdominal segments, well developed legs terminating in one or two claws. The mandibles are sharp and sickle-shaped, and there are six ocelli on each side of the head. There are three larval instars and larval development often takes 6–12 months. Pupation takes place in the soil.

In the new edition of Imms (1977) the tiger beetles (Cicindelidae) have been relegated to a subfamily within the Carabidae; there are some 2000 species of tiger beetles worldwide, many brightly colored, with long legs, prominent eyes and large mandibles; most are tropical and subtropical but a few are found in temperate regions. *Cicindela* is mostly terrestrial (Fig. 10.2) dwelling on open ground, footpaths and the like, where they run, jump, and fly for short distances. The larvae live in burrows in the ground, often a bank surface (Fig. 10.3) where they lurk at the tunnel entrance waiting for prey to pass. Species of *Cicindela* occur on sandy beaches in most parts of the world, as part of the littoral insect fauna. The genus *Collyris* (and some close relatives) is unusual in that the larvae are said to tunnel into living twigs and branches of various shrubs, with the entrance directed downwards. Several species of *Collyris* and *Neocollyris* are recorded as minor pests on tea, coffee, cocoa, cotton and jujube bushes in different parts of India and S.E. Asia. Also in S.E. Asia the black flightless tiger beetle *Tricondyla pulchripes* (Fig. 10.4) is an arboreal species that is quite common and its larvae are also twig tunnellers, and at least one species is found damaging cocoa bushes in Java.

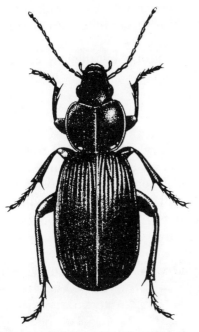

Fig. 10.1 Typical carabid beetle, strawberry seed beetle (*Harpalus rufipes*) (Coleoptera; Carabidae); from Cambridge, U.K.

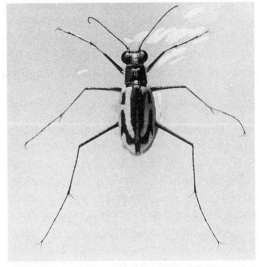

Fig. 10.2 Typical tiger beetle adult (*Cicindela anchoralis*) (Coleoptera; Carabidae); body length 14 mm (0.56 in.); South China.

Fig. 10.3 Breeding site for blue-spotted tiger beetle (*Cicindela separata*); larval burrows in earth bank; Stonecutters Island, Hong Kong; inserts of both a larva and an adult beetle.

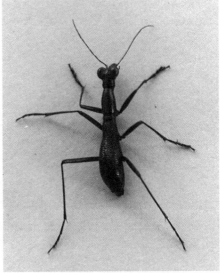

Fig. 10.4 Black flightless tiger beetle (*Tricondyla pulchripes*) (Coleoptera; Carabidae); body length 18 mm (0.7 in.); Hong Kong.

The other subfamilies in the Carabidae are not so distinctive and easy to recognize, and there is considerable dispute over their constitution, and so they will not be mentioned here. Carabid beetles in the tropics range in size from 10–40 mm (0.4–1.6 in.), and there is great diversity in coloration from black (dull and shiny) to brown with green elytra, or entirely shiny blue or green. Some are found in low vegetation, or on the ground, but many are arboreal. It has been suggested that Carabidae be divided into three different ecological groups, first, the *geophiles* which live on the ground (and away from water); second, the *hydrophiles* living on the seashores, edges of ponds, streams or swamps; and third, the *arbicoles* living on the trunks of trees, in foliage. There are so many different genera of Carabidae found throughout the tropics that it is not feasible to make specific references. The main point is that collectively they are important as part of the natural predator complex exerting pressure on many insect pest populations. One notable case was the introduction of *Pheropsophus hilaris* from India into Mauritius to control larvae of *Oryctes* (rhinoceros beetle).

Four other small families are also placed in this group, but they are all basically water beetles and of no importance agriculturally.

Dytiscidae (true water beetles)   *4000 spp.*

A worldwide family, but predominantly Palaearctic; fiercely predacious beetles, some large in size. Both adults and larvae are predatory, and species of *Dytiscus* and *Cybister* will kill small fish and frogs. In paddy fields these beetles will be mostly beneficial in that most of their prey will be insects and some of them pests. But in fish ponds the large species are regarded as pests as they will kill small fish very readily. A typical adult and a larva of a large tropical species of *Cybister* are shown in Figs. 10.5, 10.6.

Fig. 10.5 Common large water beetle (*Cybister tripunctatus*) (Coleoptera; Dytiscidae); body length 28 mm (1.2 in.); South China.

Fig. 10.6 Larva of large water beetle (*Cybister tripunctatus*); body length 50 mm (2 in.); from a fish pond in South China.

Gyrinidae (whirligig beetles) *700 spp.*

These are water surface swimmers, often gregarious, that will dive when pursued. The larvae are elongate, predacious, with long plumose abdominal tracheal gills. In paddy fields both adults and larvae will kill small rice insects. Adults are small (4–8 mm/0.16–0.32 in.), shiny and black. Pupation takes place in a cocoon attached to aerial leaves of water plants.

SUBORDER POLYPHAGA

The great majority of beetles are placed in this suborder and their grouping into superfamilies is difficult for the nonexpert to follow, and in various textbooks different systems of classification have been used. The system used in Imms (1977) is a modification of that proposed by Crowson (1968).

HYDROPHILOIDEA

These are mostly water beetles, characterized by having elongate maxillary palps functioning as antennae, and the antennae are used in respiration. They are less adapted for an aquatic life than are the Dytiscidae. There are five families in this group but only one of any importance.

Hydrophilidae (scavenging water beetles) *2000 spp.*

A large family, especially abundant in the tropics; the adults feed on decaying vegetable matter, as do most of the larvae, but *Hydrophilus* larvae are carnivorous. Quite a large number of species are terrestrial, and to be found in damp places or in rotting vegetable matter. *Dactylosternum* is found in rotting plants of banana and sisal in East Africa. In temperate regions, several species of *Helophorus* are minor crop pests: *H. nubilis* (wheat shoot beetle) larvae eat the stem of wheat (and sometimes oat) seedlings below soil level in Europe, and *H. porculus* and *H. rugosus* are the turnip mud beetles whose larvae tunnel in the roots and stems of turnips, cabbages, lettuce, and beans.

HISTEROIDEA

A group of predacious insects, comprising three families, but only one of any importance.

Histeridae *2500 spp.*

A large group of predatory beetles of compact shape, black, hard, and shining, with geniculate and clubbed antennae. A few species are brown or have red-marked elytra, and some are metallic. The elytra are terminally truncated and leave the two apical segments of the abdomen exposed. When alarmed they simulate death and closely retract legs and antennae under the body. Some live in dung, others under tree bark, and have flattened bodies; some have cyclindrical bodies and frequent tunnels in wood bored by other insects. The larvae are also carnivorous and prey on caterpillars, fly larvae, and other beetles. *Niponius* (Himalayas, Indonesia, Japan) has larvae that prey on larval Scolytinae in their tunnels. *Plaesius javanus* (Fig. 10.7) larvae and adults prey on banana

Fig. 10.7   *Plaesius javanus*
adult (Coleoptera; Histeridae).

weevil (*Cosmopolites* and *Odoiporus*) larvae, and in Fiji this species was introduced in 1914 as an early and very successful case of biological control. After a period of eight years the predator population became established and controlled the population of banana weevil adequately. *Pachylistes chinensis* preys on muscoid fly larvae in fresh cow dung and has been introduced into Fiji, Samoa, Solomons, and Australia in successful attempts to reduce the dung-fly populations—these flies are veterinary, medical, and urban pests.

STAPHYLINOIDEA

A very large group, characterized by having short elytra, and a series of special anatomical features. Mostly they are fungus feeders or predacious; only a few are phytophagous, and very few inhabit dry areas. Nine families are included; some are very small and of no agricultural consequence and are not included here.

Leptinidae (parasitic beetles)

A small group of ectoparasitic beetles, eyeless and often wingless; *Leptinus* is found in nests of *Bombus* in Europe, and in burrows of small rodents. *Platypsyllus castoris* lives on the beaver in Europe and North America, and *Leptinillus* also lives on the American beaver.

Anisotomidae (= Leiodidae)   *1300 spp.*

These were formerly placed in the Silphidae, but are distinguished by having a five-segmented antennal club. Some are fungus-feeding and the larvae live underground; adults are nocturnal. Others are scavengers, in the nests of mammals, other insects, and one group is cave-dwelling.

Silphidae (carrion beetles)   *200 spp.*

Large beetles, mostly found in the Holarctic region, and most are carrion feeders. *Nicrophorus* includes the well-known orange and black burying beetles of Europe. A few species have predacious larvae—*Phosphuga* attacks snails, and *Xylodrepa* preys on caterpillars. The one genus *Aclypea* is vegetarian and damages beet and other root crops and some cereals in Europe; adults and some larvae eat the leaves, and other larvae in the soil eat the roots (Fig. 10.8).

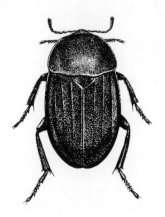

Fig. 10.8 Beet carrion beetle (*Aclypea opaca*) (Coleoptera; Silphidae).

Staphylinidae (rove beetles)   *27,000 spp.*

A very distinctive group of ground beetles with very short elytra, concealing large and well-developed wings, and they fly readily and well. Body form varies somewhat but most are elongate; some are brightly colored but most are black. The majority are small and inconspicuous, but the European *Ocypus olens* reaches 28 mm (1.12 in.) in body length. Most species are predacious, many being myrmecophilous and another group is termitophilous. The adults do not appear to show particular prey preferences except for a size preference since many of them are small. The larvae are campodeiform and very active in soil and litter. In Europe, the larvae of some species of *Aleochara* are pupal parasites of Cyclorrhaphous Diptera and are important in the natural control of some root flies (Anthomyiidae); and in Java another species attacks *Lyperosia* (stable flies). A very widespread genus in the tropics is the small black and orange colored *Paederus*; in South China, paddy fields were sometimes literally covered around the edges with *P. fuscipes,* which could also walk on the water surface; in East Africa another species is known as the "Nairobi eye-fly"—it flies to lights at night and if handled secretes a noxious fluid, and if the fluid on the fingers comes into contact with the eyes the result is extreme irritation. In Indonesia, *P. fuscipes* adults have been observed attacking stem borer eggs and moths, but mostly they prey on Cicadellidae.

In summary, it is evident that in many agricultural situations the staphylinid beetles are important as part of the natural enemy group of various crop pests.

SCARABAEOIDEA

One of the most distinctive groups of Coleoptera and one of the most important agriculturally as there are many pest species. The adults are stout-bodied, compact insects, with short clubbed antennae and toothed front tibiae; the eighth abdominal tergite forms an exposed pygidium; most are fossorial in habits. They walk rather clumsily but fly well—only a very few species are apterous. Sexual dimorphism is pronounced, and the differences may affect almost any part of the body, but more usually the head and mandibles, and some males have enormous antler-like processes protruding anteriorly (Fig. 10.13). The largest species grow to a body length (head to abdomen tip) of 80–100 mm (3–4 in.) (Fig. 10.11).

Eggs are large, relatively few in number, usually laid in the soil or leaf litter. As with most Coleoptera there are three larval instars and larval development is very slow, usually taking 1–2 years. The larvae are all very similar in appearance and are characteristically termed scarabaeiform—a typical larva is soft-bodied, fleshy, with a swollen abdomen, thoracic legs, a brown, well sclerotized head capsule and stout mandibles; the body at rest is held in a C-shaped position (see Fig. 10.17). The larvae generally lie on their back or side and seldom move more than a few centimeters, often being surrounded by food material. In many groups, both adults and larvae can stridulate, but do so rather quietly.

Passalidae (sugar beetles)   *500 spp.*

These beetles are quite large and the body is flattened and more or less parallel-sided; black or brown in color, and with elytra (which cover the abdomen) with deep, longitudinal striations. Adults are often found under loose tree bark in forests in the tropics. Both adults and larvae eat rotting wood, and both stridulate. The larvae are less crescentric than most scarabaeids. It is reported that the adult beetles actually feed the larvae, by chewing and partly digesting the rotten wood, but this is not confirmed, although adults and larvae are generally found living together. These beetles should not be confused with tree pests; if found on a tree they are only there because of the prescence of rotten wood.

Lucanidae (stag beetles)   *750 spp.*

The elytra cover the entire abdomen in this group too, but they are not longitudinally striate. The antennal club is loose. Males have enlarged mandibles; in some species they are as long as the rest of the body; the male is invariably larger than the female. The more temperate *Lucanus* has very large mandibles, but the tropical *Prosopocoilus* has a less spectacular dimorphism. Larvae develop in rotting wood (trees or roots) on which they feed; some tropical species develop in one year but the temperate *Lucanus* is thought to require four years, and others have been recorded as taking six years. Pupation takes place in a cell composed of gnawed wood fragments. These are not pest species in that they only occur in well rotten wood. Some of the adults are nectar feeders and can be seen on flowers or feeding on exuding sap from tree trunk wounds. In Indonesia, *Eurytrachelus* and *Odontolabis* (10 cm/4 in. long) are recorded from the inflorescence of coconut and *Aegus* (2–3 cm/0.8–1.2 in.) on coffee flowers.

Geotrupidae ("dor" beetles)   *300 spp.*

These beetles resemble Scarabaeids but the elytra cover the abdomen tip; the spiracles are all in the pleural membrane; the scutellum is large and obvious; and the antennae are composed of 11 segments. They are coprophilus in habits, strongly convex, black or brown in color, and part of the usual "dung beetle" complex. The adult beetle digs several burrows several centimeters deep under a patch of dung in which eggs are laid. The burrows are provisioned with pieces of dung plugged at the blind end on which a single egg is laid per plug.

Scarabaeidae (chafers; cockchafers; June beetles, etc.; white grubs, etc.)   *17,000 spp.*

A very large group, worldwide in distribution, stout-bodied, often convex, sometimes black in color otherwise they may be very brightly colored. There may be sexual dimorphism but the male mandibles are not enlarged; some spiracles lie in the abdominal sternites (of which six are visible); antennae have 8–10 segments. The larvae may be parasitized by Scoliidae (Hymenoptera)—*Megascolia azurea* is widespread from

India through S.E. Asia to China (page 534) and attacks larvae of many different Scarabaeidae found in rubbish heaps or compost. Various Scoliidae have been used in biological control projects against chafer grubs—sometimes with considerable success.

The family is composed of six large and well-defined subfamilies, and several smaller ones; some authorities prefer to regard these groups as separate families but here we are following Imms (1977) in regarding them as subfamilies. They are, however, sufficiently distinct both anatomically and biologically to warrant separate treatment in the text.

Cetoniinae (rose or flower chafers, etc.)   *2600 spp.*

Brightly colored, often metallic, diurnal insects with weak mouthparts, that feed on nectar, pollen, and soft overripe fruits. The mandibles are usually thin and incapable of biting. The adults are rather flattened dorsally, and the elytra are incurved (emarginate) level with the hind legs; in flight the elytra are only slightly raised and the wings protrude through these lateral emarginations. Adults of some species (*Protaetia* especially) can be seen flying over areas of grassland (or lawns) on sunny days looking for oviposition sites. Larvae mostly develop in turf, under grassland, feeding on roots and decaying vegetable matter, but some genera apparently prefer refuse heaps, and a few prefer dead, rotten wood. The genus *Agestrata* is atypical in that they can penetrate plant tissues; *A. orichalcea* is a large, handsome, golden-green beetle, 4 cm (1.6 in.) long (Fig. 10.9) recorded from India, throughout S.E. Asia to East China; in Indonesia the larvae are found in the growing point of *Pandanus* (screw pine) but it is not clear whether they are the primary infestation or whether they are there following damage by pyralid caterpillars (*Acara*). The adult beetle is recorded feeding on overripe wild figs in China, but in India it is said to feed on pollen stores of *Apis indica* in hives. One tribe within this group (Cremastochilini) are unusual in that both adults and larvae live in the nests of ants and termites.

A very widespread oriental genus is *Protaetia*—recorded from India to Japan. They range in color from shiny green (*P. orientalis*—Fig. 10.10) to bronze and shiny blue/black (*P. impavida*—India). A very spectacular genus in the Cetoniinae is *Goliathus* (goliath beetles)—several species occur throughout tropical Africa and Asia, with a body length

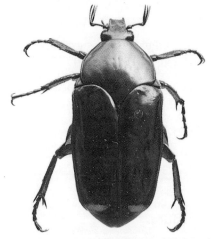

Fig. 10.10 Oriental rose chafer (*Protaetia orientalis*) (Coleoptera; Scarabaeidae; Cetoniinae).

Fig. 10.9 Large green flower chafer (*Agestrata orichalcea*) (Coleoptera; Scarabaeidae; Cetoniinae); body length 40 mm (1.6 in.); South China.

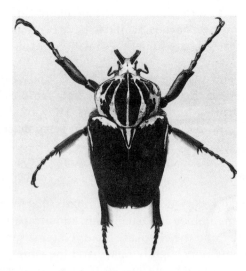

Fig. 10.11 Goliath beetle (*Goliathus go-liathus*); length 100 mm (4 in.); Mbira Forest, Uganda (Coleoptera; Scarabaeidae; Cetoniinae).

of up to 11 cm (4.4 in.); basically native to tropical rain forests where they live in the canopy and are seldom seen. Fig. 10.11 shows *Goliathus goliathus* from the Mbira Forest in Uganda; at 9–12 cm (3.6–4.8 in.) in body length this is the largest species; the larvae are said to live in rotten wood of dead trees in the forest.

Despite the weak mouthparts possessed by the adult beetles, some are recorded as pests for they destroy the anthers and pistil of flowers in their quest for nectar and pollen. Several species of *Oxycetonia* and *Protaetia* destroy flowers of fruit trees in India and Japan, especially *Citrus,* cherry, apple, peach, pear, and plum. *Protaetia* beetles have been seen eating ripe papaya in the Seychelles, and also feeding in the ear heads and tassels of maize. Other species of *Protaetia* in the larval stage are recorded damaging roots of sugarcane in Java. The genus *Pachnoda* is native to Africa and Arabia where it occurs as a large number of species; the adult beetles are recorded damaging flowers and fruits of quite a wide range of cultivated plants, including peach, apricot, cotton, citrus, roses, mango, *Acacia,* maize, sunflower, and the milky grains of sorghum and pearl millet, and many garden flowers. Larvae have been found in forest leaf litter, shallowly in the soil.

A few other genera of Cetoniinae recorded as pests of some importance in Africa and the Near East are included below—they are almost all recorded damaging the flowers of crops and other cultivated plants; one or two species are reported to damage leaves and unripe cotton bolls.

Cetonia aurata (European rose chafer)—larvae in soil eat roots of grasses and some crop plants; adults can destroy flowers; Europe (Fig. 10.12).

Diplognatha spp.—on cotton, mango, coconut, citrus, coffee and roses; East, West and South Africa.

Epicometis hirta—adults eat pollen of many flowers; Near East.

Leucocelis spp.—on coffee, citrus, apple, cotton, maize (silks); East, Central and West Africa.

Oxythyrea spp.—Near East.

Polyplastus spp.—on wheat and *Hibiscus;* East Africa.

Rhabdotis spp. on mango, castor, roses; East Africa.

Tropinota spp.—Near East.

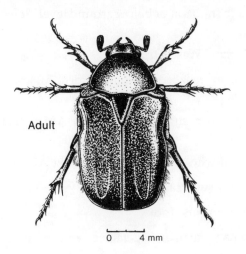

Adult

0   4 mm

Fig. 10.12 European rose chafer (*Cetonia aurata*) (Coleoptera; Scarabaeidae; Cetoniinae).

Dynastinae (rhinoceros/unicorn beetles)   *1400 spp.*

These are among the largest beetles and the most spectacular insects known; usually black in color and nocturnal in habits, often with striking sexual dimorphism. The adults usually have a "horn" on the head, either slender and recurved or bifurcated, together with a single or paired process protruding from the prothorax. In a few cases (*Oryctes rhinoceros*) both sexes have the frontal horn. They are entirely tropical (one species in southern Europe) but most abundant in the Neotropical region. Adults have well-developed mouthparts and they damage various plants by biting foliage, stems, and tubers, sometimes even making short tunnels. The larvae live in soil and rotting vegetation and some damage roots of various plants. The two crops most heavily attacked are coconut (and oil palm) and sugarcane. Box (1953) lists 65 species of Dynastinae recorded damaging sugarcane worldwide. The palms are damaged by the adult beetles boring in the crown, and the cane is damaged by adults eating into shoots at or below ground level, and the larvae eating the roots in the soil. Yams are also heavily damaged, by the feeding adults making shallow tunnels in the tubers underground. The adults could be regarded as being of two types—the smaller, rounded species (*Heteronychus*, etc.) being active burrowers that attack various crop plants underground; and the larger spectacularily horned "rhinoceros" beetles (*Dynastes, Chalcosoma, Oryctes,* etc.) decidedly more aerial and usually found on palm crowns. Some larvae live entirely in rotten wood, often preferring dead palm trunks, others almost entirely in refuse dumps, compost heaps, and piles of rotting vegetation, and a few live in grassland turf (but not so many as Melolonthinae). Some species have larvae that are apparently more adaptable and opportunistic and can be found in a diversity of habitats, and they may be pests.

The larvae of many Dynastinae are regularly parasitized by species of Scoliidae (Hymenoptera), as are other Scarabaeidae; one successful case of biological control was started in 1917 when *Scolia oryctophaga* was introduced into Mauritius from its native Madagascar, to control *Oryctes tarandus* which was causing serious damage to the sugarcane industry.

Some of the notable pests include the following:

Chalcosoma atlas (Atlas beetle)—adults damage coconut flowers and young fruits; India, S.E. Asia, Indonesia (Fig. 10.13).

Dynastes spp. (Hercules beetles)—U.S.A. and Central and South America.

Dyscinetus spp.—adults bore into sugarcane shoots at or below ground level; South America.

Heteroligus spp. (yam beetles, etc.)—on yams; West Africa.

Heteronychus arator (black maize beetle)—on maize, etc.; South Africa, Australia, New Zealand (C.I.E. Map No. A. 163).

Heteronychus consimilis (sugarcane beetle)—on sugarcane; tropical Africa (Fig. 10.14).

Heteronychus licas (black wheat beetle)—on wheat, sugarcane, etc.; East Africa.

Heteronychus pauper (sugarcane beetle)—on sugarcane; S.E. Asia.

Heteronychus poropygus—larvae damage rice roots; India.

Heteronychus sanctae-helenae—larvae damage maize, vegetables, and pastures; Australia.

Ligyrus spp.—adults damage *Xanthosoma* and larvae on sugarcane roots; West Indies, Central and South America.

Oryctes boas (African rhinoceros beetle)—adults gnaw crown of coconut and other Palmae, may destroy growing point; tropical Africa (C.I.E. Map No. A. 298).

Oryctes monoceros (African rhinoceros beetle)—on Palmae; tropical Africa, Madagascar, Mauritius, Seychelles (C.I.E. Map No. A. 188) (Fig. 10.15).

Oryctes rhinoceros (Asian rhinoceros beetle)—on Palmae; India, S.E. Asia, Taiwan, Fiji, Samoa, etc. (C.I.E. Map No. A. 54).

Oryctes spp.—about 40 species are known, quite similar in appearance, from Africa and southern Asia, most recorded feeding on Palmae, but also on sugarcane.

Papuana laevipennis (taro beetle)—adults damage taro, coconut, banana; Solomon Islands, Papua New Guinea.

Papuana spp.—several other species also damage coconut, banana; larvae attack groundnuts and various vegetables; Papua New Guinea and Solomon Islands.

Phyllognatha dionysius—damages paddy rice; India.

Prionoryctes caniculus (yam beetle)—adults eat tubers of yam underground; tropical Africa.

Scapanes australis (and spp.)—adults damage coconut and bore banana pseudo-stem; Australia, Papua New Guinea.

Strategus spp.—adults bore pineapple stems in South America, and damage sugarcane in the West Indies and South America.

Xylotrupes gideon (unicorn/elephant beetle)—adults damage foliage of coconut, rubber, sugarcane, citrus, banana, etc.; India, S.E. Asia, South China, Papua New Guinea, Australia (Fig. 10.16).

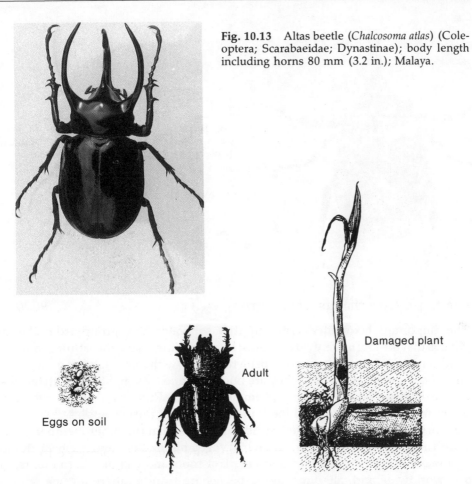

Fig. 10.13 Altas beetle (*Chalcosoma atlas*) (Coleoptera; Scarabaeidae; Dynastinae); body length including horns 80 mm (3.2 in.); Malaya.

Eggs on soil

Adult

Damaged plant

Fig. 10.14 Black cereal/sugarcane beetle (*Heteronychus* sp.) (Coleoptera; Scarabaeidae; Dynastinae).

Damage

Adult ♀   ♂

Larva

0   1 cm

Fig. 10.15 African rhinoceros beetles (*Oryctes monoceros*); Mahé, Seychelles (Coleoptera; Scarabaeidae; Dynastinae).

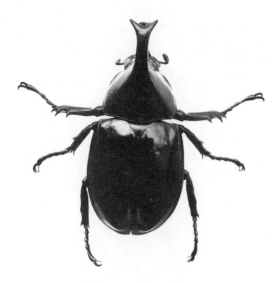

Fig. 10.16 Unicorn/elephant beetle (*Xylotrupes gideon*); body length 45 mm (1.8 in.); South China (Coleoptera; Scarabaeidae; Dynastinae).

Melolonthinae (cockchafers; chafer grubs, etc.) *9000 spp.*

The important taxonomic criterion that separates this group from the two preceding is that the labrum is evident and sclerotized (they have the labrum concealed and membranous). Most adults are dull brown in color, with a fat, rounded body; during flight, the elytra are held vertically; they are nocturnal and fly to lights at night; the claws of the hind legs are equal in size and immovable (cf. Rutelinae); they have strong mandibles and can bite quite hard food, including small, unripe, apples.

The larvae are generally more injurious to crops than the other members of this family, and some prefer to live in decaying vegetable matter (and rubbish dumps) although many live in soil and turf and eat plant roots and will bore into tubers. The group is worldwide and, although most species are tropical, there are many in the Holarctic Region, and in North America they are major pests of open grassland. Although most damage is done by the larvae, some damage is done by adults that eat leaves, flowers, and sometimes fruits.

Larval development is often protracted—in the tropics it may be as short as six months but is often about a year; in cooler climates, larval development takes 1–3 years (*Melolontha* in Europe, and *Phyllophaga* in Canada take three years).

Generally, each large country or region throughout the agricultural world has its populations of cockchafers collectively causing significant economic damage to crops, but the constitution of the cockchafer/white grub complex varies locally—some genera are confined to a single country, (for example, *Cochliotis* to Tanzania, *Dermolepida* to Australia, and *Leucopholis* to the Philippines and Indonesia). Most of the different genera are certainly restricted to individual continents so far as is known (although *Serica* is found in both Africa and India); this is, in part, because cockchafers are not migratory or long-distance fliers and most adults probably do not fly more than a kilometer or so in their adult life. Of course, a few are transported accidentally with agricultural produce, but this is seldom recorded for Scarabaeidae. There is some dispute over generic ranges in relation to taxonomy, and it has been suggested that *Holotrichia* is Holarctic in distribution, but the usual interpretation is that *Holotrichia* is the Old World genus (Asiatic), and the New World counterpart is *Phyllophaga,* of which there are many pest species, especially in Canada.

The more important and widespread pest species of Melolonthinae are listed below:

Apogonia spp.—larvae are pests of sugarcane, adults eat various plants; Indonesia.

Amphimallon spp. (European/American chafers, etc.)—pests of pastures and several crops; Europe, Asia, U.S.A. and Canada (C.I.E. Maps Nos. A. 371 & 391).

Clemora smithi—a serious pest of sugarcane; Mauritius, West Indies, Central America.

Cnemarachis spp.—West Indies, Central and South America.

Cochliotis melolonthoides (sugarcane white grub)—a devastating pest of sugarcane, but only abundant in Tanzania (Fig. 10.17).

Dermolepida spp. (sugarcane beetles/greybacks)—on sugarcane; Australia, Indonesia.

Exopholis hypoleuca and spp.—larvae polyphagous; Indonesia.

Holotrichia spp. (chafers; chafer grubs)—polyphagous on many crops; Europe, Asia, India, Indonesia (Fig. 10.18).

Lepidiota spp. (sugarcane white grubs)—in pastures, sugarcane, and many other crops; India, tropical Asia, Philippines, Australasia.

Leucopholis spp. (chafers/white grubs)—larvae polyphagous on many crops; India, Java, Philippines.

Melolontha spp. (cockchafers)—larvae polyphagous root eaters, adults bite pieces from leaves and young apples; Palaearctic (C.I.E. Maps Nos. A. 193 & 194) (Fig. 10.19).

Odontria zealandica—polyphagous; New Zealand.

Phyllophaga spp. (20) (June beetles)—pastures and crops damaged; North, Central, and South America.

Schizonycha spp. (chafer grubs)—polyphagous; tropical Africa (Fig. 10.20).

Serica brunnea (brown chafer)—polyphagous; Europe.

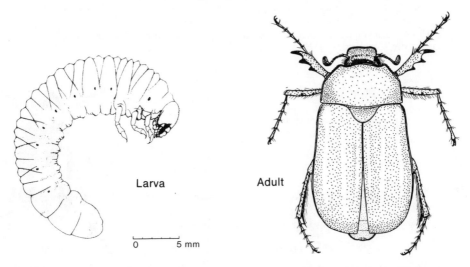

Larva   Adult

0   5 mm

Fig. 10.17 Sugarcane white grub (*Cochliotis melolonthoides*) ex Tanzania (Coleoptera; Scarabaeidae; Melolonthinae).

Serica orientalis (oriental brown chafer)—polyphagous; China.

Serica spp. (brown chafers)—polyphagous; Africa, India, and Asia.

Trochalus spp.—on various crops, and roses; East Africa.

Box (1953) lists over 80 species of Melolonthinae recorded from sugarcane worldwide.

Fig. 10.18 Large brown cockchafer (*Holotrichia geilenkenseri*); body length 12 mm (0.5 in.); South China (Coleoptera; Scarabaeidae; Melolonthinae).

Damage to young apples by adult chafers

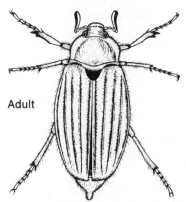

Adult

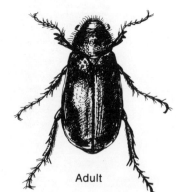

Fig. 10.19 Cockchafer (*Melolontha melolontha*) (Coleoptera; Scarabaeidae; Melolonthinae); from Cambridge, U.K.

Eggs

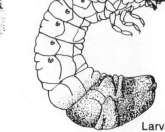

Larva

Pupa

Adult

Fig. 10.20 Small brown chafer (*Schizonycha* sp.) ex Kenya (Coleoptera; Scarabaeidae; Melolonthinae).

Rutelinae (flower beetles; June beetles; May beetles)   *2500 spp.*

The general appearance (facies) of the adults is that of the Melolonthinae, but the adults are generally brightly colored, the hind tibiae are often thickened, and the claws are mobile, long, and of unequal length. Most adults are nocturnal and hide during the day, but *Popillia* is diurnal. As with many other Scarabaeidae, most of the larvae are crop pests in that they damage plant roots by their feeding. However, the adults are also pests in that they eat leaves and flowers (see Fig. 10.22). Occasionally complete tree/bush defoliation occurs. *P. japonica* is a very serious polyphagous pest and occurs in North America in dense populations and defoliation of plants by the adult beetles is commonplace. Typical damage is shown in the Fig. 10.24 where *Popillia histeroidea* is eating a flower of *Melastoma* in China. In some countries in the Near East it has been shown that the larvae can cease their development in the absence of suitable food plants (roots) and then take three years to develop instead of the usual two, so most crop rotations are ineffective against these pests.

The emergence of the adults from the soil is strikingly synchronized, and large numbers appear simultaneously over a short period of time (sometimes a few days or a week). The common name of June/May beetles comes from the fact that at this time of the year in Europe and North America, the newly emerged adults fly to lights at night and are very conspicuous. In Australia, the Southern Hemisphere, adults emerge in December and here *Anoplognathus* are called "Christmas beetles". Some of the Melolonthinae are also called "June beetles" for similar reasons, and some species were regularly misidentified in past literature, with the result that "June Beetle" has been used for a large number of different species.

Some of the more important pest species are listed below; in some cases it is the adult beetle that is the main pest, in others the larval stage, in other cases both stages are pests.

Adoretus spp. (many) (flower beetles)—adults and larvae are polyphagous; pantropical in the Old World and now also in the U.S.A.

Anomala spp. (100+) (chafers, etc.)—a very widespread genus attacking many different plants both as adults and larvae; found throughout Africa, Asia, and the U.S.A. in both tropical and temperate regions (16 pest species recorded in Japan alone) (Figs. 10.21, 10.22).

Fig. 10.21 Green flower beetle (*Anomala cupripes*); body length 20 mm (0.8 in.); Malaya (Scarabaeidae; Coleoptera; Rutelinae).

Fig. 10.22 Green flower beetle damage to cocoa leaves; Malaya (*Anomala cupripes*—Coleoptera; Scarabaeidae; Rutelinae).

Anoplognathus spp. (Christmas beetles)—polyphagous; Australia.

Lachnosterna spp. (chafer grubs)—larvae polyphagous; U.S.A. and Canada.

Mimela spp. (chafers)—throughout Asia.

Phyllopertha horticola (garden chafer)—polyphagous; most of Europe.

Phyllopertha nazarena (Nazarene chafer)—polyphagous; Near East.

Popillia japonica (Japanese beetle)—adults polyphagous defoliators; eastern Asia, U.S.A., Canada (C.I.E. Map No. A. 16) (Fig. 10.23).

Popillia spp. (many)—adults polyphagous on many different cultivated plants (many ornamentals) throughout Africa and Asia (Fig. 10.24).

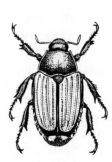

Fig. 10.23 Japanese beetle (*Popillia japonica*) (Coleoptera; Scarabaeidae; Rutelinae).

Fig. 10.24 Flower beetle adult (*Popillia histeroidea*) feeding on flower of *Melastoma;* South China (Coleoptera; Scarabaeidae; Rutelinae).

Aphodiinae   *1200 spp.*

A small group of beetles, oblong and convex in shape, small in size, with concealed labrum and mandibles. Most are found associated with dung, but some larvae feed on plant roots. No important pest species are known.

Scarabaeinae (= Coprininae) (dung beetles)   *2000 spp.*

These are the true "dung beetles", with a rounded, convex body, living entirely in dung, with membranous mandibles incapable of biting. The classical *Scarabaeus* of Egypt and its allies roll a ball of dung for some distance to a suitable location, where a burrow is dug, and in cells underground a dung plug is lodged with a single egg. In some cases, the female beetle "guards" the nest and may even tend the young. Both adults and young are coprophilous. A few genera are myrmecophilous.

In Africa the elephant dung beetle is the giant *Heliocopris* (Fig. 10.25). In Asia a similarly large species is *Cartharsius molussus,* reported in India to cover the exterior of the dung ball with clay. When first discovered by Europeans they were thought to be ancient stone cannon balls; one was recorded buried to a depth of 2 m (6 ft. 6 in.). These are the largest genera and measure up to 40–50 mm (1.6–2 in.) in body length, but some species are quite small (8–10 mm/0.32–0.4 in.).

Many species of dung beetles have been imported into Australia (as the group is poorly represented there) in an attempt to utilize the vast amounts of dung produced by the cattle and sheep, and hence to reduce the "dung" fly population which is a serious problem both agriculturally and domestically.

Fig. 10.25 Elephant dung beetle (*Heliocopris* sp.) adult; body length 65 mm (2.6 in.); Murchison Falls Park, Uganda (Coleoptera; Scarabaeidae; Scarabaeinae).

The genus *Onthophagus* causes "beetle disease" in India by invading the bowels of children; but the infestation is not damaging. In Ethiopia it was reported that adult scarabs invaded the human alimentary canal and frontal sinuses in a remote part of the country.

BUPRESTOIDEA

Buprestidae (jewel beetles; flat-headed borers)   *11,500 spp.*

A family of tropical timber beetles, with some species occurring in temperate regions. They are mostly brightly colored, some quite brilliant and iridescent, and some species are made into jewelry in S.E. Asia (Fig. 10.26). The adult body is elongate and tapering, and somewhat flattened. The larvae are quite characteristic in being legless, with a tiny head partly withdrawn into the prothorax (Fig. 10.30), and the prothorax is large, expanded, and flattened. The larvae tunnel in, or under, the bark of trees, sometimes deeply into the wood, or in the roots of trees; a few bore into stems of herbaceous plants, and a very few make leaf mines or galls. Adults are active and fly readily. Some species appear to prefer host trees that are damaged, sickly, or stressed (as do many timber beetles), but others definitely attack healthy living trees. One of the problems with timber beetles is that when infestations are evident, the trees invariably look sickly and unhealthy, and so often it is not apparent whether they are now looking sickly because of the insect damage or whether the insect damage was a consequence of the tree being sickly in the first place. There has been recent work in the U.S.A. that indicates that some *Eucalyptus* trees were attacked by timber beetles after relatively mild water stress, so it seems that the amount of stress required to make some trees vulnerable is really very low.

One of the characteristic features of larval activity is that there is no frass externally, the tunnel behind the larva is filled with tightly packed frass (Fig. 10.29) and so there are no external frass holes as there are with some Cerambycidae.

This family is striking in that there are a few genera that have very large numbers of species attributed to them; these include *Agrilus*—700 spp.; the Australian *Stigmodera* with 400; and both *Chrysobothris* and *Sphenoptera* with about 300 each. Some species of both *Agrilus* and *Chrysobothris* are found in temperate regions in Europe, Asia, and the U.S.A.; some are pests on fruit, nut, and forest trees.

Many species complete their life history in a single year, but in cooler climates two years are required. Pupation takes place in a larval tunnel just beneath the bark, and the emerging adult leaves a characteristic oval or crescentic hole, in the wood surface. The larger tropical species measure some 30–40 mm (1.2–1.6 in.) in body length, but the species in the temperate regions tend to be smaller (7–15 mm/0.3–0.6 in.), and the leaf miners smallest of all at about 3–10 mm (0.12–0.4 in.). *Sternocera aequisignata* is a common large species found throughout S.E. Asia, and made into brooches and sold in Thailand (Fig. 10.26).

The more important buprestid pests include the following:

Agrilus acutus (jute spiral borer)—on jute, okra, and other Malvaceae; India, S.E. Asia.

Agrilus auriventris (citrus borer)—on citrus; China and Japan (Fig. 10.27).

Agrilus occipitalis (citrus branch borer)—on citrus; Philippines, and Indonesia.

Agrilus spp. (700)—many species on many different hosts worldwide.

Aphanisticus spp. (sugarcane borers)—on sugarcane; Java.

Capnodis carbonaria (almond borer)—on stonefruits; Mediterranean, India, southern U.S.S.R.

Capnodis tenebrionis (peach borer)—on stonefruits; Mediterranean and southern U.S.S.R.

Capnodis spp. (fruit tree borers)—many different fruit trees (Rosaceae) bored; Africa, Europe, Asia.

Chalcophora japonica (pine buprestid)—in *Pinus;* China and Japan (Figs. 10.28, 10.29).

Chalcophora mariana (conifer buprestid)—in various conifers; Europe.

Chrysobothris spp. (300) (flat-headed borers)—attack a very wide range of hosts (deciduous trees and shrubs); worldwide (Fig. 10.30).

Chrysochroa spp.—attack a wide range of living trees; S.E. Asia.

Psiloptera argentata (silvery buprestid)—on grapevine, etc.; Near East.

Sphenoptera gossypii (cotton borer)—on cotton; N.E. Africa, India.

Sphenoptera neglecta (cotton borer)—on cotton; Africa.

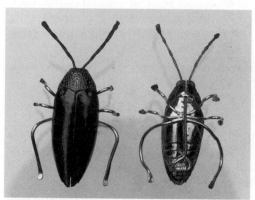

Fig. 10.26 Jewel beetles (*Sternocera aequisignata*) made into brooches; Chiang Mai, Thailand; body length 35 mm (1.4 in.) (Coleoptera; Buprestidae).

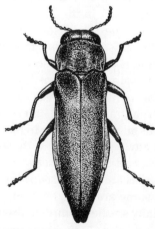

Fig. 10.27 Citrus borer (*Agrilus auriventris*) ex China (Coleoptera; Buprestidae).

Sphenoptera spp. (300)—attack a wide range of trees and shrubs (including cotton, cowpea); Africa and India.

Stigmodera spp. (400) (jewel beetles)—on many hosts; Australia.

Toxoscelus auriceps (chestnut twig borer)—on chestnut, etc.; Japan.

Trachys spp. (buprestid leaf miners)—larvae mine leaves of fruit trees, shrubs, legumes, etc.; India, S.E. Asia, Japan, and Europe (including U.K.).

Fig. 10.28 Pine buprestid (*Chalcophora japonica*); body length 35 mm (1.4 in.); from *Pinus*; South China (Coleoptera; Buprestidae).

Fig. 10.29 Pine plank bored by larva of pine buprestid (*Chalcophora japonica*) showing frass-filled tunnel (partly cleared); South China (Coleoptera; Buprestidae).

Fig. 10.30 Flat-headed borer (*Chrysobothris* sp.) adult; body length 15 mm (0.6 in.); and larva; South China (Coleoptera; Buprestidae).

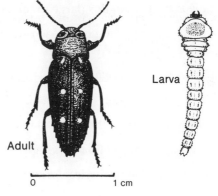

Larva

Adult

0   1 cm

ELATEROIDEA

Six families are placed here in Imms (1977), but only the Elateridae are of importance agriculturally.

Elateridae (click beetles; wireworms)   *7000 spp.*

The adult beetles have small fore-coxae and hind angles to the pronotum acute and projecting, and they have the power of leaping into the air from lying on their back using a special "click" mechanism which operates as an articulation between the pro- and mesothorax. Adults often feign death if disturbed in vegetation, but when touched they usually leap to their feet with an audible "click". Adults range in body length from 3–50 mm (0.12–2 in.); most are brown in color and cryptic in behavior, but a few tropical species are brightly metallic.

The larvae are the notorious "wireworms" of temperate agriculture. The group is worldwide, with larvae to be found in the soil, eating plant roots or boring into tubers and rhizomes. Most regions and countries have something of a wireworm problem but the pest complex is composed of different species in each region; very few species are widely distributed. Not all the larvae are strictly vegetarian; a few are wood-eating (*Melanotus*) and some species of *Athous* and *Paracalais,* etc., are carnivorous. *Paracalais* larvae prey on larger larvae boring in wood and timber in Australasia. The genus *Pyrophorus* in South America are "fireflies" and they have luminescent patches, as well as luminescent eggs and larvae. The adult beetles apparently do not feed much but may eat a certain amount of leaf material and flowers (pollen).

Development is slow and some species of *Agriotes* in Europe may require five years for complete development; and adults may overwinter in hibernation.

In Europe the largest wireworm populations are generally in pasture and grassland and most damage to cultivated crops is done when grassland is plowed and planted with potatoes, wheat, or sugar beet, etc. The group is actually most damaging to crops in the temperate regions of Europe, Asia, and North America. But damage is done in some tropical situations; Box (1953) lists some 55 species of Elateridae recorded as pests of sugarcane.

Some of the more damaging species of wireworms in the warmer parts of the world include:

Agriotes spp. (wireworms)—polyphagous; Europe, Asia, North America (Fig. 10.31).

Athous spp. (garden wireworms, etc.)—polyphagous; Europe (Fig. 10.32)

Compsosternus auratus (large green click beetle)—common species in South China.

Conoderus spp. (sugarcane wireworms)—on sugarcane roots; China, Australia, U.S.A., and West Indies.

Ctenicera (= *Corymbites*) spp.—grassland pests; Europe, U.S.A. (Fig. 10.33).

Drasterius spp.—on potatoes, sugarcane, etc.; India, Japan, U.S.A., West Indies.

Ectinus spp.—on wheat, etc.; Japan.

Heteroderes spp. = *Conoderus* spp.

Lacon spp. (tropical wireworms)—polyphagous; pantropical.

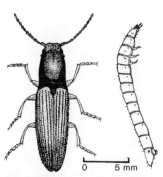

Fig. 10.31 A typical click beetle and wireworm (*Agriotes* sp.) (Coleoptera; Elateridae).

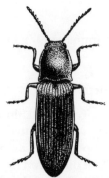

Fig. 10.32 Garden wireworm (*Athous niger*) (Coleoptera; Elateridae); from Cambridge, U.K.

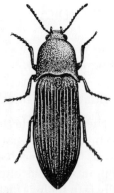

Fig. 10.33 Upland wireworm (*Ctenicera* sp.) (Coleoptera; Elateridae).

Limonius spp. (sugar beet wireworms)—on sugar beet; U.S.A.

Melanotus spp.—attack sugarcane, sweet potato, etc.; Near East, Asia, Japan, U.S.A.

Simodactylus cinnamomeus—on sugarcane; Hawaii and Fiji.

Tetralobus spp.—on coconut; East Africa.

CANTHAROIDEA

A somewhat difficult group to define—most are predacious, soft-bodied, and may show bioluminescence. There are eight or nine families placed here, but only three are of any size.

Lampyridae ("fireflies"; "glowworms") *1700 spp.*

These beetles are usually sexually dimorphic and at least one sex has luminescent organs. The adults fly freely at night and are called "fireflies", and the larvae that sit on the ground or in low vegetation are called "glowworms". The light produced is usually brightest in the female beetle. Eggs and pupae may also be luminescent to varying degrees. For the adults this light clearly brings the sexes together at night for mating, but the function in the larvae is not apparent. The adults seldom feed but the larvae are predacious on snails and slugs which they attack with their sharp, hollow, sickle-shaped mandibles. Enzymes are pumped into the body of the mollusc which is digested externally, and the liquid sucked up by the beetle larva. The larvae are clearly of value agriculturally in preying on slugs and snails.

Cantharidae (soldier beetles) *3500 spp.*

It is thought that both adults and larvae are probably predacious. Adults are often found in large numbers on flowers and in low vegetation—they are diurnal in habits. The larvae live in soil and litter and are somewhat flattened. They may well be of some importance in natural control of some insect populations.

Lycidae (net-winged beetles) *3000 spp.*

Quite a large family of diurnal beetles, mostly tropical, to be found on flowers, leaves, and under bark. Adults are often brightly colored—usually reddish brown (*Lycus* spp.)—and the elytra are often large and rather soft. Many species are apparently distasteful to birds. They are of no significance agriculturally but may often be encountered, especially on flowers. Several species of *Calochromus* are found inhabiting rotten tree stumps. Most species appear to be predacious—both adults and larvae.

DERMESTOIDEA

Four families are placed here, but only one has any economic significance; the others are myrmecophilous, or have larvae that live in slime-molds, etc.

Dermestidae (hide beetles, etc.) *700 spp.*

Small, rounded beetles, often with a median ocellus, usually covered with fine scales. In the wild state these beetles are scavengers and feed on dead animal remains (cadavers) especially on skin, hair, horns, hooves, feathers, and other forms of keratin. But now many species are urban pests and live in houses and other domestic premises such as food stores where they feed on dried meats, bacon, cheese, wool, hides, skins, furs, carpets, etc. *Dermestes lardarius* (larder beetle) (Fig. 10.34) is an important pest of stored products and feeds on a wide range of animal and plant products throughout the warmer parts of the world. *Dermestes maculatus* (hide beetle) (Fig. 10.35) is more typical of the genus and its diet is restricted to dried animal protein. In many tropical countries there is a dried fish industry as a local source of animal protein, sometimes based upon a marine fishery (Malaysia, Thailand, Hong Kong, etc.), or else on freshwater lakes and rivers (Uganda, Malawi, etc.). Traditionally the fish are sun-dried in the open air which presents many problems, especially when rainstorms arrive unexpectedly. The two main groups of insect pests attacking the drying and the dried fish are fly maggots (Muscoidea) and *Dermestes* beetles, especially *D. maculatus*. Hide beetle larvae are causing considerable damage to the dried fish industry in many parts of Africa and tropical Asia.

Two other domestic pests of importance include *Attagenus pellio* (black carpet beetle) (Fig. 10.36) and *Anthrenus scrophulariae* (carpet beetle) (Fig. 10.37) now more or

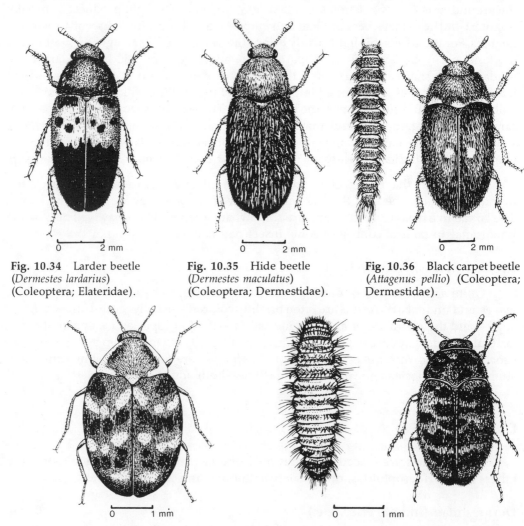

Fig. 10.34   Larder beetle (*Dermestes lardarius*) (Coleoptera; Elateridae).

Fig. 10.35   Hide beetle (*Dermestes maculatus*) (Coleoptera; Dermestidae).

Fig. 10.36   Black carpet beetle (*Attagenus pellio*) (Coleoptera; Dermestidae).

Fig. 10.37   Carpet/museum beetle (*Anthrenus* sp.) (Coleoptera; Dermestidae).

Fig. 10.38   Khapra beetle (*Trogoderma granarium*) (Coleoptera; Dermestidae).

less cosmopolitan and damaging carpets, clothing, and furs. Species of *Anthrenus* are called "museum beetles" and can be very damaging to dried natural history museum specimens; both adults and larvae are damaging but the larvae are by far the more destructive. The larvae are characteristically very hairy dorsally.

One of the very few truly herbivorous dermestid beetles is the khapra beetle (*Trogoderma granarium*) (Fig. 10.38) which is now cosmopolitan throughout the warmer parts of the world. This is a very serious pest of stored grains, and the larvae are able to penetrate intact grains. They often occur gregariously in stored produce and frequently damage far more grains than they eat. The larvae can survive under very dry conditions, and are reputed to be able to feed on grain with a moisture content as low as 2%; and they can actually go into a state of facultative diapause in the event of food shortage or adverse physical conditions, for a period of many months (or even years!). Its preference is for a hot, dry climate, and it is a serious pest in the hot, dry tropics where food has to be stored over the long dry season; famine reserves are often severely damaged as these beetles can develop under conditions intolerable to most stored products insects. Adults are wingless, and short-lived, and do not feed; dispersal is restricted entirely to human agency. Under optimum conditions—37°C (98.6°F) and 25% RH—the life cycle can be completed in as little as three weeks.

BOSTRYCHOIDEA

A group of families often associated with timber, and some species are very damaging to trees. Adults are hard-bodied, with pronotum large and hood-like and head deflexed; the larvae have soft bodies without sclerotized dorsal plates and without setae, and they have a C-shaped posture. There are four large families within the group.

Anobiidae (furniture/timber beetles) *1100 spp.*

A group of destructive timber beetles with a few stored products pests; the larvae are decidedly scarabaeiform. The group is worldwide, but some of the best-known pest species are temperate in distribution. There are four main species of economic significance.

> *Anobium punctatum* (common furniture beetle) (Fig. 10.39)—a temperate species found also in the warmer subtropical regions; very damaging to household timbers, rafters, floorboards, furniture. The larvae bore in solid wood, especially the softer woods. It is thought to be a European species, now distributed throughout Asia, temperate Australasia, South Africa, U.S.A., and Canada. The adult is small in size (2.5–4.5 mm/0.1–0.2 in.); larvae are 5–7 mm(0.2–0.3 in.)when fully grown and they tunnel extensively in the wood, usually taking about a year for development. The larvae bore also in the bark and wood of dead trees; most wood-borers have mycetomes in caecae at the anterior end of the intestine, to aid with digestion. When the female lays eggs they become infected with the microorganisms on the shell surface from the female anus. When the adults emerge they leave a round, small exit hole and a pile of frass is extruded, making the site quite conspicuous. Fig. 10.40 shows bark of *Pinus* with *Anobium* holes.

> *Lasioderma serricorne* (tobacco beetle)—a tropical species that requires a temperature above 19°C (68°F) (and more than 30% RH) for development; now com-

Fig. 10.39 Common furniture beetle (*Anobium punctatum*) (Coleoptera; Anobiidae).

Fig. 10.40 Bark of pine tree with *Anobium* emergence holes (Coleoptera; Anobiidae); only found in dead wood; Friskney, U.K.

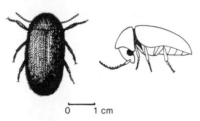

Fig. 10.41 Tobacco beetle (*Lasioderma serricorne*) (Coleoptera; Anobiidae).

Fig. 10.42 Drugstore beetle (*Stegobium paniceum*) (Coleoptera; Anobiidae).

pletely pantropical in distribution—optimum conditions are 30–35°C (86–95°F) and 60–80% RH. The larvae are quite polyphagous in diet and are able to attack intact seeds and grains as well as a very wide range of stored foodstuffs; they are particularly damaging to tobacco, both as stored leaves and as cigarettes in packets. The pupal cocoon is stuck on to a solid substrate and the adults gnaw their way out and through any wrappings in which the produce is contained; adults fly readily, but most dispersal is through trade (Fig. 10.41).

Stegobium paniceum (drugstore beetle) (Fig. 10.42)—a more temperate species, quite cosmopolitan in distribution ranging from India to Japan, Europe, and North America, and in parts of Africa. The larvae feed on almost all types of stored foodstuffs of vegetable origin, including coriander seeds, dried ginger, turmeric, and various drugs.

Xestobium rufovillosum (deathwatch beetle)—a larger (5–7 mm/0.2–0.3 in.) temperate beetle whose larvae bore in oak and other hardwoods where fungal decay is present. The adults apparently do not fly, and dispersal, such as it is, is effected by the active first instar larvae. It is recorded throughout Europe, North Africa, New Caledonia, and the U.S.A.

Ptinidae (spider beetles)   *700 spp.*

Small beetles (2–5 mm/0.04–0.2 in.) with head and pronotum narrow in relation to the rather globular elytra and body; legs and antennae elongate, giving an overall impression of a spiderlike appearance; there is some sexual dimorphism. These are stored products pests, feeding on dried material of both animal and plant origin. In the wild state they are to be found in the nests and lairs of birds and mammals in warmer regions of the world, and some are myrmecophiles. About 24 species are recorded infesting stored foodstuffs worldwide; some of the more abundant species include the following:

Eurostus hilleri (= *Pseudoeurostus hilleri*)—Europe, Asia, Canada.

Gibbium psylloides—pantropical.

Mezium americanum (American spider beetle)—cosmopolitan.

Niptus hololeucas (golden spider beetle)—temperate, West Asia and Europe.

Ptinus clavipes (brown spider beetle)

Ptinus fur (white-marked spider beetle)

Ptinus hirtellus (brown spider beetle)

Ptinus tectus (Australian spider beetle) (Fig. 10.43)—cosmpolitan.

Ptinus villiger (hairy spider beetle)

Tipnus unicolor

Trigonogenius globulus (globular spider beetle)—common in East Africa.

Many species appear to be indigenous to the Mediterranean Region and have subsequently been dispersed by trade throughout the warmer parts of the world, although a few species appear to be endemic to Australasia and to North America.

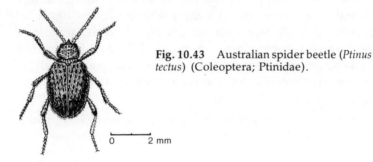

Fig. 10.43   Australian spider beetle (*Ptinus tectus*) (Coleoptera; Ptinidae).

0   2 mm

Bostrychidae (black borers; auger beetles, etc.)   *430 spp.*

Adults usually black (or brown) in color, body cylindrical in shape, with large prothorax hooded, and often spinose covering the deflexed head; antennae short but with a loose club of three terminal segments; body length 3–20 mm (0.12–0.8 in.). The group is worldwide but most frequently seen in the warmer countries. The adults bore galleries (circular in cross-section) 6–10 cm (2.4–4 in.) long in branches and trunks of trees, usually sickly or moribund trees being preferred. *Apate monachus* is one of the commonest tree borers, and the nocturnal beetles remain in the galleries during the day but emerge at night and fly quite freely. Adults live for 2–3 months and may excavate 6–8 galleries each during their lifetime. Eggs are laid in bark crevices (about 80 per female) and the larvae tunnel in the bark of the dead or dying tree—apparently eggs are never laid in living wood. In Africa, coffee and cocoa are attacked by the adult beetles but apparently breeding does not occur on either of these hosts. In East Africa it is presumed that breeding takes place in wild hosts. The larvae bore in the dead bark and tunnel longitudinally down the dead branch. Larval development takes nearly three years. The larvae have an intestinal mycetome similar to that of the Anobiidae, but in this group it appears that the microorganisms are transmitted to the eggs with the sperm at mating. These beetles do not possess cellulases and they have to use the starches and sugars present in the wood as food; they are thus unable to feed on seasoned timber. Although *Apate monachus* is a widespread pest on many woody crop plants (and sugarcane) throughout much of the tropics, it appears that breeding only occurs on a few crop plant species. In some of the other wood-boring species (*Sinoxylon, Schistocerus,* etc.) breeding takes place within the female galleries of the host tree.

In India, there are 17 species of Bostrychidae recorded attacking mango trees; six different species are recorded from coffee and from cocoa in Africa and South America.

Two members of the family are extremely important stored products pests:

Rhizopertha dominica (lesser grain borer)—formerly of South America, is now completely pantropical in distribution and occurs in heated grain stores in Europe and U.S.A. This tiny beetle (2–3 mm/0.08–0.12 in.) is a typical bostrychid in appearance (Fig. 10.46) and is a primary pest of stored grains, both as the adult and the larval stage, and also feeds on stored flours, cereal products and dried tubers such as cassava. Post-harvest cereal crop losses attributable to this pest have sometimes been very high.

Prostephanus truncatus—is the larger grain borer, only slightly larger than the lesser though, and it has recently been making news headlines as a new pest of stored products in tropical Africa (see page 32). In its native Central America, damage to stored maize has been relatively slight. But in 1975 it was accidentally introduced into Tanzania where it is now firmly established, and is spreading to adjacent countries as well as to West Africa. In Africa it is causing post-harvest storage losses in maize of 70–80% in a matter of four months, and is in addition a serious pest of stored cassava tubers. It is now rated as one of the most serious insect pests in tropical Africa.

Some of the more important Bostrychidae known as agricultural pests are as follows:

Apate indistincta (black coffee borer)—this species is restricted to East Africa, where it occasionally damages coffee bushes.

Apate monachus (black borer)—polyphagous on a wide range of woody crop plants (olive, citrus, coffee, cocoa, grapevine, etc.) and sugarcane (the adult beetle that is); found throughout Africa, Mediterranean Region, West Indies and South America (Fig. 10.44); it breeds in olive, carob, almond, peach, grapevine, and various other trees, but only in dead wood.

Apate spp. (black borers)—several other species are found on coffee and other trees; in Africa and the West Indies.

Bostrychoplites spp.—bore coffee and cocoa; tropical Africa.

Bostrychopsis spp.—recorded killing orchard trees in Australia; also on coffee in Africa, and coffee and bamboo in S.E. Asia.

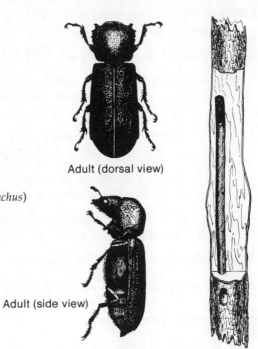

Adult (dorsal view)

Fig. 10.44 Black borer (*Apate monachus*) (Coleoptera; Bostrychidae).

Adult (side view)

Infested branch cut to show tunnel

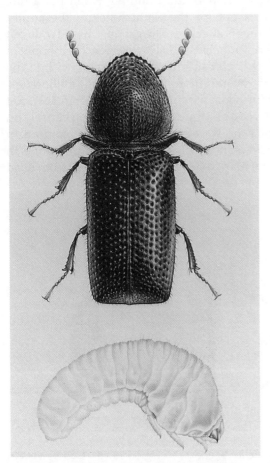

Fig. 10.45 Larger grain borer (*Prostephanus truncatus*) (Coleoptera; Bostrychidae) (I.C.I.).

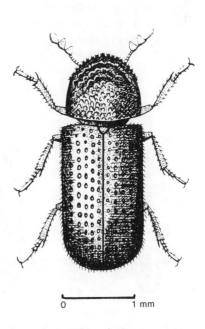

0   1 mm

Fig. 10.46 Lesser grain borer (*Rhizopertha dominica*) (Coleoptera; Bostrychidae).

Dinoderus spp. (small bamboo borers)—bore bamboo, rattan palm, mango, etc.; India, S.E. Asia, Japan, Australia.

Heterobostrychus spp.—on mango, mulberry; India.

Prostephanus truncatus (larger grain borer)—native to Central America and now established in Africa, causing heavy losses of stored maize, other grains, and cassava (C.I.E. Map No. A. 465) (Fig. 10.45).

Rhizopertha dominica (lesser grain borer)—attacks stored grains and foodstuffs; cosmopolitan in warmer regions (Fig. 10.46).

Schistocerus bimaculatus (grape cane borer)—bores grapevine stock, apple, citrus, almond, fig, etc.; Mediterranean Region.

Sinoxylon spp.—bore in bamboos, cotton stems, mango, fig, citrus, mulberry, acacia; Africa, Mediterranean, India, S.E. Asia.

Xyloperthella spp.—bore cocoa; tropical Africa.

Xyloperthodes spp.—bore *Cassia* and *Bauhinia* trees; Africa.

Lyctidae (powder post beetles) *70 spp.*

A small family, closely related to the preceding group, so-called because they tunnel the sapwood of susceptible hardwood timbers until nothing is left but a fine powder; the outer veneer remains intact except for the flight holes of the adults. Apparently only timber from broad-leafed trees is attacked. The group is worldwide but best represented in the Holarctic region. Several species of *Lyctus* are now cosmopolitan pests. The female beetle uses a long, narrow, ovipositor to insert eggs down into the lumen of a major xylem vessel at a depth of several (up to 8 mm/0.32 in.) millimeters. Apparently, sapwood with a starch content (in xylem parenchyma) of less than 3% is not used for oviposition. The larvae feed on the wood as they develop, subsisting entirely on the starch and sugars in the cell contents, and a moisture content of 6–30% is required. In warmer climates most species are probably bivoltine, but in cooler regions univoltine. In India, two species of *Lyctus* are recorded from Mango trees—but presumably from dead branches. Fig. 10.47 shows an adult *Lyctus* beetle.

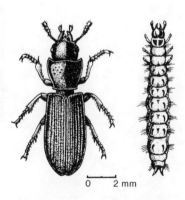

Fig. 10.47 Powder post beetle (*Lyctus* sp.) (Coleoptera; Lyctidae).

Fig. 10.48 "Cadelle" beetle (*Tenebriodes mauritanicus*) (Coleoptera; Trogositidae).

Fig. 10.49 Copra beetle (*Necrobia rufipes*) (Coleoptera; Cleridae).

CLEROIDEA

Six families are placed here but they are not of much importance agriculturally. Adults have five tarsal segments, and the larvae, which are usually predacious, have protruded mouthparts, and a pair of horny urogomphi posteriorly.

Trogossitidae   *600 spp.*

A tropical group of great morphological diversity; some live in decaying trees and prey on the larvae of wood-eating beetles, and some are fungivorous. *Tenebriodes mauritanicus* is the "Cadelle" (Fig. 10.48) to be found in stored foodstuffs—it is omnivorous in diet and feeds both on grains and processed foodstuffs, and on some of the insects to be found there. The larvae can survive for several months without food, and the adults may live for 1–2 years apparently, so continuation of infestation is ensured.

Cleridae (checkered beetles)   *3400 spp.*

A large tropical family, most adults are of graceful form and beautiful coloration; they are small to medium in size. Both adults and larvae are usually predacious on wood-boring beetles (especially Scolytidae). The genus *Necrobia* occurs on animal carcasses and skins, and is partly saprophagous in diet, but also eats the dipterous maggots found in carcasses. *Necrobia rufipes* is called the copra beetle (Fig. 10.49) for it eats stored copra in the tropics, but it is also called the red-legged ham beetle as it is a serious pest of stored ham and bacon. *N. ruficollis* is the red-shouldered ham beetle in Japan and North America.

Melyridae   *4000 spp.*

A large family of tropical beetles with filiform antennae and elongate five-segmented tarsi, and projecting fore-coxae. There is great diversity of form within the family (which according to Richards & Davies, 1977, is in need of revision) and many beetles look like Cantharidae. Adults are often found on flowers, sometimes in large numbers. The larvae resemble those of Cleridae and are thought to be predacious. A common species seen in Hong Kong is the blue-bodied *Idgia oculata* with yellow prothorax and antennae (Fig. 10.50). In Australia adult beetles of *Laius femoralis* have been reported feeding on developing rice grains.

Fig. 10.50 Adult of *Idgia oculata* (Coleoptera; Melyridae); Hong Kong.

LYMEXYLOIDEA

Lymexylidae (timber beetles)   *40 spp.*

A small group, mostly tropical in distribution, of elongate, soft-bodied beetles, whose larvae bore tunnels in hardwoods, and often have fungi associated with their infestation causing additional damage to the trees. *Melittoma insulare* is the coconut palm borer (Fig. 10.51) causing very serious damage to the palms in the Seychelles and Madagascar (C.I.E. Map No. A. 152). It was at one time regarded as the most important pest of coconut palm in the Seychelles, but chemical control was generally successful (Brown, 1954). *M. sericeum* is the chestnut timberworm of the U.S.A., and another species of *Melittoma* occurs in Australia.

In tropical Africa, India, S.E. Asia, and Australia, the genus *Atractocerus* is regularly found in forested areas where it is reputed to infest forest trees—the body of the adult is elongate and soft, with tiny elytra but large hind wings with which they fly to lights at night. The adult *Atractocerus* looks superficially like a Staphylinidae. The larvae are long and very thin, so presumably make narrow tunnels in the trees. In India, *A. reversus* is a pest species, attacking timber of the tree *Boswellia serrata*—the source of Salai wood used for making boxes and paper. The beetles infest green sappy logs soon after felling and the larvae make long, deep (narrow) tunnels in the heart of the sappy logs, and the tunnel walls are usually infected with fungi. Severely damaged timber is often unfit for commercial use. In East Africa, *A. brevicornis* bores in the trunk of cashew trees.

CUCUJOIDEA

In this group is placed a large number of species, in some 43 families, according to Crowson (1968) and followed by Richards & Davies (1977), but only about five families contain species of agricultural interest. The group is very heterogeneous and different authorities view this assemblage quite differently.

SECTION CLAVICORNIA

Adults have front coxae never projecting, tarsi never 5–5–4 in both sexes (often 5–5–5 or 3–3–3), antennae usually clubbed. Generally a difficult group to recognize.

Nitidulidae (sap beetles)   *2200 spp.*

A large family of variable form and habits; tarsi usually 5–5–5 (rarely 4–4–4); fore- and mid-coxae very transverse, abdomen usually with last one or two tergites not covered by the elytra. Many species are to be found on flowers, feeding on nectar and pollen, some being quite host specific; others are found in decaying animal matter, or in fungi, or attracted to sap and fermenting plant fluids; they are often found at injury sites on plants where sap is exuding. Species of *Cybocephalus* are predacious on scale insects (Coccoidea), aphids, and some phytophagous mites. About 16 species are pests of stored products, and *Carpophilus* (dried fruit beetles) are quite important pests of dried fruits, as well as sometimes attacking fruits still on the tree. Both adults and larvae of *Carpophilus* feed on fruits; on field crops they are seldom of real importance as pests, causing a little flower damage, and mostly feeding on sap at damage sites, but there is often fungal infection associated with their feeding. Pineapples are regularly infested with *Carpophilus* when injured, and in some instances the beetles are recorded as a nuisance in canning factories where special attention is required to keep them out of the cans. Citrus fruit may be attacked; and damaged and overripe dates are regularly attacked—the fungi carried on the bodies of the beetles have often ruined entire date crops.

Species of some importance as agricultural pests include:

Amphicrossus spp.—recorded on guava, peach, plum, pear, pomegranate; India.

Carpophilus dimidiatus (corn sap beetle)—as for *Carpophilus* spp. below.

Carpophilus hemipterus (dried fruit beetle) (Fig. 10.52)—as for *Carpophilus* spp. below.

Carpophilus spp. found on a wide range of crop plants including date palm, sugarcane, pineapple, citrus, mango, peach, fig, guava, castor, grapevine, etc., either on flowers or ripe fruits or damaged fruits, and on a wide range of stored and dried fruits; pantropical.

Haptoncus spp.—adults bore into ripe citrus fruits or infest flowers; Africa, Japan.

Meligethes spp. (blossom beetles)—in flowers of Cruciferae; Europe (Fig. 10.53).

Soronia spp.—in flowers of citrus; India.

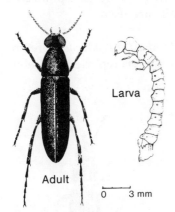

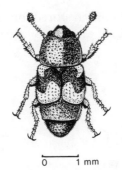

Fig. 10.51 Coconut palm borer (*Melittoma insulare*) adult and larva (Coleoptera; Lymexylidae).

Fig. 10.52 Dried fruit beetle (*Carpophilus hemipterus*) (Coleoptera; Nitidulidae).

Fig. 10.53 Blossom beetle (*Meligethes* sp.) (Coleoptera; Nitidulidae).

Cucujidae (flat bark beetles)   *500 spp.*

These small, flattened beetles often have filiform antennae, and they are often to be found under the bark on trees; many are predacious or feed on dead insects in spider webs, but some feed on plant debris, and a few are adapted to feed on stored products. *Cryptolestes* are tiny beetles, only about 2 mm (0.08 in.) in length, but are quite damaging stored products pests.

Cryptolestes ferrugineus (rust-red grain beetle)—secondary pest on stored grains; now worldwide (Fig. 10.54).

Cryptolestes pusillus (flat grain beetle)—on stored foodstuffs; Asia, North America.

Cryptolestes turcicus—often found in flour mills; throughout Asia and North America.

Laemophloeus spp.—a large genus (formerly including *Cryptolestes*); recorded on apricot and grapevine, others are grain-eating (secondary), and some predacious; India.

Silvanidae (flat grain beetles)   *400 spp.*

Sometimes this group is regarded as a subfamily of Cucujidae which they strongly resemble, but their antennae are usually clubbed. Adults usually found under tree bark, or on plants, and some are predacious (on Scolytidae and other bark borers); they are small in size (2–3 mm/0.08–0.12 in.). Two species in South America live in a state of symbiosis (mutualism) with a mealybug in hollow leaf petioles (Imms, p. 880).

The important pest species are the tropical:

Oryzaephilus mercator (merchant grain beetle) and *Oryzaephilus surinamensis* (saw-toothed grain beetle) (Fig. 10.55)—general feeders, to be regarded as secondary pests on stored grains in that they feed on fragments left by the primary pests (*Sitophilus, Ephestia,* etc.). *O. surinamensis* is more frequently found on cereals and cereal products, and *O. mercator* on oilseed products. The difference between the two species relates to the shape of the head behind the eyes—in *mercator* the temple is drawn out into a rounded point but it is flat in *surinamensis;* and also the male genitalia are used.

An additional species to be found in stored foodstuffs is *Ahasverus advena* (foreign grain beetle), but it is a fungus feeder and so not directly a pest; it is reported to prefer *Penicillium* to *Aspergillus.*

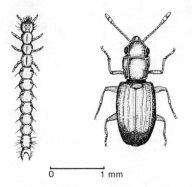

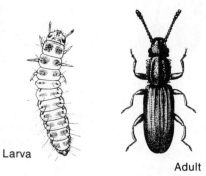

Larva

0   1 mm

Adult

Fig. 10.54 Rust-red grain beetle (*Cryptolestes ferrugineus*) (Coleoptera; Cucujidae).

Fig. 10.55 Saw-toothed grain beetle (*Oryza-ephilus surinamensis*) adult and larva (Coleoptera; Silvanidae).

Cryptophagidae (silken fungus beetles)   *800 spp.*

A small group of beetles, mostly fungus-feeders, of little agricultural interest. The many species of *Cryptophagus* are all fungus feeders, with the exception of a few found in the nests of bees and waps. *Antherophagus* adults are often to be seen on flower heads, but the larvae are in the nests of bumblebees (*Bombus* spp.). The genus *Telmatophilus* spends the whole life cycle in the head of bulrush (*Typha*) and other water plants. In Europe, one of the numerous species of *Atomaria (A. linearis)* is called the pygmy mangold beetle and in the spring the feeding adults damage young seedlings of beets and other Chenopodiaceae (Fig. 10.56).

Byturidae (raspberry beetles, etc.)

A very small family but well-known in temperate agriculture as there are four species of *Byturus* that eat the fruits of raspberry and other species of *Rubus* (Fig. 10.57).

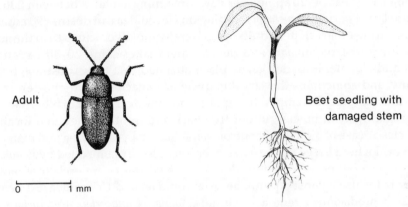

Adult

Beet seedling with damaged stem

0   1 mm

Fig. 10.56 Pygmy mangold beetle (*Atomaria linearis*) (Coleoptera; Cryptophagidae) and damaged sugar beet seedling; Cambridge, U.K.

0   2 mm

Fig. 10.57 Raspberry beetle (*Byturus tomentosus*) (Coleoptera; Byturidae) adult, and larva on a raspberry fruit; Copenhagen, Denmark.

Coccinellidae (ladybird beetles)   *5000 spp.*

Small to moderate-sized beetles, with a characteristically convex body shape and head partly concealed under the pronotum. The larvae are soft-bodied and of varying coloration, but often a dark bluish color with yellow and white spots, and with long sickle-shaped mandibles. There are usually tergal tubercules and soft spines, but in a few genera there is, instead, a white flocculent secretion (*Cryptolaemus*, etc.) over the body surface.

There are three distinct subfamilies; both adults and larvae of the Coccinellinae are carnivorous and prey on aphids, scale insects, small caterpillars, etc.; and it is thought that the Tetrabranchinae are also predacious. The Epilachninae are phytophagous and several species are agricultural pests.

Most adults, if disturbed, secrete a bitter brown fluid that is apparently poisonous to vertebrates. The characteristic color of the adults is red with black spots, but some are blackish, some brown, or yellow, with or without spots. Presumably the bright coloration and the diurnal habits are to be regarded as aposematic and a warning to predators.

Eggs are yellow and laid in batches projecting from the plant foliage—the total number laid varies with the species from about 200–800 (average). Their importance as predator of insect pests cannot be overemphasized; early work in the U.S.A. established that some larvae eat 14–20 aphids per day, consuming a total of between 200 and 500 for the entire larval period. *Hyperaspis* feeding on Coccoidea can destroy 90 adults and 3000 nymphs during its larval life. Adults are generally more voracious than the larvae. There are usually four larval instars and development takes some 20–30 days according to species and climate. Pupation takes place attached to a firm substrate, by the caudal extremity, and sometimes the larval exuviae are attached to the posterior end. Some species are migratory in temperate regions, and may overwinter in hibernation as adults, often gregariously in bushes, under tree bark and in other sheltered locations.

A classic case of biological control was achieved in California and many other parts of the world where *Icerya purchasi* has been successfully controlled by *Rodolia* and other ladybirds. Several species of ladybirds are now available commercially as biological control agents for use against Aphididae, Aleyrodidae, and Coccoidea. Some of the most important predacious genera include: *Adalia*, *Chilocorus*, *Coccinella*, *Hyperaspis*, *Menochilus*, *Rodolia* and *Scymnus*.

The Epilachninae are phytophagous and both adults and larvae are to be found on the foliage of Solanaceae, Cucurbitaceae, and some legumes, where they eat holes in the leaf lamina. *E. varievestis* is atypical in that it attacks beans (*Phaseolus* spp.) in the New World; it is a serious pest and crop losses of up to 50–60% have been recorded. The most notable pest species include:

Epilachna chrysomelina (12-spotted melon beetle)—on curcurbits, potato, cotton, etc.; southern Europe, Africa, Near East (C.I.E. Map No. A. 409) (Fig. 10.58).

Epilachna fulvosignata (eggplant epilachna)—on Solanaceae; Africa.

Epilachna hirta (potato epilachna)—on potato and *Solanum*; Africa.

Epilachna sparsa (spotted leaf beetle)—on Solanaceae and Cucurbitaceae; India, S.E. Asia, China (Fig. 10.59).

Epilachna varievestis (Mexican bean beetle)—on beans (*Phaseolus* spp.); U.S.A., Mexico, Central America (C.I.E. Map No. A. 46).

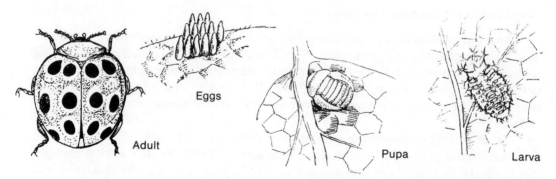

Eggs

Adult

Pupa

Larva

Fig. 10.58 Epilachna beetle (*Epilachna* sp.) (Coleoptera; Coccinellidae); ex Kenya.

Eggs

Damage

Larvae

Adult

Fig. 10.59 Spotted leaf beetle (*Epilachna sparsa*) on *Solanum* leaves; Hong Kong (Coleoptera; Coccinellidae).

Epilachna spp.—many species feed on a wide range of crops and wild hosts; Africa, Asia, U.S.A. (Le Pelley, 1959, lists 17 "species" of *Epilachna* as pests in East Africa).

Some authorities have split the total number of species between *Epilachna* and the genus *Henosepilachna*.

The oriental species *E. sparsa* is variable in coloration, and the number of spots on the elytra vary from 12–26, with the result that the different forms, in different countries, have repeatedly been referred to as different "species" according to the number of spots present, and the literature is very confusing.

E. chrysomelina in Israel has been reported to transmit squash mosaic virus in cucumber.

Lathridiidae (minute brown scavenger beetles)   *600 spp.*

A small but widespread family, all the members of which feed on fungi, particularly molds and Mycetozoa. Several species are regularly found in food stores (*Enicmus, Lathridius, Coninomus, Cartodere,* etc.) but they are apparently only feeding on fungal growths.

SECTION HETEROMERA

Beetles with tarsi 5–5–4 in both sexes usually; a large group somewhat difficult to define to the nonspecialist, containing about 33 families, most of which are small and of no particular interest agriculturally. The families of some agricultural importance are included below:

Mycetophagidae (hairy fungus beetles)   *200 spp.*

A small group of small and densely pubescent beetles associated with fungi; brown or black in color, and often with yellow or reddish spots on the elytra. Two genera are regularly found in stored products, but they are invariably feeding on fungi and not directly on the foodstuffs or grains. Several species of *Mycetophagus* and *Typhaea* are to be found, although *T. stercorea* (hairy fungus beetle) is probably the commonest, and to be found now in most parts of the world.

Tenebrionidae (darkling beetles, etc.)   *15,000 spp.*

A large group of beetles, well represented throughout the world, but most abundant in the tropics. There is some dissimilarity in the adults but the larvae are very similar in appearance, and are often referred to as "false wireworms"—the yellow mealworm (Fig. 10.62) is a good example of a typical larva. Many are ground beetles, often black in color, reminiscent of Carabidae, and often apterous, or with vestigial wings and elytra often immovable. Some are wood feeders and to be found mostly on tree trunks and bark; these usually fly quite freely. A large number of species are characteristic of desert habitats (200 species in Namibia); others live in dung, fungi, dead animals, or with stored foodstuffs. Some species of *Gonocephalum* appear to prefer the upper shore region of sandy beaches. In India *G. depressum* is a pest of grapevine, and is also said to be a vector of some fowl pinworms.

The stored products pests are very important, especially *Tribolium*, for these small, flat, reddish beetles are prolific breeders and very long-lived (a year or more) and their life cycle can be completed in about five weeks (at 30°C/86°F). They are basically secondary pests of grains, being too small to damage the intact grains, but often associated with *Sitophilus* weevils. They are most abundant in processed flours, biscuit, and other cereal products. In grain and flour, the population tends to be very dispersed; partly because the female secretes a dispersive pheromone, and partly because at high densities the adults are cannibalistic and eat eggs and young larvae. More than 100 species of Tenebrionidae are recorded feeding in stored foostuffs worldwide. Most of the stored produce pests are now totally cosmopolitan or almost so, as a result of food transportation to and from across the world. The more important genera and species of Tenebrionidae are listed below:

Alphitobius laevigatus (black fungus beetle)—cosmopolitan in warmer regions.

Alphitophagus spp. (fungus beetles)—several species are recorded.

Astylus spp. (false wireworms)—larvae feed on seed (maize, etc.) in soil prior to germination; South Africa.

Blaps mucronata (churchyard beetle)—usually found in cellars and buildings; Europe, Asia.

Gnathocerus cornutus (broad-horned flour beetle) (Fig. 10.60)—pantropical and temperate.

Gnathocerus maxillosus (slender-horned flour beetle)—pantropical only.

Gonocephalum dermestoides—adults damage stem of cotton; Africa.

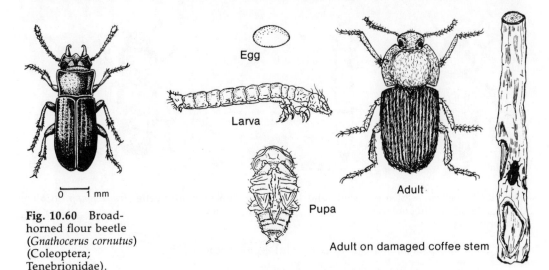

Fig. 10.60 Broad-horned flour beetle (*Gnathocerus cornutus*) (Coleoptera; Tenebrionidae).

Egg

Larva

Pupa

Adult

Adult on damaged coffee stem

Fig. 10.61 Dusty brown beetle (*Gonocephalum simplex*) (Coleoptera; Tenebrionidae).

Gonocephalum patrule (dusty radish beetle)—attack many vegetables; N.E. Africa.

Gonocephalum simplex (dusty brown beetle)—adults chew young bark on coffee bushes; larvae in soil damage seedlings of maize, cotton, tobacco, chickpea, etc.; Africa, south of the Sahara (Fig. 10.61).

Gonocephalum spp. (= *Opatrum*) (dusty brown beetles)—both adults and larvae damage a wide range of crop plants (adults on coffee, cotton, grapevine, etc.; larvae on roots of cereals and vegetables and also destroy seedlings); Africa, India, Asia.

Latheticus oryzae (long-headed flour beetle)—a tropical pest most abundant in S.E. Asia.

Macropoda spp.—damage legumes and tree seedlings; East Africa.

Mesomorphus longulus (hay beetle)—adults and larvae damage many vegetables and field crops (adults overwinter in haystacks); Near East.

Mesomorphus villiger (tobacco darkling beetle)—adults damage transplanted tobacco seedlings by gnawing the stem base, extensive replanting sometimes required; India.

Opatroides frater—damage tobacco seedlings, etc.; Near East and India.

Opatrum spp. = *Gonocephalum* spp.

Palorus spp. (depressed flour beetle, etc.)—some species are omnivorous and feed on both broken grains and on other insects; pantropical.

Phrynocolus spp.—damage wheat and sunflower seedlings; East Africa.

Rhytinota spp.—attack legumes; East Africa.

Tenebrio molitor (yellow mealworm) (Fig. 10.62)—a temperate pest in Europe, North Asia and North America.

Tenebrio obscurus (dark mealworm)—as above.

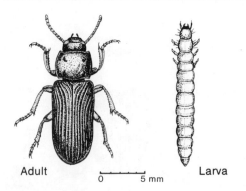

Adult   0   5 mm   Larva

Fig. 10.62 Yellow mealworm beetle and larva (*Tenebrio molitor*) (Coleoptera; Tenebrionidae).

Fig. 10.63 Flour beetle (*Tribolium* sp.) (Coleoptera; Tenebrionidae).

Tribolium castaneum (red flour beetle)—pantropical (Fig. 10.63).

Tribolium confusum (confused flour beetle)—more temperate.

Tribolium destructor (dark flour beetle)—pantropical.

Zophosis spp.—recorded damaging coffee, cotton, maize, barley, wheat, groundnut, etc.; Africa and Near East.

Alleculidae (= Cistelidae)   *1100 spp.*

A small group, worldwide, characterized by having pectinate tarsal claws, and to be found on flowers or leaves; the larvae are found in rotten wood, humus, or soil. *Omophlus lepturoides* is reported damaging potatoes, and species of *Extenostoma* and *Isomira* are pests of pasture in East Africa.

Meloidae (blister/oil beetles)   *2000 spp.*

These are largish beetles, soft-bodied, 10–35 mm (0.4–1.4 in.) long, characterized by having a deflexed head, distinct narrow neck, and long legs. They are diurnal in habits; the adults feed mostly on the inflorescence of plants (eating the pollen and the petals) and they can cause a loss of fruits, although damage is often unsightly rather than serious. Species of *Epicauta* are slightly different in that most of their damage is leaf-eating. The Leguminosae and Malvaceae are often favored hosts of some blister beetles, but others appear to prefer the flower spike of Gramineae (millets, maize, rice, etc.). One striking feature about their infestations is that often they occur in very large, very local, populations and they may completely defoliate or deflower a crop or a hedge of *Hibiscus*. The adults are sluggish, and if handled exude an acrid yellow fluid which can cause skin blistering (hence the common name). The blistering agent is cantharidin ($C_{10}H_{12}O_4$); this chemical is used medicinally to treat various urinogenital disorders. *Lytta vesicatoria* is the notorious "Spanish-fly" of southern Europe alleged to possess powerful aphrodisiacal properties. In fact, the cantharidin derived from ground dried beetles merely causes inflammation of the ureters and the urethra, and large doses are very toxic. In recent years, in both the U.K. and Europe, there have been human deaths recorded through the unwise ingestion of quantities of "Spanish-fly" at parties. In India, species of *Mylabris* are collected annually as a source of pharmaceutical cantharidin—the elytra alone are used, as apparently there is a greater concentration of the chemical in that part of the body.

The life history of Meloidae is unique in that the first instar larvae are very active, and are called triungulins, and they seek out the underground egg-pods of short-horned grasshoppers and locusts (Acrididae), or in some cases the nests of Aculeate Hymenoptera underground. Having found an egg-pod, the triungulin eats its way in and starts to eat the eggs; it then molts and becomes a soft-bodied, short-legged, inactive form which stays in the grasshopper egg capsule eating the eggs through the second and third instars. The fully grown third instar is scarabaeiform, leaves the egg-pod and nearby in the soil turns into the pseudopupal stage in which it may hibernate for a while before eventually changing into a sixth larval stage prior to pupation. In southern U.S.A. and Europe overwintering hibernation is the rule, but in the tropics presumably this does not take place. Oviposition takes place in open sandy areas which are the developmental sites for Acrididae; each female may lay 2000–10,000 eggs—presumably larval mortality rates are high.

In many tropical areas where there is a locust problem it is often found that the adult Meloidae are damaging to legume crops and others, but the larvae are very important natural predators on the local Acrididae, so the precise status of the agricultural impact of Meloidae is difficult to determine.

In the case of the bee-nest predators (*Sitaris, Meloe,* etc.), eggs are laid in the sandy areas where the bees make their underground nests; the triungulin has to catch hold of a passsing bee (usually by the hairy legs) and get carried down into the nest underground where it eats the bee eggs and then goes on to eat the stored pollen and nectar. The usual hosts are *Anthophora* and *Andrena* spp., and usually the triungulins are sealed individually inside each cell as the bees construct and provision a succession of cells. These species occur in the cooler climates and most of the beetle larvae overwinter in the pseudopupal stage but a few do develop directly, with adults emerging in the autumn of the same year.

Some of the more important and interesting species of Meloidae include:

Cissites spp.—in nests of *Xylocopa* bees in East Africa.

Coryna spp. (pollen beetles)—polyphagous on anthers and flowers of many crops (pulses, cotton, sweet potato, etc.); tropical Africa (five spp. recorded as pests in East Africa) (Fig. 10.64).

Cylindrothorax spp.—on the panicle and tassel of maize, millets, rice, and also legumes; Africa and India.

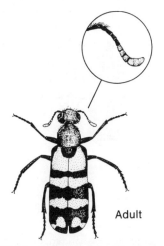

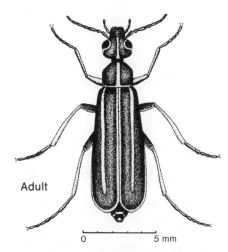

Fig. 10.64 Pollen beetle (*Coryna* sp.) (Coleoptera; Meloidae).

Fig. 10.65 Striped blister beetle (*Epicauta albovittata*) (Coleoptera; Meloidae).

Epicauta aethiops—adults eat leaves of vegetables and fodder crops; North Africa.

Epicauta albovittata (striped blister beetle)—on legumes; East Africa (Fig. 10.65).

Epicauta gorhami (striped blister beetle)—on many hosts; China.

Epicauta hirticornis—on various plants; India.

Epicauta limbatipennis—on millets and legumes; East Africa.

Epicauta ruficeps (red-headed blister beetle)—polyphagous; S.E. Asia.

Epicauta tibialis (black blister beetle)—polyphagous; China (Fig. 10.66).

Epicauta vittata (striped blister beetle)—on beans; Canada, U.S.A. and South America.

Epicauta spp. (black blister beetles)—adults eat foliage of many crop plants; seven species recorded as pests in U.S.A., and ten species in Canada.

Lytta (Cantharis) vesicator (Spanish-fly)—found in Europe, Asia, and U.S.A.

Meloe spp. (oil beetles)—larvae prey on *Anthothora* and *Andrena* bees in Europe.

Mylabris cichorii (small yellow-banded blister beetle)—S.E. Asia, China.

Mylabris phalerata (large yellow-banded blister beetle)—India, S.E. Asia, China (Fig. 10.67).

Mylabris pustulata—recorded on cucurbits; India, S.E. Asia.

Mylabris spp. (banded blister beetles)—many species recorded on many plants, most are polyphagous, on flowers of Leguminosae, cotton, other Malvaceae, sweet potato, maize, sorghum, etc.; Africa, India, and the warmer parts of Asia.

Sitaris spp.—larvae prey on *Anthophora* bees in Europe.

Zonitis spp.—Australia.

Zonitomorpha spp.—recorded on maize and sesame; East Africa.

Fig. 10.66 Black blister beetle (*Epicauta tibialis*); body length 23 mm (0.9 in.); South China (Coleoptera; Meloidae).

Fig. 10.67 Large yellow-banded blister beetle (*Mylabris phalerata*); body length 28 mm (1.12 in.); South China (Coleoptera; Meloidae).

CHRYSOMELOIDEA

A very large group of phytophagous or xylophagous beetles, apparently quite well defined taxonomically but on rather esoteric criteria. Tarsi are 5–5–5 segmented but the fourth segment is tiny and hidden in the emargination of the third which is deeply bilobed; the larvae have thoracic legs more or less developed, and there may be abdominal pseudopods. Biologically they represent several quite distinct groups.

Cerambycidae (= Longicornia) (longhorn beetles)   *20,000 spp.*

The longhorn beetles have antennae usually at least two-thirds as long as the body, often much longer, arising on strong tubercles, and capable of being flexed back over the body; all tibiae with two spurs; claws almost always simple. The family is worldwide, found wherever there are trees and shrubs, but is most abundant in the tropical rain forests. There are many species of importance, both agriculturally and as forestry pests; many shade trees and ornamentals are damaged. The host plants are typically trees, but sometimes woody shrubs, and a few herbaceous plants. The herbaceous plants are usually bored along the stem pith (cucurbits, etc.) or in the roots (carrot, etc.)—these beetles are usually quite small in size. Some species prefer dead or dying trees, selecting either dry or moist wood, but others appear to prefer to attack healthy trees. In the litera-ture there are conflicting reports in that some species are said to prefer living healthy hosts, and other accounts state that only moribund or dead trees are attacked. Attacked mature trees usually show severe injury symptoms in a year or two, and large trees can be seen to die after several years of heavy infestation by large species such as *Batocera;* branches can be killed rapidly but tree trunks take longer to be killed. Young trees and saplings can be killed in a matter of months. Recent work on timber borers in California has shown that trees can be vulnerable to beetle attack after only a moderate level of water stress; apart from the water stress the trees were healthy. Adult beetles often chew patches of tree bark, as well as leaves and shoots, and sometimes the surface of fruits. In several species, the adults are recorded actually girdling the stem on which they feed and they may kill the plant; in India *Sthenias grisator* is the grapevine stem girdler.

Viewed worldwide, some tree crops are extensively attacked by longhorn beetles although any individual plantation is probably attacked by only a few species. Figures are available from the literature for cocoa of 104 species (Entwistle, 1972), coffee 54 (Le Pelley, 1968), and in India alone Butani (1979) records 18 species attacking mango and 13 on *Citrus.*

Eggs are laid on, or into, the tree bark, and the larvae bore into the wood. Some larvae prefer to tunnel in the bark, or just under the bark, whereas others prefer sap-wood, and some the heartwood. When burrowing under the bark they sometimes ring the branch or small stem, causing rapid death. Larvae are white or yellowish, fleshy, soft-bodied, sometimes finely pubescent, and thoracic legs and abdominal pseudopods may be developed. In some species nine pairs of pseudopods are present, presumably to aid locomotion in the tunnel system. The head is small and invaginated into the prothorax. The body is generally cylindrical in shape, and the tunnel bored is circular in cross-section, but some of the bark borers tend to be somewhat flattened, and the tunnel bore oval. Larval development generally takes 1–3 years, according to species and climate. *Batocera* larvae grow as large as 8–10 cm (3.2–4 in.). Some species make frass holes as they tunnel so that frass is extruded at intervals (e.g. *Dirphya*) but many do not make frass holes, the frass being contained under the bark and in crevices; the tunnel is usually clear behind the larva, as distinct from the Buprestidae.

The most common symptom of longhorn attack is a hole in the trunk or branch with frass extruded. Pupation takes place in a cell under the bark at the end of a tunnel; the chamber is closed with a plug of frass or a wad of wood fibers. Sometimes a special operculum of calcium carbonate mixed with gummy exudate or silky material is constructed. The young adult may remain inside the pupal chamber for a while, then it bites its way to the exterior; newly emerged adults often tend to remain on the tree trunk after emergence where mating takes place. The adult emergence hole is neat and usually circular (Fig. 10.68). Adults live for 1–3 months.

The larvae that feed on cambium material (such as *Glenea* on cocoa) have a more nutritious diet and develop more rapidly and there may be as many as 3–4 generations per year. The species that bore shrubs and herbaceous plants are generally smaller in size and develop quite rapidly. *Alcidon* in South America bores in the stem of eggplant and has 3–4 generally per year; and *Apomecyna* in cucurbit stems in Africa likewise develops rapidly and has 4–6 generations annually. A few species, tiny in size, have larvae that bore into ripe seeds of cotton and berries of coffee (*Sophronica* spp.).

The smallest beetles are only 5–6 mm (0.2–0.24 in.) in body length but the largest have a body of 6–8 cm (2.4–3.2 in.) and antennae 10–14 cm (4.5–5.6 in.) long—those of the males being longer than the female.

Many beetles are brightly colored, although a few are cryptic (e.g. *Petrognatha gigas*; Fig. 10.82), and some species stridulate loudly, usually by rubbing the hind margin of the prothorax against a specially striated area at the base of the scutellum. A few species stridulate by rubbing the hind femora against the edges of the elytra. The family is regarded as being composed of three large subfamilies, and two smaller ones are sometimes recognized.

Fig. 10.68   Longhorn beetle emergence hole in pine tree trunk; hole diameter 12 mm (0.48 in.); South China (Coleoptera; Cerambycidae).

Cerambycinae

The second largest group taxonomically; adults are variable in shape and coloration; diurnal in habits; antennae often very long; head partly deflexed; prognathous to hypognathous; hind legs clearly longer than middle legs. Larvae have rudimentary thoracic legs evident and abdominal pseudopods. Some of the more notable species include:

> *Aeolesthes holosericea* (apple stem borer)—on apple, apricot, cherry, guava, mulberry, peach, pear, plum, walnut; India.

Aeolesthes spp.—on almond, apple, apricot, cherry, cocoa, peach, quince, and walnut; India and the Philippines.

Callichroma sp.—eats eggplant leaves; West Africa.

Cerambyx spp.—bore fruit and nut trees, grapevine; Europe and Mediterranean Region.

Chelidonium brevicorne—bores clove trees; Malaysia.

Chelidonium cinctum (green orange borer)—on citrus; India.

Chelidonium sinense (blue bamboo borer)—bamboos; South China.

Chlorophorus annularis (bamboo longhorn)—bores bamboos and sugarcane; S.E. Asia, Papua New Guinea, China, Japan, U.S.A. (Fig. 10.69).

Chlorophorus carinatus—bores coffee and cocoa; East Africa.

Chlorophorus varius (grape wood borer)—in grapevines; Mediterranean Region.

Gracilia minuta (rubus borer)—larvae bore dead stems of *Rubus, Corylus,* etc.; Europe.

Hexamitodera semivelutina—bores clove; Indonesia.

Hylotrupes bajulus (house longhorn beetle)—bores coniferous timbers; Europe, Mediterranean Region and South Africa.

Nothopeus spp. (blue clove borer, etc.)—in clove, roseapple; Indonesia.

Phoracantha semipunctata (eucalyptus longhorn)—a major pest of *Eucalyptus* trees in southern U.S.A., parts of Asia; a minor pest in Australia.

Plocaederus spp. (mango borers)—larvae bore mango, cocoa; India and Central Africa.

Rhytidodera simulans (mango branch borer)—mango; S.E. Asia.

Uracanthus spp.—bore citrus, *Eucalyptus,* etc.; Australia.

Xylotrechus contortus (coffee stem borer)—bores coffee, teak, *Gardenia,* etc.; India, and S.E. Asia.

Xylotrechus quadripes (coffee white borer)—bores coffee; India, S.E. Asia.

Fig. 10.69 Bamboo longhorn beetle (*Chlorophorus annularis*); body length 14 mm (0.56 in.); South China (Coleoptera; Cerambycidae).

Xylotrechus spp.—several species bore in coffee, maple, walnut, etc.; India, S.E. Asia, China and U.S.A.

Xystrocera spp.—bore coffee and cocoa in East and West Africa, and leguminous shade trees in Indonesia.

Prioninae

Adults are flattened, elongate and oval in shape, dark reddish brown usually, prothorax with sharp, well-defined lateral margins (at least basally); fore-coxae strongly transverse, not projecting; antennae usually short; nocturnal in habits. A small group whose larvae usually bore in dead, dry wood; only a few pest species are recorded.

Acanthophorus spp.—bore mango, plum, cocoa, eucalyptus, etc.; Africa and India (Fig. 10.70).

Dorysthenes hugelii—bores apple, pear, apricot, plum, peach, walnut; India.

Dorysthenes spp.—bore sugarcane, etc.; China.

Ergates spp.—*Pinus* and other conifers; Europe, North Africa, U.S.A.

Macrotoma spp.—mango, apple, walnut; India.

Megopis reflexa—bores sugarcane; Hawaii.

Stenodontes downesii—larvae bore rubber, citrus, cocoa, coffee, kapok, coconut, *Acacia*, etc.; Madagascar and Africa.

Stenodontes spp.—bore rubber; West Africa.

Fig. 10.70 *Acanthophorus maculatus* from eucalyptus tree trunk; body length 85 mm (3.4 in.); Ethiopia (Coleoptera; Cerambycidae).

Lamiinae

A very large group where many adults are dark colored, and with long antennae (male antennae often twice as long as female); head strongly deflexed, subvertical; hypognathous; hind legs scarcely longer than middle legs; larvae either legless or with vestigial legs. Many pest species are known.

Acalolepta cervina (canker grub)—bores teak trees; India.

Acalolepta rusticator—on coffee, cocoa, teak, cassava, etc.; Indonesia, Papua New Guinea, Solomon Islands.

Acalolepta mixta—on grapevine, passionfruit vine; Australia.

Alcidion deletum—larvae bore stem of eggplant; South America.

Analeptes trifasciata—on *Acacia* and *Eucalyptus* trees; East Africa.

Anoplophora chinensis (citrus longhorn)—bores citrus; S.E. Asia, China (Figs. 10.71, 10.72).

Anoplophora malasiaca (citrus borer)—bores citrus; Japan.

Anoplophora versteegi (orange trunk borer)—bores citrus; India.

Anthores leuconotus (white coffee borer)—coffee; tropical Africa (C.I.E. Map No. A. 196) (Fig. 10.73).

Apomecyna spp. (cucurbit longhorns)—larvae bore stems of Cucurbitaceae; Africa, S.E. Asia, China, and U.S.A.

Apriona cinerea (jackfruit longhorn)—larvae bore jackfruit, fig, mulberry, apple, peach, etc.; India, S.E. Asia, China.

Apriona germarii (jackfruit longhorn)—larvae bore jackfruit, fig, mulberry, apple, peach, etc.; India, S.E. Asia, China. (Figs. 10.74, 10.75).

Apriona japonica (mulberry borer)—mulberry etc.; Japan.

Batocera horsfieldi (square-spotted longhorn beetle)—walnut; India, China.

Fig. 10.71 Citrus longhorn (*Anoplophora chinensis*); body length 30 mm (1.2 in.); South China (Coleoptera; Cerambycidae).

Fig. 10.72 Damage to young *Citrus* stem (killed) by larva of citrus longhorn (*Anoplophora chinensis*); Hong Kong (Coleoptera; Cerambycidae).

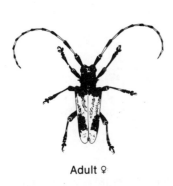

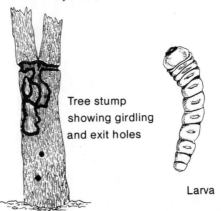

Adult ♀

Tree stump showing girdling and exit holes

Larva

Fig. 10.73 White coffee borer (*Anthores leuconotus*) ex Kenya (Coleoptera; Cerambycidae).

Fig. 10.74 Jackfruit long-horn (*Apriona germarii*); body length 48 mm (1.9 in.); South China (Coleoptera; Cerambycidae).

Fig. 10.75 Larva of jackfruit long-horn (*Apriona germarii*); South China (Coleoptera; Cerambycidae).

Fig. 10.76 White-spotted longhorn (*Batocera rubus*); body length 36 mm (1.44 in.); South China (Coleoptera; Cerambycidae).

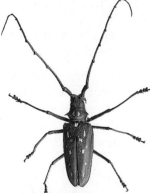

Fig. 10.77 Larva of white-spotted longhorn (*Batocera rubus*) in branch of *Ficus carica*; South China (Coleoptera; Cerambycidae).

Fig. 10.78 Red-spotted longhorn (*Batocera rufomaculata*); body length 70 mm (2.8 in.); South China (Coleoptera; Cerambycidae).

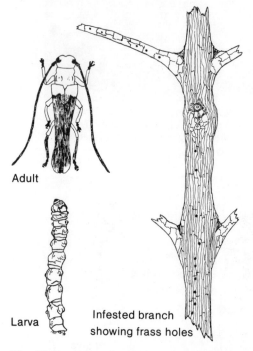

Adult

Larva

Infested branch
showing frass holes

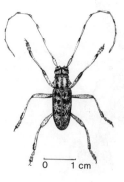

Fig. 10.79 Yellow-headed borer (*Dirphya nigricornis*) ex Kenya (Coleoptera; Cerambycidae).

Fig. 10.80 Pine sawyer (*Monochamus* sp.) (Coleoptera; Cerambycidae); body length 21 mm (0.84 in.); South China.

Batocera rubus (white-spotted longhorn)—fig, mango, jackfruit, etc.; India, S.E. Asia, South China (Figs. 10.76, 10.77).

Batocera rufomaculata (red-spotted longhorn)—this very large species bores fig, mango, guava, jackfruit, pomegranate, apple, rubber, walnut and *Sapium*, etc.; Near East, India, S.E. Asia, China (Fig. 10.78).

Batocera spp. (*c.* 8)—recorded from a similar host list, including also cocoa and kapok; India, S.E. Asia, Indonesia, Australia, China and Japan.

Bixadus sierricola—bores coffee; tropical Africa.

Ceroplesis spp. (5)—bore coffee; tropical Africa.

Dirphya spp. (yellow-headed borers)—bore coffee; East Africa, Malawi (Fig. 10.79).

Glenea spp. (cocoa stem borers)—on cocoa; Africa, S.E. Asia, Indonesia, Papua New Guinea.

Milothris irrorata (legume tree borer)—bores trunks of tree legumes; S.E. Asia.

Monochamus spp. (pine sawyers, etc.)—bore *Pinus* in Europe, Asia, Canada, U.S.A.; and coffee, cocoa, citrus, etc. in East and Central Africa (Fig. 10.80).

Olenecamptus bilobus—bores fig, jackfruit, mulberry, mango, pomegranate; India, S.E. Asia, Indonesia, China.

Paranaleptes reticulata (cashew stem girdler)—on cashew, citrus, kapok, etc.; East Africa (Fig. 10.81).

Petrognatha gigas (giant fig longhorn)—bores *Ficus natalensis* and *Chlorophora* trees in East Africa, and other *Ficus* spp. in West Africa (Fig. 10.82).

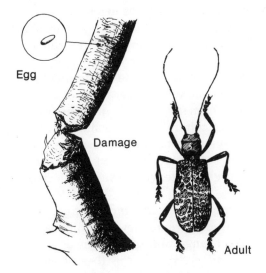

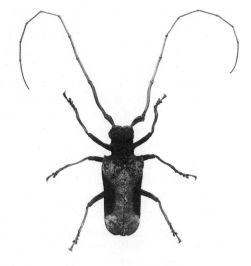

Fig. 10.81 Cashew stem girdler (*Paranaleptes reticulata*) (Coleoptera; Cerambycidae).

Fig. 10.82 Giant fig longhorn (*Petrognatha gigas*) from *Ficus natalensis;* body length 65 mm (2.6 in.); Uganda (Coleoptera; Cerambycidae).

Phytoecia geniculata (carrot borer)—larvae bore root of carrot; eastern Mediterranean.

Plagiohammus spp. (coffee borers)—bore coffee stems; Mexico and Central America.

Saperda candida (round-headed apple tree borer)—bores apple, etc.; U.S.A. and Canada.

Sophronica spp. (3) (coffee berry borers)—larvae bore ripe berries of coffee, seeds of cotton, and stems of cowpea, tobacco, and *Solanum;* East and West Africa.

Steirastoma breve (cocoa beetle)—bores cocoa stems; West Indies, Central and South America.

Sternotomis spp. (4)—larvae bore coffee stems; West Africa.

Sthenias spp.—bore almond, apple, citrus, coffee, cocoa, grapevine, jackfruit, mango, mulberry, etc.; tropical Africa and India.

Tragocephala spp.—larvae bore citrus, cocoa, coffee, cowpea, mango; Africa.

Lepturinae

A small group with larvae typically in wood in an advanced stage of fungal decay; adults often found on flowers of Umbelliferae, etc., in bright sunlight. Both evergreen and deciduous trees are infested in Europe by species of *Leptura, Rhagium,* and *Strangalia.*

Bruchidae (= Lariidae) (seed beetles; bruchids)   *1300 spp.*

The group is worldwide, but most abundant in tropical Asia, Africa, Central and South America. The adults have deeply emarginate eyes; antennae short and not capable of being flexed backwards; hind femora thickened and often toothed; elytra striate; first visible abdominal sternite as long as the following three together. They are small, stout beetles, usually grayish or brown, covered in scales, and the elytra are short and do not cover the abdomen tip. Male antennae are often serrate. The pronotum narrows anteriorly into a "neck" region.

The larvae develop only inside seeds; most hosts belong to the Leguminosae but other plants are used, especially Convolvulaceae, Malvaceae, Palmae; a total of 24 other families were recorded by Southgate (1979) though for many of these only a single plant species is associated with one beetle genus. The agricultural pests are almost entirely confined to the species attacking legumes, both pulses and forage legumes. Some 20 odd species in seven genera are crop pests on legumes worldwide. A total of 56 genera are known, placed in five different subfamilies, but three-quarters of these are referred to the Bruchinae.

There is variation in host specificity—some species are monophagous and restricted to a single genus of host, whereas at the other extreme a few are polyphagous. *Caryedon* is quite polyphagous and attacks a broad range of Leguminosae as well as plants in the Umbelliferae and Combretaceae. *Acanthoscelides obtectus* (bean beetle) is most damaging to *Phaseolus* spp. but will attack most other cultivated pulse legumes. *Bruchus pisorum* on the other hand only infests peas. Generally the different crop pest species of Bruchidae do show different host preferences within the Leguminosae.

Development only takes place within seeds. The female beetle lays her eggs usually glued to the outside of the legume pod; *Bruchus* usually lay on a young pod, but *Callosobruchus* usually go into open pods and lay directly on to the seeds themselves. The hatching larva usually bites its way into the pod through the egg base; once inside the pod it then enters a seed. Each larva usually develops within a single seed, excavating a chamber as it grows. When fully grown it makes a "window" in the seed by biting up to the test surface. It pupates inside a puparium under the "window". In large seeds, there may be several larvae in each seed, but in small seeds there will only be a single larva. The emerging adults are diurnal, fly readily, but usually do not feed; they are thus short-lived, although some have been observed taking pollen and nectar from flowers. Sometimes the adults remain in the puparium for a while and these are invariably carried into the grain stores. Many temperate species usually hibernate overwinter inside the puparium. Most adults seldom fly for more than half a kilometer and so distant field crops are not likely to be infested from local stores.

Species of *Bruchus* and *Bruchidius* mostly attack growing legumes in the field, and although many species complete their first generation actually within a store, they are all univoltine in storage, and have to return to the field and a growing crop to lay their eggs. The multivoltine *Callosobruchus* and *Acanthoscelides* can persist for many generations purely within grain stores, and so are far more important as storage pests.

The species most adapted to storage conditions, such as *A. obtectus*, usually scatter the eggs throughout the produce. Some of the bruchids in wild hosts are multiple seed feeders, and some leave the pod to pupate on the ground in leaf litter.

In addition to the pulse and forage crops damaged by these beetles, they are also important in that many species destroy the seeds of tropical tree legumes. In many parts of the world, in semiarid regions and on eroded hillsides, there have been successful plantings of various tree and bush legumes; some of which can also be used as food by browsing ungulates. The reclamation programs could be adversely affected by bruchid seed destruction. In Europe, tree lupins are being used to colonize old mine-waste dumps and, as mentioned on page 29, there is concern as they are being devastated in the U.K. by a new pest to the British Isles (lupin aphid: *Macrosiphum albifrons*)—bruchid attack could be almost as damaging in the same kind of reclamation situation in the tropics. In connection with the infestation of tropical tree legumes it is interesting to note that many Bruchidae have adapted to deal biochemically with large amounts of potentially very toxic substances that are characteristic of the foliage (and especially seeds) of these trees.

From an agricultural point of view the "field" species of Bruchidae seldom cause significant crop losses, although in some temperate situations crop losses of 25–30% have occasionally been recorded. The "storage" species usually only occasionally infest the field crop to any extent (1–2% is usual) but with continuous breeding in the tropical grain stores, the harvested pulse crops have often been recorded with damage levels as high as 80% after 6–8 months. The main "storage" pest species have their origins in populations on plants that were not human food sources, and on these wild hosts infestation levels were quite low; but on the more edible crops infestation levels escalated to the current pest proportions.

Some of the more important pest species are listed below:

Acanthoscelides obtectus (bean bruchid)—on beans (*Phaseolus* spp.) and other pulses; North, Central, and South America, southern Europe, Africa (Figs. 10.83, 10.84)

Acanthoscelides spp.—300 species known in the New World.

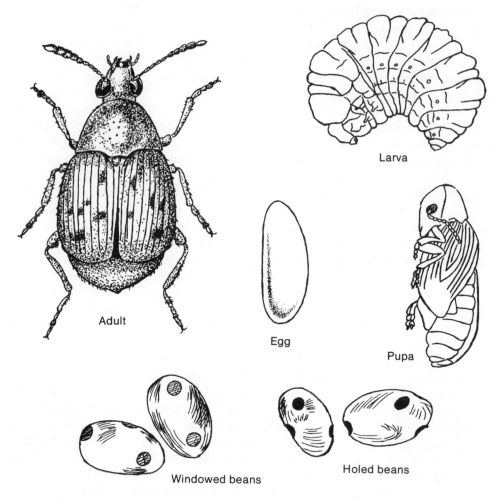

Larva

Adult

Egg

Pupa

Windowed beans

Holed beans

Fig. 10.83   Bean bruchid (*Acanthoscelides obtectus*) (Coleoptera; Bruchidae).

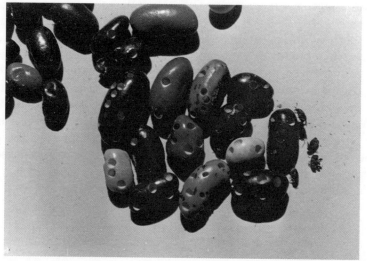

Fig. 10.84   *Phaseolus* beans bored by bean bruchid (*Acanthoscelides obtectus*); Harar, Ethiopia.

Bruchidius atriolineatus—on cowpea; West Africa.

Bruchidius bimaculatus (two-spotted beetle)—on alfalfa; Mediterranean Region.

Bruchidius quinqueguttatus (five-spotted bruchid)—on chickpea, lupins, pea, vetches; Mediterranean Region.

Bruchidius sahlbergi—in pods of Popinac; eastern Mediterranean.

Bruchidius trifolii (clover beetle)—on clovers; North Africa and Mediterranean Region.

Bruchus brachialis (vetch bruchid)—on vetches, etc.; U.S.A. and Canada.

Bruchus chinensis (Chinese pulse beetle)—on pulses, etc.; China.

Bruchus ervi (Mediterranean pulse beetle)—on lentil, etc.; Mediterranean.

Bruchus pisorum (pea beetle)—on peas (but not on dried peas in storage); cosmopolitan (Fig. 10.85).

Bruchus rufimanus (bean beetle)—on beans; Europe, Asia, Australia, U.S.A.

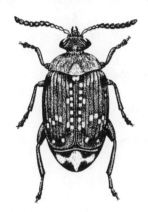

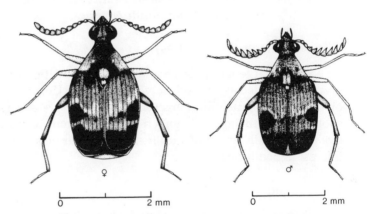

Fig. 10.85 Pea beetle (*Bruchus pisorum*) adult (Coleoptera; Bruchidae); ex Boston, U.K.

Fig. 10.86 Oriental cowpea bruchid (*Callosobruchus chinensis*) (Coleoptera; Bruchidae).

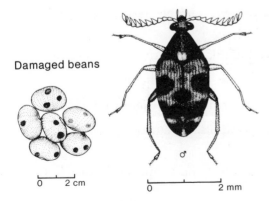

Damaged beans

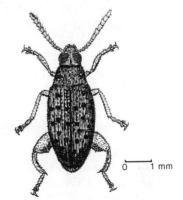

Fig. 10.87 Spotted cowpea bruchid (*Callosobruchus maculatus*) (Coleoptera; Bruchidae).

Fig. 10.88 Groundnut borer (*Caryedon serratus*) (Coleoptera; Bruchidae).

Callosobruchus analis—mostly on cowpea and mung bean; pantropical.

Callosobruchus chinensis (oriental cowpea bruchid)—on most grain legumes; pantropical (Fig. 10.86).

Callosobruchus maculatus (spotted cowpea bruchid)—on most grain legumes; pantropical (Fig. 10.87).

Callosobruchus phaseoli (cosmopolitan seed beetle)—on grain legumes;
cosmopolitan.

Caryedon serratus (groundnut borer)—attacks groundnut in the field in West Africa,
tree legumes in India, etc.; now transported widely and in food stores
throughout the world (Fig. 10.88).

Pachymerus spp.—larvae sometimes in groundnut pods, also damaging to coconuts
and nuts of oil palm; West Africa.

Specularius spp.—on cowpea, pigeon pea, and various wild legumes; Africa.

Spermophagous spp.—seeds of Convolvulaceae, Malvaceae, and some others;
Europe, Africa, and India.

Zabrotes subfasciatus (Mexican bean beetle)—mostly on beans (*Phaseolus* spp.);
Africa, Central and South America.

Chrysomelidae (leaf beetles)   *20,000 spp.*

A very large family of great diversity and with a large number of crop pests. The
group is regarded as being closely related to the Cerambycidae, and is difficult to define
on the basis of structural characters. There is no general agreement as to the number of
subfamilies, with figures ranging from 10–15. Some of these subfamilies are very distinctive anatomically and biologically, but others are less so; about nine subfamilies are of
interest agriculturally.

Criocerinae

Rather elongate, somewhat rectangular, beetles, with larvae short, thick, and fleshy,
and they feed externally on plant foliage. In some species the larvae cover themselves
with excrement, presumably for concealment. Both adults and larvae are foliage feeders
and may be found together on the leaves gregariously. Pest species include:

Crioceris spp. (asparagus beetles)—on asparagus; Europe, Asia, Africa, U.S.A. and
Canada (Fig. 10.89).

Lema bilineata (tobacco leaf beetle)—on tobacco; South Africa, South America.

Lema pectoralis (orchid leaf beetle)—on various orchids; Indonesia.

Lema spp.—on potato, yam, and various hosts; Europe, Asia, Africa, U.S.A.

Lilioceris livida (yam leaf beetle)—larvae and adults eat yam leaves; North, East and
West Africa.

Oulema melanopus (cereal leaf beetle)—on wheat and temperate cereals; Europe,
Asia, U.S.A.

Oulema oryzae (rice leaf beetle)—on rice and grasses; eastern Asia (C.I.E. Map No.
A. 260) (Fig. 10.90).

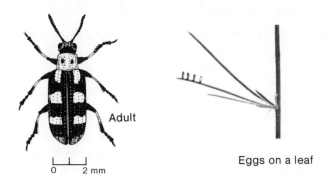

Adult

Eggs on a leaf

0   2 mm

Fig. 10.89 Asparagus beetle (*Crioceris asparagi*) (Coleoptera; Chrysomelidae; Criocerinae).

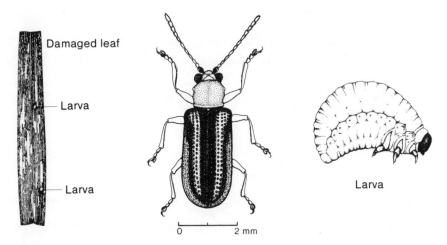

Damaged leaf

Larva

Larva

Larva

0   2 mm

Fig. 10.90 Rice leaf beetle (*Oulema oryzae*) (Coleoptera; Chrysomelidae; Criocerinae).

Oulema spp.—recorded on coffee, cotton, eggplant, potato; Africa, Asia, U.S.A.

Sagrinae

A small group of large brilliantly colored beetles (*c.* 20 mm/0.8 in. long), with very thick hind femora, to be found in the tropics. The usual body colors are shiny metallic purple, green, or black. The larvae are reported to inhabit stem swellings on *Dioscorea* and Cucurbitaceae, or root swellings on legumes. The adults eat plant foliage; both legume leaves and eggplant are recorded. They are of little importance agriculturally. *Sagra pupurea* is recorded from S.E. Asia and China; in Vietnam, *S. purpurea* was thought to be a larval pest of coffee roots, killing many seedlings below ground level, and *S. ferox* is reported on coffee in East Africa. *S. villosa* adults eat leaves of cowpeas in tropical Africa, as also do species of *Chrysolagria*. Other species of *Sagra* are minor pests on a range of crops throughout tropical Africa, India, and S.E. Asia.

Clytinae

Small, robust, beetles of a cylindrical body form, with widely separated antennae. The larvae inhabit cases composed of dried excrement, and live on plant foliage. The group is insignificant agriculturally, although a few minor pests are recorded:

Clytra spp.—recorded on cassava in Africa.

Labiostomis spp.—observed defoliating *Lezpedeza* bushes in Hong Kong.

Peploptera spp.—on *Acacia* species; East Africa.

Eumolpinae

These beetles are rounded, often oblong, mostly small in size. with head almost hidden under prothorax; generally either metallic or yellowish in color, and with prothorax narrower than the elytra. The larvae usually live in the soil and feed on plant roots, but their damage is seldom serious, except for *Colaspis brunnea* in the U.S.A. The adults eat young leaves, buds, and sometimes the surface of young fruits of a very wide range of crop plants including soft fruits, vegetables, maize, and roses; host plants of adults and of larvae are not always the same.

Agricultural pests include:

Colaspis brunnea (grape colaspis)—larvae damage roots of legumes and rice; adults polyphagous on young foliage; U.S.A. and New Zealand (Fig. 10.91).

Colaspis hypochlora (banana fruit-scarring beetle)—adults damage banana fruits; central and northern South America (Fig. 10.92).

Colasposoma acaciae (acacia leaf beetle)—on wattle; South Africa.

Colasposoma coffeae and spp. (coffee leaf beetle)—on coffee; East Africa.

Colasposoma fulgidum—on citrus; South Africa.

Colasposoma metallicum—on potato, sweet potato; India, Indonesia, China.

Colasposoma spp.—recorded on citrus, coffee, sweet potato, etc.; Africa, India, Indonesia.

Menius spp.—on cocoa, coffee, cotton, legumes, sweet potato; East and West Africa.

Pagria spp.—on beans, cowpea, and other legumes; East Africa, S.E. Asia.

Rhyparida spp.—on cotton, sugarcane, etc.; Australia.

Scelodonta strigicollis (grapevine leaf beetle)—on grapevine; India.

Syagrus spp.—adults eat leaves and larvae attack roots of *Acacia,* cotton, other Malvaceae, *Gardenia, Leucaena,* oil palm, roses; tropical Africa.

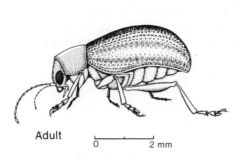

Adult   0 ⊢——⊣ 2 mm

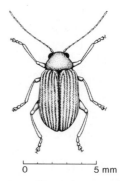

0 ⊢————⊣ 5 mm

Fig. 10.91 Grape colaspis (*Colaspis brunnea*) (Coleoptera; Chrysomelidae; Eumolpinae).

Fig. 10.92 Banana fruit-scarring beetle (*Colaspis hypochlora*) (Coleoptera; Chrysomelidae; Eumolpinae).

Chrysomelinae

Most of the species belong in this group. The adults are typically oval, convex, brightly colored, and some 3–12 mm (0.2–0.5 in.) in body length. The larvae live exposed on the plant foliage; they are short and convex, usually with a leathery, pigmented integument. Pupation takes place in the soil. There are not many species regarded as serious crop pests.

Leptinotarsa decemlineata (Colorado beetle)—adults and larvae defoliate potato and some other Solanaceae; U.S.A., Canada, Europe (C.I.E. Map No. A. 139) (Fig. 10.93); 30 other species recorded in the U.S.A., some feed only on Solanaceae.

Oidosoma spp.—on *Brassica*, sesame, and some ornamentals; East Africa.

Phaedon spp. (mustard beetles, etc.)—on Cruciferae; Europe, Asia, U.S.A. (Fig. 10.94).

Phaedonia inclusa (soybean leaf beetle)—on soybean, etc.; Indonesia.

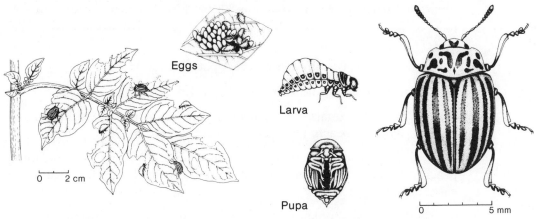

Fig. 10.93 Colorado beetle (*Leptinotarsa decemlineata*) (Coleoptera; Chrysomelidae; Chrysomelinae).

Fig. 10.94 Mustard beetle (*Phaedon* sp.) (Coleoptera; Chrysomelidae; Chrysomelinae).

Galerucinae (cucumber beetles, etc.)

A large group with many species recorded as agricultural pests. Adults tend to be soft-bodied, with normal hind femora, third tarsal segment deeply bilobed, fore-coxae globose; many are yellowish in color with dark spots or stripes; body length some 3–16 mm (0.12–0.64 in.), usually 4–8 mm (0.16–0.32 in.); generally the group is not well defined so far as the general entomologist is concerned.

Larval habits are varied, but the majority are probably soil dwellers where they feed on roots, although a number feed openly on the leaf surface together with the adults. Many of the species feed on the Cucurbitaceae, more than any other group of plants, and so they are sometimes called collectively "cucumber beetles". In North America *Acalymma* adults have been recorded as vectors of cucumber wilt disease. In some species it is the adult beetle that is the pest, in others the larva, and sometimes both stages are destructive to cultivated plants. As with some other Chrysomelidae, the adult beetle may be a pest on one crop and the larva eating the roots of another.

Some of the more important pest species include:

Acalymma spp. (cucumber beetles)—on cucurbits; adults are vectors of cucumber wilt disease; U.S.A. and Canada.

Aulacophora spp. (black/yellow cucumber beetles; pumpkin beetles)—on Cucurbitaceae; Africa, India, S.E. Asia, Australia, Japan.

Copa spp.—on cucurbits; Africa.

Diabrotica balteata (banded cucumber beetle)—found attacking many different crop plants in U.S.A., Central, and South America, but adults often prefer bean and cucurbit foliage whereas larvae often prefer roots of maize although they eat bean roots (Fig. 10.95).

Diabrotica spp. (cucumber beetles and corn rootworms)—many species are reported to occur, possibly as many as 300 worldwide, and they are serious on potato, legumes and maize in North, Central, and South America (Figs. 10.96, 10.97).

Diacantha spp.—on beans, coffee, rice, sweet potato; East Africa.

Exora spp.—on Hibiscus, sunn-hemp, etc; tropical Africa.

Megalognatha rufiventris (maize tassel beetle)—adults on maize tassels, mango flowers, avocado, peach, plum, etc.; East Africa (Fig. 10.98).

Megalognatha spp. (peach beetles, etc.)—adults damage many crops, including peach, cotton, maize, mango, plum, acacia, and beans; 16 species recorded in eastern Africa.

Monolepta australis (red-shouldered leaf beetle)—polyphagous; Australia.

Monolepta dahlmanni—polyphagous; tropical Africa.

Monolepta duplicata—polyphagous; tropical Africa.

Monolepta intermedia (four-spotted rose beetle)—on ornamentals and legumes; Ethiopia.

Monolepta punctipes (spotted leaf beetle)—polyphagous; Ethiopia.

Monolepta signata (white-spotted leaf beetle)—polyphagous adult; India, S.E. Asia, Philippines, China.

Monolepta spp. (spotted leaf beetles)—a number of species recorded from many hosts, some clearly polyphagous, including wheat, cotton, maize, plum, apple, beans, etc.; East Africa, India, S.E. Asia, and Australia.

Oides decempunctata (ten-spotted leaf beetle)—on grapevine; Asia (Fig. 10.99).

Ootheca mutabilis (brown leaf beetle)—on pulses, coffee, cocoa, cotton; Africa (Fig. 10.100).

Ootheca spp.—adults recorded on beans, cotton, coffee, sesame; East Africa.

Rhaphidopalpa spp. (Pumpkin Beetle)—on Cucurbitaceae; Africa, Mediterranean, India.

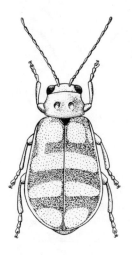

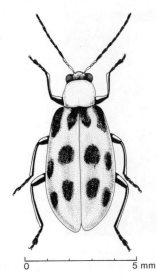

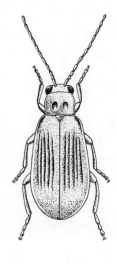

Fig. 10.95   Banded cucumber beetle (*Diabrotica balteata*) (Coleoptera; Chrysomelidae; Galerucinae).

Fig. 10.96   Spotted cucumber beetle (*Diabrotica undecimpunctata*) (Coleoptera; Chrysomelidae; Galerucinae).

Fig. 10.97   Southern corn rootworm (adult) (*Diabrotica longicornis*) (Coleoptera; Chrysomelidae; Galerucinae).

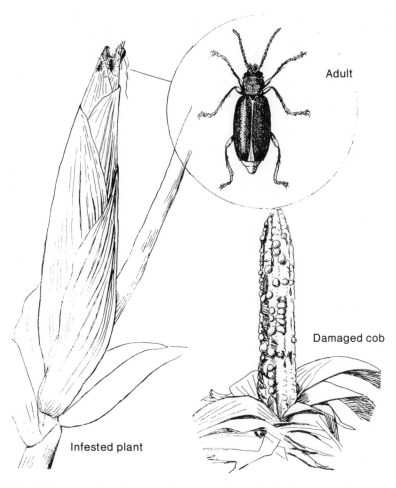

Adult

Damaged cob

Infested plant

Fig. 10.98   Maize tassel beetle (*Megalognatha rufiventris*) (Coleoptera; Chrysomelidae; Galerucinae).

Fig. 10.99 Ten-spotted leaf beetle (*Oides decempunctata*) on wild grapevine; Hong Kong (Coleoptera; Chrysomelidae; Galerucinae).

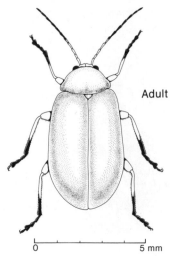

Fig. 10.100 Brown leaf beetle (*Ootheca mutabilis*) (Coleoptera; Chrysomelidae; Galerucinae).

Halticinae (flea beetles)

Mostly tiny in size (3–5 mm/0.12–0.2 in.) but *Podontia* measures 14 mm (0.56 in.); very swollen hind femora with which the adults jump considerable distances. Adults eat small holes in the leaf lamina, usually making a shot-hole effect, but sometimes they leave one (upper) epidermis intact leaving a "window". Severe damage (to seedlings usually) results from large populations of beetles; seedlings are particularly vulnerable. Adults generally prefer to jump than to fly. Most host plants are herbaceous, but some are shrubs (cotton, etc.); the genus *Aphthona* attacks the foliage of trees (apple, *Mallotus,* etc.) and large populations can be seen on leaves 4–6 m (13–20 ft.) from the ground (Fig. 10.101). More than 20 genera of flea beetles are regarded as agricultural pests on a wide range of crops throughout the warmer parts of the world.

Larval habits are varied. Many live in the soil and feed on plant roots, but they are tiny and damage is negligible; some are leaf miners and cause conspicuous damage to the plant foliage; others live inside plant stems, or in roots; and others live openly on the leaves and eat the lamina. Pupation usually takes place in the soil. In cooler regions the adults often hibernate overwinter; they resume feeding on the crop seedlings in the spring and thus can cause considerable damage, and with precision drilling of beets and some other vegetables the plant stand can be severely reduced.

Some important pest species include:

Altica cyanea (blue flea beetle)—often found on rice crops, but adults and larvae feed on the leaves of the aquatic weed *Ammania* and water chestnut (*Trapa bispinosa*); India, China.

Altica spp.—the genus is polyphagous on clovers, vegetables, grapevine, flax, rhubarb, etc.; Mediterranean Region, East Africa, India, Asia, Australia, North America.

Aphthona wallacei—adults on the shade tree *Mallotus apelta* in South China (Fig. 10.101).

Aphthona spp.—recorded on many trees (apple, *Euphorbia,* mango, peach, etc.) and also beans, cabbage, sesame, and some flowers; Europe, eastern Africa, Asia.

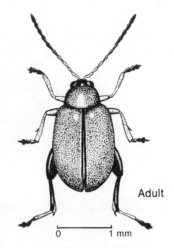

Adult

0   1 mm

Fig. 10.101   Adult flea beetle (*Aphthona* sp.) damage to leaves of shade tree *Mallotus apelta;* South China (Coleoptera; Chrysomelidae; Halticinae).

Fig. 10.102   Mangold flea beetle (*Chaetocnema concinna*) (Coleoptera; Chrysomelidae; Halticinae).

Argopistes spp. (citrus flea beetles, etc.)—on citrus in China and Japan; on olive and ornamentals in South and East Africa.

Chaetocnema spp.—on cereals, barley, maize, millet, rice, sorghum, and beet mostly, but also on sweet potato, and other Chenopodiaceae in Europe; Europe, N.E. Africa, India, Asia, Japan, U.S.A., Central and South America (Fig. 10.102).

Crepidodera ferruginea (wheat flea beetle)—on wheat and some grasses; Europe.

Epitrix aethiopica (potato flea beetle)—on Solanaceae; eastern Africa.

Epitrix integricollis (tobacco flea beetle)—on Solanaceae; eastern Africa.

Epitrix spp. (tobacco/potato flea beetles, etc.)—on Solanaceae; Australia, U.S.A., Canada.

Longitarsus belgaumensis (sunn-hemp flea beetle)—on *Crotalaria;* India.

Longitarsus gossypii (cotton flea beetle)—on cotton and some flowers); Africa.

Longitarsus nigripennis (pepper flea beetle)—larvae bore berries of pepper; India.

Longitarsus spp.—on legumes and other hosts; Europe, Asia.

Phyllotreta spp. (16+) (cabbage flea beetles, etc.)—on *Brassica* spp., and other Cruciferae, cotton, etc.; Europe, Africa, Asia, Australia, North America (Fig. 10.103).

Podagrica spp. (cotton flea beetles, etc.)—many species on cotton, *Hibiscus,* and other Malvaceae, cucurbits, and some legumes, coffee; Africa, India, S.E. Asia, China.

Podontia spp. (large yellow and brown flea beetles)—larvae and adults feed on foliage of *Spondias;* India, Malaya, Indonesia, South China (Fig. 10.104).

Prodagricomela nigricollis (citrus flea beetle/leaf miner)—an interesting pest where the adult eats citrus leaves (usually making windows), and larvae mine inside the leaves; China (Figs. 10.105, 10.106). A similar species was seen on Chinese privet but not yet identified.

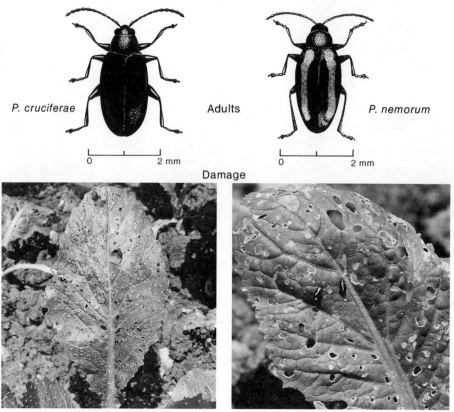

Fig. 10.103 Cabbage flea beetles (*Phyllotreta cruciferae* and *P. nemorum*) (Coleoptera; Chrysomelidae; Halticinae).

Fig. 10.104 Large brown leaf beetle (*Podontia lutea*) (Coleoptera; Chrysomelidae; Halticinae); length 15 mm (0.6 in.); South China.

Psylliodes chrysocephala (cabbage stem flea beetle)—larvae bore stems and petioles of Cruciferae in temperate regions; adults seldom seen; Europe, Canada (Fig. 10.107).

Psylliodes spp. (stem flea beetles)—larvae bore stems of Solanaceae, Cruciferae, Chenopodiaceae, etc.; Europe, Asia, U.S.A., Canada.

Sphaeroderma rubidum (artichoke beetle)—adults hole leaves, larvae mine leaves of artichoke; Europe, Mediterranean Region.

Systena spp.—adults polyphagous on many crops; U.S.A. and Canada.

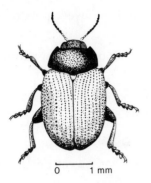

Fig. 10.105 Citrus flea beetle (*Prodagricomela nigricollis*); South China (Coleoptera; Chrysomelidae; Halticinae).

Fig. 10.106 Citrus leaf showing citrus flea beetle leaf miner (*Prodagricomela nigricollis*) tunnel mine, and "windows" eaten by adult beetles; South China (Coleoptera; Chrysomelidae; Halticinae).

Adult

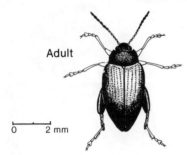

Fig. 10.107 Cabbage stem flea beetle (*Psylliodes chrysocephala*) (Coleoptera; Chrysomelidae; Halticinae).

Cassidinae (tortoise beetles)

The adults are circular or oval in shape, the body is greatly widened and flattened, and the head is concealed under the prothorax, giving an overall shield-like appearance. Many are brilliantly colored silver or golden, with the color fading after death; most are 4–6 mm (0.16–0.24 in.) long. The larvae are characteristically oval-shaped with long body spines and a tapering forked posterior to which are attached the larval exuviae; some larvae also have threads of excrement carried posteriorly.

Eggs are laid in small clusters on the underneath of a leaf, often enclosed within a papery ootheca; often about 20 eggs per ootheca. Larvae and adults are both phytophagous and feed together on the underside of mature leaves; small larvae often skeletonize and "window" part of the leaf lamina, older larvae and adults typically make small roundish holes in the lamina. Most species feed on *Ipomoea* and other Convolvulaceae in the tropics, and sometimes on some Solanaceae. Development in the tropics usually takes 4–6 weeks, and there are several generations per year. Adult females are long-lived (3–10 months) and generally lay 10–15 oothecae; total egg production per female is usually about 200.

The tortoise beetles most frequently encountered are as follows:

Aspidomorpha spp. (sweet potato tortoise beetles)—about 12 species are recorded in Africa and another 12 species in tropical Asia; feeding mostly on *Ipomoea*, other Convolvulaceae, some *Solanum*, and a few other plants; found throughout Africa, tropical Asia, and the West Indies (Figs. 10.108, 10.109).

Cassida catenata (sweet potato tortoise beetle)—recorded on sweet potato and morning glory; S.E. Asia and Indonesia.

Cassida obtusata (sweet potato tortoise beetle)—recorded on sweet potato and morning glory; S.E. Asia and Indonesia.

Cassida spp. (beet tortoise beetles)—on beet and other Chenopodiaceae; Europe, Asia, parts of Africa (Fig. 10.110).

Conchyloctenia spp. (conchylo tortoise beetles)—many species in tropical Africa, recorded feeding on foliage of *Ipomoea* and *Solanum* spp.

Laccoptera spp.—on sweet potato and other *Ipomoea* spp.; East Africa.

Metriona circumdata (green sweet potato tortoise beetle)—on sweet potato and morning glory; India, S.E. Asia, China.

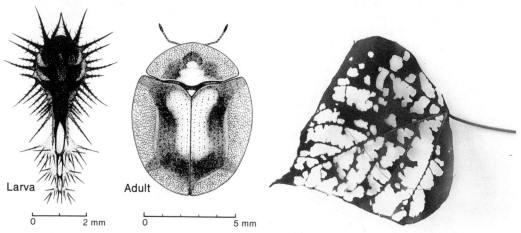

Fig. 10.108 Sweet potato tortoise beetle (*Aspidomorpha cincta*) ex Kenya (Coleoptera; Chrysomelidae; Cassidinae).

Fig. 10.109 Sweet potato leaf eaten by tortoise beetles; South China (Coleoptera; Chrysomelidae; Cassidinae).

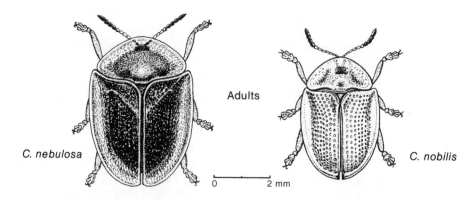

Fig. 10.110 Beet tortoise beetles (*Cassida* spp.) (Coleoptera; Cassidinae).

Hispinae (hispid beetles)

Small, elongate, beetles, mostly 4–9 mm (0.16–0.36 in.) long (some 10–16 mm/0.4–0.64 in.), often flattened, with short antennae and broad tarsi, and most have the body surface covered with spines. They mostly feed on the leaves of monocotyledonous plants such as palms, cereals, and grasses, but some dicotyledons are occasionally attacked. Adults are nocturnal in habits and fly well at night. Some larvae eat young, folded, leaves at the growing point and cause leaf scarification in long strips; other larvae are true leaf-miners and tunnel in the leaf mesophyll making blotchy mines of irregular shape. Adult feeding usually leaves longitudinal, whitish, scarring along the lamina of the leaf; adults and larvae are usually found together on the same leaves. Larval leaf mines are usually whitish when occupied but turn brown later. Pupation takes place in the larval tunnel.

Damage to crop plant hosts in the tropics is sometimes severe. The various palm hispids (as distinct from the palm leaf-miners) usually feed on the young, unopened, leaves and are especially damaging to seedlings and young plants whose leaves unfold rather slowly and so are susceptible to more damage. *Brontispa* species are often serious pests of coconut in the Pacific region. Coconut seedlings often are first attacked by *Plesispa* beetles (adults and larvae together) which feed on the newly opened leaflets causing surface stripping. Later the young palms tend to be attacked by *Brontispa* species. Both genera may be found together on palms 2–6 years of age. *Brontispa* damage is most serious on palms 2–3 years old, and infestations on 5–10 year old palms become progressively lighter. It is thought that the heart-leaves of older palms become harder and less suitable for larval development; also the leaves on older palms often open more rapidly, so the life cycle of the beetle is not completed within the shelter of the folded leaflets.

Some of the pest species in parts of S.E. Asia and the Pacific are kept partially under control through the encouragement of egg-parasites (Chalcidoidea). Pest outbreaks have often occurred when careless use of insecticides has destroyed the natural parasites.

The more notable hispid crop pests include:

Brontispa spp. (coconut hispids)—on coconut; S.E. Asia, Papua New Guinea, Solomon Islands, Pacific Islands.

Coelaenomenodera elaidis (oil palm hispid)—on oil palm; West Africa.

Dactylispa spp.—on coffee, peach, *Sterculia,* etc. (leaf mines); eastern Africa.

Dicladispa armigera (paddy hispid)—on rice and grasses; India, S.E. Asia, China (C.I.E. Map No. A. 228) (Fig. 10.111).

Dicladispa gestroi—on rice; Madagascar (C.I.E. Map No. A. 206).

Hispa spp.—on *Digitaria* and other grasses; eastern Africa.

Oncocephala spp.—on sweet potato; East Africa.

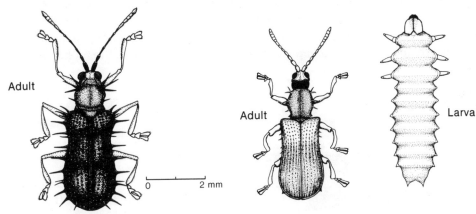

Fig. 10.111 Paddy hispid (*Dicladispa armigera*) (Coleoptera; Chrysomelidae; Hispinae).

Fig. 10.112 Rice hispid (*Trichispa sericea*); adult and larva (Coleoptera; Chrysomelidae; Hispinae).

Pleisispa spp. (palm hispids)—on Palmae; S.E. Asia.

Promecotheca spp. (palm leaf miners)—on Palmae, but mostly on native palms; S.E. Asia and the Pacific Region.

Trichispa sericea (rice hispid)—on rice; Africa (C.I.E. Map No. A. 257) (Fig. 10.112).

Wallaceana spp. (palm hispids)—on Palmae; Indonesia.

CURCULIONOIDEA

One of the largest groups of beetles, with many crop pests, showing tremendous variation in form and habits; and, in several respects to be regarded as the most highly evolved. The antennae are nearly always distinctly clubbed, and the head usually extended into a rostrum (except Scolytidae). Larvae are almost always legless—the Anthribidae larvae have thoracic legs but reduced to two segments; only the head is sclerotized; the larvae are all phytophagous.

There is dispute over the number of families to be recognized—in Imms (1977) nine families are given, but here the Scolytidae and the Platypodidae are recognized as separate families and not just subfamilies of the Curculionidae, because of their biological distinctiveness. Some of the families are small and obscure, and of little importance agriculturally, and so are not referred to here.

The Anthribidae are regarded as the evolutionary link between the Chrysomelidae and the weevils proper.

Anthribidae (fungus weevils)   *2400 spp.*

This family is tropical, mostly to be found in the Indo-Malayan region. The beetles look rather like Bruchidae in appearance, with a broad short snout, but with quite long antennae, not elbowed, but clubbed; maxillary palps normal (flexible). Most species are associated with dead wood and fungi. The genus *Brachytarsus* is predacious on scale insects (Coccoidea).

The species *Araecerus fasciculatus* (Fig. 10.113), known as either the nutmeg weevil or the coffee bean weevil, is now tropicopolitan on stored seeds and dried fruits of many kinds. In warmer countries, populations live in the field on the seeds of many wild and cultivated plants. In Hong Kong the favored host was *Nasturtium* growing wild on the hillsides. In East Africa this species tends to occur only in the dried coffee berries left on the bushes. Several other species of *Araecerus* are known as pests of cultivated plants: in Hawaii *A. laevipennis* (koa haole seed beetle), India has *A. suturalis* attacking dried fruits of custard apple and dried papaya stems, and *A. crassicornis* is the tephtosia seed weevil

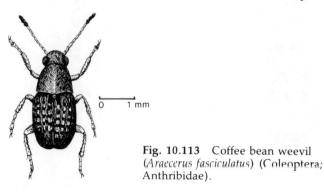

0   1 mm

Fig. 10.113   Coffee bean weevil (*Araecerus fasciculatus*) (Coleoptera; Anthribidae).

and it lives in the ripe pods of various legumes. In India *Basitropis* and *Dendrotogus* attack fruits of *Syzygium*. Several species of *Phloeobius* are recorded from coffee in East Africa.

Attelabidae (= Rhynchitinae) *300 spp.*

A small group of weevils with clubbed antennae, but not geniculate (the scape is short); maxillary palps rigid with four segments.

Some of these weevils cause "leaf roll" on various common trees in temperate regions. The female beetle makes the leaf roll by cutting the leaf lamina with her mouthparts at the end of a long pointed rostrum, and an egg is laid in the center of the roll. In some species (of subgenus *Caenorhinus*) the female cuts so as to girdle a shoot or leaf petiole to provide withered material for larval development. Most of the species in this group are Palaearctic or Nearctic in distribution; the group is of some importance to agriculture. The few noteworthy species are listed below:

Apoderus spp. (hazel leaf roller weevil, etc.)—larvae in rolled leaves of hazel, beech, birch, litchi, mango, etc.; most of Europe, India, S.E. Asia.

Attelabus spp. (oak leaf roller weevil, etc.)—on oaks, alder, chestnut, etc.; Europe.

Byctiscus spp. (poplar leaf roller weevil, etc.)—on poplars, hazel, birch, elm, etc.; Europe.

Deporaus spp. (birch and maple leaf roller weevils, etc.)—on hazel, *Acer,* beech, birch, etc.; Europe; also on mango in India.

Rhynchites bicolor (rose curculio)—flower buds destroyed on *Rosa* by ovipositing females; U.S.A.

Rhynchites (Caenorhinus) aequatus (apple fruit rhynchites)—on apple and hawthorn; Europe (Fig. 10.114).

Rhynchites (C.) caeruleus (apple twig cutter)—on apple, pear, oaks; Europe (Fig. 10.115).

Rhynchites (C.) germanicus (strawberry rhynchites)—on strawberry, raspberry, blackberry and loganberry; Europe.

Rhynchites lauraceae (laurel rhynchites)—in the mountains of Java the adults kill twigs and shoots on cinnamon and avocado (larvae not yet observed).

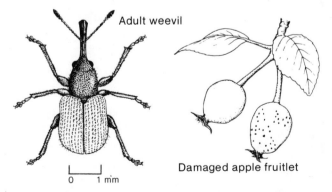

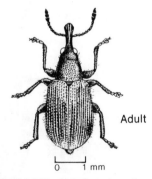

Fig. 10.114 Apple fruit rhynchites (*Rhynchites aequatus*) (Coleoptera; Attelabidae).

Fig. 10.115 Apple twig cutter (*Rhynchites caeruleus*) (Coleoptera; Attelabidae).

Brenthidae   *1300 spp.*

A tropical group of weevils with antennae often long, scape short, and antennae not geniculate and scarcely clubbed; body elongate and narrow; sexual dimorphism pronounced; size variable, but generally males the larger. They are native to tropical forests, where the larvae (so far as is known) bore in wood, and have thoracic legs. Some adults are predacious on the larvae of other wood-boring insects. None is regarded as an agricultural pest but a few are minor forestry pests.

Apionidae   · *1000 spp.*

A small group of weevils with worldwide distribution, characterized by having antennae distinctly clubbed, but not geniculate. Several genera are of importance as agricultural pests.

Apion (pod weevils, etc.)—a very large genus with hundreds of species found worldwide, but probably more abundant in cooler regions. The adults are tiny black beetles with a globular body tapering anteriorly to a long slender snout; when feeding they bite small holes in the plant foliage. Most species have larvae that develop inside seeds of Leguminosae and other plants, but some are leaf miners, and others develop inside plant stems. In the tropics the main hosts are Leguminosae, but others include cotton, cashew, jute, citrus, potato, peach, and sweet potato. A typical adult *Apion* is shown in Fig. 10.116.

Cylas (sweet potato weevils)—a truly tropical genus of importance as a pest of sweet potato. *C. puncticollis* is the African sweet potato weevil found only in tropical Africa, but is very damaging (C.I.E. Map No. A. 279) (Fig. 10.117). *C. formicarius* (sweet potato weevil) is found throughout Africa, tropical Asia, the Pacific, southern U.S.A., Central and South America (C.I.E. Map No. A. 278) where it is very damaging to this crop. Both adults and larvae may be found together in the tunnels in the tuber; the tunnels are often infected by fungi and so damaged tubers frequently rot (Fig. 10.118); tubers nearer the soil surface are generally more heavily damaged than the deeper ones.

Piezotrachelus spp. occur in eastern Africa on cowpea, peach, nug, yam, and leaves of coffee.

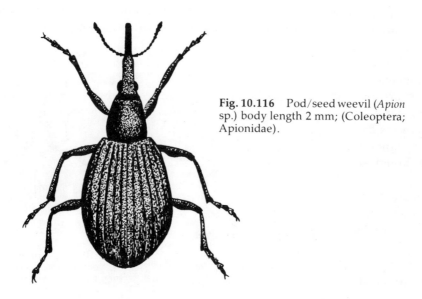

Fig. 10.116   Pod/seed weevil (*Apion* sp.) body length 2 mm; (Coleoptera; Apionidae).

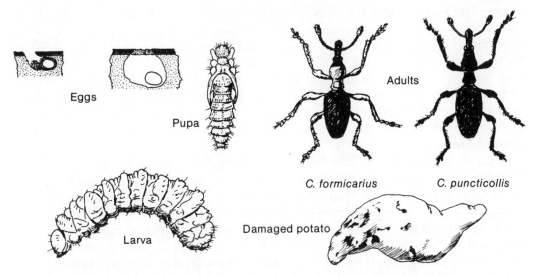

Eggs

Pupa

Larva

Adults

C. formicarius   *C. puncticollis*

Damaged potato

Fig. 10.117   Sweet potato weevils (*Cylas* spp.) (Coleoptera; Apionidae).

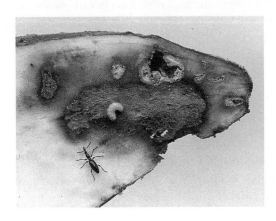

Fig. 10.118   Sweet potato tuber attacked by *Cylas formicarius,* showing adult weevil and larva; South China (Coleoptera; Apionidae).

Curculionidae (weevils proper)   *60,000 spp.*

A very large group of agricultural pests; totally worldwide. Adults have antennae geniculate; long trochanters; and many have a long rostrum bearing the biting mouth parts distally. Many subfamilies are recognized by various taxonomists but their designation is problematical and the taxonomic characters very esoteric, so they are not used in this text.

Larvae are generally legless, with a well-sclerotized head capsule, and large, well-developed mandibles. They are white, or pale bodied and usually live concealed either in the soil or inside plant tissues—they are never found feeding openly on plant foliage as are many other beetle larvae. Pupation takes place in the soil for the great majority of species, although some pupate (usually inside a cocoon) at the larval site on the plant body.

The adults often show sexual dimorphism, and the female has a longer, narrower rostrum. Many adults are black, brown, or gray in color, but some are brilliant blue or green. The color is sometimes owed to pigmented scales, and sometimes it is a fine powdery substance on the body surface. The diamond beetles, *Cypus* and *Entimus,* of Brazil are among the most brilliantly colored beetles recorded in the world.

The agricultural pests are varied in their biology and habits, and they attack all parts of the plant body. Adults generally eat plant foliage, usually leaves and young buds, sometimes bark; and the leaf-eating adults may be regarded as pests as well as the phytophagous larvae. The female weevil uses her long rostrum in order to bite deep into the host plant tissues, and the hole produced appears to be used (often) for feeding and then is often used as an oviposition site (see Fig. 10.127). Some species appear to be without males and it is assumed that reproduction is parthenogenetic. A few species are wingless, and some are winged although they very seldom fly; most adult beetles live for quite lengthy periods of time. Dispersal is either by flight or through the movement of infested produce.

Larval feeding habits, and hence their importance as agricultural pests, vary tremendously within the family; this is to such an extent that it is probably worthwhile viewing the group from this point of view, as is done below. It should be stressed that for many species the characterizations made below are constant and the insect does only that type of damage, but for others there is a variation of habits according to the host plant and cultivation conditions; some of the more polyphagous species can cause damage in several different ways so there may be considerable overlap between the different categories. Damage category designation is also difficult in that overlap is inevitable, for example leaf eaters will eat a terminal bud which is effectively a shoot; and the distinction between tree bark borers and wood borers is a fine one. The main ways in which weevil larvae and adults damage cultivated plants are reviewed below:

DAMAGE TO CULTIVATED PLANTS BY WEEVILS (*s.l.*) (Curculionoidea)

Roots

The majority of weevil larvae are soil-dwellers that eat the roots of plants by biting pieces of tissue with their biting and chewing mouth parts. Most of these species are not particularly host-specific and usually eat what is available, having very limited powers of dispersal.

Taproot—woody; a few species bore down woody trunks and end up boring inside the major woody roots (*Sipalinus*); others tend to be restricted to boring in the roots rather than stems (*Hylobius* spp.—pine weevils) but these are few in number. *Otiorhynchus* spp. and *Phyllobius* spp. will eat woody roots (and others) underground from the outside surface; also included here are *Diaprepes* spp., *Dicasticus* spp., *Pachnaeus litus* (citrus root weevil), and *Aperitmetus brunneus* (tea root weevil).

Taproot—fleshy; larvae of *Otiorhynchus* may tunnel from the outside into the interior; but *Lixus* spp. (cabbage/beet weevils) and *Listrononus oregonensis* (carrot weevil) spend the whole larval life tunneling inside the root, often having entered at the base of the stem.

Fine roots—dry soil; the larvae of the majority of species fall into this category, for example: *Phyllobius* spp. (common leaf weevils), *Myllocerus* spp. (gray weevils), *Graphognathus* spp. (white-fringed weevils), *Hypomeces squamosus* (gold-dust weevil), etc.

Fine roots—aquatic; several species are aquatic and known as "rice water weevils," and the larvae live in the soil submerged in paddy fields and eat the roots of the rice plants; *Echinocnemus oryzae* (paddy root weevil), *Hydronomidius molitor* (Indian rice root weevil), and *Lissorhoptrus* spp. (rice water weevils).

Root nodules on legumes—several species have larvae that do this but the most notable are the species of *Sitona* (pea and bean weevils) in the more temperate regions.

Stems (including tubers and rhizomes)

Stem borers—woody; the female weevil generally bites a hole deep into the stem tissues and lays an egg at the bottom, as with *Cyrtotrachelus longimanus* (bamboo weevil); many of these stems are technically woody but not very hard; also *Alcidodes* spp. (stem weevils), *Mecocorynus loripes* (cashew weevil), *Scyphophorus interstitialis* (sisal weevil), *Pissodes* spp. (conifer weevils), *Rhynchophorus* spp. (palm weevils), *Diocalandra* spp. (coconut weevils), and *Sternochetus goniocnemus* (mango weevil).

Stem borers—herbaceous; some larvae start tunneling in the leaf petiole and then continue the tunnel down into the stem; *Alcidodes* spp. (stem weevils), *Trichobaris* spp. (potato stalk borers), *Ceutorhynchus quadridens* (cabbage stem weevil), *Metamasius* spp. (sugar cane borers), *Odoiporus longicollis* (banana stem weevil), and *Lixus* spp. (cabbage/beet weevils) where the larvae may bore part of the stem before entering the root.

Shoot/twig borers—*Cyrtotrachelus longimanus* (bamboo shoot weevil), *Sternochetus goniocnemus* (mango weevil), *Pissodes* spp. (conifer weevils). *Rhynchites caerulus* (apple twig cutter) is a special case where the female cuts off a young soft twig after laying an egg in the soft tissues, and the larva feeds in the pith of the withering shoot on the ground.

Rhizome/tuber borers—larvae of *Cylas* spp. (sweet potato weevils), *Euscepes postfasciatus* (West Indian sweet potato weevil), and *Cosmopolites sordidus* (banana weevil).

Stem girdlers—adults of *Alcidodes* spp. (stem girdler weevils); larvae that bore under bark and often girdle the stem include *Pempherulus affinis* (cotton stem weevil); larvae in soil at or just below ground level feed on the bark and often girdle the stem, *Otiorhynchus* spp. (root weevils).

Stem gall makers—larvae of *Ceutorhynchus pleurostigma* (turnip gall weevil) in the stems of cultivated Cruciferae.

Seedling stem cutters—the adults of *Tanymecus* are called surface weevils and they feed on the soil surface cutting through the stems of soft seedlings.

Flowers

Flower buds—several species are referred to as "blossom weevils"; the female punctures an unopened flower bud and lays an egg inside, and the larva develops inside the bud which remains closed (capped) and it pupates there. In the cases of *Rhynchites bicolor* (rose rhynchites), *Ceutorhynchus asperulus* (red gram bud weevil), and *Anthonomus pomorum* (apple blossom weevil), the "capped" flower bud remains *in situ* on the plant, but with *Anthonomus rubi* (strawberry blossom weevil) after oviposition the female weevil girdles the flower stalk by a series of small bites and the flower bud falls to the ground and the larva inside develops rapidly (in about two weeks) feeding on the withering floral parts and receptacle.

Flower heads—larvae of *Larinus* spp. (thistle weevils) feed and develop in the flower head of thistles and other Compositae.

Pollen—adults of some species feed on pollen and occasionally also nectar in various flowers; one example is *Anthonomus grandis* (cotton boll weevil).

Seeds and fruits

Seeds—there are quite a few "seed weevils"; where the female lays an egg either directly into an ovary (clovers, etc.) or else into a young fruit and the larva has to bite its way into the developing seed (mango, etc.). Some of the best known species include: *Apion* spp. (clover seed weevils), *Apion* spp. (pod/seed weevils), *Amorphoidea* spp. (seed weevils), *Chalcodermus* spp. (tea seed weevils), *Ceutorhynchus assimilis* (cabbage seed weevil), and *Sternochetus mangiferae* (mango seed weevil), also *Araecerus fasciculatus*.

Grains—obviously these are seeds, but it is convenient to treat them separately; the species of *Sitophilus* (grain weevils) only develop inside cereal grains in field infestations, although they may be recorded from other seeds in food stores.

Fruits—the species that attack pods are included in the above category "seeds." The larvae of these species basically feed on the "flesh" of the fruit or on the fruit pulp, such as *Anthonomus grandis* (cotton boll weevil), *Baris* spp. (melon weevils), *Conotrachelus* spp. (plum curculio, etc.), *Omophorus stomachosus* (fig weevil), and *Sternochetus frigidus* (mango weevil).

Nuts—again there is little real distinction between categories, but the damage is very obvious; the larva develops inside the nut eating the kernel, and when fully grown it bites a large hole in the shell through which it leaves to pupate in the soil. Examples include *Balanogastris kolae* (kola nut weevil), *Curculio nucum* (hazelnut weevil) and *Curculio* spp. (nut weevils).

Leaves

Damage to plant leaves is mostly done by feeding adults (a number of species are known as "leaf weevils") in the vast majority of cases; some adults are diurnal and are readily seen *in situ*, but some are nocturnal and so although the damage symptoms are evident the weevils are very seldom encountered.

Leaf miners—generally rare in the Curculionoidea, but one notable species is *Rhynchaenus mangiferae* (mango leaf miner) whose larvae make small blotch mines in mango leaves. Other species of *Rhynchaenus* mine leaves of beech and other forest trees in Europe. In the U.S.A., larvae of *Odontopus calcentus* (yellow poplar weevil) mines leaves of several ornamental trees, and larvae of *Rhynchaenus rufipes* (willow leaf weevil) makes blotch mines in willow (*Salix*) leaves.

Leaf margin notching—some of the "broad-nosed" weevil adults make characteristic feeding notches around the periphery of the leaf margin, usually of quite regular shapes; including *Nematocerus* spp. (shiny cereal weevils), *Systates* spp. (systates weevils), and *Sitona* spp. (pea and bean weevils).

Leaf lamina eating—feeding adults tend to eat at the leaf margin but some will also make holes in the lamina, but usually the holes eaten do not have a consistent regular shape; such as *Dermatodes* spp., (gray weevils), *Deporaus marginalis* (mango leaf weevil), *Isanates* spp., *Myllocerus* spp. (gray weevils), *Phyllobius* spp. (common leaf weevils), and *Tanymecus* spp. (surface weevils).

Leaf folding—larvae of *Hypera* spp. (clover leaf weevils) live for some time inside the folded leaflets of clovers, where they eat the leaf lamina; and larvae and adults of *Goniopterus scutellatus* (eucalyptus weevil) are said to eat eucalyptus leaves together.

Leaf rolling—the leaf-rolling characteristic of some temperate European weevils on native trees is unusual in that in some cases the leaf roll falls to the ground and the larva is actually feeding on a wilted and decaying leaf; the female makes a series of cuts in the leaf lamina and induces a lateral rolling of the lamina and then she lays an egg inside; the leaf falls prematurely and the larva feeds on the decaying leaf on the ground. Other leaf-rolling species include *Deporarus marginatus* (mango leaf weevil) and *Apoderus* spp.

Leaf buds—adults of several species that feed on leaves may also eat buds and may destroy terminal shoots; including *Anthonomus grandis* (cotton boll weevil), *Otiorhynchus* spp. (black vine weevil, etc.), *Pissodes* spp. (conifer weevils).

Leaf petioles—larvae of several species start their lives as miners in leaf veins and petioles, and often they move down into the stem proper, such as *Alcidodes* spp. (stem weevils); *Listrononus oregonensis* (carrot weevil) larvae usually bore petioles before entering the root; *Lixus concavus* (rhubarb weevil) larvae in the U.S.A. bore extensively in the petioles.

Tree trunks and woody stems

Quite a large number of weevils are forestry pests, but many of these are not usually included within the scope of agricultural entomology.

Bark gnawers—damage is especially serious to seedlings and saplings—young plants may be killed by bark removal; adults of *Otiorhynchus* spp. (black vine weevil, etc.), *Alcidodes* spp. (stem weevils).

Bark borers—larvae of some species bore and tunnel under the bark in the vascular tissues, including *Lophobaris piperis* (pepper bark weevil), *Magdalis* spp. (bark weevils), and *Pantorhytes pluthus* (cocoa bark borer).

Living wood—trunks and branches mostly, many are cambium feeders; sometimes they bore down into the larger roots; some species prefer twigs; larvae of *Cratosomus punctulatus* (citrus borer) and *Magdalis* spp. (bark weevils, as above), bore under the bark in the vascular tissues. *Pissodes* spp. (conifer weevils) often show preference for young terminal shoots; also *Rhina* spp. (palm trunk borers) and also various larvae belonging to the family Brenthidae. *Sipalinus* bore in trunks and branches of *Pinus*, etc. The palm weevils (*Diocalandra* spp. and *Rhynchophorus* spp.) could also be placed in this category, as also could the adults of the Scolytidae, but they do not look like weevils and so have not been included in this categorization. The *Xylosandrus* spp. are the black twig borers.

Timber and dead wood borers—there are not many species in this category that are of agricultural importance, although many different species are known to attack dead forest trees and felled timber. Several species of *Euophryum*, native to New Zealand, are now widespread in temperate regions of Europe in dead oaks, willows, etc. Also in this category are placed the larvae of various species of the family Brenthidae.

Rotten wood borers—there are species to be found in decaying wood infested with fungi, and they are very seldom of any interest agriculturally. The full extent of the weevil fauna in this category in the tropics is not at present known, but the following genera are well-known in damp wood in Europe: *Eremotes, Euophryum, Pentarthrum, Rhyncolus* (pit-prop beetle, etc.), and the larvae of some Anthribidae.

Some of the more important and interesting species of weevils are listed below:

Alcidodes affaber (malvaceous stem weevil)—larvae bore stems and petioles of cotton and other Malvaceae, adults eat buds; India.

Alcidodes dentipes (striped sweet potato weevil)—adults girdle stem, larvae live in stem galls in sweet potato, cotton, woody legumes, etc.; Africa (Fig. 10.119).

Alcidodes gossypii (cotton stem-girdling weevil)—on cotton, etc.; tropical Africa.

Alcidodes spp. (stem-girdling weevils, etc.)—larvae bore in stems of beans, cotton, other Malvaceae, and young shoots of kapok, etc.; adults gnaw stems; Africa, India, and S.E. Asia.

Amorphoidea spp.—adults visit flowers, larvae develop in floral ovaries and fruits of cotton, coffee, etc.; only recorded from S.E. Asia.

Anthonomus grandis (cotton boll weevil)—on cotton; southern U.S.A., West Indies, Central and South America (C.I.E. Map No. A. 12) (Fig. 10.120).

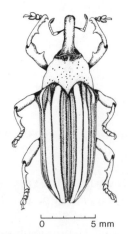

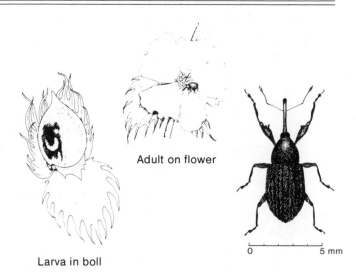

Adult on flower

Larva in boll

Fig. 10.119 Striped sweet potato weevil (*Alcidodes dentipes*) (Coleoptera; Curculionidae).

Fig. 10.120 Cotton boll weevil (*Anthonomus grandis*) (Coleoptera; Curculionidae).

Anthonomus pomorum (apple blossom weevil)—larvae inside capped buds of apple, pear, quince; Europe and Japan (Fig. 10.121).

Anthonomus pyri (apple bud weevil)—larvae inside capped buds of apple and pear; Europe.

Anthonomus rubi (strawberry blossom weevil)—larva inside unopened flowerbud; Europe.

Anthonomus signatus (strawberry blossom weevil)—as above; U.S.A. and Canada.

Aperitmetus brunneus (tea root weevil)—larvae damage taproot of tea, coffee, beans, *Brassica* spp.; East Africa (Fig. 10.122).

Balanogastris kolae (kola nut weevil)—larvae inside fruits (nuts); West Africa.

Baris spp. (melon weevils)—larvae bore cucurbit fruits in the Mediterranean and Africa, and pineapple in South America; other hosts in tropical Asia.

Blosyrus spp. (sweet potato black weevils)—adults eat leaves of sweet potato, coffee, etc.; Africa. S.E. Asia, China.

Ceutorhynchus asperulus (red gram bud weevil)—larvae in closed flower buds; India.

Ceutorhynchus assimilis (cabbage seed weevil)—larvae inside seed pods of rape and other Cruciferae; Europe, U.S.A., and Canada (Fig. 10.123).

Ceutorhynchus pleurostigma (turnip gall weevil)—larvae gall stem and root of Cruciferae; Europe (Fig. 10.124).

Ceutorhynchus quadridens (cabbage stem weevil)—larvae tunnel seedling stems and leaf petioles of Cruciferae; Europe, and U.S.A.

Ceutorhynchus spp. (turnip weevils, etc.)—several species attack turnips and radish in Canada and the U.S.A.; 52 species recorded in the U.K.

Chalcodermus spp. (plum curculio, etc.)—larvae bore inside apples, stone fruits, and nuts; U.S.A. and Canada.

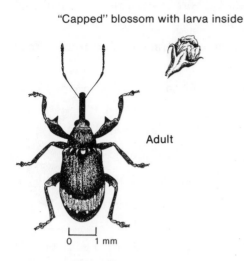

"Capped" blossom with larva inside

Adult

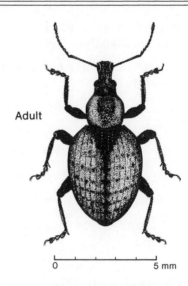

Adult

Fig. 10.121 Apple blossom weevil (*Anthonomus pomorum*) (Coleoptera; Curculionidae), and "capped" blossom; Cambridge, U.K.

Fig. 10.122 Tea root weevil (*Aperitmetus brunneus*) (Coleoptera; Curculionidae); ex Kenya.

Adult on rape flowers

Emergence hole in rape seed pod

Fig. 10.123 Cabbage seed weevil (*Ceutorhynchus assimilis*) (Coleoptera; Curculionidae); Skegness, U.K.

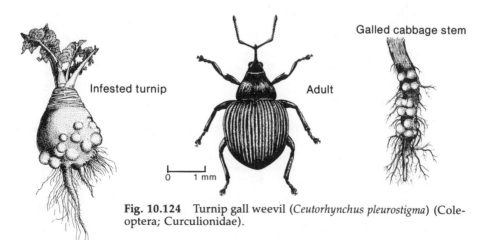

Infested turnip

Adult

Galled cabbage stem

Fig. 10.124 Turnip gall weevil (*Ceutorhynchus pleurostigma*) (Coleoptera; Curculionidae).

Cosmopolites sordidus (banana weevil)—larvae bore banana rhizome; pantropical (C.I.E. Map. No. A. 41) (Fig. 10.125).

Cratosomus punctulatus (citrus borer)—larvae bore branches and trunk of citrus; West Indies.

Curculio nucum (hazelnut weevil)—larvae inside hazelnuts and acorns; Europe (Fig. 10.126).

Curculio neocorylus (hazelnut weevil)—larvae inside hazelnuts; U.S.A.

Curculio occidentalis (filbert weevil)—larvae inside hazelnuts; U.S.A.

Curculio sayi (small chestnut weevil)—larvae inside chestnuts; U.S.A.

Curculio spp.—larvae found inside nuts and various fruits in many parts of the world.

Cyrtotrachelus longimanus (bamboo weevil)—larva inside shoot of bamboo; South China (Fig. 10.127).

Dermatodes spp. (gray weevils)—adults eat leaves of cocoa, citrus, cinchona, tea, tung; S.E. Asia, and China.

Diaprepes abbreviatus and spp.—larvae in soil eat roots of citrus, sugarcane, maize, and many other crops; West Indies.

Dicasticus spp.—larvae eat roots of tea, coffee, cotton, cassava, apple, etc.; Africa.

Diocalandra frumenti (four-spotted coconut weevil)—larvae bore all parts of the palms; East Africa, India, S.E. Asia, Pacific Region (C.I.E. Map No. A. 249) (Fig. 10.128).

Diocalandra taitense (Tahiti coconut weevil)—larvae damage coconut; S.E. Asia, Pacific Region, and now Madagascar (C.I.E. Map No. A. 248).

Echinocnemus oryzae (paddy root weevil)—aquatic larvae eat rice roots; India, Asia.

Entypotrachelus meyeri—damage tea, coffee, etc.; East Africa.

Euscepes postfasciatus (West Indian sweet potato weevil)—larvae tunnel tubers of sweet potato; Japan, southern U.S.A., West Indies, Central and South America.

Goniopterus scutellatus (eucalyptus weevil)—adults and larvae eat leaves of *Eucalyptus;* Australia, New Zealand, Africa, and South America.

Graphognathus spp. (white-fringed weevils)—polyphagous on many vegetables and field crops, larvae in soil eat roots and adults eat leaves; South Africa, Australia, New Zealand, U.S.A., and South America (C.I.E. Map No. A. 179) (Fig. 10.129); *Graphognathus leucoloma, G. minor, G. peregrinus* are the most important species.

Hydronomidius molitor (Indian rice root weevil)—aquatic larvae eat roots of rice; India.

Hylobius spp. (pine root weevils)—larvae damage roots of *Pinus* and other conifers; Japan, U.S.A.

Hypera postica (lucerne weevil)—larvae infest folded leaflets of lucerne, etc., adults eat leaves; Europe, India, parts of Asia, U.S.A. (C.I.E. Map No. A. 456).

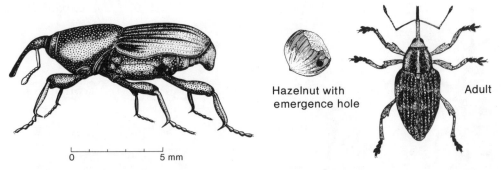

0   5 mm

Fig. 10.125 Banana weevil (*Cosmopolites sordidus*) (Coleoptera; Curculionidae).

Hazelnut with
emergence hole

Adult

Fig. 10.126 Hazelnut weevil (*Curculio nucum*) (Coleoptera; Curculionidae); Alford, U.K.

Damage

Larva

Adult ♀

0   1 mm

Fig. 10.127 Bamboo weevil (*Cyrtotrachelus longimanus*) and larva; female body length 32 mm (1.3 in.); bamboo (probably *Sinocalamus oldhami*) showing secondary branching following destruction of the apical shoot; South China (Coleoptera; Curculionidae).

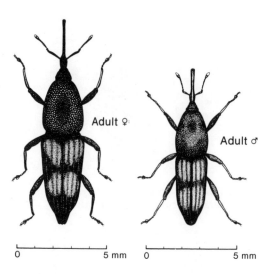

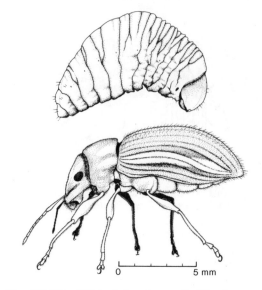

Fig. 10.128 Four-spotted coconut weevil (*Dio-calandra frumenti*) (Coleoptera; Curculionidae).

Fig. 10.129 White-fringed weevil (*Graphogna-thus* sp.) (Coleoptera; Curculionidae).

Hypera spp. (clover leaf weevils)—damage clovers, etc.; Europe, Asia, U.S.A., Canada (Fig. 10.130).

Hypomeces squamosus (gold dust weevil)—adults eat leaves of citrus, sweet potato, kapok, etc.; larvae in soil eat living roots of cotton, tobacco, etc.; India, S.E. Asia, Indonesia, Philippines, China (Fig. 10.131).

Isanates spp.—adults destroy young foliage of coffee, tea, apple, plum; East Africa.

Larinus spp. (thistle weevils)—larvae develop in flower head of thistles, safflower, and other Compositae; Europe, Asia (Israel to Japan).

Lachnopus coffeae—adults eat young leaves, buds, and fruits of coffee, and larvae in soil may eat roots; Central America.

Lissorhoptrus oryzophilus (rice water weevil)—larvae eat roots of rice in soil, and adults eat leaves; U.S.A., Canada, now Japan (C.I.E. Map No. A. 270) (Fig. 10.132).

Lissorhoptrus spp.—on rice; South America.

Listroderes costirostris (vegetable weevil)—polyphagous; U.S.A., and South America.

Listroderes obliquus (vegetable weevil)—adults damage many vegetable crops and may defoliate, larvae eat roots; Australia, New Zealand, Japan, South Africa.

Listrononus oregonensis (carrot weevil)—larvae bore root of carrot, and petioles of celery; U.S.A.

Lixus spp. (beet/cabbage weevils)—larvae bore inside stem/root of cabbage, beet, and many vegetables (also tea); Mediterranean, Africa, Asia, Indonesia, U.S.A.

Lophobaris piperis (pepper bark weevil)—larvae bore bark of pepper vines; Malaysia, and Indonesia.

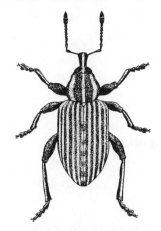

Fig. 10.130 Clover leaf weevil (*Hypera* sp.) (Coleoptera; Curculionidae).

Fig. 10.131 Gold-dust weevil (*Hypomeces squamosus*) on *Citrus* leaf; body length 12 mm (0.48 in.); Thailand (Coleoptera; Curculionidae).

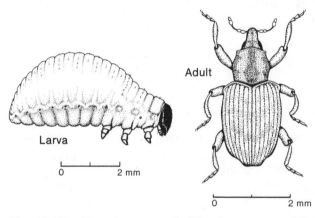

Fig. 10.132 Rice water weevil (*Lissorhoptrus oryzophilus*) (Coleoptera; Curculionidae).

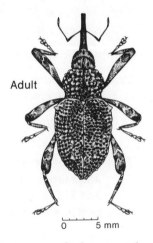

Fig. 10.133 Cashew weevil (*Mecocorynus loripes*) (Coleoptera; Curculionidae).

Magdalis spp. (bark weevils)—adults eat leaves, larvae bore under tree bark of apple, pear, plum, etc.; Europe, and U.S.A.

Mecocorynus loripes (cashew weevil)—larvae bore tree trunk of cashew; East Africa (Fig. 10.133).

Metamasius hemipterus (West Indian cane weevil)—larvae bore stems of sugarcane, stools of banana; West Indies, and Australia.

Metamasius spp. (cane/pineapple weevils, etc.)—larvae bore sugarcane, pineapple, etc.; West Indies.

Myllocerus maculosus (cotton ash weevil)—adults eat foliage of many fruit trees, larvae eat roots of cotton seedlings in soil, and many other crop plants; India.

Myllocerus subfasciatus (brinjal gray weevil)—adults eat leaves and larvae eat roots of Solanaceae; India.

Myllocerus spp. (gray weevils)—some 12–15+ species recorded as pests of sugarcane, and many fruit and nut trees; India, Asia to Japan (Fig. 10.134).

Nematocerus spp. (shiny cereal weevils)—larvae eat roots, adults eat leaves of cereals, and many other crops; tropical Africa (Fig. 10.135).

Odoiporus longicollis (banana stem weevil)—larvae bore pseudostem of banana; India, S.E. Asia to South China, Papua New Guinea (Fig. 10.136).

Omorphorus stomachosus (fig weevil)—larvae inside figs; Africa.

Otiorhynchus clavipes (red-legged weevil)—larvae eat roots, adults eat leaves, buds, etc., of many fruit trees, bushes, and strawberry; Europe (Fig. 10.137).

Otiorhynchus cribricollis (apple weevil)—adults on fruit trees, larvae in soil attack cover legumes; southern Europe, North Africa, Australia, and U.S.A. (C.I.E. Map No. A. 423).

Otiorhynchus sulcatus (black vine weevil)—polyphagous, both adults and larvae, on many crops; Europe, Asia, Australia, New Zealand, U.S.A., and Canada (C.I.E. Map No. A. 331) (Fig. 10.138).

Otiorhynchus spp. (root weevils, etc.)—adults damage buds of fruit trees, larvae in soil eat roots; Europe, Asia, Australia, New Zealand, U.S.A., and Canada.

Pachnaeus litus (citrus root weevil)—larvae damage citrus roots; U.S.A. (Florida).

Pantorhytes plutus (cocoa bark borer)—adults damage cocoa flowers, larvae tunnel under stem bark of cocoa; Papua New Guinea.

Pempherulus affinis (cotton stem weevil)—larvae bore seedling stem and may girdle stems of cotton and other Malvaceae; India and S.E. Asia.

Phyllobius spp. (common leaf weevils)—adults eat leaves of fruit trees, nuts, roses, and many other plants, larvae in soil damage pastures; Europe, parts of Asia, and U.S.A. (Fig. 10.139). In the U.K. there are ten species, five of which are pests of some importance.

Phytonomus = *Hypera.*

Pissodes spp. (conifer weevils)—larvae bore and kill terminal shoots of *Pinus* and other conifers causing "bushiness"; Asia, Japan, U.S.A., and Canada.

Protocirius colossus—larvae bore crown of sago palms, rattan, and young coconut; S.E. Asia.

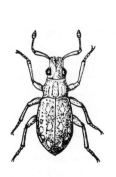

Fig. 10.134 Gray weevil (*Myllocerus* sp.) (Coleoptera; Curculionidae) (ex Butani).

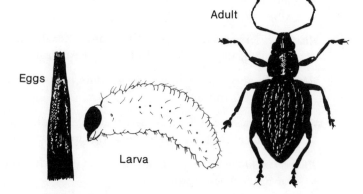

Fig. 10.135 Shiny cereal weevil (*Nematocerus* sp.) (Coleoptera; Curculionidae).

Damage

Larva & pupa   0   1 mm   Adult

Fig. 10.136 Banana stem weevil (*Odoiporus longicollis*); adult female body length 13 mm (0.52 in.); damaged stems, with larva and pupal cocoon (South China); young banana plant with stem broken following heavy weevil infestation (Penang, Malaya) (Coleoptera; Curculionidae).

Rhabdoscelus obscurus—larvae bore sugarcane and Palmae; S.E. Asia, Australia, and the Pacific Islands.

Rhina spp. (= *Rhinostomus*)—larvae bore palm trunks; East Africa, West Indies, Central and South America.

Rhynchaenus mangiferae (mango leaf miner)—larvae make leaf blotch mines on mango; adults eat young leaves; India.

Rhynchaenus spp. (leaf-mining weevils)—larvae mine leaves of temperate trees (beech, etc.); Europe, Asia, and U.S.A.

Rhynchophorus ferrugineus (Asiatic palm weevil)—on palms; India, S.E. Asia (C.I.E. Map No. A. 258) (Figs 10.140, 10.141).

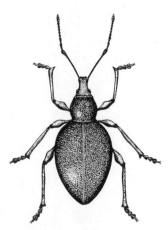

Fig. 10.137 Red-legged weevil (*Otiorhynchus clavipes*) (Coleoptera; Curculionidae).

Fig. 10.138 Black vine weevil (*Otiorhynchus sulcatus*) (Coleoptera; Curculionidae).

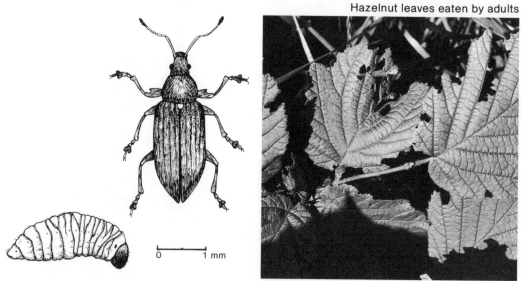

Hazelnut leaves eaten by adults

0   1 mm

Fig. 10.139 Common leaf weevil (*Phyllobius pyri*) (Coleoptera; Curculionidae) adult and larva; adult feeding damage to hazelnut leaves (Gibraltar Point, U.K.).

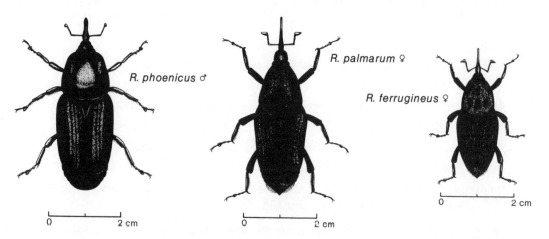

R. phoenicus ♂

R. palmarum ♀

R. ferrugineus ♀

0   2 cm

0   2 cm

0   2 cm

Fig. 10.140 Palm weevils (*Rhynchophorus* spp.) (Coleoptera; Curculionidae); adults.

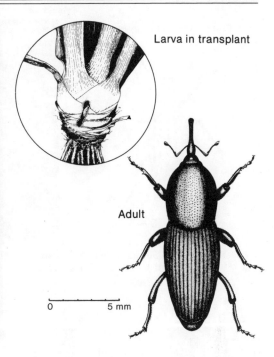

Larva in transplant

Adult

0   5 mm

Fig. 10.141 Coconut palm with crown destroyed by palm weevil larvae (*Rhynchophorus ferrugineus*) (Coleoptera; Curculionidae); Penang, Malaysia.

Fig. 10.142 Sisal weevil (*Scyphophorus interstitialis*) (Coleoptera; Curculionidae) ex Tanzania.

Rhynchophorus palmarum (South American palm weevil)—on palms; Central and South America (C.I.E. Map No. A. 259) (Fig. 10.140).

Rhynchophorus phoenicus (African palm weevil)—on palms; tropical Africa (Fig. 10.140).

Rhynchophorus spp. (palm weevils, etc.)—larvae tunnel in the crown of coconut, oil palm, and other palms; Africa, S.E. Asia, Pacific Islands (Fig. 10.140).

Scyphophorus interstitialis (sisal weevil)—larvae tunnel bole of sisal; East Africa, Java, Sumatra, southern U.S.A., Central and South America (C.I.E. Map No. A. 66) (Fig. 10.142).

Sipalinus aloysii-sabaudiae—larvae bore stem of cassava; Africa.

Sipalinus hypocrita (pine borer)—larvae bore trunk of *Pinus,* etc.; China and Japan.

Sitona spp. (pea and bean/clover weevils)—many species, adults notch leaves, larvae live in root nodules; Europe, Asia, U.S.A., and Canada (C.I.E. Map Nos. A. 372 and 437) (Fig. 10.143). In the U.K. 20 species are known, and 10 are pests.

Sitophilus granarius (grain weevil)—on stored wheat, barley, etc.; a more temperate species; cosmopolitan but at high altitudes in the tropics (Fig. 10.144).

Sitophilus oryzae (rice weevil)—on rice and stored grains; cosmopolitan (Figs. 10.144, 10.145); on maize, etc., in the field.

Sitophilus zeamais (maize weevil)—on maize both in the field and in storage, and on other stored grains and foodstuffs; cosmopolitan.

Field beans with notched leaves

Fig. 10.143   Sitona weevil (*Sitona* sp.) (Coleoptera; Curculionidae) adult, and *Vicia* beans with damaged leaves; Skegness, U.K.

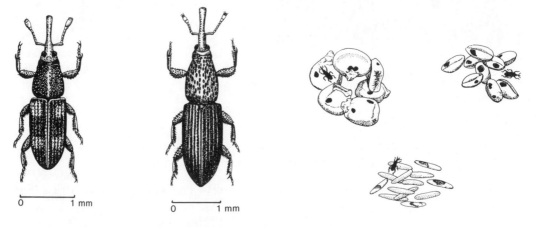

Fig. 10.144   Rice weevil (*Sitophilus oryzae*) and grain weevil (*Sitophilus granarius*) and damaged grains (Coleoptera; Curculionidae).

Fig. 10.145   Maize cob (after harvest) with field infestation of rice weevil (*Sitophilus oryzae*) (Coleoptera; Curculionidae); Alemaya, Ethiopia.

Sphenophorus spp. (billbugs)—on cereals and sugarcane; North, Central, and South America.

Sternochetus frigidus (mango weevil)—larvae bore pulp of mango fruits; S.E. Asia.

Sternochetus goniocnemus (mango twig borer)—larvae bore twigs and shoots of mango; Indonesia.

Sternochetus mangiferae (mango seed weevil)—larvae bore in seed of mango fruits; Africa, India, S.E. Asia, Australia (C.I.E. Map No. A. 180) (Fig. 10.146).

Systates pollinosus (systates weevil)—adults notch leaves and may defoliate young plants; polyphagous on coffee, citrus, tobacco, legumes, etc.; tropical Africa.

Systates spp. (20+) (systates weevils)—adults polyphagous on a wide range of cultivated and wild plants; tropical Africa (Fig. 10.147).

Tanymecus dilaticollis (southern gray/surface weevil)—adults eat leaves of many different crop plants, and may destroy seedlings by stem cutting; eastern Europe, and S.W. Asia (C.I.E. Map No. A. 357).

Tanymecus palliatus (gray/beet/surface weevil)—pest of beet, etc.; China (Fig. 10.148).

Tanymecus spp. (surface weevils)—adults damage many crops, such as maize, sorghum, sugarcane, apple, jujube, etc.; North and East Africa, India, and China.

Temnoschoita spp.—larvae bore banana stem and Palmae; West and East Africa.

Trichobaris spp. (potato stalk borers)—larvae bore stalks of potato and eggplant; U.S.A. and Canada.

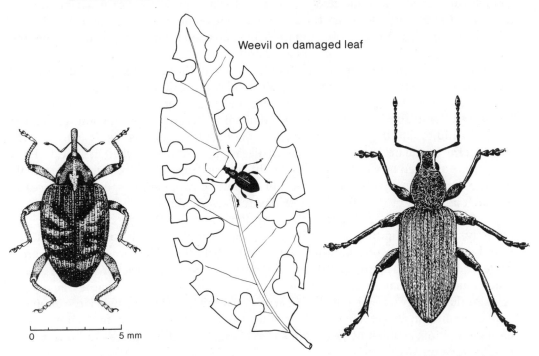

Weevil on damaged leaf

0   5 mm

Fig. 10.146 Mango seed weevil (*Sternochetus mangiferae*) (Coleoptera; Curculionidae).

Fig. 10.147 Systates weevil (*Systates crenatipennis*) (Coleoptera; Curculionidae) ex Kenya.

Fig. 10.148 Gray/beet/surface weevil (*Tanymecus* sp.) (Coleoptera; Curculionidae).

Scolytidae (bark beetles and ambrosia beetles)

In Imms (1977) these beetles are included as the Scolytinae and Platypodinae within the Curculionidae, but since they are so distinct biologically it seems preferable to regard them both as belonging to a separate family.

The adults are small (1–3 mm/0.04–0.12 in. long mostly), dark in color, cylindrical in body shape, with head deflexed and adapted for burrowing through wood and other plant tissues. As wood borers they are unusual (together with the Bostrychidae) in that it is the adults that are the primary tunnelers and not the larvae. The adult females tunnel into the sapwood under the bark to construct the breeding galleries some 2–12 cm (0.8–5 in.) long. Mating usually takes place in the tunnel, and then the female lays about 50 eggs along the tunnel. Mating sometimes takes place outside on the tree trunk prior to tunneling and in some species the male enlarges part of the female gallery and mating takes place there, or else the male may start the tunnel (having arrived first at the selected host tree). The female beetles of many species have special body cavities (mycangia) in which fungal spores or conidia of the ambrosia fungus (*Ambrosiella,* etc.) are carried, and these spores are deposited in the breeding tunnel to establish a mycelium to be used as food for the larval beetles. The mycangia of some species are cavities behind the mandibles, in others in the thorax, and some carry the fungus in the crop. The larvae feed on the starches and sugars present in the sapwood, or on the fungal mycelium present in the brood gallery. Generally the larvae make narrow tunnels at right angles to the breeding gallery, and the overall pattern of tunnels is often quite specific to each species. Some larvae feed only on the material in the wood. Other larvae feed almost entirely on the fungus—these are the "ambrosia" beetles proper; generally a tropical group for it is thought that in temperate regions the temperatures are too low for proper fungal development. The five larval instars are spent in the tunnel, but pupation often takes place in a cell near to the surface of the bark, from which eventually the exit tunnel will be made. Each infestation generally has a single entrance tunnel, but finally a large number of exit holes mark the site. Sometimes other tunnels (called "ventilation" tunnels) are made to the exterior, and these may be used for frass expulsion; sometimes frass is also ejected from the entrance tunnel.

Most of the species tunnel between the bark and the sapwood (see Fig. 10.151) of trees and woody shrubs; but some bore into solid wood, others into roots or twigs, a few into herbaceous plants, and also a few bore into fruits and seeds of palms, coffee, fir cones, etc. Some species are particularly damaging because they transmit pathogenic fungi (in addition to the ambrosia fungi) to the host tree, such as the *Scolytus* spp. that transmit the causal organism of Dutch elm disease (*Ceratostomella ulmi*) in Europe. Flight activity is governed by a combination of light, temperature and wind; when the females emerge they are positively phototactic, but this later disappears. Dispersal distances are as much as 1–30 km (⅝–20 miles) per day for several days.

Most Scolytidae are characterized by having very effective dispersal (they fly strongly) and host-finding mechanisms. A great deal of research has been conducted into their aggregation pheromones and their host-finding. The forestry pests seem to prefer to live in large aggregations on host trees that are sickly or severely stressed, so it is advantageous to have a large number of beetles converging on a suitable tree more or less simultaneously. The first beetles to arrive on a suitable tree produce large quantities of pheromones to attract other beetles to the tree. They often have to find a new tree for each generation, as most hosts do not survive long enough for more than one generation to develop.

The twig borers (*Xylosandrus* and *Xyleborus* spp.) can apparently attack healthy twigs, but they also destroy stressed seedlings, and some species invade the healthy branches via deadwood stumps left after pruning. The female lays some 20–30 eggs in a brood chamber, and usually the combined effect of adults, larvae, and fungi kill the plant distal to the point of attack. The life cycle is generally completed in about one month.

Some of the ambrosia beetles prefer recently felled timber, either with bark intact or stripped, and in general the group (Scolytidae) is more important as forestry pests than in the restricted agricultural context.

The system of reproduction in these beetles is unusual. Some species are monogamous (*Scolytes*), others bigamous (*Ips*), and some entirely polygamous (called spandrous) such as *Xyleborus* with a sex ratio of 1 : 10–60. In the latter group the male is often smaller than the female, flightless, a short-lived haploid (produced from unfertilized eggs) that does not leave the breeding gallery where it emerged.

Within the family there is considerable variation in the feeding life-styles, and these have been summarized by Beaver (1977) as follows:

Phloeophagy—feeding on the phloem/cambium layer; a very common type, mostly associated with trees in the temperate regions; *Dryocoetes, Polygraphus,* and some species of *Xyleborus.*

Herbiphagy—feeding on the soft tissues of herbs and woody twigs; an uncommon type; e.g. *Hylastinus* spp.—the clover root borers.

Spermaphagy—feeding on seeds and fruits, and sometimes also leaf-stalks (petioles) in the tropics; generally rare and confined to the tropics, including *Coccotrypes* (date stone borer) and *Hypothenemus hampei* (coffee berry borer).

Xylomycetophagy—these are the ambrosia beetles, and feed on the fungus cultivated within the breeding gallery tunnel system either under the bark of trees, or in twigs. A very common type, most abundant in the tropics, but quite plentiful in the warmer temperate regions; e.g. *Scolytus* and some *Xyleborus* and *Platypus* mostly in the tropics and in felled timber.

Xylophagy—the true wood feeders; quite rare.

Some of the more notable members of the Scolytidae are listed below:

Scolytinae

Blastophagus spp. (= *Tomicus*)—attack pines, firs, and spruces; temperate regions.

Coccotrypes carpophagus—attacks stored *Areca* nuts and *Anona* seeds; India.

Coccotrypes dactyliperda (date stone borer)—a primary pest of green fruits (dates) causing them to fall; Mediterranean Region.

Conophthorus spp. (pine cone beetles)—in pine cones; U.S.A., and Canada.

Dendroctonus spp. (pine beetles)—attack *Pinus*; U.S.A., and Central America.

Dryocoetes spp. (spruce/birch bark beetles)—attack various trees; Japan, and U.S.A.

Hylastes spp. (pine bark beetles)—in *Pinus* trunks; Japan.

Hylastinus obscurus (clover root borer)—in roots of clovers, etc.; Europe, U.S.A., Canada.

Hylesinus oleiperda (olive bark beetle)—tunnel bark of *Olea;* Mediterranean Region.

Hypoborus ficus (fig bark beetle)—on fig, mulberry, grapevine; Mediterranean Region.

Hypothenemus hampei (coffee berry borer)—adults bore ripe berries of coffee; Africa, S.E. Asia, now South America (C.I.E. Map No. A. 170) (Fig. 10.149).

Hypothenemus spp. (= *Stephanoderes*)—attack fruits, etc., of apple, cocoa, kapok, mango, sugarcane, cassava; Africa, U.S.A., West Indies, South America.

Ips spp. (pine/larch beetles)—attack various conifers; Europe, Japan, U.S.A., and Canada.

Poecilips myristicae—bore fallen pala nuts and acorns (*Quercus* spp.); Indonesia.

Polygraphus spp. (fir bark beetles)—attack firs, etc.; Japan.

Ruguloscolytus amygdali (almond bark beetle)—on almond; Mediterranean, Middle East.

Scolytus mali (large fruit bark beetle)—attacking various fruit trees; Europe, U.S.A.

Scolytus rugulosus (fruit tree bark beetle)—on many fruit and nut trees; Europe, North Africa, eastern Asia, North and South America (C.I.E. Map No. A. 392) (Fig. 10.150).

Scolytus scolytus (larger elm bark beetle)—on elms; Europe and Asia (C.I.E. Map No. A. 348) (Fig. 10.151).

Scolytus spp. (bark beetles)—many species known that attack fruit trees, nuts, and forest trees; throughout Europe, Asia, and North America.

Taenioglyptes spp.—bore walnut, *Pinus, Larix,* mulberry, etc.; Japan.

Xyleborus affinis—on guava, rubber, sugarcane; Africa, India, West Indies, Central and South America.

Xyleborus dispar (pin-hole borer)—polyphagous on many trees; Europe.

Xyleborus ferrugineus—polyphagous; pantropical (C.I.E. Map No. A. 277).

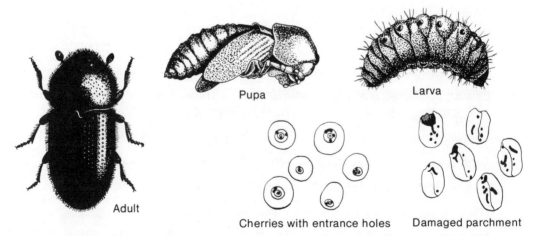

Fig. 10.149 Coffee berry borer (*Hypothenemus hampei*) (Coleoptera; Scolytidae) ex Kenya.

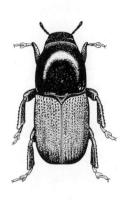

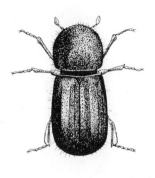

Fig. 10.150 Fruit tree bark beetle (*Scolytus* sp.) (Coleoptera; Scolytidae); body length 3 mm (0.12 in.).

Fig. 10.151 *Scolytus* breeding gallery under bark of an elm tree.

Fig. 10.152 Tea shot-hole borer (*Xyleborus fornicatus*) adult female, body length 3 mm (0.12 in.) (Coleoptera; Scolytidae).

Xyleborus fornicatus (tea shot-hole borer)—a polyphagous primary pest; Madagascar, India, S.E. Asia, Papua New Guinea (C.I.E. Map No. A. 319) (Fig. 10.152).

Xyleborus perforans (coconut shot-hole borer)—polyphagous; pantropical (C.I.E. Map No. A. 320).

Xyleborus morstatti—on coffee, avocado, etc.; Africa, and India.

Xyleborus saxeseni (fruit tree wood ambrosia beetle)—on apple, cherry, etc.; Europe.

Xyelborus sobrinus (citrus ambrosia beetle)—on citrus; Japan.

Xyleborus spp. (shot-hole borers; ambrosia beetles)—many different trees attacked (most stressed or moribund) including citrus, rubber, coffee, cocoa, tea, cotton, and grapevine; Europe, Africa, and Asia.

Xylosandrus compactus (black twig borer)—polyphagous, attacking healthy trees; pantropical (C.I.E. Map No. A. 244).

Xylosandrus morigerus (brown coffee borer)—polyphagous, primary pest on some trees and secondary pest on others; S.E. Asia (C.I.E. Map No. A. 292).

Xylosandrus spp.—on alder, *Castanopsis,* etc.; Japan.

Platypodinae

Crossotarsus spp.—on walnut, mango, etc.; India.

Platypus spp.—on mango, jackfruit, etc.; India.

11

ORDER DIPTERA
flies
c.130: 85,000

One of the largest orders, the Diptera show great diversity both morphologically and ecologically, and many are extremely specialized. One result of this extensive diversity is that this large number of species is split into a very large number of families, many of which are small (containing only a small number of species) and very specialized, which makes generalization difficult. Adults are characterized by possessing a single pair of wings, the hind pair being modified into halteres (balancing organs), and the wing venation is generally reduced. Mouthparts are suctorial, often in the form of a proboscis, and sometimes modified for piercing. Prothorax and metathorax are small and reduced, and are fused with the large mesothorax. Most species have all tarsi five-segmented. Metamorphosis is complete and there is great diversity shown between the adults and larvae, in all respects. Larvae are generally eruciform and apodous; head usually reduced, and often retracted; tracheal system varied but often amphipneustic. Pupa either free or enclosed within a puparium formed from the hardened larval cuticle.

Most adults are diurnal and feed on flower nectar, or on decaying organic matter, but many are predacious and prey on other insects, and a number are bloodsucking ectoparasites on vertebrates (some diurnal and some nocturnal). In most of the bloodsucking groups, except the Pupipara and Muscidae, it is only the female that takes blood. Because of the large number of haematophagous species, the Diptera are very important as medical and veterinary pests, and many are vectors (and often intermediate hosts) of a wide range of pathogenic organisms causing diseases in both man and domestic animals (malaria, sleeping sickness, yellow fever, etc.—see page 97). Some of the blood-sucking species are nocturnal, especially some of the mosquitoes (Culicidae), but most are diurnal. With the medical and veterinary pests, the great majority are adults, and often females only, and usually their larvae are free-living and often saprophagous, and may be aquatic. When the adult flies take a blood meal from a vertebrate host they invariably pump in saliva to prevent blood clotting prior to actually imbibing any blood—the saliva contains anti-coagulant chemicals. It is these chemicals in the saliva that cause the irritation associated with insect "bites" as they stimulate the host animal antibody reaction. In human individuals that normally have a low blood antibody titer there is usually no irritant effect with the feeding of biting flies, which is why the biting fly photographs used here invariably show the fly *in situ* on the arm or leg of the author.

In the Pupipara are several families in which the adults are permanent ectoparasites on birds and mammals, and they show interesting anatomical and behavioral modifications for this mode of life; one special adaptation being that they are larviparous ("pupiparous") and the larvae retained in the uterus are entirely fed by the female fly through special internal "milk" glands.

Eggs are usually elongate-oval in shape, whitish, and often with a characteristic surface sculpturing, and they are usually concealed. Those laid in cow dung, rotting organic matter, and wet soil, often have a flattened, tubular projection that functions as a plastron for respiration. Some mosquitoes lay eggs in groups, called egg-rafts, on the surface of the water.

The larvae show tremendous ecological diversity, unrivaled among the Arthropoda. Four families have larvae mostly phytophagous (Cecidomyiidae, Agromyzidae, Chloropidae, and Tephritidae); two are fungivorous (Mycetophilidae, Platypezidae); and several are saprophagous and scavenging (Bibionidae, Sepsidae, Phoridae, Anthomyiidae, Heliomyzidae, Scatophagidae). Predacious larvae are mostly in the Brachycera (most families), with some in the Syrphidae and Muscidae. Families with larvae truly parasitic are found in the Tachinidae, Oestridae, Pipunculidae, Bombyliinidae, Nemestrinidae, Conopidae), and also some Acalyptrates. The Diptera is second only to the Hymenoptera Parasitica in importance as a group of natural control agents of insect pests, and insect populations generally. Truly aquatic larvae are found in the Nematocera (Culicidae, Simuliidae, Chironomidae, etc.), and in the Stratiomyidae and Tabanidae of the Brachycera; but in a very large number of families some larvae live in wet soil or under semiaquatic conditions (Ephydridae, especially). Some of the aquatic larvae are filter-feeders, many are scavenging or saprophagous, a few are phytophagous, and some predacious; some are omnivorous and feed on a surface film of algae and organic detritus.

A number of larvae attack or live in the bodies of vertebrate animals, especially mammals, and this mode of life is referred to as myiasis. There is primary myiasis induced by true parasites, and secondary myiasis by saprophagous larvae infesting wounds which have become infected by microorganisms.

Anatomically the larvae of Diptera are reduced; there are no true legs, but some species have groups of body spines or pseudopods used to effect locomotion. Body segmentation is often obscure; there may be 12 body segments, three thoracic and up to nine abdominal. A few families have a well-defined head—larvae of Culicidae are referred to as "eucephalous," but most Diptera fall into the category of being "hemicephalous" with a reduced head, often partly retractable into the thorax, and the most advanced Cyclorrhapha have the head quite vestigial and are called "acephalous." Antennae are tiny structures, 1–6 segmented, best developed in a few predatory Nematocera. Mouthparts are varied, but in *Bibio* and some other Nematocera, the basic components are all recognizable, and there are functional biting mandibles. In the Cyclorrhapha the typical mouthparts have atrophied in correlation with head reduction and there has been developed a secondary feeding structure called the cephalopharangeal skeleton, being a series of black, articulated sclerites that operate in the vertical plane (in the Brachycera, the reduced mandibles operate vertically instead of the usual lateral action). Anteriorly there is a pair of mouth-hooks (mandibular sclerites) which articulate (vertically) with an H-shaped intermediate (hypostomal) sclerite, which in turn joins a large double basal sclerite (pharyngeal sclerite) in which the pharynx is lodged. The joint salivary duct passes through the hypostomal sclerite. The mouth-hooks in phytophagous species are toothed, but in carnivorous larvae the hooks are longer and sharply pointed.

The tracheal system is used as a systematic character of value. The primitive condition is peripneustic and is virtually confined to the Nematocera, with nine pairs of spiracles in some families and ten in *Bibio*. Newly hatched Cyclorrhapha are metapneustic but become amphipneustic in the second and third instars. The propneustic state is very rare; apneustic larvae are found in *Chaoborus*, Chironomidae, and some other aquatic forms. In peripneustic (and amphipneustic) larvae the first pair of spiracles is found on the prothorax and the second pair on the metathorax or first abdominal segment (never on the mesothorax). The position of the posterior pair is variable; usually on the last segment, but sometimes on the penultimate, or even the previous one. In some aquatic metapneustic larvae the two longitudinal tracheal trunks form a fine plexus of tracheae by the spiracles, and they pass to the walls of the posterior heart region. It is thought that the blood is brought into close contact with the oxygen in the tracheae and that they are functioning as a type of lung. Aquatic larvae usually develop accessory respiratory structures in the form of " blood-gills" (Chironomidae), and more usually "tracheal-gills" (Ephemeroptera, Odonata, Trichoptera, and some Diptera).

The number of larval instars varies considerably; in the Nematocera, four is general (six in Simuliidae), the Brachycera vary from five to eight, but in the Cyclorrhapha there is uniformity at three larval instars. Pupation occurs in two different ways. In most Nematocera and most Brachycera the larval exuvium is cast at pupation, but in the Stratiomyidae and some gall midges (Cecidomyiidae) the exuvium persists and encloses the pupa—when the adults emerges from the gall the exuvium is left protruding (see Fig. 11.7). In the Cyclorrhapha the larval exuvium shortens and hardens becoming smooth, shiny, ovoid, and eventually dark brown or black; the puparium is totally immobile. The whole process is very complicated and involves the formation of a pre-pupal stage, and a pharate adult is often present within the puparium for a while. In aquatic species it is often the case that the pupa has some kind of gill structure for respiration.

It is generally agreed that the Diptera comprises three suborders, the Nematocera, Brachycera, and Cyclorrhapha, and they show a progressive increase in specialization. In the following sections the small families of no agricultural interest are omitted.

SUBORDER NEMATOCERA

Adults with long, many-segmented antennae (no arista), and larvae usually having a well-developed exserted head, and mandibles that bite in the horizontal plane; pupae are obtect and free, and in some species quite active (except in some Cecidomyiidae). For more esoteric taxonomic characters refer to Imms' textbook (1977).

TIPULOIDEA

Tipulidae (crane flies; leatherjackets)   *13,500 spp.*

Adults are large in size (2–3 cm/0.8–1.2 in.); antennae quite long and sometimes pectinate in the male (but not plumose); no ocelli; legs long and deciduous; mesonotum with a V-shaped transverse suture; female has horny ovipositor; and some species have the front of the head prolonged, even forming a distinct proboscis. Larvae live in soil or water, and are soft, fleshy, metapneustic, and with fleshy retractile anal processes. The larvae of some species (Tipulinae) live in soils where they eat plant roots and are termed "leatherjackets"—they are particularly damaging to quality turf and to cereal seedlings, but are generally more important as temperate pests than tropical. Some species are adapted for life in very cold climates (e.g. snow flies—*Chionea*) and the adults have reduced wings. The larvae generally have a reduced head embedded in the prothorax and are called hemicephalous, but the mandibles are well developed. The elongate, extendable cylindrical body is gray or brown; in aquatic forms the anal processes include protrusible blood gills, but, in terrestrial forms these are greatly reduced. However, there may be retractile pseudopods on some body segments to aid locomotion through the soil. Some of the aquatic larvae are predacious on aquatic worms, others are saprophagous; one aquatic species can pierce grass roots with its posterior spiracular processes to obtain oxygen. Some aquatic larvae make silken cases in which they live, and the free-living ones usually make a silken cocoon for pupation, and in these respects they resemble Trichoptera.

In the Cylindrotominae, the larvae are green, either terrestrial or aquatic, and feed on aquatic mosses or Angiosperms; the body surface is covered with filamentous outgrowths. *Cylindrotoma* larvae feed openly on the leaves of various marsh plants, just like a caterpillar. One species of *Limonia* has larvae mining leaves of *Crytandra* in the Hawaiian Islands where it is very abundant.

Adults are diurnal, but most prefer cool, shady places and tend to avoid direct sunlight. Most are about 2–3 cm (0.8–1.2 in.) in body length, but the smallest are only 0.7 cm (0.28 in.) and the largest tropical species have a wingspan of up to 10 cm (4 in.). Mouthparts are usually reduced and scarcely functional; adults drink and take some nectar from the more "open" flowers and they may live for several weeks. The long-snouted crane flies apparently take nectar from flowers with a longer corolla. In the smaller species, mating is gregarious and swarms of males can be seen waiting for the females to emerge; this is generally accepted as being a primitive feature.

Members of the Tipulinae (long-palped crane flies) are the largest species, often brown in color; the larvae are vegetarian, eating plant roots, rotting wood, algae, moss, and probably also some small invertebrates occasionally; roots of tree seedlings may also be eaten as well as those of herbaceous plants and grasses. In this group is the Holarctic genus *Tipula* with several pest species of importance.

A more tropical species is the rather beautiful *Ctenophora* with a dark body and wings with orange banding; the larvae of this genus live in rotting wood. The similar *Tanyptera*, black and red in color, has a slim, sharp ovipositor and is thought to lay its eggs in wood that is less decayed and the larvae could be regarded as wood-borers.

The Limoniinae (short-palped crane flies) is a large group of mostly smaller species many of which have larvae that are carnivorous. The genus *Limonia* is enormous and worldwide and polydemic. Many larvae live in salt marshes and freshwater swamps and some are truly aquatic and fiercely predacious, swallowing prey almost as large as themselves. These aquatic predacious larvae are far more slender and active than their "leatherjacket" counterparts. A well-known tropical example of this group is the completely black colored crane fly *Hexatoma*, the larvae of which are thought to be aquatic and predacious; species of *Hexatoma* are known in Europe. Within this group are some fungivorous species that are recorded causing damage to Oriental mushrooms, namely species of *Ula*.

Some of the species of Tipulidae recorded as pests of agricultural crops worldwide are listed below:

Dictenidia fasciata (orchid crane fly)—larvae damage orchids; Japan.

Limonia amatrix (citrus crane fly)—larvae damage citrus roots; Japan.

Limonia nohirai (mulberry crane fly)—Japan.

Limonia spp. (small yellow crane flies)—Palaearctic and Australasia.

Nephrotoma spp. (spotted crane flies)—larvae polyphagous; Palaearctic.

Tipula sp.—recorded damaging sugar cane; Java.

Tipula spp. (crane flies; leatherjackets)—larvae polyphagous in soil; Holarctic (Fig. 11.1).

Ula fungicola (pine agaric crane fly)—larvae damages mushrooms; Japan.

Ula shiitakea (shiitake mushroom crane fly)—Japan.

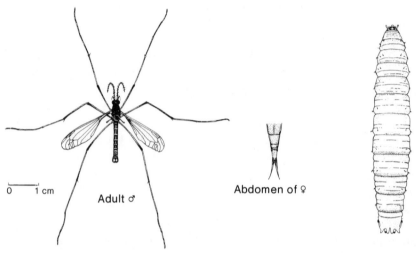

0   1 cm

Adult ♂

Abdomen of ♀

Fig. 11.1   Crane fly (*Tipula* sp.) adult male; and leatherjacket (larva) (Diptera; Tipulidae).

PSYCHODOIDEA

Psychodidae (sand flies; moth flies)   *1200 spp.*

Minute fragile flies, moth-like in appearance, with legs, body, and wings covered with coarse hairs and scales; no ocelli. Larvae are usually aquatic or saprophagous. Adults are usually found in close proximity to the larval habitat, in cool, damp, shaded locations. Some species are attracted to lights at night and often found on windows. The group is completely cosmopolitan. Taxonomists may recognize several different sub-families, but it is more general to accept two. The Psychodinae (moth flies) adults are quite stout-bodied, dark and hairy, and to be found resting on walls in cool, dark corners, especially bathrooms, cowsheds, caves, sewers, etc. (Fig. 11.2). They are generally 2–3 mm (0.08–0.12 in.) in length and the wings are held laterally at rest (like a moth) or roof-like over the abdomen. The larvae are aquatic and live in shallow, muddy water with a high organic content that is often deficient in oxygen. The head is well developed, and at the end of the abdomen is a protrusible respiratory siphon. Sewage filter beds are a favorite temperate breeding site; in the tropics, paddy fields and vegetable plots irrigated in the Chinese style and fertilized with "night soil" or animal dung may be heavily infested. A few genera are restricted to freshwater cascades and streams where *Simulium* larvae live, and *Trichomyia* live in rotting wood. Several genera are termitophilous in South America, and have the wings reduced to strap-like structures. One genus of Psychodinae is bloodsucking—*Sycorax silacea* is reputed to feed on the edible frog in France and it may also transmit a parasitic worm.

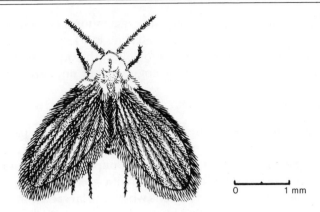

0   1 mm

Fig. 11.2   Moth fly (*Psychoda* sp.) (Diptera; Psychodidae).

The Phlebotominae (sand flies) are tiny, delicate, and hairy, and resemble minute hairy midges. The females are bloodsucking and pantropical in distribution, and 350 species are known. They are crepuscular or nocturnal in habits; they fly weakly and are easily disturbed when feeding. Many species of *Phlebotomus* are found in semiarid areas of Africa and the Mediterranean, but it is now known that they are also quite abundant in tropical rain forests. *Phlebotomus* flies are often confused with the Ceratopogonidae (biting midges), being of the same order and size, and some of the latter breed high on sandy shores in the tropics where the females bite painfully, and locally they are usually referred to as "sandflies." The bloodsucking female Phlebotominae feed on a wide range of vertebrate hosts including many reptiles and amphibians as well as birds and mammals. The piercing proboscis includes an elongated pair of mandibles distally shaped like a pair of scissor-blades and with which the host skin is cut. In humans they transmit the pathogen for "sandfly fever" in the Mediterranean Region, and they also transmit the flagellate protozoans responsible for leishmaniasis (kala-azar and oriental sore) throughout parts of Africa and tropical Asia. Sand fly larvae live in damp crevices in the soil, sometimes 20–30 cm (8–12 in.) deep, or between stones; in the absence of liquid water the larvae and pupae soon die of desiccation. The larvae are minute, with a well-developed head capsule, and bear long caudal bristles; fully grown larvae measure 5–8 mm (0.2–0.32 in.). It can be said that *Phlebotomus* larvae are actually aquatic, feeding on organic debris and animal dung, in crevices where condensation of atmospheric moisture results in drops of water being present. They breed in rodent burrows, termite nests, and around human habitations. In East Africa, one of the most frequently sampled sites for monitoring *Phlebotomus* populations was the entrance tunnels to the large *Macrotermes* termite mounds—presumably the larvae are present lower in the depths of the nest.

CULICOIDEA

A group of water midges—the larvae are all aquatic, smallish in size to tiny, and some of the adults are important as bloodsuckers of the Vertebrata.

Dixidae (pond midges)   *200 spp.*

A small cosmopolitan group, formerly placed in the Culicidae; adults resemble small crane flies, but are without the U-shaped thoracic suture. There are only two genera: *Dixa* which is worldwide, and *Neodixa* in New Zealand. Larvae are like those of

Anopheles and usually live on aquatic vegetation just below the water surface; sometimes they are actually above the waterline but kept wet by surface tension—they are filter feeders using mouth-brushes to collect the organisms. The adults are totally non-biting; the group is interesting ecologically but not agriculturally.

Culicidae (mosquitoes)   *3000 spp.*

Very slender flies, with long legs, long antennae (plumose in males; pilose in females) and an elongate piercing proboscis; wings with fringing scales posteriorly and along the veins; no ocelli. Larvae and pupae aquatic and active; larvae distinctly metapneustic. A worldwide group but with greatest diversity in the tropics, although extremely abundant in Arctic regions where the larvae are very important ecologically in the food webs of all the nesting birds. The larvae generally are important in freshwater food webs; some euryhaline species are important in saline and brackish pools on both rocky shores and salt marshes/mangroves. The adults are extremely important economically as vectors of disease organisms; some of the more important diseases transmitted by biting mosquitoes include malaria, yellow fever, dengue, filariasis, etc. (for further information see page 96). The most important genera medically include the following:

Aedes—many species are conspicuously banded black and white; many urban species for they are adapted to breed in tiny bodies of water such as rain butts, tin cans, broken bottles, footprints in the mud, etc.; in the wild they breed in tree holes and leaf axils (pineapple, banana, etc.). *A. aegypti* is the notorious yellow fever vector. Most countries and regions have a local species of *Aedes* causing trouble as a common domestic/urban species; in Hong Kong it is *A. albopictus* (Fig. 11.3).

Culex—a very large genus, completely worldwide, and with many subgenera and also subspecies recorded. The species *C. fatigans* is virtually tropicopolitan and is the vector of many pathogenic organisms; in many different countries it is called the brown house mosquito (Fig. 11.4). It is a typical mosquito in appearance; the larvae are found in pools and ponds and at the edge of slow-moving streams.

Anopheles—the malarial mosquitoes; worldwide in distribution, but more restricted to the warmer parts of the world (Fig. 11.5).

There are four well-established subfamilies; the Culicinae (*Culex, Aedes,* etc.), Anophelinae, Toxorhynchinae, and the Chaoborinae (glassworms)—the latter two are not bloodsucking but of interest ecologically.

Fig. 11.3 Black and white house mosquito (*Aedes albopictus*); Hong Kong (Diptera; Culicidae).

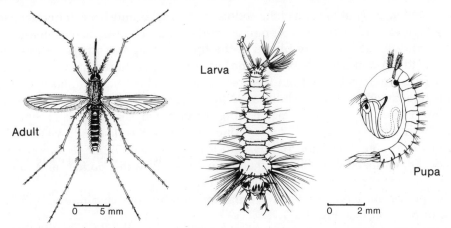

Fig. 11.4 Brown house mosquito (*Culex fatigans*) adult, larva and pupa; Hong Kong (Diptera; Culicidae).

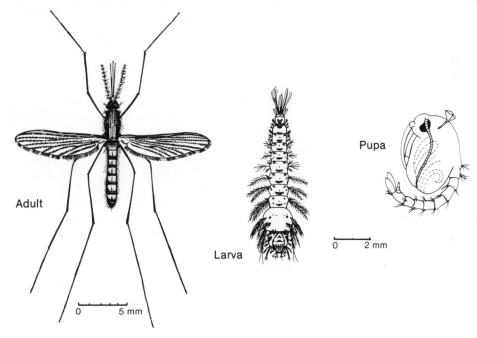

Fig. 11.5 Malarial mosquito (*Anopheles* sp.) adult, larva, and pupa; Hong Kong (Diptera; Culicidae).

The overall life history is similar within the family but details vary quite considerably according to the species concerned. Eggs are laid on, or near, the surface of water (fresh, brackish, or saline; stagnant or moving). Females of *Aedes* and some *Anopheles* lay eggs singly on the water surface. *Culex, Mansonia,* and *Theobaldia* lay batches of eggs up to 300 or more, stuck together in a raft on the water surface. Most of the eggs have small flotation chambers for buoyancy.

The larvae have a well-developed head with effective eyes, and the chewing mouthparts usually supplemented by a pair of lateral feeding brushes; they feed mostly on diatoms, microscopic algae, and detritus. But a few are predacious and have the mouth brushes replaced by stout spines used to seize the prey. In the case of the pelagic glassworms (*Chaoborus*) the larvae have prehensile antennae used to seize the prey— which are invariably larvae of other mosquitoes. The thoracic segments are fused to form a single rounded mass; at the end of the abdomen the anal somite bears four tracheal gills and long bristles. The respiratory siphon is on the penultimate segment (eighth) of the abdomen, and the spiracles are placed at the tip of the siphon. *Anopheles* are surface feeders and the siphon is very short, and the larvae tend to lie horizontally just under the water surface, whereas the bottom-feeding *Culex* has a long siphon and spends considerable time traveling from the bottom of the water body where it feeds to the surface for gaseous exchange—at the surface it hangs head-down. *Chaoborus* is exceptional in having a body almost transparent and is apneustic—it is pelagic in mid-water in larger bodies of freshwater (lakes and ponds) and it has two pairs of air-sacs internally for buoyancy. Other species of Chaoborinae occur in small water bodies, and one even occurs in the water inside pitcher plants in Australia.

The pupae are also aquatic and quite mobile, but the respiratory siphons are thoracic in position (see Fig. 11.5).

The different species of Culicidae tend to be rather specific in their choice of water body for development, as chosen by the female for oviposition. Some prefer large bodies of open water, others only lay in small streams or ponds that are overhung with vegetation, some require the pond to be choked with vegetation, some want hill streams. Oxygen concentration, temperature, light/shade, organic content, and water salinity are all decisive factors in influencing oviposition. A number of tropical species are adapted to breed in small temporary pools—generally these have an accelerated life cycle. A few live in tree holes, leaf axils, cut bamboos, and even in pitcher plants. Many of the species that live in ephemeral water actually lay eggs before the arrival of the water; the eggs are laid in soil or debris in suitable hollows, the embryos develop until ready to hatch and then remain in diapause until they are immersed in water (when the rains come) when the young larvae emerge very rapidly. Many of these small bodies of trapped water are referred to as "container habitats" as distinct from "ground pools"; ground pools are typically the most ephemeral. Apart from the Chaoborinae whose larvae practice body surface respiration, there are a few other species who do not have to visit the water surface for respiration—these are species of *Mansonia* which live in swamps (choked with vegetation) and they can insert their modified siphons into the root tissues of the plants for their oxygen supply.

The adults are mostly crepuscular or nocturnal, although some species are active in daylight—these being the forest species of *Aedes* and *Sabethes* whose natural habitat is cool and dark under the forest canopy; their natural hosts are forest monkeys and birds, hence *Aedes* species being the vectors of the pathogens known as "arborviruses" (usually the term is restricted to "arbo-viruses," meaning "arthropod-borne viruses," but both definitions occur in the literature, and both are appropriate). Adults of many species in the tropics hibernate to avoid the dry season, and in temperate regions to avoid the winter; the Arctic species usually pass the cold winter period as diapausing eggs. The mouthparts are extended into a long proboscis used for piercing and sucking. Males feed on plant juices, but the females require a blood meal for egg maturation, in addition to their basic diet of nectar and plant juices. They take blood from a wide range of Vertebrata, but each species usually has its preferred range of hosts. Mammals, birds, reptiles, and amphibians, as well as humans, are all attacked by a wide range of mosquitoes, and even some fish such as mudskippers which emerge from the water are attacked. A few species of mosquito can apparently lay eggs without first having a blood meal, but these are very few. Some adults are very long-lived and several months is a regular life-expectancy, even to a year or more.

An interesting genus of tropical mosquitoes is *Toxorhynchites*; the adults are large, blue and white mosquitoes to be found in rain forests and occasionally domestic premises where their presence often causes alarm needlessly; their proboscis is sharply

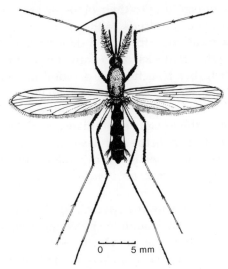

Fig. 11.6 Predatory mosquito (*Toxorhynchites splendens*); Hong Kong (Diptera; Culicidae).

bent to one side at about half its length (Fig. 11.6). These adults are not bloodsuckers, and they are harmless to humans, as they feed only on nectar from flowers; the larvae, however, are predacious and they usually live in the tree holes, etc., where they prey on larvae of *Aedes*. So *Toxorhynchites* species should be regarded as highly desirable insects that restrict the numbers of some of the forest/urban pest species of mosquitoes. They are also found in pitcher plants where they prey on whatever mosquito larvae are to be found. In Trinidad, there is an active campaign to use *Toxorhynchites* in a biological control program to reduce the numbers of *Aedes* breeding in inaccessible container habitats, apparently with considerable success.

Simuliidae (black flies; turkey-gnats; buffalo-gnats) *1450 spp.*

Small, stoutly built flies, black in color, legs short, wings broad with thickened anterior veins and remainder faint, and a large anal lobe; antennae short (11-segmented); no ocelli; and males holoptic. The larvae live in running water, often in cascades attached to rocks, stones, or vegetation by the anal extremity; spiracles closed; and pupae also in the same locations fastened to the rocks. The larvae are filter-feeders. Some species breed in clean, stagnant, water which is well aerated by having dense aquatic vegetation.

A small family but worldwide in distribution in freshwater streams, especially the faster flowing streams. Many species appear to need a blood meal for egg maturation—only the females suck blood—but not in all species. Often in this group the preferred hosts are birds rather than mammals. This explains partly why there are so many *Simulium* in the Arctic tundra of Europe, Asia, and Iceland, for these regions are renowned for the vast numbers of waterfowl (ducks, geese, and waders) that breed here in the short Arctic summer. *Simulium* species are equally abundant in the U.S.A. and Canada. In the Arctic regions the breeding birds, especially the young, are feeding on both the aquatic insect larvae and the adult flies (including of course *Simulium*!). Males fly to lights at night and may be trapped at considerable distances from where they bred; 100–300 km (60–200 miles) have been recorded, particularly when aided by the wind. Females are mostly diurnal, but sometimes fly at night.

Although these flies are important disease vectors (see page 97), probably most problematic is nuisance damage due to their bloodsucking. They are found in dense swarms in some locations, and hundreds or thousands may attack a single person or host-animal simultaneously, actually forming a dense black cloud around the victim; they are very persistent in their attacks, often crawling over the head and face in large numbers. Livestock are disturbed by actual loss of blood and the harassment which puts them off feeding to such an extent that cattle have been known to weaken and die. Parts of the Sudan, Rumania, Iceland, etc., are virtually uninhabitable by either humans or domestic livestock owing to the seasonal abundance of *Simulium* flies.

In Africa, *S. damnosum* transmits the filarial worm *Onchocerca*—causing filariasis or onchocerciasis. In humans, the parasitic worm usually settles in the eye and causes "river blindness." Other species transmit *Onchocerca* worms in Central and South America.

Egg-laying techniques vary with the species; some females lay on aquatic vegetation above the waterline, other just drop their eggs singly into the water, some enter the water to lay the eggs on submerged objects. One reason for the success of *S. damnosum* is that the larvae can develop in a wide range of aquatic habitats from torrents and cascades to almost stationary water in the lower river reaches. Each female may lay up to 500 eggs. *S. neavei* is remarkable in that the larvae live and develop in the branchial chamber, or on the body integument of freshwater crabs (*Potamon niloticus*); other species are now recorded living on the bodies of mayfly and dragonfly nymphs in Africa in a state of phoresy. There are usually 6–7 larval instars. The larva has a terminal proleg bearing a circlet of tiny hooks with which it attaches to the substrate; the head hangs downstream using the mouth-brushes for trapping food particles. Larvae are often so numerous on aquatic vegetation or under or on the sides of stones that they resemble a sheet of mucus. There is also a thoracic sucker with hooklets, formed by the fusion of a pair of pseudopods; rectal blood gills protrude from the anus. Larvae are quite mobile and move freely, but when fully grown they spin a silken cocoon attached to the surface on which they are living, and in here they pupate. The pupae have long branched respiratory filaments (gills) which usually protrude from the cocoon.

Worldwide, many species are known and are often placed in a series of subgenera. Some of the more important species are listed below:

Simulium arcticum—on birds in the Arctic Circle.

Simulium columbaschense—attacking domesticated animals; the Balkans.

Simulium damnosum—attacking humans and domestic animals; tropical Africa.

Simulium griseicolle ("Nimitti")—on turkeys, other birds, donkeys and humans; N.W. Africa.

Simulium indicum (potu fly)—the Himalayas and India.

Simulium maridionale (turkey-gnat)—on turkeys, chickens; U.S.A.

Simulium neavei—Africa.

Simulium occidentalis—on turkey, etc.

Simulium pecuarum (buffalo-gnat)—on buffalo and cattle; U.S.A. and Canada.

Simulium venustum—on ducks, etc.

Ceratopogonidae (biting midges) *4000 spp.*

Small or tiny flies, some species scarcely more than 1 mm (0.04 in.) in body length, gnat-like; male antennae plumose, female pilose; mouthparts adapted for piercing and sucking. The group is mostly Holarctic, but species are found on all continents. The adults are bloodsucking or predatory in some way; the female has a pair of mandibles like elongate scissors in the piercing proboscis (similar to those in *Phlebotomus* and *Simulium*); on other insects it is thought the mandibles operate scissor-like and cut the integument, but on people the blades are used in a stabbing manner. *Culicoides* females feed on vertebrate blood, but *C. anophelis* is interesting in that it pursues mosquitoes that have had a blood meal and it robs them of some of the blood by piercing the abdomen. Females of *Forcipomyia* (a very large genus) feed mostly on other insects such as adult Neuroptera, Lepidoptera, and some Diptera, and caterpillars of moths and some sawflies. Sometimes the midge takes blood very delicately from wing veins rather than piercing the body; they may be seen clinging to the wings of *Chrysopa* and other insects. *Lasiohelea* is a warm blood feeder and readily attacks humans—it is thought to breed in organic debris or soil, and can be found in forests and wooded areas, along with *Forcipomyia* species. Many genera feed on nectar, pollen, and plant juices, and some attack small, non-biting midges scarcely much larger than themselves. One strange genus is *Atrichopogon* that feeds on the blood of blister beetles (Meloidae) despite the presence of cantharidins in the beetle blood.

Probably the most important genus is *Culicoides* which contains a large number of species whose females suck blood from vertebrates, including humans. The feeding bite is painful, often followed by the formation of a red inflamed disc and swelling; and sometimes a clear blister develops. The adults are so small as to be scarcely visible to the unaided eye and in Africa one local name is "No-see-ums." The larvae are to be found in aquatic habitats around the edge of ponds and streams, and they are quite long and wormlike; the larvae are apneustic and obtain oxygen by diffusion from the surrounding water. In addition to freshwater habitats the larvae also inhabit brackish or salt water. In many parts of the tropics, *Culicoides* abound in mangroves where the adults feed in small swarms and the larvae live in the mud. In the Seychelles, on certain beaches, *Culicoides* breed in the wet sand at the top of the shore; the biting females drive the sunbathers off these beaches unless the sand is sprayed with insecticide. However, it is not feasible to be too dogmatic when describing breeding habits, some *Culicoides* are quite terrestrial and some *Forcipomyia* larvae are aquatic.

A few species transmit parasites in livestock (see page 95), but the nuisance value of the family is only associated with their irritating bites. And some species of *Forcipomyia* are recorded as being important pollinators of cocoa in Indonesia.

Chironomidae (lake midges; bloodworms) *5000 spp.*

These are delicate, gnat-like flies with conspicuous antenna (plumose in the male, pilose in the female); head small and often concealed; mouthparts usually poorly developed and most species do not feed in the adult stage; forelegs elongate; anterior wing veins more strongly marked than posterior veins; wings scaleless; larvae aquatic and apneustic. The adults are renowned for spectacular mass mating flights but otherwise are not very interesting. The larvae, however, play vital roles in most freshwater food webs and are very important ecologically; a few have been mentioned as damaging crops. They are most numerous in ponds and lakes, but a few can be found in slow-flowing streams and rivers. Most larvae live in the mud at the bottom of ponds where they construct soft tubes and they feed on the plankton and organic debris there.

Feeding is usually a form of filtration, but sometimes the mud surface is scraped. Some species feed entirely on algae and a few actually tunnel into stems of water plants, using their mandibles for biting the plant tissues. The subfamily Tanypodinae are all carnivorous and predacious on other insect larvae (mostly mosquito larvae) which they eat using quite well-developed mandibles. Benthic larvae are usually red in color (hence "bloodworms") as they contain hemoglobin as a respiratory pigment. Each larva has two pairs of elongate blood gills on the eleventh segment and two pairs of papilla-like anal gills. The surface dwelling and pelagic larvae (and predacious larvae) are generally greenish for they have hemocyanin as the respiratory pigment. It is generally accepted that hemoglobin is more efficient as a respiratory pigment than hemocyanin, and it has been developed in insect larvae and other aquatic invertebrates that live under conditions of low oxygen tension. Stream species of Chironomidae are usually greenish in color. It used to be thought that all chironomid larvae were to be found in shallow water but, in recent years, study of the group has revealed a far greater ecological diversity; some species are found in the benthos of very deep lakes. The Clunioninae include several genera that are maritime and feed on marine algae, and have apterous adults; larvae have been recorded from considerable sublittoral depths (down to 20 m/65 ft. or more). *Clunio* in Europe has winged males, but the females are apterous. Other species are abundant in salt lakes in Africa and Asia. Members of the Diamesinae live in icy Arctic waters and glacier streams in some mountainous regions. *Pontomyia* is a truly pelagic insect; larvae are recorded on the foliage of the sublittoral *Halophila ovalis*, living in mud tubes and feeding on diatoms; the adult female is apterous and wormlike and remains inside her larval exuvium on the *Halophila*; the male is winged and can run about on the sea surface and also dives beneath the surface. Species of *Polypedilium* live in shallow temporary pools in North Africa and, as reported in some classical papers by the late Professor Hinton, the larvae can withstand almost complete desiccation; when dehydrated they can survive for years in a state of cryptobiosis waiting for the next period of rain when the pools would fill. Some species have larvae that attach themselves to nymphs of mayflies and other aquatic insects (Odonata, Coleoptera, Diptera, etc.), and snail shells as commensals, and a few are truly parasitic and prey on mayfly nymphs and aquatic snails. A few larvae are not aquatic and are to be found in soil, animal dung, compost, rotting wood, and the like.

Eggs are laid in a mass inside a gelatinous envelope, often in a "string"; most egg masses float on the water surface. A few species are parthenogenetic.

Pupation usually take place either in the larval tube or at least in the same microhabitat as the larva.

Adults are short-lived and weak fliers and may form vast mating swarms over the water surface, usually in the evenings. In parts of Africa, especially Lake Victoria, the swarms are so extensive and dense as to appear cloud-like. At times the swarms are netted and the midges compressed into blocks of "midge butter" and used for culinary purposes. In regions where aquarists are abundant, there may be a local trade in the harvesting of "bloodworms" from local ponds and river edges for sale as food for aquarium fishes. In Hong Kong and South China (and parts of S.E. Asia) this collecting practice is augmented by the actual rearing of *Chironomus* larvae (Fig. 2.12) in shallow concrete pools and pans where they are cultivated in organically rich water and feed on the thick planktonic "soup." In many parts of S.E. Asia, bloodworms may be harvested from the edge of fishponds as part of the "multicropping" practice of rearing ducks and fish together using the same ponds. In regions of field plot irrigation, as practiced in S.E. Asia, there may be large numbers of chironomid larvae in the water and extensive populations can be found in paddy fields where the organic content of the water is high. Most species to be found in paddy fields are totally harmless, and serve as food for the fish that may be there, but a few species are recorded as being polyphagous and damaging to the rice plants.

The few pest species recorded, include the following:

Chironomus oryzae (rice midge)—recorded on recently transplanted rice seedlings, damaging the stem bases; S.E. Asia from Malaya to Japan.

Chironomus plumosus—reported grubbing up roots of very young plants; Hungary.

Chironomus thummi—larvae eat leaves of rice seedlings; Camargue region of southern France.

Cricotopus bicinctus—larvae eat and cut leaves of rice seedlings; Hungary and California (U.S.A.).

Cricotopus trifasciatus—larvae gnaw stems and leaves of rice seedlings; Camargue (France).

Hydrobaenus macleayi—recorded damaging early sown beans in organic enriched soils in Australia when germination was delayed by cold conditions.

Spaniotoma furcata—reported by Oldroyd (1964) as being an occasional pest in greenhouses where the larvae damage seedlings.

Stenochironomus nelumbus (lotus lily midge)—larvae eat lotus leaves; Japan.

BIBIONOIDEA

Bibionidae (fever flies; March flies)   *700 spp.*

A small group of quite large, robust flies, mostly black in color (but a few species are very small). Adults have short antennae; large wings; male eyes holoptic; ocelli present. Larvae are found in soil, particularly soils with a high organic content; they are elongate and somewhat wormlike, but with a definite exerted head capsule and well-developed mouthparts. They often occur in large numbers, both as larvae in the soil and swarms of adults on flowers or on the vegetation generally. Swarming adults can be a nuisance, but larvae in large numbers can be damaging to plant roots and to seedlings—they will eat more or less any vegetable matter with their biting and chewing mouthparts. They are probably most abundant in temperate parts of the Holarctic region, but are well represented in Australasia, and in some more tropical regions such as India and Indonesia.

The few pest species recorded include:

Bibio spp.—known in the U.S.A. as March flies, and in the U.K. *B. marci* is St. Mark's fly; both species swarm in very large numbers in the spring; larval damage is reported from a very wide range of field and vegetable crops, but damage levels usually low.

Dilophus spp. (fever flies)—they also swarm in very large numbers in the spring, and also in the autumn in some regions; swarms in urban areas may be annoying but the flies can be of importance as fruit tree pollinators; Europe.

Plecia spp.—recorded as abundant in vegetable plots and gardens in Java, but seldom damaging to growing plants; also recorded in Japan.

MYCETOPHILOIDEA

In some books the Sciaridae are regarded as a subfamily within the Mycetophilidae, but the general trend seems to be to regard it as a separate family.

Mycetophilidae (fungus midges)   *2000 spp.*

Small flies with elongate antennae and legs, widely distributed throughout the world in both temperate and tropical regions. The larvae are smooth and vermiform with a dark head capsule, and they live gregariously in fungal fruiting bodies (mushrooms, etc.) and rotting vegetable matter. A very few species are predacious, and the remarkable genus *Planarivora* is endoparasitic in land planarians. The larvae have labial glands modified to produce silk, and webbing is often found associated with their infestations. In New Zealand the famous "glowworm cave" at Waitomo has the cave ceiling covered with luminous larvae of *Arachnocampa luminosa* suspended in webs. Other luminous larvae are known in parts of Asia.

The few pest species belong to the genera *Mycetophila* and *Exechia* and are often referred to collectively with some Cecidomyiidae as "mushroom midges." The *Mycetophila* larvae are often sufficiently abundant to destroy entire mushroom cultures, although usually they occur as mixed infestations with Sciaridae and some Phoridae, as well as some Cecidomyiidae.

Sciaridae (dark-winged fungus gnats)   *1000 spp.*

Small dark midges, widespread in distribution, with species adapted to a very wide range of conditions and climates; the larvae live in soil, feeding on almost any type of organic debris. The group is in somewhat of a state of taxonomic confusion—many early records of *Sciara* were apparently misidentifications and did not belong to this genus, so there is doubt about many early records. Adults fly to lights at night and in many rain forest areas of the tropics light-trap catches may be very high.

The larvae are recognizable (together with Mycetophilidae) by being wormlike but with a black head capsule, and biting mouthparts. The larvae of *Sciara militaris* are unusual in that they are recorded as "marching" in long rows across the floor in rain forests and sometimes coffee plantations in Indonesia; the line of larvae may be 2–3 cm (0.8–1.2 in.) wide and 2–3 m (6–10 ft.) in length. Occasionally *Sciara* breed in urban or domestic situations and have been recorded as causing a nuisance, both as larvae and as adults inside the buildings.

Mostly Sciaridae are to be regarded as minor pests; although recorded damaging a wide range of field and vegetable crops in many parts of the world, damage levels are usually slight. But in greenhouses, there may be serious damage; and mushroom cultures may be sometimes destroyed by the feeding larvae. In both Europe and North America there are forms of potato tuber scab and rot that are definitely thought to be due to damage by feeding larvae of *Pnyxia scabiei*, etc. Damage is usually done to the plant roots but *Bradysia tritici* also eats the stem of wheat seedlings. A few species are recorded as leaf miners and a species of *Bradysia* in Hong Kong makes leaf galls.

Some of the more important pests are listed below:

Bradysia tritici (moss fly)—larvae damage wheat and other cereals; Europe.

Bradysia spp. (mushroom/sciarid flies)—in mushroom cultures; cosmopolitan.

Lycoriella spp. (mushroom/sciarid flies)—in mushroom cultures; Europe.

Pnyxia scabei (potato scab gnat)—damage potato tubers; Europe, Asia, U.S.A.

Sciara spp.—previously regarded as a large and widespread genus, but of late interpreted more narrowly, and many early records now regarded with uncertainty; cosmopolitan.

Cecidomyiidae (= Itonidae) (gall midges)   *4500 spp.*

Tiny, delicate midges; antennae long and moniliform with conspicuous whorls of setae; wings with few veins and no crossveins. The larvae are also tiny, and like the adults may be red, orange, gray, or yellow in color; some larvae are quite elongate, others shorter and more rounded; all have a reduced head; most possess a stout, sternal spatula under the thorax thought to be used for digging in the ground prior to pupation. Some larvae are able to contort their body suddenly to leap into the air; this is said to be done by locking the anal crotchets on to the extremity of the spatula and then releasing the tension.

Most larvae are phytophagous and many induce galls in the plants in which they live, hence their common name of "gall midges"; the study of plant galls is often referred to as cecidology. Saliva produced by the feeding larvae apparently induces the host plant to form gall tissue. For the average field entomologist the group is taxonomically very difficult. The family is easy enough to recognize but the genera and species require an expert taxonomist for identification. Generally, the field entomologist relies on the body color of the insect together with host plant identity and the type of gall formed to arrive at an approximate identification.

The group actually shows greatly diversified habits and has been interpreted as follows:

1. Zoophagous species. These mostly prey on Homoptera and mites; but some prey on other Diptera (both larvae and pupae); a few are parasitic on Aphididae and other small Homoptera.

2. Saprophagous species. Found in decaying vegetable matter, in fungi, in dung, with some species living in the excrement of caterpillars.

3. Phytophagous species. These can be further subdivided:
   i. Plant feeders making no galls—cereal and grass midges, in the flower heads of Gramineae; others in the flower heads of Compositae and others.
   ii. Plant gall inquilines—living inside galls induced by some Hymenoptera, Coleoptera or Diptera (Tephritidae, and other Cecidomyiidae).
   iii. True gall formers (cecidogenous species)—the great majority of species occur in this category.
      a. Leaf and leaflet semi-galls—*Dasineura* spp. on legumes, rose, violet, etc.
      b. Leaf galls proper—longan gall midge, etc. (Fig. 11.7).
      c. Shoot and bud galls—pea midge, hawthorn shoot midge, etc.
      d. Seed and fruit galls—sorghum midge, sesame gall midge, etc.
      e. Stem/twig galls—*Asphondylia morindae* on *Aporusa chinensis* (Fig. 11.8).
      f. Root galls—only a few species, on roots of ornamentals and aerial roots of some orchids; some adults appear to be subterranean.

Some of the predacious species are of importance as agents of natural control for certain insect pest species. A few of the more notable include:

Fig. 11.7   Longan gall midge (genus and species indeterminate) leaf galls; South China (Diptera; Cecidomyiidae).

Fig. 11.8   Stem/twig galls of *Asphondylia morindae* on *Aporusa chinensis*; Hong Kong (Diptera; Cecidomyiidae).

Acaroletes spp.—prey on spider mites (Tetranychidae).

Aphidoletes spp.—prey on a wide range of Aphididae.

Arthrocnodax spp.—attack many different phytophagous mites.

Coccodiplosis spp.—prey on Coccoidea in many parts of the world.

Diplosis spp.—prey on both mealybugs and scale insects (soft and hard); worldwide.

Lestodiplosis spp.—prey on phytophagous mites belonging to several families, and also on Psyllidae, Aleyrodidae, and some small Coleoptera.

Phenobremia spp.—recorded attacking many Aphididae in many parts of the world.

Schizobremia spp.—prey on Coccoidea; worldwide.

Most of the predacious larvae feed rather in the manner of syrphid larvae, in that they pierce the body of the bug and suck out all the fluid body contents.

A number of species of Cecidomyiidae, especially some seed-inhabiting ones, are of importance in that they can destroy plant weeds and some have been used successfully in biological control programs. A good example of this is *Zeuxidiplosis giardi* introduced into Australia to help control St. John's wort *(Hypericum)*, by destroying buds, shoots, flowers, and leaves.

The family is large and there are very many species (many hundreds) to be encountered, mostly on plants, and most species are quite specific and to be found only on one plant (genus or species). Many are found on wild hosts but a large number will be seen on cultivated plants, although not many are serious pests. To give an indication of the abundance of gall midges generally, it can be noted that for four dominant genera the numbers of species recorded for the U.K. and for the U.S.A. are as follows: *Asphondylia*—U.K. 15 spp., U.S.A. 55 spp.; *Contarinia* (blossom and fruit midges)—72 spp. and 32 spp.; *Dasineura* (flower and leaf midges)—136 spp. and 95 spp.; *Lasioptera* (fruit and stem galls)—9 spp. and 74 spp. The plant groups most attacked by gall midges appear to be the Leguminosae and the Gramineae, but most plant families are attacked by some midges. Sorghum midge, saddle gall midge, hessian fly are major pests; and below is presented a selection of the more important midge pests of cultivated plants:

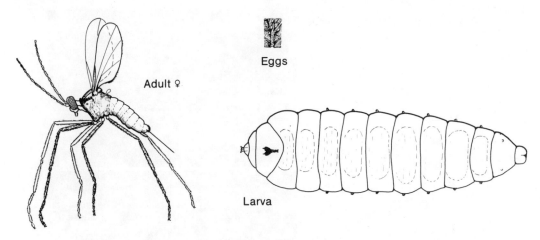

Fig. 11.9 Sesame gall midge *(Asphondylia sesami)* (Diptera; Cecidomyiidae).

Asphondylia sesami (sesame gall midge)—in sesame pods; East Africa, India (Fig. 11.9).

Asphondylia websteri (alfalfa gall midge)—on alfalfa; U.S.A.

Contarinia citri (citrus blossom midge)—in citrus flowers; Mediterranean, Mauritius, India, and China.

Contarinia gossypii (cotton gall midge)—in cotton flowers; West Indies.

Contarinia humuli (hop strig midge)—in hop flower heads; Europe.

Contarinia johnsoni (grape blossom midge)—in grape flowers; U.S.A.

Contarinia lycopersici (tomato flower midge)—in tomato flowers; Hawaii, West Indies.

Contarinia mali (apple blossom midge)—in apple flowers; Japan.

Contarinia merceri (coxsfoot/foxtail midge)—in grassheads; Europe.

Contarinia medicaginis (lucerne flower midge)—in alfalfa flower heads; Europe.

Contarinia nasturtii (swede midge)—distort leaves of Cruciferae; Europe.

Contarinia pisi (pea midge)—in shoots or pods of pea; Europe (Fig. 11.10).

Contarinia pyrivora (pear midge)—destroy fruits of pear; U.S.A., Europe.

Contarinia sorghicola (sorghum midge)—larvae in sorghum ovaries; Africa, Japan, Australia, U.S.A., South America (C.I.E. Map No. A. 72) (Fig. 11.11).

Contarinia tritici (yellow wheat blossom midge)—on wheat (also barley); Europe, Asia to Japan (C.I.E. Map No. A. 182).

Contarinia spp. (flower midges)—larvae in flowers of legumes, grasses, cereals, some vegetables, fruit trees, and bushes; Europe, Asia, North America.

Dasineura affinis (violet leaf midge)—roll violet leaves; Europe.

Dasineura brassicae (brassica pod midge)—in swollen siliqua of Cruciferae; Europe.

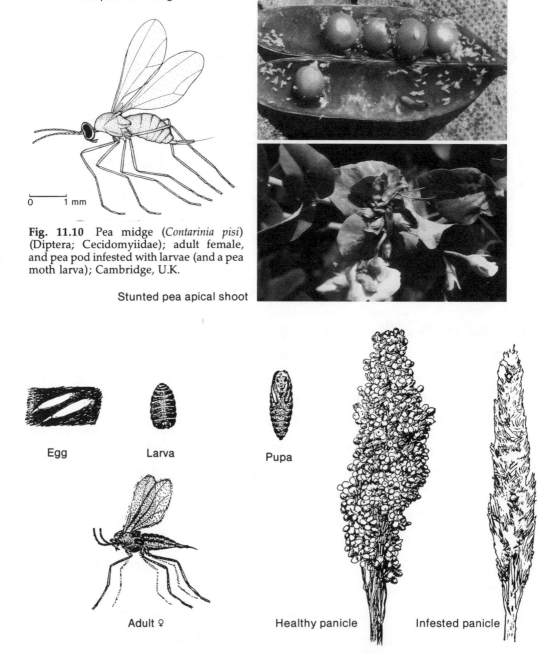

Pea pod with midge larvae inside

0   1 mm

Fig. 11.10 Pea midge (*Contarinia pisi*) (Diptera; Cecidomyiidae); adult female, and pea pod infested with larvae (and a pea moth larva); Cambridge, U.K.

Stunted pea apical shoot

Egg   Larva   Pupa

Adult ♀   Healthy panicle   Infested panicle

Fig. 11.11 Sorghum midge (*Contarinia sorghicola*) (Diptera; Cecidomyiidae).

Dasineura coffeae (coffee flower midge)—in flowers of coffee; tropical Africa.

Dasineura crataegi (hawthorn button-top midge)—apical shoot gall; Europe.

Dasineura leguminicola (clover seed midge)—in red clover flowers; Europe, U.S.A., Canada.

Dasineura mali (apple leaf midge)—curl leaf margins; Europe and Japan.

Dasineura mangiferae (mango blossom gall midge)—in mango flowers; India.

Dasineura oleae (olive leaf midge)—larvae gall olive leaves; Mediterranean, Ethopia.

Dasineura trifolii (clover leaf midge)—larvae fold leaflets; Europe, U.S.A.

Dasineura spp. (leaf midges)—larvae fold leaves, infest buds, of forage legumes, fruit trees, and bushes, flowers, etc.; throughout Europe, Asia and North America.

Diplosis spp. (mulberry midges)—three species attack mulberry in Japan.

Erosomyia indica (mango flower midge)—in mango flowers; India.

Erosomyia mangiferae (mango blister midge)—blister leaf galls; West Indies.

Eumarchalia gennadii (carob midge)—larvae damage pods of carob; Mediterranean.

Geromyia pennisiti (millet grain midge)—in millet heads; India, Africa.

Haplodiplosis marginata (saddle gall midge)—gall stems of barley and wheat; throughout Europe and the U.K.

Japiella medicaginis (lucerne leaf midge)—larvae fold leaflets; Europe, U.S.A.

Mayetiola destructor (Hessian fly)—larvae gall stems of wheat, cereals, grasses; Holarctic (C.I.E. Map No. A. 57).

Mayetiola spp. (barley/oat stem midges, etc.)—several species on different temperate cereals; also Holarctic.

Mycophila spp. (mushroom midge complex)—bore mushrooms; cosmopolitan.

Neolasioptera murtfeldtiana (sunflower seed midge)—in sunflower ovaries; U.S.A.

Orseolia oryzae (Asian rice stem gall midge)—on rice; India, S.E. Asia, China (C.I.E. Map No. A. 171) (Fig. 11.12).

Orseolia oryzivora (African rice stem gall midge)—on rice, Africa (C.I.E. Map No. A. 464).

Prolasioptera berlesiana (olive fruit midge)—in fruits of olive; Mediterranean.

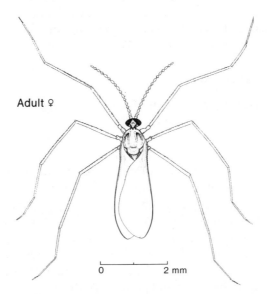

Adult ♀

0   2 mm

Fig. 11.12   Asian rice stem gall midge (*Orseolia oryzae*) (Diptera; Cecidomyiidae).

Resseliella occuliperda (red bud borer, of rose)—in buds of *Rosa*; Europe.

Resseliella theobaldi (raspberry cane midge)—gall raspberry stems; Europe.

Rhopalomyia chrysanthemi (chrysanthemum gall midge)—galls leaves, stems of chrysanthemums; U.S.A. and Japan.

Sitodiplosis mossellana (orange wheat blossom midge)—in ear of wheat, etc.; Europe, Asia, North America (C.I.E. Map No. A. 183).

Stenodiplosis panici (millet gall midge)—larvae inside grains of common millet; central Asia.

Thomasiella sorghivora (sorghum stem midge)—larvae inside sorghum stems; West Africa.

Thomasiniana oleisuga (olive bark midge)—larvae kill twigs of olive; Mediterranean Region.

SUBORDER BRACHYCERA

This group contains 16 families, recognized partly by wing venation and short, porrect palpi; the antennae are somewhat variable but generally three-segmented and shorter than the head. The arrangement of the families is still disputed. The larvae are generally rather wormlike, with a small retractile head and vertical biting mouthparts, and typically amphipneustic. The pupa is often spiny, and usually free in the soil.

TABANOIDEA

Rhagionidae (snipe flies)   *500 spp.*

A small group of predacious flies that feed on other insects, with carnivorous larvae in soil, leaf litter or freshwater. One or two species are bloodsucking and attack humans. *Chrysophilus ferruginosus* is native to S.E. Asia where the larvae prey on larvae of banana stem weevil *(Odoiporus)* and *Rhabdoscelis* in sugarcane; this species of fly has now been introduced into other countries for the biological control of banana weevils.

Tabanidae (horse flies; clegs)   *3500 spp.*

A large family of worldwide distribution, of flies of medical and veterinary importance. The adults are stout-bodied flies, medium to large in size, with very large eyes; often gray or brown in color. They are diurnal in habits and fly mostly in warm sunny weather. Male flies feed mostly on nectar and plant juices, and are frequently to be found on flowers; females will also feed thus but prefer vertebrate blood. The usual hosts are mammals, reptiles, and amphibians—birds are seldom attacked. On people and domestic animals they feed in a somewhat invidious manner in that they fly straight on to the body of the host and sit conspicuously while they insert the proboscis and feed. They are often so intent on the feeding that they can be swatted without difficulty, and can even be picked off by hand. Males have lost their mandibles and so are quite unable to pierce with their proboscis. The actual "bite" is not really painful, but the general effect induced in the host tends to be alarm and panic. Cattle often suffer reduced weight gains or loss of milk yield if plagued by tabanids.

It is thought that the Tabanidae evolved in South America, where most species still occur. The larvae are often said to be aquatic but in reality few are; most cannot live submerged for long and they take atmospheric oxygen through a posterior respiratory siphon. They may be best described as being semiaquatic, found in wet soil at the sides of ponds and rivers; a few are even intertidal in wet sand. Many larvae, including those of *Chrysops,* feed on vegetable debris in the soil. Larvae of *Tabanus, Haematopota,* and some other genera, are carnivorous and feed on other insects, small worms, and crustaceans.

The three best-known genera are *Tabanus, Haematopota,* and *Chrysops.*

Tabanus spp. (horse flies, etc.)—these are large, dark-colored flies with clear wings; found totally worldwide; most typical of open woodland and savanna country (Fig. 11.13).

Haematopota spp. (clegs)—generally slightly smaller in size, with speckled wings, and eyes marked with zigzag banding; they are most abundant in Africa and parts of Asia, and often seem to prefer the edges of forest or woodland clearings; the genus is almost extinct in North America.

Chrysops spp. (deer flies)—these are in life very beautiful flies with dark-banded wings and emerald colored eyes (somewhat spotted) (Fig. 11.14) that fade rapidly after death; most abundant throughout Africa, Asia, and Europe, and also North America, in open woodland (characteristically deer country).

Most of the injury inflicted on people and domestic animals by feeding Tabanidae is harassment and loss of blood, but they are important mechanical vectors of various parasitic organisms. The spongy labella of the proboscis holds a considerable amount of uncongealed blood, and, since feeding flies are often disturbed, when they move immediately to another host or on to another part of the body of the same host they are very effective mechanical transmitters. They regularly transmit anthrax and anaplasmosis in this manner. There are three pathogens especially associated with tabinid vectors, namely, *Trypanosoma evansi,* tularemia, and *Loa loa.*

Trypanosoma evansi is a blood parasite of camels, which also attacks horses, dogs, cattle, and domestic buffalo, causing the disease "surra." Transmission is mostly by Tabanidae, but it is purely mechanical. This disease is often fatal, and losses can be considerable.

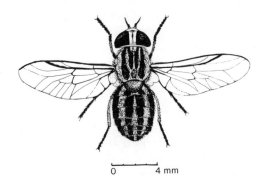

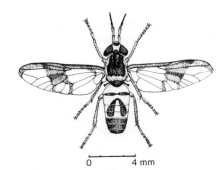

Fig. 11.13 Common brown horse fly (*Tabanus* sp.) of South China; body length 15 mm (0.6 in.) (Diptera; Tabanidae).

Fig. 11.14 Deer fly (*Chrysops* sp.); South China (Diptera; Tabanidae).

Tularemia is a febrile disease caused by the bacterium *Pasturella tularensis,* sometimes fatal, occurring in both Eurasia and North America. Essentially a disease of rodents and lagomorphs, trappers become infected through handling and skinning these animals, or eating the flesh insufficiently cooked. But mechanical transmission by ticks and Tabanidae is common—the main vector in the U.S.A. is *Chrysops discalis.*

Loa loa is the only parasite cyclically (biologically) transmitted by Tabanidae. This nematode is found throughout Africa in rain forest areas. Loasis is usually associated with painful swellings of the joints, especially wrists and ankles. This filarial nematode is found in monkeys and humans where it lives subcutaneously. The biological vector is the tabanid fly *Chrysops*—four species of canopy-dwelling *Chrysops* are involved; two diurnal in habits and two crepuscular or nocturnal. In the sleeping monkeys, microfilariae pass into the peripheral circulation for ease of transmission.

Stratiomyidae (soldier flies)   *1300 spp.*

Medium-sized flies; body somewhat flattened; some are metallically colored, others with white, yellow, or green markings; the distinctive antennae (just longer than head width) have the third segment annulated; generally poorish fliers; often to be found in large numbers on flowers of Umbelliferae and herbage in damp situations. The group is found uniformly distributed throughout the world in all regions. The larvae are sometimes quite elongate, to be found in soil, mud, or dung; terrestrial larvae are generally scavengers and the aquatic ones more carnivorous. In Australia, *Altermetoponia rubriceps* is abundant in soil and regularly damages roots of sugarcane in Queensland; accidentally introduced into New Zealand, this species injures grass roots and causes pasture deterioration. In East Africa, *Ptecticus elongatus* is recorded from papaya fruits and also the roots of carrots.

At the time of adult emergence the numbers of flies in the herbage may be very high, but the group is not really of any importance economically (Fig. 11.15).

ASILOIDEA

Therevidae   *700 spp.*

A small family about which little is known. It is thought that most adults are predacious on other insects. Adults are somewhat like small Asilidae but densely pubescent and with nonprehensile legs, and nonprotruding eyes. Most larvae live in the soil, they are elongate and smooth, and some are predacious on other insect larvae, including wireworms. Clearly these insects form part of the large complex of predacious insects involved sometimes in the natural control of some pest species.

Nemestrinidae   *250 spp.*

A small, well-defined family predominantly found in the warmer regions of the world. Adults are quite large flies, without bristles; often with a long proboscis; wing venation distinctive with many veins parallel to the hind margin, and some with a "primitive" network of crossveins. The adults are often to be seen hovering over flowers taking nectar. They are most abundant in hot, dry, tropical areas. The larvae of the species whose life history is known are parasitic on grasshoppers, locusts, and some beetle larvae (Scarabaeidae). A few of the more important parasites are as follows:

Fig. 11.15 Soldier fly (*Stratiomyia* sp.) (Diptera; Stratiomyiidae).

Fig. 11.16 Robber fly (*Philodicus javanicus*); body length 20 mm; South China (Diptera; Asilidae).

Hirmoneura spp.—in tunnels of Cerambycidae in tree trunks; and also on *Amphimallon* larvae in the soil.

Neorhynchocephalus spp.—attack grasshoppers (Acrididae) in the New World.

Symmictus spp.—attack *Locusta* and *Schistocerca gregaria* in Africa and Asia.

Trichopsidea spp.—parasitize Australian plague locust (*Chortoicetes terminifera*) in Australia.

Asilidae (robber flies)   *5000 spp.*

A distinctive group of predatory flies; many are quite large and have an elongate tapering body, and long, strong, prehensile legs with large paired hooks and large pulvilli. The head is transverse, with large, well-spaced eyes, and the body is bristly. The group is totally worldwide, but some of the largest and most striking species are tropical. The stout, ventrally projecting, horny proboscis is used for sucking the body fluids from the insect prey which are captured on the wing. They do not feed on vertebrates but may bite if handled. Most species have lost the mandibles and sometimes even the maxillae, the piercing part of the horny proboscis is the single unpaired stylet formed from the hypopharynx.

The prey taken is varied and includes a number of very aggressive insects as well as distasteful ones such as Danainae butterflies, aculeate Hymenoptera, and Odonata. Prey size is as always an important factor; large Asilidae can be seen carrying small dragonflies and damselflies, and conversely small robber flies can be seen as the prey of large dragonflies. A typical adult robber fly is shown in Fig. 11.16. A few species are typical of sandy shores in all different parts of the world, both tropical and temperate.

The larvae live in soil, sand, rotten wood, or leaf litter; some are predacious and some are scavengers. In Java, the larvae of *Philodicus javanicus* (Fig. 11.16) are recorded preying on larvae of *Adoretus* and *Phyllophaga*. Larvae of *Laphria* have been found in the tunnels of Cerambycid larvae in trunks of *Pinus*. The pupae are characteristically very spiny about the head, thorax, and abdomen.

The agricultural value of Asilidae is difficult to assess for although they eat many crop pests they also destroy a number of beneficial insects.

Bombyliidae (bee flies)   *4000 spp.*

Medium-sized flies, stout-bodied, often darkly colored and densely pubescent; wings often dark, legs long and slender, and a distinctive long proboscis. Adults are usually to be seen hovering over flowers before alighting to feed on the nectar; at rest the wings are usually held outspread. Many species appear to mimic bees or wasps in their general appearance. The group is widespread but best represented in Africa and other parts of the tropics.

The larvae live in soil and are parasitic on solitary bees and fossorial wasps. Young larvae are elongate and slender, with a very small head, and undergo hypermetamorphosis when they become shorter and flattened.

A few of the more notable and most frequently encountered genera are:

Anthrax—many species are known, parasitic on Noctuid larvae or pupae (including *Spodoptera* in Africa), Aculeate Hymenoptera, eggs of Orthoptera (Acrididae), and some are hyperparasites on dipterous and hymenopterous parasites of Lepidoptera; cosmopolitan (Fig. 11.17).

Bombylius—parasitize solitary bees; cosmopolitan.

Comptosia—the dominant genus in Australia—is thought to attack the egg-pods of grasshoppers and locusts.

Ligyra—many species known; *L. tantalus* in S.E. Asia is recorded parasitizing *Compsomeris* (Hymenoptera, Scoliidae) larvae preying on chafer grubs (Coleoptera, Scarabaeidae) in rubbish dumps and compost heaps.

Systoechus—parasitize egg pods of locusts in Africa, Asia, and Australia.

Systropus—parasitize larvae of *Parasa* (Lepidoptera, Limacodidae); Africa and parts of Asia.

Thyridanthrax—parasitic on pupae of *Glossina* (tsetse flies) in Africa.

Fig. 11.17 Small bee fly (*Anthrax* sp.); body length 6 mm (0.24 in.); South China (Diptera; Bombylidae).

EMPIDOIDEA

Empididae (dancer flies)   *3000 spp.*

Smallish, bristly flies, with a horny proboscis for piercing their prey, which are other insects—mostly small Diptera. Some species have mating swarms that are quite conspicuous.

Larvae are elongate, with a tiny, retractile bead, and live in soil or damp humus, and a few are aquatic—the few species studied are all carnivorous in habits.

Dolichopodidae (long-legged flies)   *4500 spp.*

Quite a large group; these flies mostly occur in wet places, among low herbage; some genera are found on seashores. The adults are predacious on small, soft-bodied insects, but some are regularly seen on flowers where they are apparently taking nectar.

Larval habits are varied, some are aquatic, some live in soil or humus, others live among the flotsam on seashores, and some prey on larvae of wood-boring Coleoptera. It is thought that probably most of the larvae are predacious.

SUBORDER CYCLORRHAPHA

These are regarded as the more advanced members of the Diptera. Adults have three-segmented antennae (four in the Platypezidae) with a subterminal arista; wing venation is usually reduced. Larvae have a vestigial head, and some species are usually referred to as "maggots"; there are usually only three larval instars. Pupae are exarate, and pupation takes place within a puparium. The group is usually subdivided into two sections.

SECTION ASCHIZA

Adults with frontal suture absent; lunule either absent or indistinct; ptilinum absent; wing cell Cu elongated and extending more than halfway to wing margin. A number of small families are not of apparent economic importance and are omitted from this text.

PHOROIDEA

Phoridae (scuttle/carrion flies; humpbacked flies)   *3000 spp.*

A somewhat anomalous group of flies, tiny to small in size (1–6 mm/0.04–0.24 in.); grayish or yellowish in color; active runners; of a humpbacked appearance; antennae apparently consisting of one large segment (which conceals the others) and bearing a long arista; wings often vestigial or absent. Wing venation is distinctive; the longitudinal veins are strong but short, compressed into the first half of the leading edge; the other veins faint and usually unbranched. Most adults are to be found among decaying vegetation, but some occur in the nests of termites and ants. The group is of great taxonomic and evolutionary interest, but only of slight economic interest. Some species are pests of cultivated mushrooms, and others are parasitic on other insects and of some importance as natural control agents.

Larvae are maggot-like and are distinctive in having rows of tiny spinelike processes along the abdomen (somewhat like larvae of *Fannia*—the lesser housefly); in *Megascelia* these processes are fringed. This is thought to be an adaptation to living in a semifluid medium.

Fully-winged phorids are widespread and abundant everywhere, and sometimes recorded in domestic situations in large numbers; they are apparently typical cave-dwellers. The main breeding places where larvae can be found in large numbers are dead snails, animal corpses, fungal mycelia, and rotting vegetation. The species associated with urban habitats are sometimes involved in myiasis when eggs or larvae are accidentally ingested with contaminated fruit, vegetables, or meat. Oldroyd (1964) recalled a record of a European in Burma passing larvae, pupae, and adult flies in his stools for a period of over a year, under conditions indicating that a colony of phorids was living and breeding inside his intestine. One interesting genus is *Wandolleckia* whose two or three species live in the mucus on the upper surface of the foot of the giant African snail *(Achatina)*—they are visible as small white specks.

Some of the species of Phoridae recorded as being of some economic importance or of particular ecological interest are as follows:

Aphiochaeta xanthina—larvae reported to cause myiasis in livestock in India.

Apocephalus spp.—larvae are endoparasitic in *Camponotus* ants in the tropics.

Megaselia matsutaki (pine agaric humpbacked fly)—Japan.

Megaselia scalaris—a parasite of *Amsacta* caterpillars in India; also recorded causing human intestinal myiasis.

Megaselia spp. (3+) (mushroom scuttle flies)—serious pests of cultivated mushrooms in the U.K. and in Europe.

Plastophora spp.—larvae are endoparasitic in *Solenopsis* ants in parts of the tropics.

Termitoxeniinae—several species spend the whole of their life cycle inside nests of termites, and the adults are so modified as to be scarcely recognizable as Diptera.

SYRPHOIDEA

Pipunculidae (big-headed flies)   *600 spp.*

Tiny flies (1–3 mm/0.04–0.12 in. long) with a large spherical head and huge eyes; wings much longer than abdomen, and they hover with great ease; ovipositor horny and exerted. Adults most frequently seen hovering over flowers or in low herbage.

The larvae are endoparasitic in Homoptera Auchenorrhyncha, especially Cercopidae, Cicadellidae, and Fulgoroidea. The egg is laid on the host insect and the larva penetrates into the host abdomen where it grows until it fills most of the abdomen. Pupation takes place in the soil. In India *Pipunculus annulifemur* is of some importance as a parasite of mango hoppers (*Idiocerus* spp.).

Syrphidae (hover flies, etc.)   *5000 spp.*

Moderate-sized flies with brightly colored markings; many mimic wasps and bees in appearance; most feed on nectar and pollen and can be seen hovering over flowers or settled and feeding, especially on flowers of Compositae and Umbelliferae. Many species are of importance as flower pollinators. It is a large and sharply-defined family, of considerable importance agriculturally. The basic body color is usually black, blue, or metallic, with stripes or spots of yellow, or else the body is plump, dark, and hairy and resembles a small bumblebee. The group is worldwide and abundant.

The larvae are quite diverse in habits, and there are some anatomical modifications associated with their life styles. Larvae have a very reduced head; body cuticle tough and leathery; segmentation obscure, but apparently 11 somites are present. Being blind and legless they differ from most other predacious insects, but they are apparently quite successful predators. Eggs are laid in the vicinity, if not in the midst, of the prey on infested plants so the larvae do not have to search far. The prey is seized and impaled on the mouth hooks and the body fluids sucked out. The effectiveness of syrphid larvae as predators is shown by reports that one well-developed larva usually eats 50–60 aphids per day (400–500 per larval lifetime); they are recorded as taking aphid nymphs at a rate as rapid as one per minute when actively feeding. The pupal puparium is typically stuck on to foliage by cement under the tapering caudal segments.

The diversity of larval life styles are indicated below:

1. Phytophagous larvae
   a. Feeding externally on pollen (maize)—*Mesogramma*
   b. Feeding internally in bulbs—*Merodon* and *Eumerus*.

2. Fungivorous larvae—*Cheilosia, Platychirus,* some *Melanostoma* spp.

3. Saprophagous larvae (scavengers)
   a. Living in dung—*Rhingia* (cow dung)
   b. Living in tree holes—*Callicera, Xylota*
   c. Polluted water and mud—*Eristalis, Helophilus,* etc.
   d. Rotting vegetation—*Platychirus,* some *Melanostoma* spp.
   e. Nest scavengers with ants, termites, and Aculeate Hymenoptera—*Microdon, Volucella,* etc.

4. Carnivorous larvae
   a. Aphidophagous larvae —*Syrphus, Melanostoma, Xanthogramma, Scaeva, Episyrphus* (India), *Dideopsis, Ischiodon* (Indonesia), and many others.
   b. Coccidophagous larvae.
   c. Psyllidophagous larvae—*Baccha* in India.
   d. Caterpillers and sawfly larvae—*Dasyrphus,* some *Syrphus* and *Scaeva* species.

5. Larvae causing accidental intestinal myiasis—*Eristalis*

The aquatic larvae of *Chrysogaster* found in swamps have posterior spiracles extended into sharp spines and they are used to pierce the stems of aquatic grasses underwater to obtain oxygen; most of the other aquatic forms are "rat-tailed maggots."

As already stressed, Syrphidae are economically important for the adults that polli-nate flowers and the predacious larvae that destroy aphids and other crop pests; but a few phytophagous species can be serious pests of cultivated plants. It is arguable whether these bulb flies are really phytophagous or saprophagous in that they cause bulb damage and hence rotting or else are more commonly recorded attacking damaged bulbs. It is sometimes thought that *Merodon* are more clearly primary pests and that *Eumerus* may more often be secondary pests. The pest species recorded are more typical of warmer regions although they may be recorded in some numbers in cooler temperate regions in the summer.

Eumerus amoenus (Mediterranean lesser bulb fly)—Mediterranean Region.

Eumerus chinensis (Chinese bulb fly)—China and Japan.

Eumerus figurans (ginger maggot)—U.S.A.

Eumerus flavicinctus—larvae found in damaged cassava, taro, and other tubers in Indonesia.

Eumerus okinawensis—recorded from Okinawa and Japan.

Eumerus strigatus (small narcissus flies, etc.)—recorded throughout Europe, Asia, and North America (Fig. 11.18).

Eumerus tuberculatus (small narcissus flies, etc.)—recorded throughout Europe, Asia, and North America (Fig. 11.18). Several other species are known in Europe, but they are probably more saprophagous.

Merodon equestris (large narcissus fly)—Europe, Japan, New Zealand, Tasmania, and North America (C.I.E. Map No. A. 120) (Fig. 11.19).

Merodon geniculata (Mediterranean narcissus bulb fly)—Mediterranean Region.

These flies are recorded mostly from the bulbs of Liliaceae and Amaryllidaceae (particularly *Narcissus,* snowdrop, and onions), but the maggots are also found in some root crops, especially in damaged roots.

Mesogramma polita (corn-feeding syrphid fly)—larvae eat pollen of maize flowers. U.S.A.

Fig. 11.18 Lesser bulb fly (*Eumerus* sp.) (Diptera; Syrphidae).

Fig. 11.19 Large narcissus fly (*Merodon equestris*) (Diptera; Syrphidae).

SECTION SCHIZOPHORA

Adults with frontal suture and lunule distinct; ptillinum always present; wing cell Cu short and vestigial (except Conopidae). This group of Diptera has been subject to taxonomic reappraisal in recent years, and in the latest edition of Imms (Richards & Davies, 1977) the previously used categories of Calyptratae, Acalyptratae, and Pupipara are discarded. There is, however, still considerable dispute as to the precise content of the Schizophora, and the two extremes are as few as 47 families and as many as 66 families. Richards & Davies (1977) include 48 major families, but some of these are not included in the present text as they are of no particular agricultural importance. Many of the taxonomic criteria defining these families are quite esoteric and such characters are not listed here for obvious reasons; for any such detail students are recommended to refer to Richards & Davies (1977).

Lonchaeidae   *300 spp.*

A small family where the larvae are mostly scavengers in rotting vegetation or dung. A few are phytophagous and make galls on grasses, others are recorded in damaged or overripe fruits in Africa and parts of Asia. *Lonchaea aristella* is recorded as being very damaging to fig fruits in the Mediterranean Region, and *Silba (= Lonchaea) sibbosa* in Java is a primary pest of *Citrus* flower buds, the larvae eating out the flower buds which fall to the ground. The larvae are able to jump when disturbed.

Cryptochaetidae

A small, tropical group of flies whose larvae are endoparasitic in Coccoidea. A well-known species is *Cryptochetum iceryae* which parasitizes *Icerya* (cottony cushion scale). This group may be placed in the Drosophilidae (see below).

LAUXANIOIDEA

## Braulidae (bee "lice")											*8 spp.*

These small, wingless flies resemble Hippoboscidae, and are found clinging to the queen and worker honeybees. The larvae tunnel in the wax combs, but apparently cause little damage. The main genus is *Braula,* and *B. coeca* (bee "louse") is found throughout Europe, Asia, and Japan in local beehives.

DROSOPHILOIDEA

## Drosophilidae (vinegar flies; small fruit flies)							*2500 spp.*

Mostly small, yellowish flies, 3–4 mm long (0.12–0.16 in.), with red eyes. They are attracted by various by-products of fermentation and frequently found on damaged ripe and overripe fruits. Most species are actually fungivorous and feed on the yeasts and other fungi present rather than directly on the fruits themselves.

Some species are leaf miners *(Scaptomyza)*, others are predacious, or rather the larvae are predacious, on small Homoptera. *Acletoxenus* larvae prey on larvae and pupae of *Aleurocanthus* and *Trialeurodes* and other Aleyrodidae in tropical Asia. In India *Gitonides* is predacious on the sugarcane mealybug, and *Cryptochaetum* is an important internal parasite of Coccoidea; *C. iceryae* was imported from Australia into California in one of the classic cases of biological control to control *Icerya purchasi.*

The genus *Drosophila* now contains over 1000 species and is well known from genetic experimentation through the world as both a teaching aid and for research material. In the Hawaiian Islands there has been explosive evolution of *Drosophila* leading to the development of 400 different species (Fig. 11.20).

According to Kalshoven (1981) it is possible that *Drosophila lurida* may be a primary pest of *Citrus* and durian fruits in Indonesia—the larvae are found in groups under the peel and as they grow they penetrate deeper into the fruit which then falls prematurely with a rotting pulp. *Zaprionus* spp. are recorded from many different fruits in eastern Africa but they are probably not primary pests. *Drosophila suzukii* is the cherry drosophila of Japan. *Drosophila* flies are sometimes a considerable urban nuisance when attracted to stored fruits, vinegars, etc. Laboratory studies have shown that the adult flies carry spores of *Aspergillus* and other yeasts and transmit the fungi to damaged but otherwise uninfected grapes and other fruits.

The tiny larvae are distinctive in that they have a protrusible, long, terminal respiratory siphon formed from the posterior spiracles.

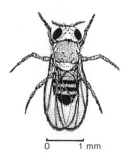

Fig. 11.20 Vinegar fly (*Drosophila* sp.) (Diptera; Drosophilidae).

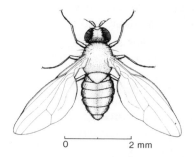

Fig. 11.21 Rice leaf miner (*Hydrellia griseola*) (Diptera; Ephydridae).

Ephydridae (shore/brine flies)   *1500 spp.*

Small, black or darkly colored flies that inhabit marshy places, swamps, mangroves, seashores, and alkaline lakes; adults only 2–3 mm (0.08–0.12 in.) in body length. Some species of *Hydrellia* and *Scatella* are found well away from water, usually on grasslands and lawns in Australia. The larvae are densely covered with a fine pubescence and have a pair of respiratory tubes emerging from the terminal anal siphon; the abdomen bears eight pairs of conspicuous pseudopods armed with hooks. The puparium has an elongate respiratory siphon whose apex rests on the water surface; but the pupae of *Notiphila* take oxygen from the plant roots, pierced by posterior spiracular spines. Most larvae are phytophagous, eating either water plants, or browsing on algal mats on seashores and salt-marshes; but some are saprophagous, and a few are carrion feeders. A few species are parasitic—*Timerina* feeds on the eggs of spiders. *Psilopa petrolei* is unusual in that the larvae live along the edges of pools of crude petroleum in California, feeding on trapped insects and possibly also bacteria.

This is a large and extremely versatile family of flies, of great interest ecologically, with a few species of some importance as agricultural pests. Some of the more interesting species are listed below:

Ephydra macellaria—recorded damaging paddy rice in France, Spain, Hungary, Egypt, and parts of central Asia.

Ephydra riparia (large shore fly)—usually found in salt-marshes and on seashores around Europe, and has been recorded as a minor pest of rice in the Camargue region of southern France; also recorded throughout Europe, Africa, Asia, North and Central America.

Ephydra spp. (shore flies, etc.)—mostly recorded infesting water plants in marshy areas and seashores; Europe, Asia, North and Central America.

Hydrellia griseola (rice leaf miner)—larvae mine leaves of rice and temperate cereals (wheat, barley, etc.); Europe, North Africa, Malaysia, Japan, Korea, U.S.A. and South America (Fig. 11.21).

Hydrellia philippina (rice whorl maggot)—larvae mine rice leaves; India, S.E. Asia, Philippines, Papua New Guinea, Indonesia, and Japan.

Hydrellia sasakii (paddy stem maggot)—recorded from India through to Japan.

Hydrellia tritici (wheat leaf miner)—larvae mine leaves of wheat; temperate Australia.

Hydrellia spp. (pondweed leaf miners)—many species occur as miners in watercress, *Potamogeton* pond weeds, duckweeds, and a range of grasses and meadow plants; Europe, North Africa, Asia, North America, and South America.

Notiphila spp. (rice stem/root maggots)—larvae in stems and roots of aquatic plants including reeds, water lilies, *Potamogeton* spp., and paddy rice; Europe, Asia, North America, and some in South America.

Psilopa petrolei (petroleum fly)—this is a large, widespread genus, but this species occurs in California (U.S.A.) and the larvae live on the edges of pools of crude petroleum.

Scatella spp. (leaf miners, etc.)—a very large genus, very widespread, with most larvae found in organic mud, sewage filter beds, etc., but some are leaf miners in various plants; one species mines leaves of Chinese cabbage in South China.

NOTHYBOIDEA

A small group of rather obscure, small families, some of which are allied to the Muscoidea, and at least one in the past was included with the Drosophilidae; but there are a couple of species of particular importance as pests of cultivated plants.

Psilidae (carrot fly, etc.)   *170 spp.*

Small flies, with a rounded, rather soft body, and globular head with wide ocellar triangle; wing venation is distinctive and various body bristles are not developed. The larvae are all phytophagous, and the group is basically Holarctic in distribution.

Two species are of particular importance as pests:

Psila nigricornis (chrysanthemum stool miner)—larvae mine in the stool of chrysanthemum plants, and sometimes in the root of carrot and lettuce; Europe, including the U.K., and also in Canada.

Psila rosae (carrot fly—U.K.; carrot rust fly—U.S.A.)—a very serious pest on carrot, parsnip, celery, and also on parsley, hemlock, and various other wild

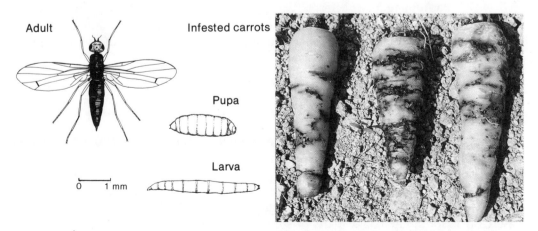

Fig. 11.22   Carrot fly (*Psila rosae*) (Diptera; Psilidae) instars and damage; Cambridge, U.K.

Umbelliferae, where the larvae tunnel into the root. Most tunnels are superficial but some go into the root cortex, and most tunnels are infected with fungi, causing rots; young plants are usually killed, after initial wilting; found in Europe and parts of Canada and the U.S.A. (C.I.E. Map No. A. 84) (Fig. 11.22).

MUSCOIDEA (= Calyptratae)

Scatophagidae (= Cordiluridae)   *500 spp.*

A small group, mostly phytophagous (leaf and stem borers), but some are predacious in the larval stages, and others are saprophagous. The yellow dung fly *(Scatophaga stercoraria)* larvae live in cow dung and the adults prey on other flies that come to feed or oviposit on the cow dung.

Anthomyiidae (root flies, etc.)   *1200 spp.*

A small group formerly included within the Muscidae, but now generally regarded as being distinct; the main taxonomic criterion being that in the wing the vein $Cu_1 + 1A$ reaches the edge of the posterior wing margin, though sometimes faint distally. The adults look just like houseflies, some 5–7 mm (0.2–0.28 in.) in body length. The larvae are mostly phytophagous. Many live in soil and eat plant roots, but a few are leaf miners, some are stem/shoot borers, some live on the seashore in seaweed wrack, and a few are fungivorous. Biologically and ecologically the members of these two families cannot really be separated, and some plant infestations are mixed.

The adults are mostly nectar-feeders and may be found on a wide range of flowers where they may be important pollinators. In Europe, the life cycle of several species appears to be correlated with the sequential flowering of a series of wild hedgerow Umbelliferae used as food sources.

The larvae can be categorized according to their feeding habits, thus:

1. Seaweed feeders—on mudflats and seashores—*Fucellia.*

2. Fungivorous larvae—*Mycophaga.*

3. Root maggots
   a. Phytophagous species; larvae bore into intact roots and subterranean stems, especially bulbs, and they also bore into large cotyledons, etc. of seeds and seedlings—*Delia* spp.
   b. Saprophagous species; larvae are often associated with the former species and feed on the damaged tissues, or feed on dead organic matter, including crop residues in the soil and going on to damage roots of living plants— *Delia platura, Pegohylemyia,* etc.

4. Cereal shoot flies—larvae bore into shoots of young cereal and grass plants and destroy the growing point, causing a "dead-heart"—*Delia coarctata, Delia flavibasis, Phorbia securis,* etc.

5. Leaf miners—larvae make large blotch mines in the leaves of Chenopodiaceae, Solanaceae, etc.—*Pegomya.*

The species most important as crop pests tend to be more temperate in distribution, either Palaearctic or Holarctic, but some do occur in North Africa, East Africa, Near East, China, South Japan, and parts of southern U.S.A.

Field identification of root maggots is usually very difficult because of the large number of saprophagous species of Anthomyiidae and Muscidae that live in soil rich in humus or organic debris (such as rotting crop residues). Some of these species are secondary pests in that they can feed on roots already damaged by the primary pests.

The genera *Delia* and *Hylemya* have recently been reappraised taxonomically in Europe, and many of the crop pest species formerly regarded as being in *Hylemya* (or other genera) are now placed in *Delia*, so the literature is confusing. As an indication of the relative abundance of the different genera, in the U.K. at the present time there are 34 species of *Delia*, 5 of *Hylemya*, 25 of *Pegohylemyia*, 41 of *Pegomya*, and 9 of *Phorbia*.

The species of Anthomyiidae encountered as pests of cultivated plants with some regularity are listed below:

Delia antiqua (onion fly)—larvae bore bulbs of onions and other Amaryllidaceae; Holarctic, but in Israel they develop during the cool "winter" period (C.I.E. Map No. A. 75) (Fig. 11.23).

Delia brunnescens (carnation maggot)—on carnation; U.S.A.

Delia coarctata (wheat bulb fly)—larvae cause dead-heart in winter wheat, rye, barley, and some grasses; Europe and West Asia (C.I.E. Map No. A. 115).

Delia echinata (spinach stem fly; carnation tip maggot)—Europe, Japan, U.S.A.

Delia flavibasis (barley fly)—larvae "dead-heart" barley, wheat, maize, bulrush, millet, and some grasses; southern Europe, and much of Africa (Fig. 11.24).

Delia floralis (turnip maggot)—larvae attack seedlings and roots, of Cruciferae, tobacco, etc.; Europe, China, Canada, U.S.A.

Delia hirticrura (Mediterranean onion fly)—on onions; Mediterranean Region.

Delia pilipyga (turnip maggot)—attacking Cruciferae; China and Japan.

Delia planipalpis (cruciferous root maggot)—on Cruciferae roots; Canada.

Delia platura (bean seed fly; corn seed maggot)—larvae bore sown seeds and seedlings of beans, maize, onions, tobacco, cotton, marrow, cucumber, peas, lettuce and crucifers; cosmopolitan but less abundant in the tropics (C.I.E. Map No. A. 141) (Fig. 11.25). Often in a "bean seed fly" infestation there are several species involved.

Delia radicum (cabbage root fly)—a serious pest of Cruciferae, attacking the roots; Europe, parts of Asia, North America (C.I.E. Map No. A. 83) (Fig. 11.26).

Hylemya cerealis (wheat stem maggot)—larvae in wheat, oats, etc.; Canada.

Pegohylemyia fugax—saprophagous on cruciferous roots; Europe, Canada.

Pegohylemyia gnava (lettuce seed fly)—recorded in parts of Europe.

Pegomya dulcamarae (potato leaf miner)—larvae mine potato leaves; Japan.

Pegomya hyoscyami (mangold fly; beet leaf miner, etc.)—larvae blotch mine leaves of sugar beet, spinach, mangels (Chenopodiaceae); Europe, Asia, and now North America. It is reported that a race attacks Solanaceae.

Pegomya rubivora (loganberry cane maggot)—larvae in shoots of *Rubus;* Europe, U.S.A.

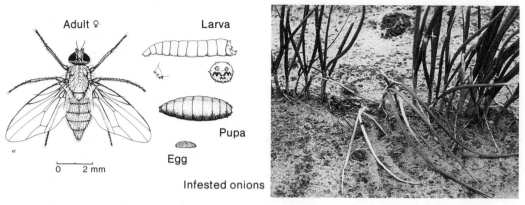

Fig. 11.23 Onion fly (*Delia antiqua*) (Diptera; Anthomyiidae); all stages, and infested salad onions; Sandy, U.K.

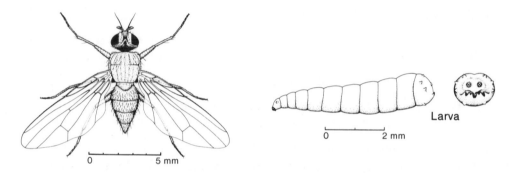

Fig. 11.24 Barley fly (*Delia flavibasis*) larva and adult (Diptera; Anthomyiidae).

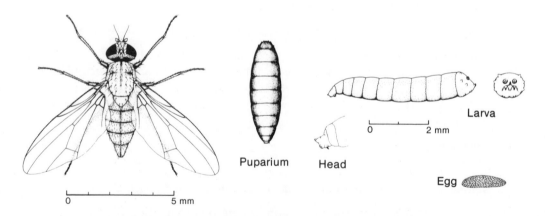

Fig. 11.25 Bean seed fly (*Delia platura*) (Diptera; Anthomyiidae).

Phorbia securis (late wheat shoot fly)—larvae cause dead-heart in wheat and sometimes barley; much of northern Europe.

Muscidae (houseflies, etc.) *3900 spp.*

Small to quite large flies, usually quite dark in color, thorax often longitudinally striped gray; wing vein $Cu_1 + 1A$ does not reach the wing margin; many species have a

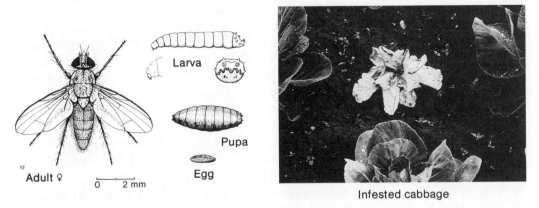

Fig. 11.26 Cabbage root fly (*Delia radicum*) (Diptera; Anthomyiidae); all stages, and damaged cabbage plant; Sandy, U.K.

general resemblance to the housefly. The current interpretation of this family, according to Richards & Davies (1977), is far more narrow than formerly, with Anthomyiidae, Fanniidae, and Glossinidae removed, although a number of groups formerly placed in the Anthomyiidae are now put into the Muscidae.

It seems fairly clear that the Muscoidea originated from a compost-feeding ancestor, but they are an ancient group and some have diversified into plant-feeders, others into dung and carrion feeders, with some even blood-feeding from warm-blooded vertebrates. As a group, the Muscoidea shows great ecological diversity and adaptability and many species have become alarmingly synanthropic through the world.

In many of the groups of flies dealt with so far in this text, it has been larval feeding that has been important physiologically as that stage has been responsible for the accumulation of protein for the insect life history completion. Protein is invariably needed at some stage in order for egg development to proceed; it can be obtained by the larva and stored in the body for later use, or else it can be obtained by the female, through, for example, a blood meal. And also the larvae were important ecologically and economically as the pest organisms affecting humans, domestic animals and a wide range of plants. But, in the Muscidae, the adults assume predominance in the urgency of their feeding. Many species still take nectar from flowers but there has been a gradual dependence on humans and their domestic animals, seeking sweat, urine, dung, and in some cases blood. The basic muscoid mouthparts are spongelike and adapted for mopping up liquids through the tubular pseudotrachea on the end of the labellum. Most flies in this group are attracted to wounds and body punctures, and some can increase the blood flow by rasping the skin with their hardened, prestomal teeth. In a few cases the whole labium has become stiffened, narrowed, until it forms a stiletto as in the *Glossina* species.

Most members of this group lay large numbers of eggs—*Musca domestica* lays up to 1000 eggs either singly or in batches, at a rate of 100–150 per day. They are laid on fresh horse dung, human, cow, or poultry dung, decaying vegetable matter, and garbage, generally including decaying foodstuffs, meats, and animal carcasses. Under optimum conditions (33–35°C/91–95°F) the life cycle from egg to adult can be completed in eight days (at 10–15°C/50–59°F) it takes 40–50 days).

Adults are quite long-lived (2–10 weeks) and may fly 3–4 km (2–2½ miles) from their site of emergence, and have been recorded dispersing up to 8 km (5 miles). The adults feed at the site where oviposition takes place, as well as on exposed food in domestic situations. They have a habit of periodical regurgitation as well as defecating on the food source, and the adult flies thus have a great propensity for transmission of human diseases by microorganisms. Most of this transmission is purely mechanical, either through feces, or vomit, infected mouthparts, or bacteria and fungal spores attached to the legs or feet. Larvae feeding on infected material will produce infective adults. The list of pathogenic organisms transmitted by *Musca domestica* and the other synanthropic members of the Muscidae includes viruses, bacteria, fungi, protozoan parasites, nematodes, tapeworms, and other arthropods—for further details see page 97.

The importance of adult flies as veterinary pests consists mostly of the irritation and nuisance of the flies on the bodies of domestic livestock. Harassed beef cattle fail to increase in weight at the proscribed rate, and dairy cattle suffer a milk yield decrease. The few species of Muscidae that actually suck blood can transmit blood parasites and pathogens in both humans and domestic livestock. Some of the flies with prestomal teeth can actually break the skin surface, but the less well-endowed species cannot penetrate intact skin but they can break open scabs and enlarge the edges of wounds.

The larvae of many species of Muscidae are occasionally involved in human myiasis—intestinal, urinogenital, aural, traumatic, and nasopharyngeal myiasis, but this is not a common occurrence.

Some of the adult flies are of importance as flower pollinators, but generally they are not so important in this respect as the Calliphoridae.

The family Muscidae is split into five (or more) usefully grouped subfamilies, as follows:

MUSCINAE

Adults mostly fluid suckers, a few with prestomal teeth to rasp skin for bloodsucking; larvae in soil, rotting vegetation or dung; most are saprophagous, some predacious, some initially saprophagous turning predacious. It includes *Musca, Muscina, Hydrotaea, Ophyra, Morellia,* etc.

PHAONIINAE

Adults nectar and fluid feeders; larvae mostly phytophagous, but some may prefer decaying vegetable matter. Including *Atherigona, Helina, Phaonia,* etc.

LIMNOPHORINAE AND COENOSIINAE

Larvae are aquatic in either freshwater or marine habitats, and predacious on soil invertebrates.

Limnophora spp. (shore flies)—mostly found in freshwater.

Coenosia spp. (salt-marsh flies)—mostly in salt marshes or littoral on seashores.

STOMOXYINAE

Adults are bloodsuckers with piercing and sucking mouthparts; larvae live in moist dung and moist rotting organic material; *Haematobia, Stomoxys.*

The number of species of agricultural importance is quite limited, as not many of the Muscidae *(sensu stricta)* are truly phytophagous; the main pest species are listed below:

Atherigona exigua (rice seedling fly)—larvae make dead-hearts in rice and other tropical cereals; S.E. Asia.

Atherigona orientalis—found throughout the Old World tropics and has now also spread to the New World tropics; it is equally abundant in urban areas and the wild. The larvae are scavengers and recorded from rotting fruits and vegetables, kitchen refuse, dung, and carrion; its abundance and readiness to enter houses and crawl on food makes the adult the third most important vector of fecal pathogens.

Atherigona oryzae (rice stem fly)—in rice and other cereals; India, S.E. Asia, Australasia, China (C.I.E. Map No. A. 411) (Fig. 11.27).

Atherigona shibuyai (sugarcane stem maggot)—Japan.

Atherigona soccata (sorghum shoot fly)—larvae cause dead-hearts in a wide range of tropical and temperate cereals, and various larger grasses; Mediterranean, Africa, Near East, India and parts of S.E. Asia (C.I.E. Map No. A. 311) (Fig. 11.28).

Atherigona spp. (shoot flies, etc.)—many species are known throughout Africa, Asia and Australasia, and a large number feed in stems of cereals and grasses, causing seedling "dead-hearts"; a few species are saprophagous and found secondarily in rotting fruits (Fig. 11.29).

Helina spp.—recorded from fruits of *Coffea* and Cucurbitaceae; eastern Africa.

Muscina spp.—several species regularly found with root maggot infestations.

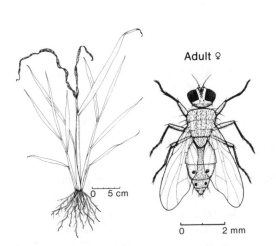

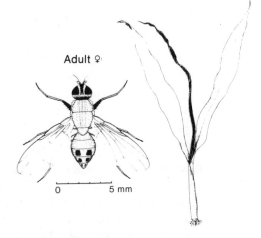

Fig. 11.27 Rice stem fly (*Atherigona oryzae*) (Diptera; Muscidae).

Fig. 11.28 Sorghum shoot fly (*Atherigona soccata*) (Diptera; Muscidae).

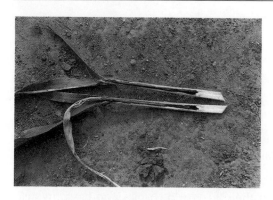

Fig. 11.29 Sorghum seedling cut open to show damage by shoot fly larva (*Atherigona soccata*); Alemaya, Ethiopia (Diptera; Muscidae).

Phaonia corbetti—larvae are recorded eating flowers of Nipa palm in Indonesia.

The Muscidae recorded as domestic/urban pests (i.e. medical and hygienic) of humans and their livestock are listed below:

Haematobia (= Lyperosia) exigua (buffalo fly)—adults are bloodsuckers and transmit bovine and equine diseases in Asia and Australasia; larvae live in dung.

Haematobia (L.) irritans (horn fly)—adults tend to cluster round the horns of cattle; they are bloodsuckers and important pests of horses and cattle; Europe, Asia, North America, now Australasia.

Hydrotaea spp. (sweat flies, etc.)—a Holarctic genus whose species attack people and domestic livestock; adults suck blood, and eye, nose exudates, but can also rasp the skin to cause a blood flow; larvae live in soil and are saprophagous/predacious.

Morellia spp. (sweat flies, etc.)—larvae live in cow dung; adults are attracted to sweat and mucus on people and livestock and will suck blood from bite sites of *Stomoxys* and tabanids; said to be the intermediate host for the eye worm of cattle *(Thelazia rhodesii)* in Europe, Asia, and Africa; a cosmopolitan genus.

Musca autumnalis (face fly)—native to Europe, Africa, and Asia, now established in North America (since *c*. 1952); adults hibernate in large clusters and cause a domestic nuisance; in warm weather, adult females cluster around the face of humans and livestock (often as many as 50–100 flies per animal are not uncommon) and cause great nuisance and harassment.

Musca crassirostris (bloodsucking housefly)—an obligate bloodsucker attacking cattle and some other domestic livestock, but people are only rarely bitten; the adult mouthparts are basically similar to *M. domestica* but the proboscis is more bulbular and with strong prostomal teeth, with which it scratches the skin or scabs; breeding occurs in cow dung; widespread in the Mediterranean Region, Africa, and S.E. Asia.

Musca domestica (housefly)—the species is totally cosmopolitan, as a series of regional subspecies, adapted for life in cooler and hotter locations; synanthropic generally, with adults invading buildings, and larvae in dung, refuse heaps, crew-yards, chicken houses, etc.; very abundant and very important as vectors of many fecal pathogens; larvae sometimes cause myiasis (Fig. 11.30).

Musca sorbens (bush eye fly)—larvae develop in dung; adults are strongly attracted to eyes, nose, and mouth regions, and to sores; widespread in Africa, S.E. Asia, Australasia and the Pacific Islands (in some respects replacing *M. domestica*).

Musca vetustissima (Australian bush fly)—a very important domestic and cattle pest in Australia (sometimes called *M. sorbens*), breeding in cattle dung, and causing great irritation to humans and cattle.

Muscina stabulans (false stable fly)—larvae in dung, carrion, rotten fruits, or bird nests, predacious on other larvae (or nestlings); cosmopolitan.

Ophyra spp. (tropical houseflies)—several species are pantropical; adults are very important as vectors of fecal pathogens in warm climates; larvae feed in nests, carcasses, dung, rotting vegetable matter; the most important species of pests in the tropics are *O. aenesens, chalcogaster,* and *nigra.*

Passeromyia spp.—larvae suck blood from nestling birds in Africa and Australia.

Stomoxys calcitrans (stable fly/biting housefly)—a cosmopolitan species thought to be native to Africa and tropical Asia; a serious pest of cattle—sucking blood, transmitting pathogens and parasites, and harassing the livestock; only occasionally biting people; mouthparts modified for piercing skin and sucking blood; larvae live in dung, organic refuse, or rotting vegetable matter (Fig. 11.31).

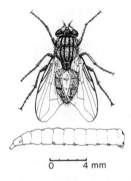

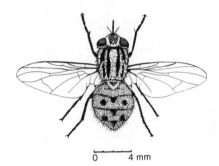

Fig. 11.30 Housefly (*Musca domestica*) adult and larva (Diptera; Muscidae).

Fig. 11.31 Stable fly/biting housefly (*Stomoxys calcitrans*) (Diptera; Muscidae).

Fanniidae   *260+ spp.*

These flies are now generally separated from the Muscidae as a distinct family. The larvae live in dung and other decaying organic matter and have a characteristic shape, being broad and flat with paired segmental outgrowths—thought by some authorities to be associated with life in a semiaquatic habitat.

Fannia canicularis (lesser housefly)—in houses the adults are often seen circling or hovering in midair in the center of the room; very abundant in most domestic premises, but they seldom alight on humans or their food; a cosmopolitan species; larvae are pests in chicken deep-litter houses, and are commonly associated with human urinogenital myiasis (Fig. 11.32).

Fannia scalaris (latrine fly)—larvae live in latrines, dung, fungi, animal and bird nests; recorded causing myiasis; cosmopolitan.

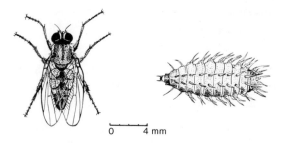

Fig. 11.32   Lesser housefly (*Fannia canicularis*) adult and larva (Diptera; Fanniidae).

Fannia spp.—some 220 species are known (according to Pont, in Smith, 1973) and ten species are listed as being the most likely to be encountered in medical or hygiene studies. Some species (like the two above) are cosmopolitan, but some are confined to Africa or South America, or to other regions.

Gasterophilidae (horse bot flies, etc.)   *18 spp.*

Quite large, very hairy flies, often partly yellow and resembling bumblebees, mouthparts reduced and useless; sometimes this group is regarded as a subfamily of the Oestridae. The larvae are intestinal parasites of Equidae, and some other grazing mammals. The best known example is *Gasterophilus intestinalis* (horse bot fly). The furry, yellow, bee-like female deposits her eggs singly, stuck to hairs on a horse, usually on the legs, or on other parts that can be reached by the horse's mouth. During oviposition the female hovers with a sound that disturbs horses in the same way that the warble fly affects cattle. Each female lays up to 1000 eggs. The larvae develop inside the eggs but do not emerge until licked by the horse—then the young larvae emerge and quickly penetrate the tongue epithelium, pass along it down into the esophagus and into the stomach. The larvae use their mouth-hooks to anchor themselves to the stomach wall and they absorb food material from the stomach contents, so strictly speaking they are kleptoparasites in that they are stealing some of the horse's food and not feeding on the horse itself. The larvae are air-breathers and the posterior spiracles have opening and closing mechanisms to take advantage of air bubbles in the stomach; the larva also has hemoglobin as a respiratory pigment. The larvae bear rows of small black spines on each body segment to facilitate their boring. Fully grown larvae detach themselves from the stomach wall and pass to the exterior in the feces—pupation occurs in the soil. Little serious direct damage is done to the host but clearly there is some upset, and the horse is frightened by the sound of the female fly hovering. Horse bots are common and widespread and most Equidae are infected, often quite heavily, in all parts of the world but probably more in tropical than the temperate regions.

The main species concerned are:

Cobboldia spp. (elephant bot flies)—in the stomach of elephants in Africa and India; working elephants have been incapacitated and some even died when the bot infestation blocked the stomach.

Gasterophilus haemorrhoidalis (nose bot fly)—eggs laid on the lips of the host, larvae develop in stomach and temporarily in rectum.

Gasterophilus inermis—eggs laid on cheeks of host.

Gasterophilus intestinalis (horse bot fly)—eggs laid on legs and hatch after licking; larval development in the stomach.

Gasterophilus nasalis (throat bot fly)—eggs laid under the chin, larvae enter mouth to develop in stomach or anterior duodenum.

Gasterophilus nigricornis—eggs laid on cheeks or nose and larvae develop in stomach.

Gasterophilus pecorum—most common and most pathogenic species in this genus; eggs laid on grasses and develop in the host stomach after eating.

The hosts for *Gasterophilus* are Equidae (horses, zebras, donkeys, mules) and very rarely humans; heavily infested animals may die, partly of intestinal irritation and obstruction of the alimentary canal.

Cyrostigma spp. (rhinoceros bot flies)—very large species that live in the stomach of different species of rhinoceros.

Oestridae (warble or bot flies)   *34 spp.*

A small family of stout, furry flies, often quite bee-like in appearance; with larvae that are endoparasitic in mammals. Larvae are more frequently encountered than the adults; generally each species attacks a closely related group of hosts.

The life history of *Hypoderma* is generally typical of most species in this family. Eggs are laid on hairs on the flanks, legs, and feet of cattle; after about 4–5 days the eggs hatch and the larvae bore into the skin. The larvae bore beneath the skin for several months until they reach the esophageal region where they remain a while. Eventually they come to lie under the skin along the back. The skin is pierced and the posterior spiracles are in contact with the air. The fully developed larvae inhabit quite large swellings (warbles) along each side of the spinal column of the host. Eventually the larvae work their way out through the skin, fall to the ground, and pupate in the soil. The injuries to the cattle include perforation of the hide, reduced milk yield, general deterioration of the meat, and lowered rate of weight gain—economic losses can be very high.

The larvae are stout, cylindrical, and not tapering, and usually covered with many small spicules. Mouthhooks are present in the first instar larvae, which are active burrowers but later they are often reduced and atrophied. They feed on serous exudates from the host tissues which are increased by the irritation caused by the larvae. These flies attack wild Bovidae and Cervidae mostly, but now are serious pests of domestic cattle, sheep, and goats, etc., and most of the fly species here mentioned are occasionally found attacking people causing myiasis of one type or another. The main species of economic importance are as follows:

Crivellia silenus (goat warble fly)—sheep and goats are often heavily attacked; North Africa, Mediterranean Region, and warm temperate Asia.

Gedoelstia spp. (antelope nasal bot flies)—throughout Africa these are found in most types of antelope; the larvae penetrate the eye orbit through a vein, and eventually lodge in the frontal sinuses. In wild hosts the infestation is tolerated, but in sheep (and occasionally humans) first instar larvae (they develop no further) may cause blindness.

Hypoderma spp. (ox warble flies)—found throughout Europe, Asia, and North America in a series of species; larvae attack cattle (Bovidae) and deer

(Cervidae) and regularly cause human myiasis which results in great discomfort and occasional paralysis.

Oedemagena tarandi (reindeer warble fly)—infests reindeer and caribou in Eurasia and North America.

Oestrus ovis (sheep nostril bot fly)—widespread in many parts of the world; a larviparous species, the larvae live in the nasal passages and frontal sinuses of sheep, goats, and their wild relatives.

Rhinocephalus purpureus (horse head maggot)—larvae live in frontal sinuses and nostrils of horses—fatalities are quite common; Europe, Asia, Africa.

Cuterebridae (rodent bot flies)

A small group of parasitic flies characterized by having an elongate scutellum and reduced postnotum (postscutellum); there are only a few genera in this family.

Cuterebra spp. (rodent bots)—a New World group that attack rodents, rabbits, and hares, and in some cases also infests cats, dogs, monkeys, and humans.

Dermatobia hominis (human bot fly)—larvae recorded attacking many different mammals and some birds, as well as people. A very serious pest of cattle in Central and South America. An unusual pest in that the female fly seizes a mosquito (particularly *Psorophora*) or else another Muscoid fly, and attaches her eggs to the body of the recruited vector; when the vector alights on a human host (or otherwise) the body warmth triggers eclosion and the larvae quickly bore into the skin where they cause warble-like swellings.

Calliphoridae (blow flies; bluebottles; greenbottles, etc.)   *1100 spp.*

Stout, hairy-bodied flies, often metallic blue or green in color; dorsal surface of abdomen usually without bristles; antennal arista markedly plumose usually for its whole length. The group is completely worldwide, but may be most abundant in the Holarctic region.

Most species are saprophagous carrion or dung eaters and many are synanthropic, but there has been a gradual evolution towards parasitism in this group. The adults take liquid food, most feed on nectar from flowers, and some are very important pollinators of the more open-flowered crops. *Calliphora* species are reared in most temperate countries to provide maggots for fishing bait, and this is quite a sizeable industry now. In some country areas the blow fly pupae are also sold to farmers; if *Brassica* crops for seed are enclosed within a fly-proof cage of netting and the blow fly pupae introduced, the emerging adults will pollinate the crop plants enclosed. Most adults live for about a month with food.

So far as crops are concerned, this group of flies is only beneficial in its pollinating behavior. But from medical, hygienic, and veterinary points of view these flies are quite important pests. Adults are attracted to blood and wound exudates, but feed only on oozing liquids—direct damage is done by the larvae, although indirect distribution of some pathogens can be effected by the feeding adults. Some of the larvae have graduated from feeding on wounds on animal bodies to actually penetrating living flesh, causing cutaneous myiasis and in some cases the larvae are now totally obligatory parasites. Larvae are identified largely by the characteristics of the posterior spiracular plate (as illustrated in Smith, 1973, p. 311 and 312).

Some larvae have been used as surgical maggots. Sterile larvae of various Calliphoridae, especially *Lucilia sericata,* have been introduced into human wounds where they fed on the necrotic tissues and bacteria and cleaned the wound completely without invading the living tissues. Maggot therapy has been practiced by certain tribes in the hills of Burma and Yunnan (China) and some Australian aborigines since ancient times. It was first noticed in the West in the sixteenth century and utilized in the American Civil War. In World War I it was observed that not only were some of the maggot-infested wounds kept clean but they did not develop infections, as did many of the treated wounds. Later it was established that the *Lucilia* larvae secrete allantoin which is responsible for the enhanced healing. Maggot therapy was quite widely practiced until the development of the sulfa drugs and antibiotics in the 1940s.

Most species are oviparous and the females search out suitable oviposition sites by using smell and taste mostly, and the adults are quite long-lived (30–40 days on average). Pupation typically takes place in the soil or surface litter. There is so much diversity of habits within the group that generalization is difficult, and it is more useful probably to consider the more important species individually, albeit briefly.

Auchmeromyia luteola (Congo floor maggot)—found in Africa south of the Sahara; the larvae are bloodsucking and feed nocturnally on sleeping humans (on the floor), hiding during the day in crevices and litter; four other species occur in Africa but apparently do not attack humans.

Bengalia spp.—a tropical genus where the species live in termite nests, and marauding adults snatch eggs, pupae, and food from marching driver ants.

Calliphora spp. (blowflies; bluebottles)—several species are strongly synanthropic; the genus is most abundant in the temperate Holarctic Region and cooler parts of Australia. Some species are associated with "sheep strike" in Australia. Adults are important pollinators (Fig. 11.33).

Chrysomya—is an abundant and widespread genus of large, blue/green flies, and in the Old World tropics is the equivalent of the more temperate *Lucilia* and *Calliphora* (but they are not entirely replaced). It is also the Old World equivalent of the New World screwworm (*Cochliomyia*). Several species are involved in human myiasis as well as cattle and sheep.

Chrysomya bezziana (Old World screwworm)—the larvae are obligate parasites in wounds, and, unlike the other members of this genus, never develop in carrion or the like. Females oviposit (and feed) in open wounds, however slight; a scratch or tabanid feeding site is sufficient. Untreated wounds may

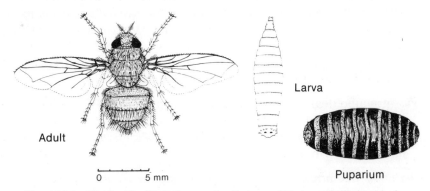

Adult   Larva

0   5 mm   Puparium

Fig. 11.33 Bluebottle (*Calliphora* sp.); all stages (Diptera; Calliphoridae).

quickly prove fatal. Hosts include humans, cattle, horses, and all types of mammalian domestic livestock. It is widespread in Africa and tropical Asia, but not Australia.

Chrysomya mallochi (steel blue blowfly; Australian screwworm)—mostly recorded as a secondary myiasis-producer, but it may be a primary pest in northern regions. The related *C. rufifacies* (hairy maggot blowfly) apparently preys on the primary maggots (*Lucilia*) involved with "sheep-strike," as well as competing with them.

Chrysomya megacephala (tropical latrine fly)—a very abundant species in both Oriental and Australasia regions, extending to Mauritius but not to Africa. Adults equally attracted to sugary foods and exposed organic materials (including feces and carrion) and are important in spreading fecal pathogens. Larvae mainly in fecal matter and carrion; occasionally a secondary infestation in wound myiasis in humans, cattle, or sheep (Fig. 11.34).

Chrysomya spp.—several species cause secondary myiasis in sheep in Australia, and several others are of some importance in different parts of Africa.

Cochliomyia hominivorax (primary screwworm)—larvae are obligate parasites feeding in living tissues; they can enter intact skin but prefer to invade through a wound or scratch; they congregate in groups and excavate a cavity; tissue destruction can be serious or even fatal. This species was one of the first used to demonstrate autocide—control of a pest through the sterilized male technique.

Cochliomyia macellaria (secondary screwworm)—a more widely distributed species, extending from Canada to the southern tip of South America. Mostly responsible for secondary myiasis of existing wounds—it is not an obligate parasite. It appears to be the more primitive of the two *Cochliomyia* species described; it is said to develop in half the time required for *C. hominivorax* and to outnumber it by almost 600 to one.

Cochliomyia (= *Callitroga*) spp. (screwworms)—a New World group causing primary or secondary myiasis in cattle, sheep, goats, and occasionally humans; now invaded North Africa, but reportedly exterminated.

Cordylobia anthropophaga (tumbu/mango fly)—eggs are laid in large batches (200–300) in sand contaminated with urine or feces in shady places. The larvae attach themselves to a host when available and burrow into the skin where they develop in boil-like lesion. Hosts are humans, dogs, rats, monkeys, and some antelopes. In East Africa it is called "mango fly" following its practice of egg-laying on the washing laid to dry on the leaf litter under mature mango trees. The eggs hatch when the clothing is worn and the larvae bore into the skin of the wearer. Infants may be seen with several hundred small boil-like lesions. Found in Africa south of the Sahara.

Lucilia (greenbottles)—stout-bodied flies, a little smaller and less bristly than *Calliphora* (Fig. 11.35). Larvae of several species will infect wounds and invade body cavities of humans and domestic animals. *Phaenicia* is now regarded as a junior synonym of *Lucilia*.

Lucilia caesar (common greenbottle)—cosmopolitan, but a scavenging species, synanthropic in habits, and not of particular importance.

Fig. 11.34 Tropical latrine fly (*Chrysomya mega-cephala*); body length 8 mm (0.32 in.); South China (Diptera; Calliphoridae).

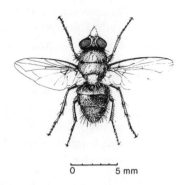

Fig. 11.35 Greenbottle (*Lucilia* sp.) (Diptera; Calliphoridae).

Lucilia cupripa (green sheep blowfly)—introduced into Australia but now responsible for about 80% of sheep strikes in that country; also a major pest in South Africa. Essentially a carrion fly, it is attracted to the soiled fleece around the anus where oviposition takes place. The young larvae first feed on the fecal matter present in the soiled fleece but later attack the skin. Parasitized sheep often refuse to feed and may die. In Australia in 1969/70 it was estimated that sheep-strike losses amounted to $28 million.

Lucilia sericata (sheep maggot fly)—widespread throughout Europe, Asia and North America where it is often quite damaging to sheep, but it is quite an innocuous species in Australia apparently. Larvae of this species have sometimes been used in the treatment of human osteomyelitis.

Phormia—adults are smaller than *Calliphora;* they are Holarctic in distribution.

Phormia regina (black blowfly)—now introduced into Hawaii and Australia where although a scavenger it is attracted to wounds and may be involved in sheep-strike.

Pollenia rudis (cluster fly)—a Holarctic species mentioned here to illustrate some of the diversity shown by the family Calliphoridae—the larvae parasitize earthworms of the genus *Allolobophora*.

Protocalliphora—a Holarctic group whose larvae live in bird nests (mostly Passeriformes) where they suck blood and may actually kill the nestlings.

Rhinophorinae—a small group that parasitize terrestrial Isopoda (Crustacea).

Stomorina lunata (locust blowfly)—in Africa this species has larvae that prey on the eggs of locusts, including the three main pest species (desert, migratory, and red locust); and for some locust populations it can be a major controlling factor.

Sarcophagidae (flesh flies)   *2500 spp.*

Large, grayish flies with a black-striped thorax and a gray/black checkered abdomen; with antennal arista plumose basally but bare distally; bristles usually present on distal part of abdomen. Only a few genera are in this family but many species. Most species are larviparous.

Sarcophaga spp. (flesh flies)—many species; a worldwide genus; most very similar in appearance (Fig. 11.36). Larvae develop in carrion, excrement, or rotting vegetable matter; some parasitize Orthoptera, Lepidoptera, and other insects, snails and other invertebrates. A few species attack vertebrates and even cause myiasis in humans. Female flies are larviparous and usually deposit 40–80 first instar larvae each. Many species are clearly synanthropic. The larvae are large and have the hind spiracles nearly vertical and deeply sunk in a posterior cavity. Throughout the world many different species are recorded as involved in human wound and intestinal myiasis; larvae passing through the human intestine cause discomfort and pain, and sometimes nervous disorder.

Theria muscaria—a species that parasitizes snails.

Wohlfartia magnifica (screwworm)—a widespread species throughout the warmer parts of the Palearctic; larvae are obligate parasites of warm-blooded animals (and are never found in carrion); hosts include humans, horses, cattle, buffalo, donkeys, sheep, goats, pigs, camels, dogs, poultry, and geese. In humans the larvae may be deposited in the ear, eyes, or nose; they cause extensive destruction of healthy tissues, and sometimes result in death.

Wohlfartia nuba—causes wound myiasis in humans and domestic animals in Africa, from Senegal through the Sahara region to Karachi.

Wohlfartia vigil—is found in North America.

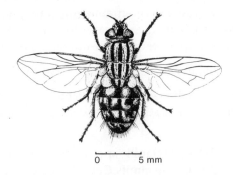

Fig. 11.36   Flesh fly (*Sarcophaga* sp.) (Diptera; Sarcophagidae).

Tachinidae (parasitic flies)   *8000 spp.*

A large group, completely cosmopolitan, which spend their larval lives as internal parasites in other insects, spiders, woodlice (Isopoda) or centipedes (Chilopoda). The adults are of moderate size, gray or brown in color, usually rather bristly, arista usually bare; post-scutellum small and not convex; second abdominal sclerite (sternite) with its sides visible, lying above those of the tergites. The classification of this group is apparently contentious and yet to be finally resolved. Identification is difficult and best left to the taxonomic experts, but the host–parasite catalogue produced by C.I.B.C. is of great assistance to the agricultural entomologist. The flies frequent flowers where they take nectar—generally their habits are rather similar in this respect.

Larval habits and their mode of parasitism vary quite considerably, and various attempts have been made to assign the family into groups on this basis. Briefly, the different approaches are as follows:

1. Eggs laid on the body of the host; the ovoid, basally flattened egg is cemented to the skin of the host, and the larva bores directly into the host body (*Gymnosoma*, etc.). One of the most common methods. In some cases they are almost viviparous (ovoviviparous) as the larva hatches immediately on oviposition and is deposited on to the body of the host (*Exorista, Plagia,* etc.). The average number of eggs laid per female is about 100.

2. Eggs laid directly into the body of the host insect—the female has a special spine-like oviscapt with which the egg is deposited inside the host body (*Alophora, Compsilura, Ocyptera,* etc.).

3. Eggs laid on the food plant of the host—many species practice this; they lay large numbers of eggs (2000+), tiny in size and dark in color; the eggs are ingested with the food and develop internally in the host insect (*Gonia, Sturmia,* etc.); such eggs may remain viable on the foliage for 6 weeks or more.

4. Eggs laid on soil surface in areas frequented by the hosts; practiced by quite a number of species; the female lays a large number of eggs (up to 13,000) and they hatch almost immediately into migratory larvae that bore their way into the first suitable host encountered (*Dexia,* etc.).

Many species have adopted larviposition presumably as a means of increasing survival chances, and the tiny larvae are deposited either on to the host insect or sometimes on to the soil. In these cases, the eggs are incubated inside the female uterus which is richly supplied with trachea.

The parasitic larvae generally have a respiratory problem inside the body of the host and most have devised a method of maintaining contact with the tracheal system of the host. The larvae live inside a sheath formed as an ingrowth of the integument, or an outgrowth of the trachea, formed by the host as a hostile reaction against the parasite. Fully grown larvae bore their way out of the host body and pupate in the soil. In a small host there is usually only one larva, but in large caterpillars and beetle grubs there may be a dozen or more. When rearing caterpillars of large emperor moths (Saturniidae) it is often a disappointment when the caterpillar body suddenly collapses and a dozen or more large tachinid maggots emerge.

The Tachinidae are extremely important in their role of parasites of many crop pests, both as natural control agents and in deliberate biological control projects. The hosts most frequently attacked are Lepidoptera (larvae), Coleoptera (adults and larvae), Orthoptera (adults and nymphs), and larval Hymenoptera, including many sawflies; some Hemiptera and some larval Diptera are also attacked. Some Tachinidae are very host-specific; for example, *Wagneria* spp. only attack larvae of Noctuidae and *Macquartia* spp. are restricted to leaf beetle (Chrysomelidae) larvae. But many are somewhat polyphagous, and extreme cases are recorded in the New World *Compsilura concinnata* with more than 100 host species, and the European *Zenillia nemea* recorded from 15 different families of Lepidoptera as well as the common earwig (Dermaptera).

A few notable examples are listed below:

Bogosia rubens—attacks *Antestiopsis* species on coffee in Africa.

Dexia rustica—parasitizes larvae of *Melolontha* (cockchafers) in the soil.

Exorista sorbilans—has as host the coffee giant looper (*Ascotis* spp.) in Africa.

Lydella grisecens—is a parasite of the corn borers (*Ostrinia* spp.) in maize.

Siphona spp.—parasitize leatherjackets (*Tipula* spp.) in turf and soil.

Theresia spp.—parasitize sugarcane borer.

Wagneria spp.—parasitize larvae of Noctuidae (Lepidoptera).

Zenillia nemea—recorded from many different caterpillars (Lepidoptera).

One case of a beneficial insect destroyed by tachinid larvae is the silkworm (*Bombyx mori*) in Japan parasitized by *Sturmia sericaiae,* which lays its eggs on the leaves of mulberry. As already mentioned most of the Saturniidae and allied moths are regularly parasitized as larvae by the Tachinidae.

Glossinidae (tsetse flies)   *22 spp.*

This small family of bloodsucking flies consists of a singe genus (*Glossina*) confined to tropical Africa (south of the Sahara), and a single record from southern Arabia. Adults are grayish brown or darker, varying in size from 6–14 mm (0.24–0.56 in.) long, and with a distinctive protruding needlelike proboscis sheathed by palpi; antennal arista is plumose with feathered hairs. At rest, the flies sit with the wings folded scissor-like over the back; the wing tips extending slightly beyond the end of the abdomen.

Previously placed as a subfamily within the Muscidae, these flies are now regarded as being quite distinct. Fossil species are recorded from the U.S.A. A total of 22 extant species are recognized in tropical Africa south of the Sahara. These flies are of great importance as vectors of the *Trypanosoma* species that cause sleeping sickness in humans and nagana in animals; many parts of Africa are still largely unpopulated because they lie within the "fly-belts" (tracts where the flies are particularly abundant). All species of *Glossina* are bloodsuckers and feed on vertebrates only (humans, other mammals, birds, and reptiles)—they take no food other than blood. Most species show a measure of host preference, but none is exclusive to one host species. Although they occur throughout tropical Africa, the distribution of tsetse is very uneven and patchy, often they seem to prefer areas of bush or patches of forest which are warm, damp, and shaded.

These flies practice larviposition and the female deposits a fully developed larva that sluggishly burrows into the soil under the leaf litter and than pupates immediately. The larvae are fed by special glands that open into the uterus; larvae are produced successively at an interval of a few days. Damp, shady areas are preferred for larviposition. Ecologically, the 22 species can be grouped into five different habitats ranging from rain forest (including mangroves and swamp forest) through forest/savanna to open savanna woodland.

It is thought that all 22 species of *Glossina* are capable of transmitting the sleeping sickness Trypanosomes, although some species have been more important than others. The flies are pool-feeders in that the labella teeth at the end of the proboscis tear the capillaries in the tissues under the skin, producing a pool of blood from which the liquid is sucked up by a muscular pharyngeal pump. Most mosquitoes, for example, are capillary feeders but some also practice pool-feeding. Serological tests show that most species will feed on a wide range of hosts, including humans, but varying levels of discrimination are shown; thus some species clearly prefer antelopes, or pigs (Suidae), rhinoceros, bushbuck, or birds, or reptiles. Generally, people are not regarded as one of the most preferred hosts.

The trypanosomes are transmitted via the salivary secretions at the time of feeding. Some trypanosomes develop within the insect proboscis (e.g. *Trypanosoma vivax*), others develop initially in the insect midgut and complete development in the proboscis (e.g. *T. congolense*); others have a different cycle in that initial development occurs in the midgut, then the trypanosomes pass via the proboscis and hypopharyngeal tube into the salivary glands where development is completed, and the infective forms are then injected into the vertebrate host at feeding (e.g. *T. brucei, gambiense,* and *rhodesiense*). Although some mechanical transmission of trypanosomes is usual, the insect is generally both vector and intermediate host.

The two most important pest species (groups) are as follows:

Glossina morsitans (group)—transmit the more localized *Trypanosoma rhodesiense* (Rhodesian sleeping sickness); they are regarded as game tsetse, endemic to open savanna woodland, and the population is dependent upon wild game being available; they are responsible for the trypanosome disease Nagana—a fatal disease of cattle and horses. Game extermination projects have been attempted in areas to control these flies.

Glossina palpalis (group)—vectors of the widespread *T. gambiense* which is the main cause of sleeping sickness; more of a waterside forest species, either in the equatorial rain forest or else in the local patches of gallery forest that fringe streams and lakesides in arid areas.

Hippoboscidae (louse flies; keds; etc.)   *200 spp.*

A small family of specialized ectoparasites that live on the body of birds and mammals, in the feathers and pelt. Adults have a characteristic tough and flattened (depressed) body; head sunk into an emargination of the thorax; palpi forming a sheath for the proboscis; antennae lying in a depression on the face; legs short and stout, with strong claws; either winged or apterous. The group is worldwide and host-specificity is marked.

The piercing proboscis is formed from the stem of the labium, and the labella are reduced and insignificant. Both sexes feed entirely on blood from the host. Some species are vectors of hematozoa that parasitize the host bird or mammal.

Together with the following two families, this group used to be referred to as the Pupipara; the egg is retained inside the "uterus" of the female fly and the larva there is nourished by a protein and fat-rich fluid secreted by the modified accessory glands ("milk glands") which empty into the "uterus" near the mouth of the larva. A fully developed larva is produced by the female. In some species it is then carried awhile adhering to the end of the abdomen. During this time the puparium is formed and the white soft prepupa becomes a black, hard, and shiny puparium which is then dropped by the female, either in the nest of the host, or to the ground, or is stuck on to the hair of the host. In many species, puparia are deposited at 2–4 day intervals over a period of several weeks or even months.

Many of the bird flies are fully winged and they fly strongly—most have to be able to search for a host and then fly on to it. The bird flies that parasitize Hirundines have reduced, nonfunctional wings, and the puparia are deposited in the enclosed nest, so the teneral adults do not really have to search for a host. Some of the mammal flies are totally wingless and then the puparium is stuck on to the hairs in the pelt (e.g. sheep ked) and the entire life cycle is spent on the body of the host. Deer keds (*Lipoptena*) are winged and fly on to the host, but then the wings are shed and the flies become flightless. Other mammal flies such as *Hippobosca* remain fully winged and mobile. Many tropical birds, and some mammals, are quite heavily infested with hippoboscids but few species are of any economic interest, although many are of considerable ecological interest.

Some of the more notable species of Hippoboscidae include:

Allobosca crassipes (lemur fly)—found only on lemurs on Madagascar.

Hippobosca spp.—a cosmopolitan, fully-winged genus with several different species on different hosts; horses, cattle, dogs, and other carnivora, camels, antelopes, and ostrich.

Lipoptena spp. (deer keds)—a Holarctic genus; with wings initially, but shed after finding a host; on deer mostly, but also on goats.

Melophagus ovinus (sheep ked)—cosmopolitan; apterous parasite on sheep, and sometimes goats; other species on chamois and antelopes in central Asia (Fig. 11.37).

Ornithomyia spp. (bird louse flies)—several species, mostly Holarctic; on a wide range of wild bird hosts; fully winged flies (Fig. 11.38).

Ortholfersia spp. (wallaby flies)—on wallabies and kangaroos in Australia.

Pseudolynchia canariensis (pigeon louse fly)—now a cosmopolitan species; a pest of domestic pigeons, and a pest of pigeons reared for food in the Orient.

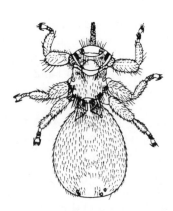

Fig. 11.37 Sheep ked (*Melophagus ovinus*) (Diptera; Hippoboscidae) (ex Hutson).

Fig. 11.38 Bird lousefly (*Ornithomyia avicularia*) (Diptera; Hippoboscidae) ex blackbird; Gibraltar Point, U.K.

Nycteribiidae (bat flies)   *250 spp.*

A small family of extremely specialized insects that live ectoparasitically in the fur of bats. A worldwide group, but mostly in the Old World tropics, especially around the Indian Ocean. Totally wingless, the adult flies have the head folded back at rest into a groove on the thorax; eyes either vestigial or absent; legs elongate with stout claws; an eversible spiny comb (ctenidium) at the anterior end of the thorax (Fig. 11.39). Fully developed larvae are deposited on to hard surfaces where the bats roost in caves or hollow trees, the female having to leave the host for a short while.

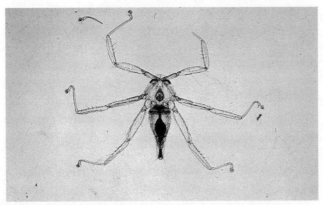

Fig. 11.39　Bat fly (*Leptocyclopodia ferrarrii*) (Diptera; Nycteribiidae) from local fruit bat; Hong Kong.

Streblidae (bat flies)

Another small family, ectoparasitic on bats, found throughout the tropics and subtropics. Adults have the head not deflexed; eyes when present are small; wings can be well developed, small, or vestigial, or absent. Most Old World species have wings and fly well, and a cylindrical body with thorax almost globular; the group is more diverse in the New World. *Ascodipteron* is unusual in that the female is very reduced and burrows into the body of the host—wings and legs are shed and she assumes a flask-like shape, scarcely recognizable as an insect.

DIOPSOIDEA

Diopsidae (stalk-eyed flies)　　　　　　　　　　　　　　　　　　*150 spp.*

A very small family of characteristic appearance in that the eyes are normally borne on long, lateral stalks. Adults are often found in large numbers near streams. Most species occur in Africa and the Oriental region, but a few are found in New Guinea and North America. Larvae are either saprophagous or phytophagous. At least five species of *Diopsis* in Africa are recorded damaging paddy rice. *D. thoracica* (stalk-eyed borer) (Fig. 11.40) is regarded as the most important pest species and it has been recorded attacking sorghum also. The larvae feed inside the leaf sheath on the central shoot, above the meristem region, causing a "dead-heart"; on older plants they eat the flower head before it emerges. Pupation takes place within the leaf sheath or stem.

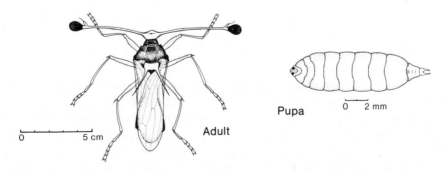

Fig. 11.40　Stalk-eyed borer (*Diopsis thoracica*) (Diptera; Diopsidae) adult and pupa.

SCIOMYZOIDEA

Several small families are placed here, including the kelp flies (Coelopidae) and the Sepsidae—yet another group with larvae living in dung.

Sciomyzidae (snail-killing flies)   *500 spp.*

A small group, worldwide, all of which are predacious upon terrestrial or fresh-water snails and other molluscs. Some larvae are able to kill up to a dozen different snails during their short lifetime; some are referred to as parasitoids and a few are saprophagous. The group is of interest as they are efficient killers of snails, some of which are important pests.

Sciomyza attack land snails and are usually specific to one snail species as prey. Some *Tetanocera* prey on slugs, and other species attack aquatic bivalves of the family Sphaeriidae. *Sepedon macropus* has been introduced into Hawaii in an attempt to control the snail vector of *Fasciola gigantea* (liver fluke) which is an important pest of cattle. Several species have been studied extensively in the hope of using them to limit populations of the snail intermediate hosts of Schistosomes.

ANTHOMYZOIDEA

A number of small families are included here but most are of no economic importance.

Opomyzidae (cereal and grass flies)   *40 spp.*

A very small family whose larvae are phytophagous in grass shoots or occasionally leaves, and can also be damaging to cereal crops though they are mostly found in Europe and Asia. According to Oldroyd (1964) most damage is done to cultivated grass crops and cereals grown on a large scale, as is typical of the U.S.S.R. and eastern Europe; larval damage is usually the destruction of the apical shoot as a typical "dead-heart". Several species of *Opomyza* and *Geomyza* are collectively referred to as "cereal and grass flies".

AGROMYZOIDEA

A large assemblage of families, some of which are not clearly established; in Imms (1977) a total of 23 families are included, but only a few are of economic importance.

Agromyzidae (leaf miners)   *2000 spp.*

A large family of tiny flies whose larvae are mostly leaf miners in a very wide range of plants including many cultivated species. About 75% of the species known are actually leaf miners; the other types of host-plant damage done are mentioned below. These flies are mostly quite small, darkly colored, and may be recognized by having the costal vein interrupted where the Sc (or R_1) runs out, and the female having an oviscap. Identification of adults is very difficult, and one major taxonomic character is the male genitalia. The world authority on Agromyzidae is K. A. Spencer who has written a monograph on the species of economic importance (Spencer, 1973), as well as a whole series of faunistic monographs for most regions of the world. For most field entomologists, an approximate identification can be arrived at by knowledge of the host plant and the life style of the larva now that Spencer has summarized these data.

Eggs are laid in the host plant tissues by means of the female oviscap, or in twigs usually through a lenticel. The larvae are small maggots and typically make a long, winding mine (tunnel) in the leaf lamina, one larva per tunnel; the tunnel appears whitish owing to light reflection from the epidermis with air trapped in the tunnel beneath. Larval fecal pellets are not conspicuous as they are deposited along the edges of the tunnel—most leaf-mining caterpillars leave a central dark line of fecal pellets. Some species make blotch mines, but this is generally more typical of other leaf-mining Diptera. Agromyzid larvae can be easily recognized in that the prothoracic spiracles are placed close together and not in the usual lateral positions. Pupation usually takes place at the end of the mine, and two spiracular horns protrude through the leaf epidermis, usually on the underside of the leaf. A few species pupate in the soil.

According to Spencer (1973) some 150–260 species are regularly associated with cultivated plants (including trees—ornamental, timber, and shade) in 25 genera. The different parts of the plant body attacked by the larvae are summarized below:

1. Leaf miners (70%)—Gramineae (*Cerodontha*)
   —Herbaceous plants (*Liriomyza* and *Phytomyza*)
   —Shrubs (*Tropicomyia*)
   —Trees (*Agromyza, Japanagromyza,* and *Phytomyza*)

2. Stems (seedlings, etc.) (15%)—*Ophiomyia*
   (also leaf petioles)

3. Cambium borers (twigs and trunks) (3%) of apple, *Acer, Salix, Prunus, Betula,* etc.
   —*Phytobia*

4. Twig galls (3%)—*Hexomyza*

5. Roots (3%)—Vegetables (*Melanagromyza fabae, Napomyza carotae,* etc.)
   (*Liriomyza braziliensis* in potato tubers in South America
   —Trees (*Phytobia*)

Host specificity ranges from complete polyphagy to quite restricted monophagy on a single genus of plant (for example, *Camellia*). Some crops are attacked by different species of Agromyzidae in different parts of the world—some of the pests are allopatric. But some pests are cosmopolitan, and others sympatric, so any one crop may be attacked by several different species of Agromyzidae in any one location. The crops most attacked by these insects belong to the families Leguminosae, Gramineae, Solanaceae, Compositae,Chenopodiaceae, Cruciferae and Cucurbitaceae. The crops most heavily attacked by Agromyzidae are mostly temperate, but most are also grown in the Mediterranean and other subtropical regions.

The primary cause of damage is by larval feeding, but sometimes the females regularly make a series of punctures in the leaf epidermis prior to oviposition and for feeding purposes. These feeding punctures on the leaves or petioles of high quality ornamentals result in considerable crop depreciation and loss of sales. As many as 100 feeding punctures in one leaf of spinach have been observed in the U.S.A. (California) and the leaf was virtually destroyed. According to Spencer (1973) the more primitive species of Agromyzidae feed internally in tree trunks, twigs, roots, stems, and seed heads of herbaceous plants. The boring of seed pods of legumes or seed-heads of Compositae, etc., is seen as a modification of internal stem boring. Leaf mining can result in the death of the leaf, and seedlings may be destroyed, and crop losses may ensue if infestations are heavy. Light infestations are very numerous and frequently encountered on a wide range of crops, but identification is seldom easy for many different insects are leaf miners.

Some of the more notable pest species of Agromyzidae to be found worldwide include:

Agromyza ambigua (cereal leaf miner)—mine leaves of temperate cereals; Europe, North America.

Agromyza oryzae (rice leaf miner)—in rice leaves, Java, Japan, eastern Siberia.

Agromyza spp. (28)—attack a wide range of herbaceous plants; worldwide.

Cerodontha spp. (9) (cereal leaf miners)—larvae mine leaves of Gramineae, and some other monocots (*Iris*); worldwide.

Hexomyza spp. (6) (gall flies)—larvae gall twigs of many temperate trees (*Salix, Tilia, Betula,* etc.); Holarctic.

Liriomyza brassicae (cabbage leaf miner)—larvae mine leaves of Cruciferae; cosmopolitan (Fig. 11.41).

Liriomyza braziliensis—larvae bore potato tubers; South America.

Liriomyza bryoniae (tomato leaf miner)—polyphagous leaf miner; Europe, West Asia.

Liriomyza chinensis (onion leaf miner)—onion leaves mined; Malaya, China, Japan.

Liriomyza huidobrensis—totally polyphagous; U.S.A. (Hawaii, California), South America.

Liriomyza sativae—polyphagous on Cucurbitaceae, Solanaceae and Leguminosae; U.S.A., Central and South America.

Liriomyza trifolii (serpentine leaf miner)—polyphagous on many hosts; North and South America, now introduced into the U.K.

Liriomyza spp. (16) (vegetable leaf miners, etc.)—attack a very wide range of vegetables, other field crops, and ornamentals; worldwide.

Melanagromyza chalcosoma—larvae in pods of pigeon pea, cowpea; eastern Africa.

Fig. 11.41 Cabbage leaf miner (probably *Liriomyza brassicae*) (Diptera; Agromyzidae) in leaf of Chinese cabbage; South China.

Melanagromyza fabae—larvae hollow stem and root of broad bean; Europe.

Melanagromyza obtusa (bean pod fly)—larvae inside pods of pigeon pea; India, S.E. Asia.

Melanagromyza sojae (bean fly)—eggs laid on leaves and larvae bore along veins, down leaf petiole into the stem and often into the root; hosts include pigeon pea, beans (*Phaseolus*), and other legumes; Africa, Near East, India, S.E. Asia, Australia.

Melanagromyza vignalis—in pods of cowpea, *Glycine*, etc.; Africa.

Napomyza carotae (carrot root miner)—larvae mine root of carrot; Europe (not U.K.).

Ophiomyia phaseoli (bean fly)—on beans (*Phaseolus* spp.), larvae mine seedling stems and leaf petioles of older plants; Africa, Asia, Australasia (C.I.E. Map No. A. 130) (Fig. 11.42).

Ophiomyia spencerella (East African bean fly)—on beans; only in eastern Africa.

Ophiomyia spp.—larvae mine roots of soybean and other plants; Japan.

Phytobia spp. (cambium borers)—fifty species known but only host data for eight; larvae bore in wood of twigs and tree trunks (e.g. cricket bat willow—*Salix* sp.) and the wood is ruined commercially; mostly a Holarctic group.

Phytomyza horticola (vegetable leaf miner)—polyphagous; cosmopolitan in Old World (Fig. 11.43).

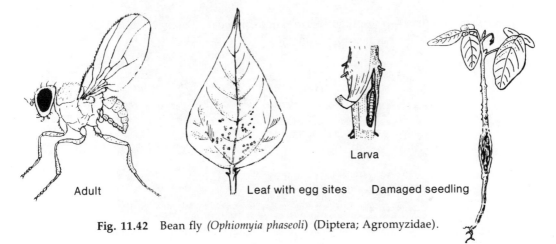

Adult   Leaf with egg sites   Larva   Damaged seedling

Fig. 11.42 Bean fly *(Ophiomyia phaseoli)* (Diptera; Agromyzidae).

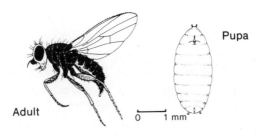

Pupa

Adult   0   1 mm

Fig. 11.43 Vegetable leaf miner (*Phytomyza horticola*) adult and pupa (Diptera; Agromyzidae).

Phytomyza ilicis (holly leaf miner)—blotch mine in holly; Europe, Canada.

Phytomyza syngenesiae (chrysanthemum leaf miner)—on different hosts; Europe, Australia, North America (C.I.E. Map No. A. 375).

Phytomyza spp. (now said to be *Chromatomyia*) (leaf miners)—in many hosts; worldwide in temperate regions.

Tropicomyia atomella (pea leaf miner)—on tea and other plants; India, Sri Lanka, Thailand, Taiwan.

Tropicomyia flacourtiae—polyphagous in leaves of tea, coffee, citrus, cotton, etc.; East, West, and South Africa, Madagascar.

Tropicomyia pisi (pea leaf miner)—larvae mine leaves of pea, white clover, and probably other legumes; Australia (Queensland).

Tropicomyia polyphyta—polyphagous leaf miner recorded from twenty different families (tea, coffee, citrus, legumes, etc.); Australia.

Tropicomyia styricola—on several plants; Japan.

Tropicomyia theae (tea leaf miner)—on tea; Seychelles, India, Sri Lanka, Japan (Fig. 11.44).

Tropicomyia vigneae (legume leaf miner)—larvae mine leaves of legumes; Africa, India, Indonesia.

Fig. 11.44   Tea leaf miner (*Tropicomyia theae*) (Diptera; Agromyzidae); Mahé, Seychelles.

Chloropidae (gout/frit flies, etc.)   *2000 spp.*

A large family of tiny flies (2–3 mm/0.08–0.12 in.), characterized by having the costal vein interrupted at R_1 or more proximally, vein Sc scarcely visible except as a fold, vein 1A and base of M_4 not developed, cell Cu_2 absent; ocellar triangle very large; eyes are often green (hence family name). These flies fall into two large groups—the Chloropinae which are mostly striped yellow and black and the Oscinellinae which are typically black. Because of their tiny size these flies mostly escape notice, and their presence is most likely to be observed in their effects on the infested plants. However, in the tropics the eye flies, *Siphunculina* (S.E. Asia) and *Hippelates* (New World), swarm and settle on the faces of people and various animals in their search for moisture, and they can be very irritating. These flies have terminal spinous pseudotracheae on the labella, which are apparently used to rasp and may make incisions in the eye conjunctiva. It appears that the pathogenic organisms carried on their bodies are responsible for conjunctivitis and various skin disorders in the Orient and parts of the tropical New World. They are pests of some medical and veterinary importance partly by their nuisance value and as mechanical transmitters of bacterial conjunctivitis, yaws, bovine mastitis, streptococci, and other pathogens. Their transmission of other diseases is strongly suspected but not actually proven as yet. The larvae live in polluted soil and are saprophagous.

A few species have larvae that are parasitic or predacious; *Thaumatomyia* is noteworthy in that adults sometimes swarm in houses in warm temperate regions, and the larvae are predacious on lettuce root aphid (*Pemphigus bursarius*).

The vast majority of the members of this family are, however, phytophagous and most feed in the stems of grasses and close relatives in the Gramineae; several are pests of cereals and grass crops grown for seed. Eggs are laid on the leaves of cereal or grass seedlings and the young larvae bore into the terminal shoot which typically enlarges and becomes gall-like (see Fig. 11.45). Sometimes the flower head or panicle is destroyed, as would be expected, but sometimes the ear (in barley) develops but fails to grow away from the ensheathing leaf. With gout fly in barley the lower stem swells but with *Meromyza* the first generation larvae cause "dead-hearts" and the second generation damage the ear within the sheath; these species are mostly temperate.

Some of the more important pest species include:

Chlorops mugivorus (wheat stem maggot)—Japan.

Chlorops oryzae (rice stem maggot)—Japan.

Chlorops pumilionis (gout fly)—larvae swell stem of barley seedlings, etc.; Europe and probably Asia (Fig. 11.45).

Chlorops spp. (40 species recognized in North America alone)—larvae gall various grasses, both wild and cultivated; Europe, Asia, North America.

Meromyza americana (wheat stem maggot)—U.S.A.

Meromyza nigriventris (wheat stem maggot)—Japan.

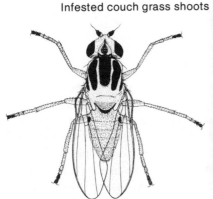

Infested couch grass shoots

Fig. 11.45 Gout fly (*Chlorops* sp.) adult and larval shoot galls on couch grass; (Diptera; Chloropidae); Skegness, U.K.

Meromyza orientalis (wheat stem maggot)—Japan.

Meromyza saltatrix (grass fly)—U.K. and Europe.

Oscinella frit (frit fly)—attacks temperate cereals; Holarctic region.

Oscinella inaequalis—destroys rice seedlings; Malaysia.

Oscinella pusilla—recorded in China and Japan.

Oscinella soror (American frit fly)—on cereals; U.S.A.

Oscinella spp. (cereal and grass flies)—28 species recorded in North America; on many species of grasses and cereals; Europe, Asia, and North America.

In eastern Africa, a number of Chloropidae have been reared from damaged cereal plants, especially sorghum and maize, and they were regarded ostensibly as pests. Several genera were involved but the two most frequently were *Elachiptera* and *Scoliophthalmus*. However, Oldroyd (1964) states that larvae of *Elachiptera* are saprophagous and feed on rotting plant material, leaf mold, and rotten wood; thus their presence in cereal shoots and stems should be regarded as probably only secondary in that they are most likely feeding on tissues damaged by other pests. The importance of the Chloropidae as pests of tropical agricultural crops is as yet not fully resolved.

Conopidae   *800 spp.*

An anomalous group, somewhat divergent evolutionarily; adults vary in size up to 25 mm (1 in.) long; with a long proboscis; abdomen with a nipped-in waist, appearing long and club-like and looking rather wasp-like. Eggs are laid on the body of the host insect; the larva develops internally inside the host abdomen, ultimately filling it completely, whereupon it pupates there.

The most striking genera include:

Conops—parasitizing *Bombus, Sphex, Vespa,* etc.

Myopa—on *Andrena, Bombus, Vespa.*

Physocephala—parasitizes the bees *Apis, Bombus,* and *Xylocopa.*

Stylogaster—in South America the adults can be seen hovering over army ants (*Anomma, Eciton*). It is not clear whether they lay their eggs on the cockroaches and muscoid flies disturbed by the ants or whether they drop eggs that are picked up by the ants.

Piophilidae (skippers)   *6 spp.*

A small group of small flies with saprophagous habits; adults have the costal vein interrupted well before R_1 and Sc is complete. Most larvae live in carrion. One species is of economic importance—*Piophila casei* (cheese skipper) (Fig. 11.46), thought to have originated in Europe but now cosmopolitan; the larvae develop in cheese and fatty ham and bacon. In hams the larvae may hollow out internal cavities without any indication on the surface. Owing to the misguided epicurean habit of eating some cheeses when infested with maggots there have been many cases of intestinal myiasis by larvae of *P. casei*. The larvae survive in the human intestine quite well, and may lacerate the gut wall with their mouth-hooks. They will also feed on hides and furs. Many of the other species of *Piophila* live in carrion; generally in corpses at a more mummified stage—the larvae are capable of eating the last fragments of dried connective tissues and tendons.

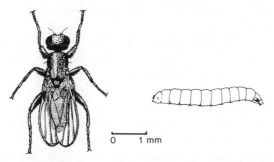

Fig. 11.46 Cheese skipper (*Piophila casei*) (Diptera; Piophilidae).

Tephritidae (= Trypetidae) (fruit flies; picture-winged flies)   *4500 spp.*

A large group of moderately sized flies, completely worldwide in distribution but more abundant in the tropics and warmer regions. Distinctive anatomical features include two small breaks in the costal vein, on the anterior margin of the wing above the second costal cell, and a short, fine, subcostal vein (Sc) in the second costal cell, sometimes elbowed to R_1. The fly body is often quite brightly colored, and the color pattern is an important taxonomic character. Wings are often spotted (maculate) or marbled, and the flies often walk slowly and wave their wings back and forth. The female has a stout, distinctive, horny ovipositor (oviscapt), and in some New World species it may be very elongate. Males typically have a comb (pecten) of bristles on the edge of the third abdominal tergite, and a pair of large, oval, shining spots on the fifth tergite. In some species of *Ceratitis* the male fly has a characteristic triangular expansion at the end of the antennal arista. Adult flies feed on sugary solutions, ripe fruits, honeydew, and with food they can live for 5–6 months, which accounts in part for their importance as crop pests.

With only a few exceptions, the larvae are phytophagous and feed on growing plant tissues and they may be grouped according to their precise life-style, as follows:

1. Inside flower heads of Compositae—*Tephritis*, some *Urophora*, and *Acanthiophilus helianthi*.

2. Inside fleshy fruits—*Ceratitis, Dacus, Anastrepha*, etc.

3. Leaf miners—*Philophylla, Euribia* (*Urophora*).

4. Stem miners/borers—*Paroxyna misella*.

5. Gall formers on various parts of the plant body, both twigs of shrubs and in herbaceous plants (thistles, knapweed, etc.)—*Procecidochares*, some *Urophora*, *Chaetorellia jaceae*.

The damage to agricultural crops is done by the larvae that feed inside developing fruits, etc.; infested fruits often fall prematurely, and may also be associated with fungal infections (especially with *Ceratitis*). The female fly uses her oviscapt to deposit eggs under the skin of young, ripening fruits; it is thought that often eggs are lain in a crack in the rind rather than actually piercing the skin of the fruit. The larvae develop inside the fruit, which may or may not fall prematurely; eventually the large larvae (maggots) leave the fruit and pupate in the soil. Probably the single most important species of fruit fly is *Ceratitis capitata* (Mediterranean fruit fly—now called medfly). It attacks a wide range of commercial fruits throughout the Mediterranean Region, most of Africa, parts of Australia, and is now also introduced into Hawaii, Central, and South America. To date, establishment has been successfully denied in the U.S.A. despite regular accidental introductions in Florida, California, and Texas. Each female (which with food can live for 5–6 months) lays 200–500 eggs, in small batches. After 2–3 days the eggs hatch and the larvae tunnel into the pulp of the fruit. There is often fungal contamination of the oviposition site and subsequently in the larval tunnels (Fig. 11.54), but not always. There are usually 10–12 whitish–yellow maggots per fruit, but occasionally more than 100 have been recorded. Usually only one female oviposits in any one fruit, and it is thought she leaves a pheromone trace which inhibits other females (but these observations only apply to one species to date). The three larval instars usually only take 10–14 days for development. Fully grown larvae either fall from the infested fruit to the ground, or else the fruit falls prematurely and they emerge to pupate in the soil. Pupation takes 8–14 days, and the whole life-cycle in the tropics generally takes 30–40 days, according to temperature (and sometimes host), and there may be 8–10 generations per year in suitable locations.

Ceratitis is essentially a subtropical genus of more than 100 species, and it is thought that high temperatures kill the immature stages; it is endemic to the Mediterranean Region and Africa. Dacus is the truly tropical genus of fruit flies, although it does also inhabit subtropical areas; it is most abundant throughout tropical Asia, Africa, and the Pacific Region. Both of these genera are gradually extending their geographical distributions throughout the warmer parts of the world through the agency of careless distribution of agricultural produce. This Afro-Asiatic genus has 500 species and 30–40 are known or potential pests. Rhagoletis has 50 species, and is widespread in temperate and subtropical regions; generally with only 1–2 generations per year and a more restricted host range. Anastrepha is the New World genus of Tephritidae of agricultural importance, with 150–200 species known, native to Central and South America, and two species now established in southern U.S.A.

Many of the crop pest species are polyphagous and may attack a wide range of cultivated fruits; a number are oligophagous and more restricted, and a few are recorded from only a single host (apple maggot, papaya fruit fly). Some species in wild hosts are quite restricted in their choice of host plants. Members of many plant families are used as hosts, but the most important are the Compositae (buds and flower heads), Rosaceae (fruits), Curcurbitaceae (fruits), and Rutaceae (fruits). In many crop situations, it is common to find fruits locally infested with larvae of several species at the same time.

Fruit flies are important crop pests partly because the larvae, which do the actual damage, are virtually safe from insecticide applications as the eggs are laid actually into the fruit tissues. Thus control measures have to be directed at the adult flies. The adults are long-lived (up to 5–6 months) and quite motile but their longevity associated with frequent feeding, enables them to be attacked by using poison baits. Present control measures used for fruit fly infestations consist of a multifaceted program involving sterile insect (male) release method (S.I.R.M.) whereby male flies are rendered sterile and released into the field population; destruction of infested and fallen fruits; bagging of ripening fruits to prevent oviposition; destruction of females using spot-baits of protein hydrolysate with added insecticide; and male annihilation by using traps baited with sex pheromones or sex attractants (methyl eugenol, sold as "Cu-lure", will attract males of most species of Dacus and Ceratitis).

In the warmer parts of the world, fruits such as peach, citrus, mango, and cucurbits, etc., may often be found to be infested by a complex of fruit fly larvae involving several different species. Usually each individual fruit is only attacked by one species of fly, but the crop as a whole may be infested by several species. In many areas, fruit fly infestations may be limiting factors in local agricultural practices. From personal experience it has been evident that melon cultivation in Uganda (near Kampala) is futile in that almost all fruits produced were heavily infested with tephritid larvae. Similarly in South China, guava trees are grown only as cover on eroded hillsides in most areas as the fruits are totally infested with *Dacus* larvae. In India, mango and melons both have six species of Tephritidae recorded infesting the fruits; worldwide they both probably have about 20 recorded pest species, but many of these are not likely to be important.

Several species of Tephritidae are so important as fruit pests, and because they are somewhat restricted (or were restricted!) to certain geographical regions they have become the subject of international quarantine legislation and also local national legislation. As these pests are so easily transported in infested fruits and other ways, it is a constant battle to limit their geographical spread. In 1974, *Dacus dorsalis* became established in California and in 1975 in northern Australia. Medfly has on several occasions been accidentally introduced into the U.S.A. (Florida, Texas, California) but each time prompt action by the U.S.D.A. and local authorities has achieved extermination of the pest enclave. However, it seems quite likely that this pest will eventually spread into regions in the U.S.A.; especially since it is now established in Mexico.

Some of the more notable species of Tephritidae encountered as pests of cultivated plants include:

Acanthiophilus helianthi (safflower budfly)—in buds of safflower and sunflower; Europe, Near East, Ethiopia, India.

Adrama determinata (tea seed fly)—attacks tea plants; S.E. Asia.

Anastrepha fraterculus (South American fruit fly)—polyphagous; Central and South America (C.I.E. Map No. A. 88).

Anastrepha ludens (Mexican fruit fly)—polyphagous; Central America, now southern U.S.A. (C.I.E. Map No. A. 89).

Anastrepha obliqua (Caribbean fruit fly)—polyphagous; West Indies, Central and South America (C.I.E. Map No. A. 90); now in southern U.S.A.

Carpomyia vesuviana (jujube/ber fruit fly)—in fruits of *Zizyphus*; India, N.E. Africa.

Ceratitis capitata (Mediterranean fruit fly; Medfly)—totally polyphagous in cultivated and wild fruits; Africa, Near East, Australia, Hawaii, Central and South America (C.I.E. Map No. A. 1) (Fig. 11.47).

Ceratitis catoirii—on Mauritius and Reunion only (C.I.E. Map No. A. 226).

Ceratitis coffeae (coffee fruit fly)—in coffee cherries; Africa (Fig. 11.48).

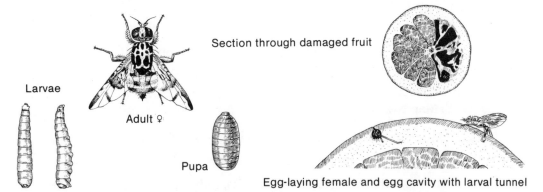

Section through damaged fruit

Larvae

Adult ♀

Pupa

Egg-laying female and egg cavity with larval tunnel

Fig. 11.47 Medfly (*Ceratitis capitata*) (Diptera; Tephritidae); ex Kenya.

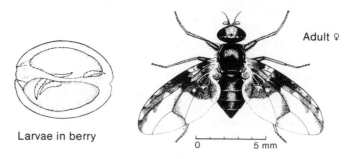

Fig. 11.48   Coffee fruit fly (*Ceratitis coffeae*) (Diptera; Tephritidae); ex Kenya.

Ceratitis cosyra (mango fruit fly)—larvae in fruits of mango, peach, etc.; Africa, south of the Sahara (Fig. 11.49).

Ceratitis rosa (natal fruit fly)—in peach, citrus, and many other fruits; Africa, south of the Sahara (C.I.E. Map No. A. 153) (Fig. 11.50).

Chaetorellia jacea (knapweed/safflower fly)—in flower head of safflower and knapweed (Compositae); Europe, Mediterranean Region.

Many pest species of *Dacus* should now be regarded as belonging to the genus *Batrocera* according to Wood (1989).

Dacus bivittatus (pumpkin fruit fly)—in various fruits; Ethiopia.

Dacus ciliatus (Ethiopian fruit fly)—in cucurbits, citrus, etc.; Africa and India (C.I.E. Map No. A. 323).

Dacus citri (Chinese fruit fly)—larvae in various fruits; China.

Dacus cucumis (cucumber fly)—in cucurbit fruits; Australia.

Dacus cucurbitae (melon fly)—larvae polyphagous in fruits of cucurbits, citrus, cotton, sunflower (head), etc.; East Africa, India, S.E. Asia to China and Japan, New Guinea (C.I.E. Map No. A. 64) (Fig. 11.51).

Dacus depressus (pumpkin fruit fly)—in fruits of Cucurbitaceae; Japan.

Dacus dorsalis (oriental fruit fly)—a polyphagous major pest; tropical and subtropical Asia, Indonesia, Hawaii (C.I.E. Map No. A. 109) (Fig. 11.52). A close relative now established in Surinam (South America).

Dacus latifrons (Malaysian fruit fly)—in various fruits; Malaysia.

Dacus musae (banana fruit fly)—larvae in bananas; Australia (Queensland).

Dacus oleae (olive fruit fly)—only recorded from *Olea* spp.; Mediterranean, South and East Africa (C.I.E. Map No. A. 74) (Fig. 11.53).

Dacus tryoni (Queensland fruit fly)—larvae polyphagous; Australia (Queensland) (C.I.E. Map No. A. 110).

Dacus tsuneonis (Japanese fruit fly)—polyphagous; China and Japan.

Dacus vertebratus (melon fly)—larvae totally polyphagous; eastern Africa.

Dacus zonatus (peach fruit fly)—in many fruits; India (C.I.E. Map No. A. 125) (Fig. 11.54).

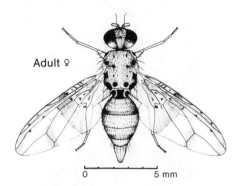

Fig. 11.49 Mango fruit fly (*Ceratitis cosyra*) (Diptera; Tephritidae).

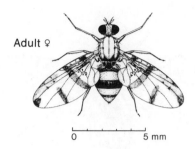

Fig. 11.50 Natal fruit fly (*Ceratitis rosa*) (Diptera; Tephritidae).

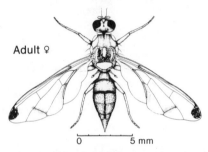

Fig. 11.51 Melon fly (*Dacus cucurbitae*) (Diptera; Tephritidae).

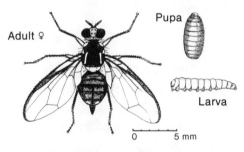

Fig. 11.52 Oriental fruit fly (*Dacus dorsalis*); adult female, larva, and pupa (Diptera; Tephritidae).

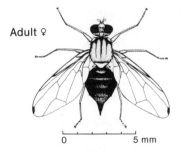

Fig. 11.53 Olive fruit fly (*Dacus oleae*) (Diptera; Tephritidae).

Fig. 11.54 Peach fruit infested with larvae of *Dacus zonatus*; India (Diptera; Tephritidae).

Euriba (Urophora) zoe (chrysanthemum blotch miner)—in chrysanthemum leaves; Europe.

Paroxyna misella (chrysanthemum stem fly)—in chrysanthemum stems; Europe.

Pardalaspis cyanescens (solanum fruit fly)—in fruits of Solanaceae; Madagascar (C.I.E. Map No. A. 140).

Pardalaspis quinaria (Rhodesian fruit fly)—polyphagous; Africa (C.I.E. Map No. A. 161).

Philophylla heraclei (celery fly)—larvae mine leaves of celery, parsnip, lettuce, parsley, etc.; Europe, North America (Fig. 11.55).

Platyparea poeciloptera (asparagus fly)—larvae mine asparagus; continental Europe.

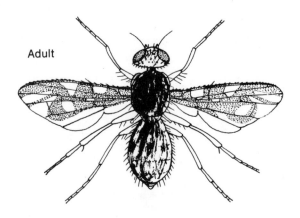

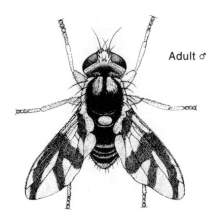

Fig. 11.55   Celery fly (*Philophylla heraclei*) (Diptera; Tephritidae).

Fig. 11.56   Apple maggot adult (*Rhagoletis pomonella*) (Diptera; Tephritidae).

Rhacochleana japonica (Japanese cherry fruit fly)—in cherries, etc.; Japan.

Rhagoletis cerasi (European cherry fruit fly)—larvae in cherries, *Prunus,* and *Lonicera;* Europe and western Asia (not U.K.); occurs as two separate races, northern and southern (C.I.E. Map No. A. 65).

Rhagoletis cingulata (cherry fruit fly)—in cherries; U.S.A. and Canada (C.I.E. Map No. A. 159).

Rhagoletis completa (walnut husk fly)—in walnut fruits; native to the U.S.A. (C.I.E. Map No. A. 337).

Rhagoletis fausta (black cherry fruit fly)— in Canada and U.S.A. (C.I.E. Map No. A. 160).

Rhagoletis indifferens (western cherry fruit fly)—western parts of the U.S.A.

Rhagoletis mendax (blueberry maggot)—in blueberry fruits; U.S.A.

Rhagoletis pomonella (apple maggot)—in apple, etc.; Canada, and U.S.A. (Fig. 11.56).

Staurella camelliae (camellia fruit fly)—in fruits of *Camellia;* Japan.

Strauzia longipennis (sunflower maggot)—in sunflower heads, etc.; U.S.A.

Toxotrypana curvicauda (papaya fruit fly)—in fruits of papaya; U.S.A. (? India).

Trupanea amoena (lettuce fruit fly)—infests lettuce; Europe, Japan.

Trypeta trifasciata (chrysanthemum fruit fly)—in chrysanthemum; Japan.

Zonosemata electa (sweet pepper maggot)—in *Capsicum* fruits; Canada, U.S.A.

Many of the world's most noxious weeds are members of the family Compositae for a diversity of different reasons. In recent years a number of host-specific Tephritidae have been employed in biocontrol programs to reduce the numbers of these weeds in Australia and other parts of the world. Some notable successes have been achieved, and it is likely that flies of this family will be used more widely in the control of weeds belonging to the Compositae.

Otitidae   *500 spp.*

This is a somewhat heterogeneous assemblage of flies that may be lumped together for convenience, but they could also be split into seven different families. Lumped together it is a large family—many species with mottled wings, the female with a flattened horny oviscapt, and with a general resemblance to the Tephritidae; but taxonomic differences include the costal vein being usually unbroken, and vein Sc meets the costa at an acute angle, and is not abruptly elbowed distally. Some larvae are phytophagous (in melons, etc.), other are saprophagous, some are parasitic on chafer grubs (*Popillia, Phyllophaga,* etc.), and *Platystoma* larvae may attack bulbs. None is particularly important economically, so it is convenient to leave the group under the general heading of the Otitidae.

12

ORDER LEPIDOPTERA
moths and butterflies
97: 120,000

A very large group of easily recognized insects with a large number of pest species. The adults have two pairs of wings, membranous and usually large; wings and body appendages clothed with broad scales; mandibles mostly absent but vestigial in a few primitive species; mouthparts represented by a long, coiled, suctorial proboscis formed by the maxillae. Larvae are eruciform, peripneustic, usually with eight pairs of limbs, and are termed caterpillars. Pupae are usually inside a cocoon or an earthen cell.

In more primitive species, the wings are equal-sized and separate (as in swift moths; Fig. 12.1) but, in most cases, the hind pair is smaller and they overlap with the front wings to function as a large single pair. Some of the faster-flying species have an arrangement of special bristles (frenulum) on the hind wing which fit into a fold on the forewing and hold the wings together for greater flight efficiency.

The typical biting mouthparts of primitive insects, which are still retained in the most primitive Lepidoptera, are generally replaced by a suctorial proboscis developed basically to take nectar from flowers, and other liquids. The proboscis is formed from extensions of the paired galea of the maxillae—the greatly elongate galea have internal concavities which when pressed together form the food canal. In the larger Sphingidae, the fully extended proboscis can be as long as 25 cm (10 in.). The fruit-piercing moths (page 496) have a short, stout proboscis with terminal spiny teeth with which they can pierce the rind of an unripe *Citrus* fruit, and some species of the related *Calpe* can pierce the skin of a rhinoceros or buffalo.

The vast majority of species are moths, most of which are drab in color and nocturnal in habits, whereas the brightly colored butterflies are diurnal, and thus far more conspicuous than their numbers warrant. The antennae are a character by which to separate moths from butterflies—most moths either have thin, filiform antennae or large feathery ones, but butterflies have long thin antennae with a terminal club. Butterflies also rest with their wings held vertically over the body, whereas moths at rest hold their wings more laterally and horizontally. Skippers (Hesperiidae) tend to be somewhat intermediate in this respect. Butterflies use a combination of sight and scent for sex location, but with moths being nocturnal they rely entirely on scent, so generally the males have antennae that are far more feathery (pectinate) than the females. In the Saturniidae, the males are reputed to be able to locate a virgin female by her sex pheromones at distances up to 5 km (3 miles) at night in the dark (downwind).

The larval stages, termed caterpillars, are very similar in general structure but vary enormously in detail from family to family. The body is elongate, soft, distinctly segmented, with a well-defined head capsule. There are three pairs of thoracic legs and a varying number (usually four) of soft, fleshy prolegs on the abdomen, and a pair of terminal claspers. The prolegs bear rows of small hooks (crochets) with which they grip the substrate. There is usually a pair of spiracles per segment, located laterally and very conspicuous. Body bristles vary greatly in extent, and chaetotaxy is usually a good taxonomic feature. Most caterpillars are phytophagous and feed on plant foliage, but some are saprophagous and feed on dried animal material (clothes moths, etc.) and cadavers; a few are predacious on other insects; several species prey on the lac insect (*Laccifer lacca*) in India, and several species of Noctuidae are readily cannibalistic under crowded conditions.

Eggs of Lepidoptera are of two main types: upright and spherical or fusiform, with the axes either equal or the vertical axis the longer, or a flattened ovoid shape with its long axis horizontal, usually with reduced sculpturing, and stuck firmly on to the plant surface. The tall or globular eggs usually have a distinctive sculpturing with longitudinal ribs. A few species drop eggs at random in vegetation, but most lay the eggs carefully on the chosen host plant on which the larvae will feed. Some eggs are laid singly (e.g. *Papilio* spp.; Fig 12.71), some in small groups (e.g. banana skipper; Fig. 12.73), and some in batches of several dozens or even up to 100. The average total number of eggs laid is often high—some cutworm moths lay up to 2000 eggs. In some cases the duration of the egg stage may be very long, as hibernation may occur at this stage; this practice is more common in temperate regions.

Larval development is typically quite rapid, except for the wood-boring species, and 5–6 larval instars are general although in some cases there may be either fewer or more. Under crowded conditions or if food is scarce the larvae are able to develop more rapidly and often pupate while still rather small. It is the larval stage that generally does the damage to cultivated plants and so is of special importance to the agricultural entomologist. Some caterpillars may be recognized at sight; in other cases the appearance of the caterpillar, type of damage, and the host plant identity, may enable an approximate identification to be made. With quite a large number of species the caterpillar appearance indicates the family to which it belongs, but it has to be reared into an adult before final determination can be made. Some caterpillars are solitary and difficult to find, others may be gregarious and extremely obvious. They are clearly preyed upon by many birds, other animals, and other insects, and so have evolved different systems for protection.

1. *Concealment:* shown very clearly by the leaf rollers, leaf miners, and tortricids that web together adjacent leaves, spinning silk for the actual fastening of the leaves. The stem borers and fruit borers can be regarded as belonging to this category. One extreme group includes the bagworms (Psychidae; Fig. 12.6) where the larvae manufacture portable cases in which they live. In the Lymantriidae and *Yponomeuta*, the larvae live gregariously in foliage covered by an extensive silk webbing (tent) which affords partial concealment and a certain measure of physical protection. Some of the leaf-eating caterpillars, and most cutworms, achieve concealment by being nocturnal—during the day the larvae lie concealed in the foliage or leaf litter/soil, and they emerge at night to feed.

2. *Protective resemblance:* the most striking examples of this are the larvae of the Geometridae that sit on twigs in the foliage during the day, so that they closely resemble twigs themselves (Fig. 12.80). It is reported that some looper caterpillars are able to change body coloration somewhat in relation to the background foliage, though most are brown in color to imitate brown twigs.

3. *Warning coloration:* some caterpillars feed on poisonous plants and are able to accumulate various toxins in their body tissues which makes them at least highly distasteful to predators and, in some cases, quite toxic and even lethal. These species are generally brightly colored in contrasting red/yellow/black patterns, and they are conspicuous in their behavior also. Some species have long, conspicuous bristles covering the body which have urticating properties and they are generally avoided by most species of predators. Other species have repugnatorial glands by which an offensive odor is released which often serves to protect the caterpillars.

Pupation may take place in the plant foliage or, as with many Noctuidae, Geometridae, and Sphingidae, the caterpillars may burrow into the earth and construct an earthen cell in which pupation takes place. Generally most butterfly pupae are exposed but often protectively colored, although Hesperiidae are usually enclosed in a fold or roll of leaf tissues. Many species spin a cocoon of silk lodged in a fold of the host plant. In some cases the silk production is so extensive that the long, single, silken thread can be unwound and used for commercial silk production; the Saturniidae are sometimes referred to as the giant silkworm moths, and the Bombycidae are the silkmoths. If diapause or hibernation occurs it is most likely to take place during the pupal stage; but, as already mentioned, it may occur in the egg stage and some Noctuidae and temperate Arctiidae overwinter as hibernating caterpillars.

Because of the wide range of life-styles adopted by caterpillars, and also their great importance as agricultural pests, it is sometimes convenient to think of them collectively according to the type of damage inflicted and the life-style of the insect. A few species (mostly Pyralidae) have larvae that are aquatic and feed on foliage of submerged rice and other hydrophytes, but these are mostly of ecological interest. Another few species live in the nests of bees (*Apis, Bombus,* etc.) where they feed mainly on the wax of the brood combs. The bloodsucking *Epipyrops* prey on bugs of the genus *Pyrops* in the Orient.

Because of the vast size of the Order Lepidoptera, the great diversity of habits shown by the larvae (and also adults), and the many types of damage done to plants and human possessions, it is worthwhile looking at the group from the point of view of larval habits and diversity. These are only broad categorizations that would lend themselves to extensive subdivision.

LARVAL HABITS/DAMAGE IN THE LEPIDOPTERA

LEAFWORMS (most families)

A very broad and general term to include all caterpillars found on the foliage of plants and eating the leaf lamina. Three basic types of feeding damage occur: skeletonization when the epidermis of one side or the other is eaten together with some other tissues (some Noctuidae in early instars do this); lamina holing is practiced by some species (diamondback moth, etc.); lamina eating from the edge. The most important families include Sphingidae (hornworms), Arctiidae ("woolly bears"), many Noctuidae, Papilionidae (swallowtails), Pieridae (white butterflies), Lymantriidae (tussocks), Geometridae (loopers), and some species from many other families. Heavy infestations can result in complete defoliation, even of large trees.

BAGWORMS (Psychidae; and some Microlepidoptera)

Bagworms are leaf lamina feeders and so are a specialized group of "leafworms" in practice. But in the Coleophoridae and Heliozelidae some species are involved in leaf mining but also have larval cases during part of their life history. The common case-bearing clothes moths (Tinaeidae) should not be confused with these phytophagous caseworms.

LEAF MINERS (Microlepidoptera; some Gelechiidae, etc.)

These are tiny caterpillars that live and feed between the two epidermal leaf surfaces, sometimes making a tunnel mine and sometimes a blotch mine. Extensive damage can destroy the leaf as a photosynthetic structure, and it may be shed prematurely; heavy infestation can sometimes result in partial defoliation. This life-style is typical of many of the Microlepidoptera, usually for the whole of the larval life. Some Gelechiidae have larvae that start as leaf miners, then they pass down the petiole into the stem and, in the case of potato tuber moth, they often end up tunneling in the tubers/roots.

LEAF FOLDERS/ROLLERS (Tortricoidea; Pyralidae; Hesperiidae)

Medium-sized caterpillars that fold, roll, or cut and roll the leaf lamina, fixing it in position by silken threads, and eating the folded lamina from within the shelter constructed. Skipper caterpillars feed on monocotyledenous plants, and on grasses they fold the leaf longitudinally. On trees and bushes a heavy infestation can result in virtual defoliation and a reduction in crop yield.

FRUITWORMS/BOLLWORMS (Tortricoidea; Lycaenidae; some Pyralidae; some Noctuidae)

A large and diverse group whose caterpillars bore into the fruits of a wide range of plants. Pods and nuts are bored to eat the seed inside; in fleshy fruits either the fruit itself or the seeds inside are eaten. Citrus, apples, and other fruits are seriously damaged, as are nuts and many legumes. Many attack cotton bolls and such is the importance of the crop that these species are known as bollworms. Typically a single caterpillar ruins a whole fruit; they are difficult to control for once inside the fruit they are more or less inviolate as well as already causing damage. Usually there is a short period of time between hatching from the egg and actually penetrating the fruit, during which time the first instar larvae are vulnerable to attack with contact insecticides. Sometimes the eggs are laid on leaves or foliage so that the larvae have some distance to travel to reach the fruits (pea moth). But some, such as pink bollworm, lay their eggs on the fruit surface and the caterpillar bores through the egg base directly into the fruit.

BUDWORMS (Tortricoidea)

This name is sometimes applied collectively to the Tortricidae in the U.S.A. Some tortricids have larvae that develop inside a large bud on various woody shrubs and trees. Larval feeding destroys the growing point of the shoot so damage can be quite serious.

STEM BORERS IN CEREALS (some Pyralidae; some Noctuidae mostly)

Because of their abundance and economic importance it is preferable to regard cereal stem borers separately. Also, ecologically, woody stem (tree trunk) borers are different. Larvae of Pyralidae are smaller than those of Noctuidae and both bore inside the stems of Gramineae; the larger caterpillars only being found in the stems of larger plants. Eggs are usually grouped under the leaf sheath but sometimes are laid on the stem or on the base of a leaf. Some species prefer to feed on the folded leaves for a while before entering the stem. Stem borers in seedlings result in the destruction of the growing point and the formation of a "dead-heart", often followed by tillering. But in larger plants the boring caterpillar hollows out part or all of the internode of the stem, sometimes then moving to another internode or even another stem on the same plant. In sugarcane the stem is solid and usually only one internode is damaged per larva. These pests are usually attacked on young plants by dusting or spraying down the funnel—the insecticide lodges inside the leaf sheath where it is particularly effective in killing the tiny caterpillars.

STEM BORERS IN HERBACEOUS PLANTS (some Gelechiidae; Pterophoridae; Sesiidae, etc.)

Stems of dicotyledenous plants, some erect herbs, climbers and creepers, are bored by smallish caterpillars. Some Gelechiidae start as leaf miners and end as stem borers, and even go down into the roots (and tubers) underground. These insects are interesting but seldom of economic importance, although a few Gelechiidae are serious pests; they can sometimes be attacked using pesticides with a penetrant action.

TREE TRUNK BORERS (Sesiidae; Cossidae; some Pyralidae; Metarbelidae)

These wood borers are medium to large in size, and they tunnel in tree trunks, branches, and even roots, and damage can be extensive. Development is slow; many species in the tropics are univoltine and in temperate regions larvae can take 2–3 years to develop. Once inside the tree, the caterpillars are difficult to kill; usual recommendations require individual treatment by poison injection into the tunnel which is then blocked by a plug of inert material. Infested branches can be cut off and burned. Metarbelidae are somewhat different in that the caterpillar feeds on the bark of the tree or shrub, but has a resting tunnel bored deep into the heartwood at a fork; while feeding on the bark at night it is sheltered under a silken web covered with frass and fecal pellets; pupation takes place in the depths of the resting tunnel.

ARMYWORMS (some Noctuidae only)

Some species of *Spodoptera*, *Mythimna*, etc., have vast reproductive powers and sometimes they occur in large numbers and behave gregariously. After destroying a crop locally they will "march" *en masse* to a new location seeking food. When there are armyworm outbreaks these more or less polyphagous caterpillars can cause devastation. In Africa, the activities of the International Red Locust Control Organization for Central and Southern Africa (I.R.L.C.O.-C.S.A.) has recently expanded the scope of its activities to include African armyworm monitoring and control. For smaller outbreaks mechanized ground spraying is sufficient; for larger scale outbreaks aerial spraying is really required.

CUTWORMS (some Noctuidae)

Some species of *Agrotis, Euxoa, Spodoptera, Noctua,* etc. have caterpillars with a different life-style. The early larval instars are often on the host-plant foliage, sometimes quite gregariously, but as they develop they leave the plant and descend to the ground becoming nocturnal in habits. They spend the daytime sheltering in leaf litter, or actually under the soil, and at night they come to the surface to feed. Typical damage includes cutting through the stems of seedlings (hence "cutworms") and they may eat part of the fallen plant body. Sometimes a single large cutworm will destroy a row of seedlings in one night. They also eat large holes in tubers and root crops from ground level to a depth of 5–10 cm (2–4 in.) in the soil. In temperate regions some cutworm species remain in the soil as large caterpillars overwinter and do not pupate until the following spring; the adults of some of these species are strongly migratory. Several important pest species regularly invade Canada from the U.S.A. each year, and similar invasions take place in parts of Asia and in Europe.

DOMESTIC (URBAN) PESTS (some Gelechiidae; Pyralidae; Tineidae; Oecophoridae)

In stored products, and in domestic situations, there are two broad types of moths whose caterpillars are very damaging. They tend to be most abundant in the tropics and warmer parts of the world, but are common in heated premises in the colder temperate regions. These two basic types are as follows:

1. Grain/flour/dried fruit eaters—several species of Pyralidae, belonging to the genera *Ephestia, Pyralis, Plodia,* and others; the genus *Sitotroga* (Gelechiidae), members of the Oecophoridae (house moths), and some others; they are basically phytophagous and mostly polyphagous.

2. Hair/wool/fur/carpet eaters—these are the clothes moths belonging to the family Tineidae (especially the genera *Tinea* and *Tineola,* Fig. 12.4) whose larvae have special enzymes in their alimentary canal that enable them to digest keratin. In the wild these species feed on dried corpses and cadavers, especially on dried skin (leather), fur, feathers, and the horns of Bovidae and antelopes (see Fig. 12.3); several species of Pyralidae also feed on these materials in the wild, and on owl pellets.

ADULT MOTH PESTS (Noctuidae, Ophiderinae; a few Geometridae; Pyralidae)

A small number of tropical moths have a proboscis that is short, stout, and terminally toothed, and it is used to pierce fruits to suck the juice; and, in a few species (*Calpe* spp.) in S.E. Asia, to suck blood from large ungulates at night. The fruit-piercing moths are mostly species of *Achaea, Othreis* (Fig. 12.119) and *Ophiusa,* although there are others. Also in S.E. Asia is a small, silver geometrid called *Problepsis* (Fig. 12.83) known locally as the "eye moth" for it is attracted to animal fluids and will visit mammals (including humans) to suck tears from the eyes; some closely related species also suck blood from wounds on the body but they are not able to pierce intact skin in the manner of *Calpe*. Some Pyralidae (Pyraustinae) are recorded taking eye fluid from mammals in S.E. Asia (see page 458).

Some of the categories referred to above are not completely exclusive so far as some species are concerned. Some widely distributed species of Noctuidae (especially some *Spodoptera*) apparently vary in their behavior and they may be leafworms or cutworms, and occasionally act as armyworms, and sometimes they climb extensively to feed on the terminal fruiting parts of the plant. The precise reasons for the behavioral flexibility of these "species" is not known as yet. With many Noctuidae it is usual for the first (and maybe second) instar larvae to be gregarious on the host foliage and they skeletonize the leaf surface in patches. As the caterpillars develop they become solitary (usually), and often change from diurnal to nocturnal in habits. Some Noctuidae even become cannibalistic under crowded conditions in the final instar.

The early instars of caterpillars are relatively easy to kill with insecticides, but the last couple of instars (5th and 6th usually) are very difficult to kill as they are far less susceptible to these poisons. The first instar caterpillars of groups such as Noctuidae are tiny, having just hatched from the egg, but the final instar (5th or 6th) is large, and these big caterpillars do most of the damage (by their eating); the final instar typically eats about 80% of all the food consumed during larval development. For leaf worms this point is of some importance, although most plants can lose quite a lot of leaf cover without any ill-effects. But, of course, the various borers have already done considerable damage long before becoming fully grown. Some caterpillar infestations are not noticed before the larvae are well developed and large enough for individual caterpillars to be conspicuous. There are many records of infestations of virtually fully-grown caterpillars being sprayed—this is basically a waste of time, for the damage has already been done, and the caterpillars are on the point of pupation and at that stage of development are extremely difficult to kill. Also empirical evidence has shown that according to the vagaries of population dynamics it is quite unlikely that the succeeding population of caterpillars in that location will be very large.

The classification of the Lepidoptera has, in the past, been varied, and often based on characters of little taxonomic value. The division of the Rhopalocera (butterflies) and Heterocera (moths) has long since been superseded, and likewise the divisions of Microlepidoptera and Macrolepidoptera which were founded only on the distinction of size. But these obsolete terms are still encountered in the literature and occasionally still used. In the present work, the term "leaf-mining Microlepidoptera" has been used as it is a convenient grouping of somewhat primitive tiny moths whose larvae make leaf-mines, difficult for the nonspecialist to identify, and of limited economic importance. In textbooks with a crop orientation it is convenient to group these small families together when they all have larvae that mine in leaves and produce somewhat similar mines and damage symptoms (Hill, 1985).

As indicated by Richards & Davies (1977), the constitution of the suborder Ditrysia is still the subject of much dispute and their arrangement of families is regarded as a conservative one. They stress that the generic classification is increasingly being based on features of the genitalia, most of which will not be at all obvious to most field entomologists.

SUBORDER ZEUGLOPTERA

A small group of the most primitive of the Lepidoptera, characterized by having functional mandibles in the adult stage, the lacinea developed and the galea not haustellate.

Micropterygidae   *c. 40 spp.*

A small family, widely distributed, of tiny, diurnal moths; an ancient primitive group with clear affinities to the Trichoptera; adults with well-developed mandibles and no proboscis. The group is of great importance phylogenetically. The larvae are said to feed on Bryophyta or detritus, and to have many jointed legs. Adults are attracted to flowers where they feed on pollen. They are thought to be relatively abundant in tropical rain forest regions.

SUBORDER DACNONYPHA

Reduced, nonfunctional mandibles are present, both wings with very similar venation; galea more or less haustellate; and the larvae are apodous. Five families are placed here by Richards & Davies (1977) but four are very small groups and not mentioned here.

Eriocraniidae

A small, primitive family; the adults have lost the lacineae; mandibles are present but vestigial; the galea form a short proboscis. The larvae are apodous leaf miners in Angiosperms and the seeds of Gymnosperms. The group is mostly Holarctic and Australian.

SUBORDER MONOTRYSIA

The female moth has two genital openings and an anus on, or behind, sternite 9, or with a cloacal opening on the same segment. They tend to be either quite large moths with similar venation in both wings, or very small moths with reduced venation on the hind wings; larvae sometimes apodous.

HEPIALOIDEA

Venation of both wings similar; female moth with two genital openings on segment 9.

Hepialidae (swift moths)   *300 spp.*

Small to large moths; antennae very short; mouthparts vestigial; wing coupling of jugate type (jugal lobe elongate and resting upon hind wing); tibial spurs absent. The group is widely distributed, but best developed in Australia and New Zealand. The larvae are elongate white caterpillars that live in soil and eat fine roots, or they bore into the roots of both fleshy and woody textures. In parts of S.E. Asia and Australia the larvae are essentially tree trunk (and root) borers, although some species appear to prefer shrub stems. The largest species in Australia have a wing span of 18 cm (7.2 in.), and many species there are brightly colored with patches of red and green, and sometimes metallic markings. Several genera in Australia are the ecological equivalent of the European *Hepialus* and the polyphagous larvae live in vertical tunnels in the soil and eat plant roots, occasionally coming to the surface at night and eating foliage. A single female of *Abantiades magnificus* lays more than 18,000 eggs; *Hepialus* in Europe lay up to 800 eggs, which are dropped singly on to the ground in arable and pasture areas. The European *Hepialus* appear to be ecologically grassland insects and most crop damage is recorded in recently plowed grassland; the adults are crepuscular.

A few species of Hepialidae recorded as doing damage to crops include:

Aeratus spp. (16) (trunk borers)—larvae bore in trunks and stems of *Eucalyptus, Acacia,* and *Lantana; Australia.*

Abantiades spp. (11)—larvae in soil eat roots of *Eucalyptus* spp.; Australia.

Endoclita sericeus (trunk borer)—larvae bore in trunk and stems of cocoa, tea, *Cinchona,* and various legume bushes; India and Java.

Endoclita signifer (grape tree borer)—larvae bore grapevine stem; Japan.

Gorgopsis libania (swift moth)—larvae are pasture pests; South Africa.

Hepialiscus sordida (tuber borer)—larvae bore tubers growing deep in the soil (*Alocasia, Dioscorea,* etc.); Indonesia.

Hepialus humuli (ghost swift moth)—polyphagous larvae in soil eat roots of many plants; Europe and Near East (Fig. 12.1).

Hepialus lupulinus (common swift moth)—polyphagous larvae in soil eat roots of many plants; Europe and Near East.

Hepialus pharus—larvae eat roots of sugarcane; Guatemala.

Oncopera spp. (12)—larvae recorded as pasture pests, and damaging some field and vegetable crops in many parts of Australia and Tasmania.

Oxycanus spp. (44)—polyphagous soil and pasture pests; Australia, New Zealand.

Palpifer spp. (tree borers)—larvae attack various trees; Japan.

Sahyadrassus malabaricus—larvae bore stems of tea bushes in parts of India.

Sthenopsis spp. (tree borers)—larvae attack various trees; U.S.A.

Fig. 12.1 Ghost swift moths, male and female (*Hepialis humuli*) (Lepidoptera; Hepialidae); Cambridge, U.K.

NEPTICULOIDEA

Small moths with reduced wing venation, especially in the hind wing; the female has a short, fleshy ovipositor; larvae are mostly leaf miners; the species in this group retain a number of primitive features.

Nepticulidae (= Stigmellidae) (pygmy moths, etc.)

A quite large group, rather primitive, of moths tiny in size, about the smallest Lepidoptera known, with a wingspan of 3–10 mm (0.12–0.4 in.). The antennal scape is characteristically enlarged and forms an eyecap dorsally. The group had only recently been the object of intensive scientific study, and it appears that this worldwide family is probably both very extensive and abundant. For example in Mackerras (1970) it is stated that only 15 species have been described in Australia, but that the group is abundant and there is probably a total well in excess of 200 species in Australia. Since the revision of the U.K. fauna there are now 100 species recorded from the British Isles. The vast majority of species are placed in the genus *Stigmella*, and *Nepticula* is now regarded as a junior synonym. The moths are quite diurnal and to be seen running over the foliage of the food plant, and flying erratically. The larvae are apodous but have some pairs of leg-like swellings on many body segments. Most larvae are leaf miners, but some tunnel in other parts of the plant body such as buds, stem, and bark, and also leaf midribs and leaf petioles. Eggs are laid singly on the foliage by the female moth, and the hatching larvae tunnel directly from the egg into the plant body. The leaf mines often start as linear tunnels (gallery mines) but end in the form of a blotch mine. Some species only make gallery mines, others only blotch mines. In a regional fauna it appears that the nature of the frass track within a mine can be diagnostic. The mines are usually made in the pallisade layer of the leaf—these are called "upper surface mines" but some species make "lower surface mines" and some eat the whole mesophyll so that the final mine resembles a "window." Pupation usually occurs outside the plant in the leaf litter, inside a cocoon of silk, but some species do pupate inside the mine and some on the plant foliage.

A very wide range of plants is attacked, many wild species and many cultivated ones. A few of the recorded species of Nepticulidae recorded as pests are listed below—they are now all placed in the extensive genus *Stigmella*.

Stigmella anomalella (rose leaf miner)—on roses; throughout Europe.

Stigmella aurella (rubus leaf miner)—larvae mine leaves of bramble, cultivated blackberry, loganberry and raspberry; throughout Europe, North Africa, Near East (Fig. 12.2).

Fig. 12.2 Rubus leaf miner (*Stigmella aurella*) (Lepidoptera; Nepticulidae) in blackberry leaf; Skegness, U.K.

Stigmella gossypii (cotton leaf miner)—larvae mine cotton leaves; U.S.A.

Stigmella juglandifoliella (pecan serpentine leaf miner)—U.S.A.

Stigmella malella (apple pygmy moth)—linear leaf miners on apple; Europe, including Italy.

Stigmella pomonella (apple leaf miner)—blotch mines on apple; throughout Europe.

Stigmella spp. (many)—recorded mining leaves of trees, shrubs, and ornamentals (including pear, plum, *Rubus,* etc.); in many parts of Europe, Asia, India, Australia, and North America; in the U.K. there are 70 species known.

Opostegidae

A small family, almost worldwide in distribution, formerly placed in the Lyonetidae. The adults have the scape of the antennae enlarged to form an extensive eyecap (much larger than in the Nepticulidae) and they have a greatly reduced wing venation. The group is little known and larvae have seldom been found, but the few that are known are leaf miners or else they tunnel in bark or the rind of stems of herbaceous plants—they are totally apodous, cylindrical, and very slender. A cocoon is said to be spun in the soil. The moths are more nocturnal than the other Microlepidoptera and are seldom seen during the day, most specimens being taken at light traps. Most of the species known at present are placed in the genus *Opostega.*

INCURVARIOIDEA

Small moths with reduced venation in the hind wing; female with sclerotized ovipositor; first antennal segment not expanded.

Heliozelidae (shield bearer moths)   *100 spp.*

A small family but worldwide in distribution (but absent from New Zealand); they are characterized by having the scales on the head depressed (as in many Tineoidea). The larvae are apodous leaf miners, and some also mine petioles, or leaf veins. Frass is retained in the tunnel. When full grown the larvae cut portable cases from the leaf lamina in which they descend to the ground to pupate. Sometimes the pupal case may be attached to the branches or trunk of the tree. Host plants are usually oak, birch, alder in Europe, and Rosaceae in the U.S.A. The adults are diurnal and may be seen on flowers, and they fly in sunshine; several species are metallic.

Coptodisca splendoriferella (resplendent-shield bearer)—is a minor pest of apple, wild cherry, and other Rosaceae in the U.S.A.

Heliozela prodela—mine in young, terminal leaves of *Eucalyptus* in Australia.

Incurvariidae

A small group of several hundred species, worldwide in distribution; rather primitive moths, tiny in size, some are brightly metallic, and they fly in sunshine. Males have very long antennae and large eyes; females have a well-sclerotized ovipositor with which they insert eggs into small pockets cut into the host plant, or another plant.

The larvae are either apodous, or only with thoracic legs; some are initially leaf miners which later feed in small portable cases; some are stem borers (*Lampronia*) and others are detritus feeders and they construct portable cases from particles of soil and detritus (leaf litter).

Several species of *Lampronia* are pests of raspberry, currants, and *Rosa* in Europe and Asia—they bore the buds and shoots; other species gall twigs of *Betula* and other trees.

Prodoxidae

The yucca moth (*Tegeticula yuccasella*) is remarkable in its symbiotic (mutualistic) relationship with yucca plants—the female pollinates the flowers of the plant and also lays eggs in the ovaries so that the larvae feed on developing seeds.

Tischeriidae (trumpet miners)

A small group but widely distributed, most abundant in America, but some species found in Europe, India, and South Africa. The larvae are leaf miners, and make a shallow mine just under the upper epidermis; the mine is a blotch mine lined with silk, and the frass is ejected through a small hole in the cuticle. The larvae have reduced thoracic legs and no prolegs, and tend to be flattened. The group was formerly included in the Lyonetidae. Pupation takes place within the mine, and the pupal exuvium extrudes after emergence.

A few pest species are known, but none is a serious pest.

Tischeria malifoliella (apple leaf trumpet miner)—mines leaves of apple; U.S.A.

Tischeria marginea (rubus leaf miner)—mine leaves of *Rubus*; Europe, Near East, North Africa.

Tischeria ptarmica (syzygium leaf miner)—larvae mine leaves of jujube; India.

Tischeria sp. (peach leaf miner)—China.

Tischeria spp. (chestnut leaf miners)—Europe.

SUBORDER DITRYSIA

Females have a copulatory pore on sternite 8 and an egg pore on sternite 9; forewing without jugum or fibula; hind wing often with frenulum; venation reduced. Pupa adecticous and obtect; larvae usually with crotchet-bearing abdominal prolegs. Some 97% of lepidopterous species are placed in this suborder; there is tremendous diversity within the group and the detailed classification is far from settled.

TINEOIDEA

A large and diverse group, open to different interpretations, and based on a number of rather esoteric characters. Adults sometimes have the head clothed in rough scales; antenna sometimes with an "eyecap"; haustellum not scaly.

Tineidae (clothes moths, etc.)   *2400 spp.*

A worldwide family of many species. However, names in the literature may lead to confusion as the family name was formerly applied to a far greater number of species;

the concept of this family has been progressively reduced; the present interpretation is based upon that in Heath & Emmet (1985). Adults have the head covered with rough scales; proboscis short or absent; labial palpi porrect and maxillary palpi usually long and may be folded. Kloet & Hincks (1972) uses five subfamilies, the most important of which is the Tineinae, the larvae of which are mostly feeders on dried animal material, in the wild on dried flesh, skin, feathers, fur and hair, the horns of Bovidae and antelopes (see Fig. 12.3), and regurgitated pellets from birds of prey; the larvae are able to digest keratin. In urban situations these species are clothes moths and the like. Some species with a more varied diet are common in bird nests, and other members of the family are to be found in bracket fungi, lichens, dead wood. The Nemapogoninae larvae are phytophagous or fungivorous, and feed on stored grains, nuts, and different types of fungal bodies, especially bracket fungi. A small number of species are important urban pests, but in households and food stores a number of other Tineidae are occasionally found in small numbers.

The more important species in this family are as follows:

Ceratophaga—now includes some species formerly placed in *Tinea; C. vastella* is probably the most important species with larvae feeding on some dried fruits, dried animal matter and horns of antelopes and cattle; found throughout tropical Africa (Fig. 12.3)

Demobrotis spp. (tobacco moths)—larvae damage dried tobacco leaves; India.

Monopsis spp.—larvae feed on various dried animal products, bird nests, owl pellets, fox scats, etc.; mostly Holarctic but some in Africa, India, and Australasia.

Nemapogon granella (corn moth)—polyphagous on stored grains, dried fungi, nuts, fruits, etc.; probably Palaearctic initially but now quite cosmopolitan.

Setomorpha rutella—a general polyphagous tropical stored products pest; pantropical.

Spatularia mimosae—a minor pest of leguminous seeds in storage; Indonesia.

Tinea pellionella (case-bearing clothes moth)—larvae feed on wool, hair, feathers, and also stored vegetable products; Holarctic, also Australia and New Zealand (Fig. 12.4).

Fig. 12.3 Buffalo horns covered with pupal exuviae and larval remains of *Ceratophaga vastella* (Lepidoptera; Tineidae); Uganda.

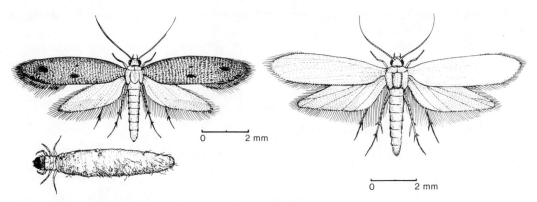

Fig. 12.4 Case-bearing clothes moth (*Tinea pellionella*) (Lepidoptera; Tinaeidae).

Fig. 12.5 Common clothes moth (*Tineola bisselliella*) (Lepidoptera; Tinaeidae).

Tinea translucens (tropical case-bearing clothes moth)—tropical Africa.

Tineola bisselliella (common clothes moth)—a serious pest on woolen fabrics, furs, and other animal fibers, and dried fishmeal, etc.; now completely worldwide (Fig. 12.5).

Trichophaga tapetzella (carpet/tapestry moth)—a widespread species feeding on coarser hair, feathers, and other animal material; its importance as a domestic pest in Europe and North America has declined since the cessation of the use of horsehair in furniture; other species occur in India.

Psychidae (bagworm moths) *800 spp.*

Quite a large group and widespread, but most abundant in the tropics and subtropics. They have evolved in an unusual manner; males are strong and swift fliers but the females are among the most degenerate of all Lepidoptera. The male has virtually no proboscis and thus cannot feed; its lifespan is short and its only activity is to seek out a female for mating. There are a few species with winged and normal females, but most are apterous and many have no antennae or legs and no functional mouthparts. Females wait for males lying in their larval "bag", and after mating the eggs are laid within the bag; most females live for 1–2 weeks. Some are parthenogenetic and males of these species are known from some parts of their geographical range but not from others. Many of the males are diurnal and can be seen searching for females in the sunlight. Australia has a rich fauna of Psychidae and many are primitive species with winged females.

The larvae live in portable cases made of silk and often ornamented externally with twigs or pieces of plant foliage. The front end of the case is open and the head and thorax of the caterpillar extrudes while feeding; the rear end of the case is open for the expulsion of fecal pellets and later for the emergence of the male and the mating of the female moth. Larval development is often protracted—in cooler climates almost a year is usual (but 2–3 years have been recorded), but in the tropics 2–4 months is more general. These insects have a considerable reproductive potential as each female may lay 2000–3000 eggs, which are given a measure of protection within the female bag. Oil and coconut palms in S.E. Asia are often attacked by very large numbers of bagworms—in Sabah (1966) infestations of 300–500 larvae per frond were recorded; the result was of course complete defoliation.

The crops most attacked by bagworms tend to be trees, palms, and bushes grown as plantation crops or ornamentals, especially Palmae, tea, coffee, cocoa, and some evergreen ornamentals (*Cupressus, Thuja,* etc.), and leguminous trees such as *Acacia, Bauhinia,* etc. Some of the better known pest species include:

Acanthopsyche spp.—on coconut, banana, tea, coffee, mango, cocoa, pomegranate; East Africa, India, Sri Lanka.

Amatissa spp.—on tea, oil palm, citrus, banana, etc.; India, S.E. Asia.

Amicta spp.—on tamarisk, etc.; Near East.

Amictoides sp. (grass bagworm)—on grasses; South China.

Canephora asiatica (mulberry bagworm)—Japan.

Clania spp.—a large genus recorded on acacia, citrus, coffee, grapevine, tea, oil palm, pomegranate, *Thuja, Bauhinia,* etc.; East Africa, India, S.E. Asia, Australia, China, and Japan (Fig. 12.6).

Crematopsyche pendula (oil palm bagworm)—on oil palm, plum, etc.; India, S.E. Asia, Australia.

Criocharacta amphiactis (grass bagworm)—on grasses; South Africa.

Cryptothelea spp.—polyphagous on cocoa, coffee, pepper, etc.; East Africa, S.E. Asia, Indonesia.

Cymnelema plebigena—on grapevine; South Africa.

Eumeta spp.—on cocoa; West Africa.

Hyalarcta spp.—polyphagous; China, S.E. Asia, Australia.

Kophene cuprea—on banana; India.

Kotochalia junodi (wattle bagworm)—polyphagous on *Acacia,* cocoa, coffee, castor, pigeon pea; West and South Africa.

Mahasena corbetti (coconut case caterpillar)—on coconut, oil palm, banana, citrus, kapok, *Derris,* and *Cupressus;* S.E. Asia, Indonesia, Papua New Guinea.

Metisa plana (oil palm bagworm)—mostly on oil palm; S.E. Asia.

Monda spp.—on coffee, etc.; East Africa.

Fig. 12.6   *Clania* bagworms on *Thuja* tree; Hong Kong (Lepidoptera; Pychidae).

Oiketicus spp.—on cocoa, coffee, *Hibiscus,* etc.; East Africa.

Psyche casta (persimmon bagworm)—on persimmon; Japan.

Pteroma spp.—on tea, cocoa, coffee, oil palm, and leguminous trees; India, S.E. Asia, Papua New Guinea.

Gracillariidae (blotch leaf miners)   *1000 spp.*

A large and cosmopolitan family of tiny moths whose larvae are mostly blotch miners in the leaves of trees and shrubs. The adults have narrow, long-fringed wings; head smooth-scaled; antennae without eyecaps; and they are crepuscular or nocturnal. One feature of distinction is that the larvae practice a form of hypermetamorphosis. In the early instars the larva is a sap-drinker, and it mines epidermal cells of leaves or sometimes tender bark. The mandibles at this stage are modified for the cutting of the walls of sap-filled cells (but not for chewing) as the larva then sucks out the cell sap. The later stages have the mandibles altered so that they are the more normal biting and chewing type; the differences are examined in considerable detail in Heath & Emmet (1985).

There are two main subfamilies with distinctly different life histories, as shown below. The Gracillariinae generally have two sap-drinking instars, followed by two phases involving several instars in the tissue-eating stage; during the first phase they continue to mine, and in the second phase they feed externally; there are some exceptions though. The Lithocolletinae have three sap-drinking instars, and continue as leaf miners through the two final instars that constitute the tissue-eating phase. In many Gracillariidae pupation takes place in a smooth, membranous cocoon often spun on the surface of a leaf. In the Lithocolletinae pupation occurs in the mine, either with or without a cocoon. The mine consists initially of an epidermal gallery and is later extended into an epidermal tentiform blotch with internal silk spinning. After larval hypermetamorphosis leaf parenchyma is eaten and the blotch mine extends to about one cm (0.4 in.) square. Most Gracillariinae leave the mine in the third or fourth instar and feed in a fold of leaf margin.

There is now more information available about this group, since the publication by Heath & Emmet (1985); clearly in temperate regions the family is well represented but the full extent of its occurrence throughout the tropics is as yet not at all clear.

Some of the recorded pest species of Gracillariidae are listed below, but the two large genera *Gracillaria* and *Phyllonorycter* together include more than 450 species.

Gracillariinae

Acrocercops astourota (pear leaf miner)—China.

Acrocercops bifasciata (cotton leaf miner)—larvae mine leaves of cotton; Africa.

Acrocercops spp. (leaf miners)—on coffee, legumes, castor, sorghum; East Africa.

Acrocercops spp. (8) (leaf miners)—on litchi, coffee, mango, macadamia, sapota, *Syzygium;* India, Sri Lanka, Java, Australia.

Callisto denticulella (apple leaf·miner)—mines apple leaves; Europe.

Caloptilia soyella (soybean leaf roller)—Japan.

Caloptilia theivora (tea leaf roller)—roll tea leaves; Sri Lanka, Japan.

Caloptilia spp.—on many ornamentals; Europe and North America.

Cryphiomystis aletreuta (coffee leaf miner)—on coffee; East Africa.

Cuphodes dispyrosella (persimmon leaf miner)—Japan.

Gracillaria spp.—on many ornamentals; U.S.A.

Lithocolletinae

(*Lithocolletis* is now regarded as a junior synonym of *Phyllonorycter*.)

Marmara elotella (apple bark miner)—mine young bark on apple trees; U.S.A.

Marmara pomonella (apple fruit miner)—mine under skin of apple fruits; U.S.A.

Phyllonorycter spp. (20) (apple/plum/oak leaf miners, etc.)—larvae mine leaves on apple, plum, oaks in Europe, India and Asia; and on Malvaceae and Leguminosae in Australia.

Spulerina astaurcta (pear bark miner)—Japan.

Spulerina spp. (cocoa pod husk miners)—larvae "scribble" on cocoa pods; West Africa.

Phyllocnistidae (leaf miners)   *50 spp.*

At present a very small family containing only a single genus, worldwide in distribution; formerly included in the Lyonetidae but later placed in the Gracillariidae. The adults are tiny and resemble *Leucoptera*. The apodous larva has an adaptation for sap feeding and is also dimorphic; the second form occurs only in the last larval instar which has atrophied mouthparts and does not eat, and its only function is to spin a cocoon. Heavily mined leaves may be killed. The few British species are associated with willows (*Salix* spp.) but the tropical species attack other groups. A few pest species are recorded but only the first is both widespread and abundant.

Phyllocnistis citrella (citrus leaf miner)—larvae mine leaves of *Citrus;* Africa and Asia (C.I.E. Map No. A. 274) (Fig. 12.7).

Fig. 12.7 Citrus leaf miner (*Phyllocnistis citrella*) in leaf of grapefruit; South China (Lepidoptera; Phyllocnistidae).

Phyllocnistis diaogella—abundant on *Phyllanthus* and *Breynia* in eastern Australia.

Phyllocnistis toparea (grapevine leaf miner)—mine leave of grapevine; India, Japan.

Phyllocnistis spp. (leaf miners)—on various common trees and grapevine; Europe, India, much of Asia and Japan.

Lyonetiidae (leaf miners)

A large family, quite heterogeneous, included by Richards & Davies (1977) in the Gracillariidae but generally regarded as a distinct family. The moths are very small and narrow-winged with quite a diversity of detailed structure in the family as at present designated, and several subfamilies are recognized. The main genera, as listed below, are all quite cosmopolitan. The larvae are leaf miners (making blotch mines), some throughout their larval life, but some larvae in their fourth instar eat one epidermis and thus skeletonize the leaf. Pupation takes place in a white silken cocoon either under a living leaf or on a dead leaf in litter.

Some of the better known pest species include:

Bedellia gossypii (cotton leaf miner)—larvae mine cotton leaves; Australia.

Bedellia ipomoeae (sweet potato leaf miner)—Fiji.

Bedellia orchilella (sweet potato leaf miner)—U.S.A.

Bedellia somnulentella (sweet potato leaf miner)—on *Ipomoea* and other Convolvulaceae; cosmopolitan in temperate regions.

Bedellia spp. (sweet potato leaf miners)—on sweet potato and *Ipomoea*; Africa.

Bedellia spp. (convolvulus leaf miners)—on Convolvulaceae; Europe and Asia.

Bucculatrix pyrivorella (pear leaf miner)—Japan

Bucculatrix thurbiella (cotton leaf perforator)—on cotton; U.S.A., Central and South America.

Crobylophora spp. (coffee leaf miners)—on coffee; Central Africa.

Leucoptera coffeella (coffee leaf miner)—larvae blotch mine coffee leaves; Central and South America (C.I.E. Map No. A. 315).

Leucoptera malifoliella (pear leaf blister moth)—Europe and China.

Leucoptera spp. (coffee leaf blotch miners)—blotch leaf mines on coffee (attacked leaves usually shed); tropical Africa (C.I.E. Map No. A. 316) (Fig. 12.8).

Lyonetia clerkella (apple/peach leaf miner)—Europe, China, and Japan.

Lyonetia prunifoliella (plum leaf miner)—Europe, Asia Minor, Japan.

Opogona glycyphaga—larvae in stem of sugarcane, and fruit (skin) of banana; Australia (Queensland).

Opogona sacchari (sugarcane stem miner)—in stem of sugarcane and fruit skin of banana; Africa, Seychelles, Mauritius, Canary Islands.

Opogona spp. (sugarcane leaf miners)—in sugarcane; S.E. Asia, Africa, Hawaii. Sometimes placed in a separate family—Hieroxestidae (Heath & Emmet, 1985).

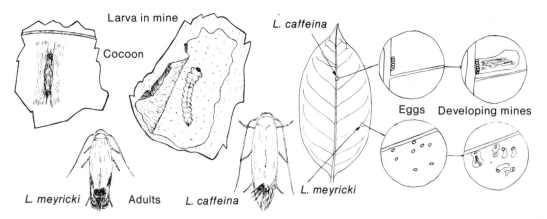

Fig. 12.8 Coffee leaf blotch miners (*Leucoptera* spp.); ex Kenya (Lepidoptera; Lyonetiidae).

YPONOMEUTOIDEA

Sesiidae (= Aegeriidae) (clearwing moths)   *1000 spp.*

A large group, almost completely worldwide in distribution, and characterized by the adults having part of the wings without scales (hence "clearwing"); antennae often dilated or knobbed; the forewings are very narrow owing to reduction of the anal region. The family is characteristic of the Northern Hemisphere; in Australia, most species are found in Queensland. Adults are diurnal and quite wasp-like in appearance. The larvae tunnel in wood of trees and shrubs, or in rootstocks of legumes, and a few in vines (stems of creepers). Pupation takes place in the larval gallery, and the pupal exuvium is often found protruding from the emergence hole. Larval development of the wood borers is generally protracted; the temperate tree borers such as *Sesia* in poplar and willow trees have a three-year life cycle. The shrub-boring, smaller, *Synanthedon* can usually complete their development in one year. The tropical sweet potato and cucurbit vine borers generally only require a couple of months to complete larval development as the food source is richer and the higher temperature accelerates development.

Infected trees are sometimes so extensively bored by the caterpillars that they die; when shrubs are attacked only the distal part of the bored stem dies, and can be removed by pruning. Quite large numbers of eggs are laid by some species—up to 1400 have been recorded for one female. All clearwings are colonial in habit; often one or two trees are selected for colonization and heavily infested (sometimes killed) and neighboring trees left uninfected. It is sometimes thought that stressed trees are preferred as hosts but this is not certain. Infestations can be recognized by several features: (a) presence of old emergence holes, (b) presence of frass at the base of the tree trunk, or on the bark, (c) pupal exuvine left protruding from holes in the bark. Adults often emerge in the early morning and may be seen sitting on the bark for a while after emergence.

The genus *Ceritrypetes* in West Africa is unusual in that it is reputed to have predacious larvae that attack *Ceroplastes* scales.

Some of the more important pest species to be found include:

Conopis hector (cherry tree borer)—Japan.

Conopis sp. (camphor clearwing)—larvae bore camphor trees; South China.

Melittia spp. (squash vine borers)—U.S.A.

Nokona regale (grape clearwing)—Japan.

Paranthrenopsis constricta (rose clearwing)—larvae bore stems of *Rosa;* Japan.

Pennisetia marginata (raspberry crown borer)—larvae bore canes of *Rubus;* U.S.A.

Sannina uroceriformis (persimmon borer)—U.S.A.

Sanninoidea exitiosa (peach treeborer)—U.S.A.

Sesia spp. (hornet clearwings)—larvae bore trunk of trees (oak, poplar, willow, birch, etc.); Europe, Asia, U.S.A. (Fig. 12.9, 12.10).

Synanthedon bibionipennis (strawberry crown moth)—U.S.A.

Synanthedon dasysceles (sweet potato clearwing)—larvae bore vine of sweet potato; East Africa (Fig. 12.11).

Synanthedon myopaeformis (apple clearwing)—on apple; Europe, Asia Minor, U.S.S.R.

Synanthedon pictipes (lesser peachtree borer)—U.S.A.

Synanthedon pyri (apple bark borer)—U.S.A.

Synanthedon salmachus (currant clearwing)—larvae bore stems of *Ribes;* Europe, Asia, Australia, New Zealand, Canada, and the U.S.A. (Fig. 12.12).

Fig. 12.9 *Sesia* larval tunnels in trunk of poplar tree (Lepidoptera; Sesiidae); Skegness, U.K.

Fig. 12.10 Adult hornet clearwing moth (*Sesia apiformis*) (Lepidoptera; Sesiidae); ex poplar tree trunk, Cambridge, U.K.

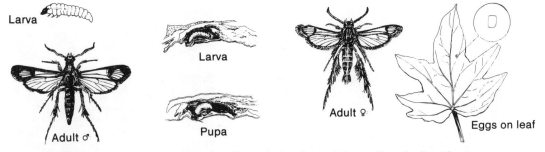

Fig. 12.11 Sweet potato clearwing (*Synanthedon dasysceles*); ex Uganda (Lepidoptera; Sesiidae).

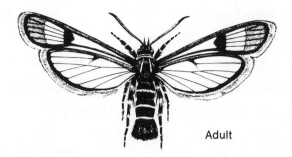

Adult

Fig. 12.12 Currant clearwing moth (*Synanthedon salmachus*) (Lepidoptera; Sesiidae); Cambridge, U.K.

Synanthedon vespiformis (yellow-legged clearwing)—larvae in trunk of walnut, chestnut, etc.; Europe and West Asia.

Vitacea polistiformis (grape root borer)—larvae bore vine rootstock; U.S.A.

Yponomeutidae (small ermine moths)   *800 spp.*

A small group but often locally abundant; the temperate species tend to be small and mostly quite drab, although *Yponomeuta* has silvery white forewings with tiny black spots; the tropical species are often brightly colored and larger. There is considerable diversity shown in the life history of these species, and the group is rather heterogeneous, with generalization somewhat difficult.

There is a number of pest species of some interest, including some of great importance to particular crops.

Acrolepiopsis spp. (leek moths)—specific to *Allium;* throughout Europe, Asia to Japan, and Hawaii (C.I.E. Map No. A. 405).

Argyresthia spp. (fruit moths)—attack temperate fruit trees in Europe and Asia to Japan.

Plutella xylostella (diamondback moth)—a major pest of cultivated Cruciferae; completely cosmopolitan (C.I.E. Map No. A. 32). (Fig. 12.13).

Prays citri (citrus flower moth)—Europe, Asia, India, Philippines, Australasia.

Prays endocarpa (citrus rind borer)—larvae mine fruit rind; India, Indonesia, Malaysia.

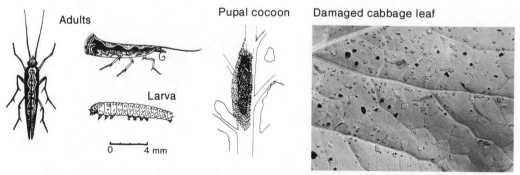

Adults   Larva   Pupal cocoon   Damaged cabbage leaf

0   4 mm

Fig. 12.13 Diamondback moth (*Plutella xylostella*) (Lepidoptera; Yponomeutidae).

Adult

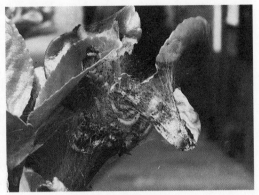

Larvae inside silken web

Fig. 12.14   Small ermine moth (*Yponomeuta* sp.) (Lepidoptera; Yponomeutidae); adult on apple leaf; Skegness, U.K.; and larvae in silken web in laurel bush.

Prays oleae (olive moth)—on *Olea;* Mediterranean, South Russia, South Africa (C.I.E. Map No. A. 123).

Yponomeuta spp. (small ermine moths)—gregarious larvae in silk web defoliate apple, cherry, plum, hawthorn, etc.; Europe, Mediterranean, Asia, Australia, North America; several very closely related species are involved, with somewhat differing host preferences (Fig. 12.14).

GELECHOIDEA

This group of moths have the haustellum more or less densely scaled at the base.

Coleophoridae (= Eupistidae) (case-bearer moths)   *400 spp.*

Small, narrow-winged moths, mostly Holarctic in distribution. May be recognized by their sitting with the antennae held porrect (protruding straight in front); in color most are drab and unicolorous. Larvae are leaf miners in their first instar, some remain leaf miners, but most become case bearers and feed externally. The leaf miners eat the entire leaf mesophyll, leaving just the upper and lower epidermis intact so that the mine resembles a window. The cases are made of silk and vary in size from about 5–8 mm (0.2–0.32 in.) long, according to the species. Many species overwinter as larvae and pupate in early summer in Europe. Most species pupate inside a case attached to a branch or the tree trunk, or in leaf litter. Most species are placed in the large genus *Coleophora* and the pest species generally attack fruit and nut trees, and some ornamental trees (elm, birch, etc.) in temperate Europe, Asia, and North America. However, *C. ochroneura* in Australia and New Zealand feeds on the flower head of white clover and may seriously reduce seed yield. Many species are to be found on wild Chenopodiaceae. Several species of *Acrobasis* attack nut trees in Europe and in North America.

Scythridae

A small cosmopolitan family, best represented in southern Europe and South Africa. Larvae live in silken tubes or a web in the foliage of various plants. One notable pest species is *Syringopsis temperatella* (cereal leaf miner) which attacks temperate cereals (wheat and barley) in the Near East. The larvae make extensive mines, sometimes causing quite serious damage, especially in times of drought.

Oecophoridae   *3000 spp.*

Quite a large family, very well developed in Australia. The adults have broad hind wings, and antennae usually with a basal pecten. The larvae feed in seed heads or spun leaves, or in decaying wood. There are not many crop pest species. A few species are cosmopolitan domestic pests known as "house moths" and the larvae feed on a wide range of vegetable matter and also on carpets, organic debris, and many types of animal material (leather, etc.), but not usually on clothing.

A few pest species of note include the following:

Anchonoma xeraula (grain moth)—on stored grains and foodstuffs; Japan.

Depressaria spp. (parsnip/carrot moths)—in seed head of cultivated Umbelliferae; Europe, Asia, North America (Fig. 12.15).

Endrosis sarcitrella (white-shouldered house moth)—cosmopolitan domestic pest (Fig. 12.16).

Hofmannophila pseudospretella (brown house moth)—cosmopolitan domestic species (Fig. 12.17).

Promalactis inonisema (cotton seedworm)—Japan.

Psorostica spp. (citrus leaf rollers)—India, and Japan.

Adult

Fig. 12.15   Parsnip moth (*Depressaria pastinacella*) (Lepidoptera; Oecophoridae).

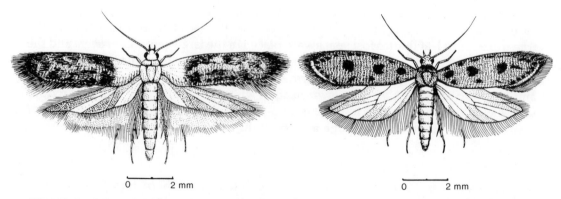

0   2 mm

Fig. 12.16   White-shouldered house moth (*Endrosis sarcitrella*) (Lepidoptera; Oecophoridae).

Fig. 12.17   Brown house moth (*Hofmannophila pseudospretella*) (Lepidoptera; Oecophoridae).

Cosmoptergidae (= Momphidae) (fringe moths) *1200 spp.*

A widely distributed family of small moths with narrow wings, whose larvae have varied habits. Some authorities refer to the group as the Momphidae, and some have the Cosmopteryginae as a subfamily. Some larvae are leaf miners, some feed in seeds or shoots, and some prey on scale insects (Coccoidea).

Some of the phytophagous species recorded on crop plants include:

Anatrachyntis simplex—larvae destroy sorghum grains; India.

Batrachedra spp. (spike borers)—larvae attack flowers of coconut; S.E. Asia.

Blastodacna atra (apple pith moth)—larvae bore shoots of apple; Europe.

Cosmopteryx spp. (leaf miners)—larvae make blotch mines in the leaves of a wide range of plants (several in sugarcane); India, and S.E. Asia.

Microcolona spp.—larvae mine leaves of guava; India.

Pansepta teleturga (cocoa webworm)—larvae eat bark and bore stems; Papua New Guinea.

Pyroderces argyrogrammus—larvae in flower head of safflower; Near East.

Pyroderces rileyi—larvae attack opened cotton bolls and eat seeds; U.S.A.

Pyroderces simplex—larvae attack seeds in open cotton bolls, also coffee berries, maize cobs, and bulrush millet heads; India and Africa.

Sathrobrota simplex—larvae mine leaves of mango, peach, pomegranate; India.

Blastobasidae

A very small group, with most of the species placed in the genus *Blastobasis*. Larvae feed on vegetable rubbish, or seeds, and some are parasites of scale insects (Coccoidea).

Xyloryctidae (= Cryptophasidae)

Small- to medium-sized nocturnal moths, quite widely distributed; well developed as a group in Australia, and to a lesser extent in India and South America. The larvae are generally concealed in shelters or tunnels in bark or wood; some feed on lichens under a shelter of silk and debris.

Cryptophasa spp.—these large species (wingspan 75 mm/3 in.) have larvae that feed on bark or bore in the trunk of many trees (*Acacia*, citrus, fig, apple, plum, etc.); Australia.

Neodrepta luteotactella (macadamia twig girdler)—eastern Australia.

Neospastis sinensis—on tea; south China.

Nephantis serinopa (black-headed caterpillar)—larvae scrape epidermis of leaflets of date and coconut palms; India.

Odites spp. (tube caterpillars)—on oil palm in S.E. Asia and coffee in East Africa.

Procometis spoliatrix—larvae eat foliage of litchi; India.

Gelechiidae   *4000 spp.*

A large group with 400 genera, found throughout the world. The moths are small, with a trapezoidal forewing, narrower than the hind wing; posterior margin of hind wing sinuate; antennae rarely with basal pecten. Larval habits varied; a number bore in fruits, seeds, shoots, or stems; some are leaf miners, leaf rollers or leaf tiers, and a few make galls; a very few are predacious on Homoptera.

A number of important pest species belong to this family. Pink bollworm is an interesting pest in that the larvae usually bore into the cotton boll immediately after hatching, sometimes even directly from eggs laid on the boll; thus the larvae are almost impossible to affect with insecticides and pest control efforts have to be directed towards the adult moths rather than the causal stage—the caterpillars. The species of *Phthorimaea* and *Scrobipalpa* that are serious pests of Solanaceae usually start as leaf miners and then bore down the petiole into the stems; in the case of potato they continue tunneling right down into the tubers underground. The extent of the damage done by these tunneling caterpillars depends in part on the nature of the crop—whether potato, tomato, tobacco, capsicum, eggplant, etc. The potato tuber moth is especially damaging to that crop as it can develop successfully in stored tubers.

Anarsia lineatella (peach twig borer)—larvae bore shoots and enter fruits of peach, plum, apricot, almond, mango; Europe, North Africa, Near East, India, China, Canada, and U.S.A. (C.I.E. Map No. A. 103).

Anarsia spp.—on mango, jujube; India.

Aristotelia fragariae (strawberry crown miner)—U.S.A.

Brachmia triannulella (sweet potato leaf-folder)—on sweet potato; Japan.

Chelaria spp. (mango twig borers)—larvae bore mango twigs; India.

Compsolechia spp. (cherry gelechiid, etc.)—Japan.

Dichomeris ianthes (alfalfa leaf tier)—web alfalfa leaves; Japan, U.S.A.

Holcocera pulverea—predacious on lac insect; India.

Keiferia lycopersicella (tomato pinworm)—on tomato; U.S.A., Central and South America.

Lecithocera spp.—on coffee (shoots) and cotton (fruits); West Africa.

Pectinophora gossypiella (pink bollworm)—larvae bore bolls of cotton and other fruits of Malvaceae; pantropical (C.I.E. Map No. A. 13) (Fig. 12.18).

Pectinophora scutigera (Queensland pink bollworm)—larvae attack cotton and other Malvaceae; Australasia and Hawaii (C.I.E. Map No. A. 14).

Pectinophora spp.—larvae bore fruits (seed capsules) of Malvaceae; pantropical.

Phthorimaea operculella (potato tuber moth)—on potato and Solanaceae; almost completely cosmopolitan throughout the warmer parts of the world (C.I.E. Map No. A. 10) (Fig. 12.19).

Scrobipalpa heliopa (tobacco stem borer)—mostly on tobacco; India, S.E. Asia.

Scrobipalpa ocellatella (beet moth)—larvae bore in root of sugar beet and mangles; Mediterranean Region and Middle East.

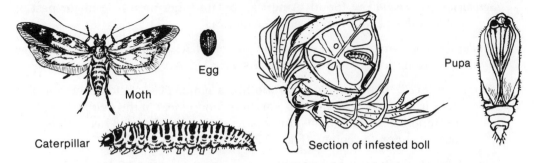

Fig. 12.18 Pink bollworm (*Pectinophora gossypiella*) (Lepidoptera; Gelechiidae).

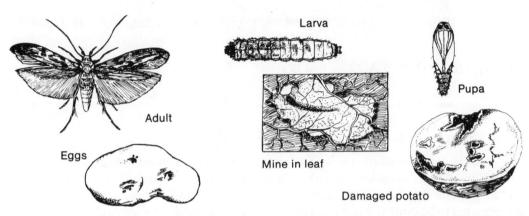

Fig. 12.19 Potato tuber moth (*Phthorimaea operculella*) (Lepidoptera; Gelechiidae).

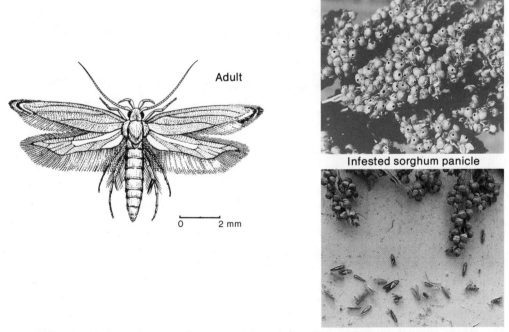

Fig. 12.20 Angoumois grain moth (*Sitotroga cerealella*) and infested sorghum panicle; Ethiopia (Lepidoptera; Gelechiidae).

Scrobipalpopsis·solanivora (South American potato tuber moth)—a major pest of potato tubers; South and Central America.

Scrobipalpa spp.—many species feed on Chenopodiaceae, either boring the stems or webbing foliage and flower heads, etc.; U.S.A.

Sitotroga cerealella (Angoumois grain moth)—a major pest of stored grains, also dried fruits; may attack maize, sorghum, and wheat in the field; completely cosmopolitan throughout warmer parts of the world, and in heated stores in cooler temperate regions (Fig. 12.20).

Stegasta bosqueella (rednecked peanut worm)—pest of groundnuts; U.S.A.

Stomopteryx (= *Biloba*) *subsecivella* (groundnut leaf miner)—South Africa, India, Sri Lanka, S.E. Asia, and China.

Tildenia inconspicuella (eggplant leaf miner)—larvae mine eggplant leaves; U.S.A.

COPROMORPHOIDEA (Alucitoidea)

These moths do not have the haustellum scaled.

Orneodidae (= Alucitidae) (many-plume moths)

A small family of distinctive appearance in having both wings cleft into six or more, narrow, plume-like divisions, densely fringed. The larvae are stout-bodied, cylindrical and bristly, and they tunnel in shoots, buds, and flower stalks causing galls. The pupae are reported to be quite distinct from the Pterophoridae and to be more like those of the Tinaeidae and Pyralidae. The group is of no particular importance agriculturally.

Carposinidae

Another small family, well represented in Australia and Hawaii, the larvae of which are mostly bark tunnelers, but a few bore into fruits, and a few are crop pests.

Carposina fernaldana (currant fruitworm)—U.S.A.

Carposina niponensis (peach fruit moth)—Japan.

Meridarchis spp.—larvae bore fruits of jujube (*Syzygium*) and olive; India.

COSSOIDEA

Cossidae (goat moths; carpenter moths; leopard moths)

A group of worldwide occurrence, particularly abundant in southern Africa and Australia. Some species are large with a wingspan of 8–18 cm (3–7 in.), but a few are quite small. Generally regarded as an ancient and primitive group. Adults are fast-flying and nocturnal, and may come to lights at night; antennae are usually pectinate in the male (only the basal half in the Zeuzerinae) and simple in the female. The larvae are wood borers in the trunks and branches of trees and shrubs, and also the pith of reeds; the larger species take 2–4 years for larval development but most tropical species develop within the one year. The larvae have a large, sclerotized prothoracic plate, small head, and very large mandibles with which they tunnel in the wood or pith. First instar larvae of *Zeuzera* and *Xyleutes,* etc. (Zeuzerinae) spin quantities of silk and are alleged to disperse aerially carried on the silken threads in the wind; but the extent of this practice (recorded in Indonesia) is not known. Pupation takes place at the end of a larval tunnel in a cocoon of silk, and the emerging adult leaves the pupal exuvium protruding from the exit hole (see Fig. 12.23). Some species lay very large numbers of eggs; in Australia, the female of *Xyleutes durvillei* is recorded laying more than 18,000 eggs.

When the larvae are tunneling, frass is extruded from the tunnel at intervals through a small hole; this is one of the main methods of recognizing a cossid infestation. Sometimes there is copious sap exudation from the frass holes. Trees heavily attacked by cossid larvae suffer extensively, and may be killed.

Some of the recorded pest species include:

Azygophleps spp.—in stems or roots of various hosts, including woody legumes (*Sesbania*); East Africa, India.

Coryphodema tristis (trunk borer)—in apple, grapevine, etc.; South Africa.

Cossula magnifica (pecan carpenterworm)—bores pecan, etc.; U.S.A.

Cossus cossus (goat moth)—polyphagous pest of woody plants (trees mostly); Europe, Asia.

Cossus japonicus (oriental goat moth)—polyphagous in trees, etc.; Japan.

Eulophonotus myrmeleon (cocoa stem borer)—larvae bore stems of cocoa, coffee, cola, etc.; tropical Africa (Fig. 12.21).

Paropta spp.—larvae in trunks and roots of olive, fig, grapevine, etc.; Near East.

Phragmataecia spp. (reed leopard, etc.)—larvae bore stems and roots of Gramineae (reeds, sugarcane, *Pennisetum,* etc.); parts of Europe, Asia, Africa, India and S.E. Asia.

Prionoxystus spp. (carpenter worms)—U.S.A.

Xyleutes capensis (castor stem borer)—larvae in stems of castor and *Cassia;* East Africa (Fig. 12.22).

Xyleutes spp.—many species are known (69 in Australia alone) boring in many different trees and bushes; Africa, Asia, and Australasia.

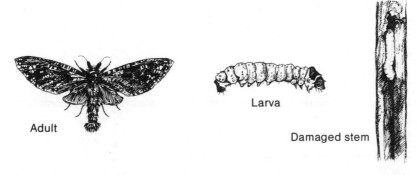

Fig. 12.21   Cocoa stem borer (*Eulophonotus myrmeleon*) adult and larva (Lepidoptera; Cossidae).

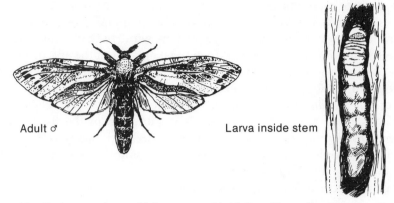

Fig. 12.22   Castor stem borer (*Xyleutes capensis*) adult and larva (Lepidoptera; Cossidae).

Fig. 12.23   Leopard moth (*Zeuzera* sp.) adult; body length 23 mm (0.92 in.); South China (Lepidoptera; Cossidae).

Zeuzera coffeae (red coffee borer)—polyphagous borer, in citrus, coffee, etc.; India, S.E. Asia, China (C.I.E. Map No. A. 313).

Zeuzera leuconotum (oriental leopard moth)—polyphagous borer; Japan

Zeuzera pyrina (leopard moth)—polyphagous tree borer; Europe, Asia, North Africa, North America (C.I.E. Map No. A. 314).

Zeuzera spp. (leopard moths)—polyphagous borers in Asia through to Papua New Guinea (Fig. 12.23).

ZYGAENOIDEA

A group with the proboscis usually atrophied, and maxillary palps vestigial or absent. A number of obscure families are placed here, along with several of economic importance.

Metarbelidae (wood/bark borers)

A small group confined to tropical Asia and Africa. The nocturnal moths slightly resemble Cossidae, but have a more reduced wing venation. In some older publications this group was regarded as a subfamily of the Cossidae. The larvae feed on the bark of trees and shrubs under a webbing of silk and frass particles. They have a refuge hole deep into the heartwood usually at a fork or where a branch has developed; during the day the larvae hide in the refuge hole (where later they also pupate) and come out at night to feed on the bark. Small trees and bushes are easily girdled and they may die; large trees may suffer multiple infestations which is clearly damaging to the health of the tree; 15–30 larvae on a large tree is not uncommon.

Most species are polyphagous, and the trees and bushes attacked include: cocoa, citrus, guava, mango, litchi, *Ficus,* mulberry, jackfruit, *Acacia, Hibiscus, Syzygium,* pomegranate, and other forest and ornamental trees. The pests are several species of each of the following four genera.

Indarbela spp. (bark borers)—found throughout S.E. Asia, from China to India and N.E. Africa (Fig. 12.24).

Metarbela spp. (bark borers)—on *Hibiscus,* etc.; most of Africa.

Salagena spp. (citrus/podocarpus borer, etc.)—larvae bore in citrus, *Podocarpus, Albizia,* and *Gossypium;* throughout tropical Africa.

Tetragra spp.—South Africa.

Adult

Larval tunnel on acacia tree

Fig. 12.24 Wood borer moth (*Indarbela* sp. nr *disciplaga*); body length 20 mm (0.8 in.); and larval tube on trunk of *Acacia* tree; South China (Lepidoptera; Metarbelidae).

Limacodidae (= Cochlididae) (stinging and slug caterpillars)

A widespread family, best represented in the tropics; adults are medium-sized, stout-bodied, rather hairy, and with broad wings; color is sometimes a drab brown, but often bright green and brown; some species show sexual dimorphism. As a group they are basically forest insects, and most are nocturnal. Eggs are flat and scalelike, laid in groups on the undersurface of leaves. The larvae show considerable morphological diversity from genus to genus. One group of larvae is called "stinging caterpillars" and have conspicuous spine-bearing scoli, the spines are often urticating, and in a few cases poisonous and quite painful. Larvae of *Setora nitens* cause serious problems in tea plantations in Indonesia as the laborers picking the tea are fearful of being "stung" by the larvae. The other group of larvae have a broad sluglike body with reduced head and legs and obscure segmentation—viewing the caterpillar from above, it is usually not possible to discern which is the head end. Many of the larvae are brightly colored, although green and yellow are probably the predominant colors.

Pupae are enclosed inside a rounded hard cocoon stuck on to the tree trunk or a leaf.

The larvae are often quite polyphagous, and as a group they tend to be associated with monocotyledenous plants (especially Palmae), although tea and other Theaceae are quite heavily attacked. Although the group is important in tropical agriculture, the larvae and pupae are normally heavily parasitized by Hymenoptera and Tachinidae and very often population control is achieved by microorganisms (bacteria, fungi, viruses). Microbial parasitism is typically at a high level when conditions are damp, so most pest outbreaks occur during the dry season. Population fluctuations are normally very dramatic and frequent with these insects.

Within the group there is quite a large number of genera and the literature is rife with conflicting names. The genus *Thosea* is a large one, found throughout S.E. Asia; and it is reported that *Latoia* is to be regarded as a junior synonym of *Parasa* which is now the largest genus in the family, widespread throughout the world except Europe.

A number of species are of importance as agricultural pests by their defoliation. On palms in S.E. Asia, populations are often very large and defoliation is not uncommon. Coffee, cocoa, tea, coconut, and oil palm each have about a dozen species of Limacodidae recorded attacking their foliage on a worldwide basis. Some of the recorded pest species of importance are listed below:

Belippa lateana—on coffee, cinchona, etc.; India.

Cania spp. (stinging caterpillars)—on oil palm, banana, tea, coconut; India, S.E. Asia, and Japan.

Chalcocoelis spp. (gelatine caterpillars)—polyphagous on coconut, cocoa, citrus, camphor, etc.; India, S.E. Asia, China, Australia.

Cheromettia lohor (gelatine caterpillar)—on oil palm, tea, coffee, banana, cocoa, kapok, *Derris,* etc.; throughout S.E. Asia.

Cheromettia sumatrensis (gelatine caterpillar)—mostly on oil palm; S.E. Asia.

Contheyla rotunda—on coconut on the west coast of India.

Darna catenata—on coconut and sago palms; Sulawesi, Papua New Guinea.

Darna trima—small spiny larvae on coconut, oil palm, citrus, coffee, cocoa, banana, etc. (quite polyphagous); throughout S.E. Asia.

(*Latoia* = *Parasa*)

Natada spp.—on mango, etc.; India.

Narosa spp.—on coffee, castor, etc.; Malaysia.

Niphadolepis alianta (jelly grub)—on coffee, tea, castor; eastern Africa (Fig. 12.25).

Niphadolepis bipunctata—also on coffee; Tanzania.

Oxyplax ochracea (slug caterpillar)—polyphagous on grasses, coconut, etc.; from India through S.E. Asia to China.

Parasa indetermina (stinging rose caterpillar)—on roses, etc.; U.S.A.

Parasa lepida (blue-striped nettlegrub)—totally polyphagous; Africa, India, S.E. Asia, Japan (C.I.E. Map No. A. 363) (Fig. 12.26).

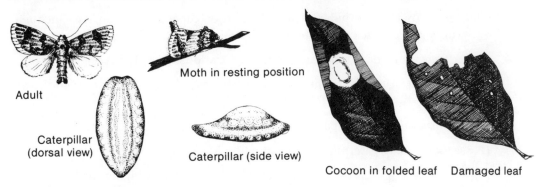

Fig. 12.25 Jelly grub (*Niphadolepis alianta*) on coffee; ex Kenya (Lepidoptera; Limacodidae).

Fig. 12.26 Blue-striped nettlegrub larva, and adult (*Parasa lepida*); wingspan 38 mm (1.4 in.); Hong Kong (Lepidoptera; Limacodidae).

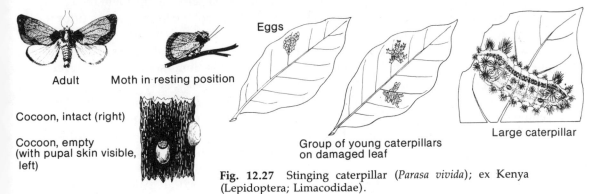

Fig. 12.27 Stinging caterpillar (*Parasa vivida*); ex Kenya (Lepidoptera; Limacodidae).

Parasa repanda—on almond and apricot; India.

Parasa vivida (stinging caterpillar)—on coffee, cocoa, groundnut, tea, castor, cotton, sweet potato, etc.; tropical Africa (Fig. 12.27).

Parasa spp.—recorded from sugarcane, *Ficus*, banana, coffee, citrus, cocoa, oil palm, etc.; Africa, India, and S.E. Asia.

Ploneta diducta (stinging caterpillar)—on oil palm, coconut, cocoa; S.E. Asia.

Scolopelodes unicolor—polyphagous, spiny, green-blue larvae on cocoa, kapok, etc.; S.E. Asia.

Setora nitens (nettle caterpillar)—polyphagous on coconut, oil palm, tea, cinchona, coffee, cocoa, etc.; throughout S.E. Asia.

Sibine stimulans (saddleback caterpillar)—on various trees; U.S.A.

Spatulicraspeda castaneiceps (humped slug caterpillar)—on tea, castor; India, Sri Lanka.

Susica himalayana—on oil palm: S.E. Asia.

Thosea asigna (slug caterpillar)—on oil palm; Malaysia.

Thosea asperiens (green nettle slug caterpillar)—a severe stinging form found on tamarind, sorghum, finger millet, and various legumes; India.

Thosea sinensis (slug caterpillar)—polyphagous on coffee, tea, oil palm, coconut, pepper, etc.; India, S.E. Asia, to China.

Thosea spp.—several species recorded from Leguminosae, castor, coffee, sorghum, millets, coconut, oil palm, pepper, banana, sugarcane, etc.; Africa, India, S.E. Asia, China.

Trichogyia albistrigella (slug caterpillar)—on Palmae; Indonesia.

Zygaenidae (burnets, foresters, etc.)   *400 spp.*

Sometimes called "day-flying moths," these insects are medium-sized, often brightly colored, and to be seen flying slowly during the daytime. A few species resemble butterflies quite closely. The larvae have stout bodies and often reduced legs, and are somewhat sluglike; often conspicuously colored and sometimes the cuticle is covered with wartlike projections called verrucae, bearing spines. The larvae live quite exposed (often gregariously) on herbaceous plants and palms. The pupae are typically enclosed inside an elongate, tough, membranous cocoon fastened to the plant foliage or some firm substrate and the mobile pupa forces its way out of the tough cocoon prior to the emergence of the imago. The coloration of both larva and adult is clearly aposematic (i.e. warning coloration) and toxins have been found in the body at all stages of development, although the larvae do not usually feed on toxic plants. Some species typically feed on herbaceous legumes and then pupate on tall grass stems.

The group is worldwide, although absent from New Zealand and poorly represented in the Pacific region. The subfamily Zygaeninae is characteristic of the Palaearctic region where the metallic green "foresters" (*Ino,* etc.), and the red-brown "burnets" (*Zygaena*) are conspicuous and common; the other subfamilies are tropical.

Artona catoxantha (coconut moth)—larvae defoliate coconut palms (sometimes completely, with a total loss of yield for 1½ years); throughout S.E. Asia, Philippines, Papua New Guinea, and Fiji.

Astyloneura spp.—on sorghum, *Cissus,* etc.; East Africa; on sugarcane in Java.

Chalconycles catori—on coconut; West Africa.

Charidea homochroa—on pea and *Cissus;* East Africa.

Eterusia spp.—pests of tea and apple; India and Sri Lanka.

Levuana iridescens—larvae defoliate coconut palms in Fiji, but now kept under control by an introduced tachinid fly (*Ptychomyia remota*); also attacks sugarcane in Fiji.

Zygaena spp. (burnet moths)—a Palaearctic genus feeding on legumes, Labiatiae, Compositae, etc.

Epipyropidae (parasitic moths)   *c. 30 spp.*

A very small family of very small moths, whose larvae feed ectoparasitically on Cicadidae, Cicadellidae, and some Fulgoroidea. The group is actually worldwide in all major zoogeographical regions but is best represented in Australia.

Eggs are laid singly, or in small groups, on the host plant—they are flat and disc-shaped. The larvae are very active and seek a host insect in the plant foliage. They are unusual in having narrow, double-toothed mandibles; also, on the head is a very large spinneret. In South China it was shown by Marshall (1970) that young larvae of *Epipyrops anomala* were usually found under the hind wings of the host (*Pyrops candelaria*) whereas larger larvae were dorsally on the abdomen. The larva extrudes silk over the body of the host; when dried the silken threads serve as anchorage for the larva which clings to them using the crochets on the abdominal prolegs. Large larvae were seen to feed by making wounds on the host abdomen from which tissue was scooped. Young larvae feed from the wing veins where they take blood. There is typically only one large larva per host insect. Fully grown larvae leave the host insects and pupate on the host's food plant inside a mass of silken material.

Several quite important crop pests are regularly parasitized by *Epipyrops* species, including:

Idioscopus spp. (mangohoppers) (Cicadellidae)—on mango; India and S.E. Asia.

Pyrilla perpusilla (Indian sugarcane leafhopper) (Lophopidae)—on sugarcane; India.

Pyrops candelaria (Fulgoridae)—on litchi and longan trees in China.

CASTNIOIDEA

Castniidae   *160 spp.*

These are brightly colored day-flying moths that resemble either skippers or nymphaline butterflies, found mostly in tropical America, Indo-Malaysia, and Australasia. The larvae bore within the stems of plants and a couple of species are recorded as crop pests in the Neotropical region.

Castnia daedalus—the larvae tunnel in coconut palms in South America (British Guiana, etc.); feeding mainly between the base of the fronds and the trunk; severe damage may kill the palm.

Castnia licoides—a major pest of sugarcane in Trinidad and the more northern countries in South America; they tunnel through the entire stem system from tip to root, and also bore in banana and coconut palms.

TORTRICOIDEA

This large group of small, dark moths has presented taxonomic problems for many years, and in some publications as many as ten or a dozen different families have been proposed. The present British and Australian interpretation is for just the two families referred to below, with a distinct series of subfamilies and tribes which will not be used in the present text as the major taxonomic criteria are somewhat esoteric for general use.

The adults are small in size but with wide wings with short hair-fringes. The group is more representative of temperate regions than of tropical, but it is worldwide and there are many species in what are regarded as the subtropical parts of the world.

Eggs are oval and flattened, and not obvious; usually laid in small groups. The larvae live concealed, usually inside a rolled or folded leaf, fruit or stem, root, flower or seed pod. The caterpillars are quite elongate and slender, with some bristles, and the full complement of abdominal limbs. The crochets on the prolegs are usually bi- or tri-ordinal and arranged in a complete circle. If exposed the caterpillars will wriggle violently, often moving swiftly backwards, and they fall off the host plant hanging by a silken thread. This behavioral habit results in larvae often being spread from crop to crop on the clothing of workers. Some larvae actually disperse in their first instar on silken threads carried by the wind and air currents. Most larvae are pale green with a dark head, but there is some variation in color, which in a few cases has been demonstrated to result from different host plants. Almost all tortricid larvae are solitary for all their life—gregarious infestations are very rare and recorded only from Australia. The pupa has two rows of spines on most abdominal segments, and it is extruded from the cocoon prior to emergence of the imago; pupation takes place usually in the larval feeding site. For details of this group as represented in Britain see Bradley, Tremewan & Smith (1973).

Tortricidae [including Olethreutidae (= Eucosmidae)]   *4000 spp.*

The two groups placed in this family have differences in some aspects of wing venation. The family is large and worldwide in distribution. They are basically insects of deciduous forests, most abundant in warm temperate regions but also common in the subtropical parts. The larvae attack forest trees, the understory shrubs, climbers, and ground flora in almost equal proportions. Some species have specialized in using herbaceous legumes as host plants.

Because of the numbers of tortricids that attack trees in the temperate regions of Europe, Asia, North America, etc., and the similarity of many larvae, it has become the practice among fruit-growers to refer collectively to the fruit pest species (of which there are many) as fruit (tree) tortricids. To indicate the importance of this group with regard to fruit cultivation, in Europe, for example, some 25 species are recorded feeding on apple, 25 on *Prunus*, 15 on *Corylus* (hazelnut), and 10 on strawberry. Most of the larvae of Tortricidae are quite polyphagous, as distinct from the Cochylidae who are either mono- or oligophagous; but a few are more specific and are restricted to a single host genus such as *Rhyaciona* on *Pinus*, or *Cydia nigricana* the pea moth.

There is sufficient diversity of larval habits that it may be worthwhile to view the group from this aspect, as shown below:

1. Shoot (stem) borers—*Enarmonia hemidoxa* (pepper top shoot borer)
   —*Eucosma nereidopa* (coffee tip borer)
   —*Rhyaciona* spp. (pine shoot borers)
   —*Tetramoera schistaceana* (gray sugarcane borer)

2. Fruit borers—*Cryptophlebia* spp., and *Cydia* spp.

3. Seed pod borers—*Cryptophlebia* spp.
   —*Cydia nigricana* (pea moth)

4. Bud borers—*Cydia pulverula* (fig bud borer)
   —*Gretchena bolliana* (pecan bud borer)
   —*Spilonota ocellana* (eye-spotted bud moth)

5. Leaf folders/tiers/webbers—the majority of species are placed in this category.
   —*Archips* spp., *Cnephasia* spp., *Cacoecimorpha* spp., *Eucosma* spp., *Spilonota* spp., *Cydia leucostoma*, and *Homona coffearia*.

6. Dead-leaf eaters—*Clepsis consimilana* (in Europe, eats dead leaves of apple, *Lonicera*, *Crataegus*, etc.)
   —several species in Australia feed on dead *Eucalyptus* leaves.

Some species are important forest pests in the North Temperate Zone, and many are pests of ornamentals, and shade trees; but the most notable belong to the very large complex of species that attack fruit trees, both in warm, temperate regions and in the subtropics. Below are listed some of the more important pest species to be found in most regions of the world; but it should be remembered that there has been great taxonomic confusion over this group and in the older literature (e.g. Evans, 1952) the nomenclature used bears little resemblance to the names being used at present, and many sources have been used in the present compilation. Doubtless the taxonomic purists will find nomenclatorial errors in the list below; this is regrettably almost inevitable:

Acleris comariana (strawberry tortrix)—on strawberry; Holarctic Region.

Acleris spp. (fruit tree tortricids)—polyphagous; Europe, India, Asia, North America.

Acleris spp. (spruce budworms)—on spruce, larch, *Tsuga*, etc.; U.S.A. and Canada.

Acropolitis rudisana—polyphagous; Australia.

Adoxophyes orana (summer fruit tortrix)—larvae on top fruit; Europe, Asia to Japan.

Adoxophyes privatana—a polyphagous leaf tier; India, S.E. Asia to Borneo.

Ancylis spp.—on strawberry in the U.S.A.; jujube in India.

Archips occidentalis (citrus leaf roller)—polyphagous on citrus, cocoa, coffee, cotton, onion, etc.; Africa.

Archips spp. (fruit tree tortricids, etc.)—ten+ species, some polyphagous, web leaves and damage the surface of fruitlets; throughout Europe, India, Asia to Japan, and North America (Fig. 12.28).

Argyrotaenia spp. (leaf rollers)—on citrus, pecan, apple, etc.; Europe, Asia Minor, India, U.S.A., and Canada.

Cacoecimorpha pronubana (carnation leaf roller)—polyphagous; Europe, North and South Africa, U.S.A. (C.I.E. Map No. A. 340) (Fig. 12.29).

Choristoneura spp. (plum tortrix, etc.)—polyphagous on fruit trees, etc.; Europe, Asia Minor to Japan. Some species are important forestry pests.

Cnephasia spp. (omnivorous leaf tier, etc.)—several species on many crops throughout Europe, Asia to Japan, North America (Fig. 12.30).

Crocidosema plebejana—on Malvaceae; Africa, S.E. Asia, Australia.

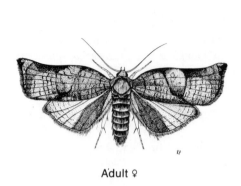

Adult ♀

Fig. 12.28 Fruit tree tortrix (*Archips podana*) adult moth (Lepidoptera; Tortricidae). Larval leaf webbing; Skegness, U.K.

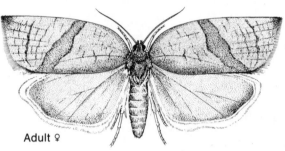

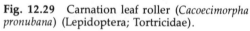

Adult ♀

Adult ♀

Fig. 12.29 Carnation leaf roller (*Cacoecimorpha pronubana*) (Lepidoptera; Tortricidae).

Fig. 12.30 Omnivorous leaf tier (*Cnephasia longana*) (Lepidoptera; Tortricidae).

Cryptophlebia leucotreta (false codling moth)—polyphagous, boring cotton bolls, citrus fruits, etc.; Africa, south of the Sahara (C.I.E. Map No. A. 352).

Cryptophlebia ombrodelta (macadamia nut borer)—polyphagous in fruits, nuts, and large legume pods; Africa, India, S.E. Asia, Japan, Australia (C.I.E. Map No. A. 353).

Cydia delineata (hemp leaf roller)—Yugoslavia.

Cydia funebrana (plum fruit maggot)—larvae in fruit of plum, etc.; Europe, Asia.

Cydia leucostoma (tea flushworm)—on tea shoots; India and Indonesia.

Cydia molesta (oriental fruit moth)—larvae bore many fruits; southern Europe, North Africa, China, Japan, Australia, U.S.A., South America (C.I.E. Map No. A. 8) (Fig. 12.31).

Cydia nigricana (pea moth)—larvae inside pea pods; Europe, Asia, Japan, U.S.A., Canada (C.I.E. Map No. A. 421) (Figs. 12.32, 12.33).

Cydiu pomonella (codling moth)—cosmopolitan wherever apples are grown (C.I.E. Map No. A. 9) (Fig. 12.34); also in pear, *Prunus*, etc.

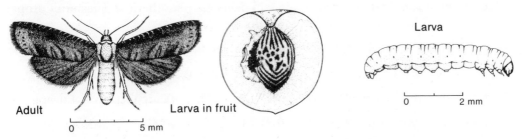

Adult   Larva in fruit

0   5 mm

Larva

0   2 mm

Fig. 12.31   Oriental fruit moth (*Cydia molesta*) (Lepidoptera; Tortricidae).

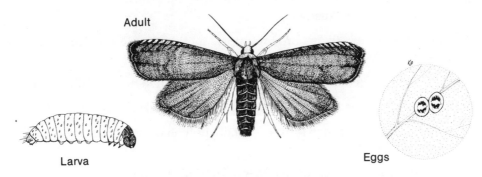

Adult

Larva   Eggs

Fig. 12.32   Pea moth (*Cydia nigricana*) (Lepidoptera; Tortricidae).

Fig. 12.33   Pea moth larva inside pea pod eating the seeds (typical damage); Skegness, U.K.

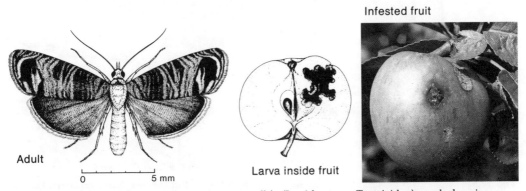

Infested fruit

Adult

Larva inside fruit

Fig. 12.34   Codling moth (*Cydia pomonella*) (Lepidoptera; Tortricidae), and showing infested fruit; Skegness, U.K.

Cydia prunivora (pear tortrix)—larvae in pears, etc.; southern and eastern Europe, South and West Asia (C.I.E. Map No. A. 422).

Cydia ptychora (African pea moth)—larvae inside pea pods; most of Africa.

Cydia sinaria (hemp moth)—larvae are shoot tiers; Pakistan, Thailand.

Cydia spp. (bud moths)—larvae bore buds of *Ficus, Pinus,* etc.; Asia, Europe.

Ditula angustiorana (a fruit tree tortrix)—Europe, Asia Minor, North America.

Enarmonia hemidoxa (pepper top shoot borer)—bores pepper vines; India, Indonesia.

Epinotia lantana (lantana tortrix)—larvae used as biological control agents to destroy lantana in Australia.

Epiphyas postvittana (light brown apple moth)—larvae in apples; Europe, Australasia, Hawaii, U.S.A.

Eucosma nereidopa (coffee tip borer)—larvae bore coffee twigs; Africa (Fig. 12.35).

Gretchena bolliana (pecan bud moth)—larvae bore buds of pecan; U.S.A.

Homona coffearia (tea/coffee tortrix)—totally polyphagous; India, S.E. Asia, China, Japan (Fig. 12.36).

Homona nubiferana—on citrus, cocoa, groundnut, etc.; Indonesia.

Leguminivora glycinivorella (soybean pod borer)—S.E. Asia, Japan.

Lobesia botrana (European grape berry moth)—larvae in grapes; southern Europe, eastern Africa, Japan.

Matsumuraeses spp. (bean podworms)—larvae bore bean pods; Japan.

Merophyas divulsana (alfalfa leaf tier)—tie alfalfa shoots; Australia.

Olethreutes spp. (fruit tree tortricids/coffee berryworms, etc.)—several species attack different plants in different ways; Europe, N.E. Africa, U.S.A.

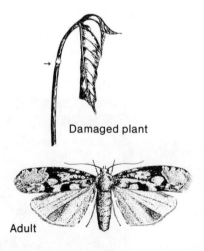

Fig. 12.35 Coffee tip borer (*Eucosma nereidopa*) (Lepidoptera; Tortricidae).

Fig. 12.36 Litchi leaves rolled by tortricid larvae, probably *Homona coffearia* (Lepidoptera; Tortricidae); South China.

Pandemis spp. (fruit tree tortricids, etc.)—many plants attacked; Europe, Asia to Japan, and N.E. Africa.

Ptycholoma lecheana (a fruit tree tortrix)—Europe, Asia Minor.

Rhyacionia spp. (pine shoot moths)—larvae bore shoots of *Pinus;* Europe, Asia, North America.

Spilonota ocellana (eye-spotted bud moth)—polyphagous; Holarctic (C.I.E. Map No. A. 415) (Fig. 12.37).

Spilonota spp. (42)—some are leaf tiers on *Eucalyptus* in Australia, others in Asia and Japan.

Suleima helianthana (sunflower bud moth)—larvae bore buds of sunflower; U.S.A.

Tetramoera schistaceana (gray sugarcane shoot borer)—larvae bore shoot of sugarcane; Mauritius, S.E. Asia, Indonesia, China and Japan.

Tortrix capensana (apple leaf roller)—pest of apple; southern Africa.

Tortrix dinota (brown tortrix)—on coffee, citrus, groundnut, cotton, guava, etc.; Africa.

Tortrix viridana (green oak tortrix)—on oaks, etc.; Europe, Asia.

Tortrix spp.—on citrus, coffee, cocoa, etc.; Africa and Asia.

In the U.K., *Laspeyresia* Hubner, 1825 (nec. R. L., 1817) is now regarded as a junior synonym of *Cydia,* as also is *Grapholita.*

Fig. 12.37 Eye-spotted bud moth (*Spilonota ocellana*) (Lepidoptera; Tortricidae).

Cochylidae (= Phaloniidae) (May be regarded as a subfamily of Tortricidae)

A small group of tortricoids, mostly Holarctic in distribution, characterized by having no vein Cu_2 in the forewing. The larvae feed inside plant tissues, usually inside stems, flowers or seed heads, and, as already mentioned, most species are either monophagous or oligophagous.

A few pest species are known, and some occur in the warmer parts of southern Europe and southern U.S.A.

Aethes spp.—larvae in flower heads of carrot, parsnip, and other Umbelliferae in Europe.

Eupoecilia ambiguella (grape cochylid)—larvae bore stems of grapevine, *Prunus,* and *Ribes;* Europe and Japan.

Lorita abornana (chrysanthemum flower borer)—U.S.A.

Phalonia epilinana—larvae on hemp; China.

Phtheochroa hospes (banded sunflower moth)—U.S.A.

PYRALOIDEA

A large group of small or medium-sized moths (usually regarded as Microlepidoptera despite some being quite large), with long and slender legs, and a generally delicate appearance; front wings elongate or triangular in shape, hind wings broad; maxillary palps are usually present, labial palps are usually held together porrect and are very conspicuous. The larvae have varied habits but most live in concealment; they are active and wriggle violently when disturbed; they are slender in form, pale in color, sometimes spotted, and with few setae. The prolegs bear a more or less complete circle of biordinal crochets. The pupa is never protruded from the cocoon at emergence.

Thyrididae (window-winged or leaf moths)

Small to largish moths, apparently forming a clear-cut, pantropical group, closely related to the Pyralidae. Most moths have a pale translucent area on the wings. The larvae feed in flower heads, on seeds, or burrow in stems and twigs where they cause gall-like swellings. Only a few are recorded as pests.

Striglina cancellata (chestnut thyridid)—Japan.

Striglina clathrata—attacks *Bridelia* trees in Uganda.

Striglina suzukii (tea thyridid)—Japan.

Striglina thermesoides—a polyphagous spinner and leaf roller of Leguminosae; S.E. Asia.

Striglina sp.—recorded on cocoa in Papua New Guinea.

Pyralidae (snout moths)

A very large group of moths, widely distributed throughout the world, but most abundant in the tropics and warmer regions, with the characteristics of the superfamily. In the past, this group was often regarded as being composed of five separate families: the Crambidae, Galleriidae, Phycitidae, Pyralidae, and Pyraustidae, and sometimes even more. The present interpretation of the group is to regard these five taxa, and a few other small groups, as being subfamilies within the family Pyralidae. In the present work the subfamilies recognized are: Crambinae, Schoenobiinae, Scopariinae, Nymphulinae, Cybalomiinae, Peoriinae, Epipaschiinae, Pyralinae, Phycitinae, Gallerinae, Glophyriinae, Odontiinae, Evergestiinae, and Pyraustinae, although the taxonomic criteria for the recognition of most of these groups are very esoteric (Goater, 1986).

The literature contains a vast number of generic names, and many synonyms are encountered in older texts. Basically many of the larvae are very similar in appearance and some groups of adult moths show striking morphological similarity. Many important pest species have been shunted from genus to genus in the past, during different taxonomic reappraisals, and there are also many published records that appear to be misidentifications which further confuse the situation.

This group has species adapted to many diverse terrestrial and aquatic habitats. Some are well adapted for aquatic life and the larvae actually have gills in the larger instars. Many specialize in feeding on dried plant and animal material and are important pests of stored produce and foodstuffs. Many species are phytophagous and a large number are important pests of cultivated crops. A few species are predacious on other insects, and in some cases the adult moths regularly visit large mammals to suck lachrymal fluid from the eyes. Most larvae produce silk and feed in concealment. Leaf rolling and leaf folding with the aid of strands of silk are both widely practiced, with the caterpillar hidden inside the roll. Many are associated with Gramineae and are very important stalk borers of sugarcane and tropical cereals. Worldwide, pyralid stalk borers are very serious pests of sugarcane, rice, and also in the New World on maize, sorghum, and millets, etc. In the Old World probably the most important maize and sorghum stalk borers are the Noctuid genera *Busseola* and *Sesamia,* which have not spread into the New World yet. The most important stalk borers are the genera *Chilo, Diatraea,* and *Scirpophaga* (= *Tryporyza*), and there are a few very important species in each genus, and many less important species.

The most important pests in the different subfamilies are summarized below:

Crambinae (grass moths; sodworms; stalk borers, etc.)

Grass moth larvae feed on grasses; they live in silken tubes in the crown of grass tussocks; in temperate regions they may be very damaging to pastures. The stalk borers tunnel in stems of Gramineae.

Agriphila spp. (grass moths; sodworms)—Europe, Asia, U.S.A., and Canada.

Chilo orichalcociliella (coastal stalk borer)—larvae bore cereal and grass stems; parts of tropical Africa (especially coastal East Africa).

Chilo partellus (= *zonellus*) (spotted stalk borer)—larvae in cereal and grass stems; East Africa, India, Thailand (C.I.E. Map No. A. 184) (Figs. 12.38, 12.39).

Chilo polychrysus (dark-headed rice borer)—larvae in rice, maize, and other cereals; India, and S.E. Asia (Fig. 12.40).

Chilo sacchariphagus (sugarcane stalk borer)—larvae bore sugarcane; Madagascar, Mauritius, S.E. Asia, South China (Figs. 12.41, 12.42).

Chilo suppressalis (Asiatic rice stalk borer)—larvae in rice, some millets, and some grasses; Spain, India, S.E. Asia to Japan and Australasia (C.I.E. Map No. A. 254) (Fig. 12.43).

Chrysoteuchia spp. (grass moths)—larvae in grass sods; western Europe (Fig. 12.44).

Coniesta ignefusalis (millet stem borer)—larvae bore stem of millet; West Africa.

Crambus spp. (grass moths; sod webworms)—pasture pests in Europe and North America.

Diatraea crambidoides (southern corn stalk borer)—in maize and sorghum; southern U.S.A.

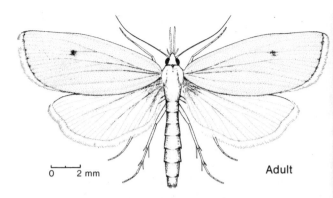

0   2 mm   Adult

Fig. 12.38 Spotted stalk borer (*Chilo partellus*)
(Lepidoptera; Pyralidae; Crambinae).

Fig. 12.39 Damage to sorghum stems by spotted
stalk borer (*Chilo partellus*); Alemaya, Ethiopia.

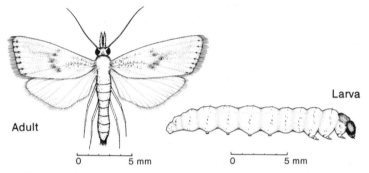

Adult   Larva

0   5 mm   0   5 mm

Fig. 12.40 Dark-headed rice borer (*Chilo polychrysus*) (Lepidoptera; Pyralidae;
Crambinae).

Fig. 12.41 Sugarcane stalk showing emergence
holes of *Chilo sacchariphagus*; South China (Lepidoptera; Pyralidae; Crambinae).

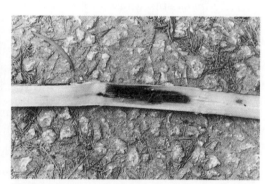

Fig. 12.42 Tunneling damage to sugarcane stalk
by sugarcane stalk borer (*Chilo sacchariphagus*);
South China (Lepidoptera; Pyralidae; Crambinae).

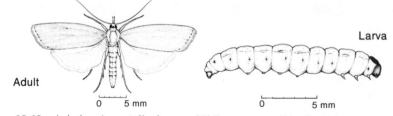

Adult   Larva

0   5 mm   0   5 mm

Fig. 12.43 Asiatic rice stalk borer (*Chilo suppressalis*) (Lepidoptera; Pyralidae;
Crambinae), and larva.

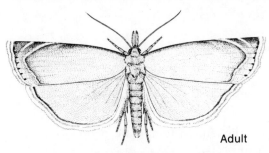

Adult

Fig. 12.44 Grass moth (*Chrysoteuchia culmella*) (Lepidoptera; Pyralidae; Crambinae).

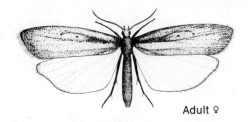

Adult ♀

Fig. 12.45 Sugarcane stalk borer (*Eldana saccharina*) adult female (Lepidoptera; Pyralidae; Crambinae).

Diatraea saccharalis (American sugarcane borer)—larvae bore sugarcane, maize, sorghum, rice, etc.; southern U.S.A.

Diatraea spp. (30+)—larvae bore stems of sugarcane, maize, sorghum, and other cereals; southern U.S.A., Central and South America.

Eldana saccharina (sugarcane stalk borer)—larvae bore sugarcane and cereal stems, also cassava stems; tropical Africa (Fig. 12.45).

Epina dichromella—larvae bore rice stems; Sri Lanka.

Hednota spp. (Australian grass-webbers)—larvae on native and introduced grasses in Australia.

Zeadiatraea spp.—Larvae bore maize, sorghum and sugarcane; Central and South America.

Schoenobiinae

Many adults are white in color and rest during the day in grasses, etc.; some species are truly aquatic.

Scirpophaga (= *Schoenobius*, = *Tryporyza*)—larvae bore stems of rice and various grasses.

Scirpophaga incertulas (yellow paddy stem borer)—in rice and grasses; India, S.E. Asia, China, Japan (C.I.E. Map No. A. 252) (Fig. 12.46).

Scirpophaga innotata (white paddy stem borer)—in rice and grasses; India, S.E. Asia, North Australia (C.I.E. Map No. A. 253).

Scirpophaga nivella (sugarcane top borer)—larvae destroy growing point of sugarcane; India, S.E. Asia, China, Japan.

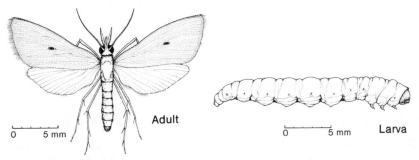

Adult

0   5 mm

0   5 mm   Larva

Fig. 12.46 Yellow paddy stem borer (*Scirpophaga incertulas*) (Lepidoptera; Pyralidae; Schoenobiinae).

Odontiinae

Noorda spp. (moringa budworms/leafworms)— on *Moringa* in India, and mango in Indonesia.

Scopariinae

Scoparia spp.—larvae feed on mosses and lichens in temperate regions (except New Zealand); not important as crop pests.

Nymphulinae (aquatic forms)

Many larvae adapted for aquatic life; respire cutaneously when small but may have gills in later instars.

Acentria nivea—larvae live in freshwater vegetation in Europe.

Elophila nympheata (China mark moth)—on rice in Italy and Hungary.

Parapoynx stagnalis (rice caseworm)—on rice foliage; pantropical (C.I.E. Map No. A. 176) (Fig. 12.47).

Parapoynx vittalis (smaller rice caseworm)—on rice in Japan.

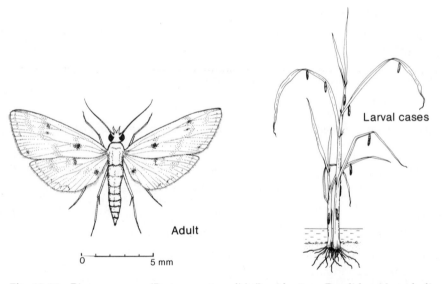

Fig. 12.47   Rice caseworm (*Parapoynx stagnalis*) (Lepidoptera; Pyralidae; Nymphulinae).

Peoriinae (= Anerastinae)

A small group with the proboscis vestigial.

Emmalocera umbricostella—larvae bore rootstock of sugarcane in Taiwan.

Maliarpha separatella (white rice borer)—on rice and grasses; causes "whiteheads"; Africa, parts of India and Asia, and S.E. Asia (C.I.E. Map No. A. 271) (Fig. 12.48).

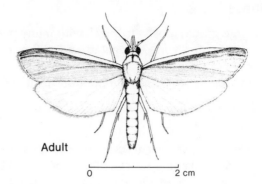

Fig. 12.48 White rice borer (*Maliarpha separatella*) (Lepidoptera; Pyralidae; Peorünae).

Adult

0   2 cm

Polyocha depressella—larvae bore the rootstock of sugarcane in India.

Saluria inficita—larvae bore stems of finger millet and foxtail millet; India.

Epipaschiinae

Orthaga euadrusalis (mango webworm)—larvae web shoots of mango; India, S.E. Asia.

Orthaga mangiferae—on mango; India.

Orthaga spp. (mango webworms)—larvae web and eat foliage of mango; India.

Orthaga spp. (40)—in Australia these species feed on fallen leaves which they spin together.

Stericta albifasciata—larvae recorded webbing shoots of avocado in Trinidad, and on mango in India.

Cybalomiinae

Hendecasis duplifascialis (jasmine bud borer)—bores flower buds of jasmine; West Africa through Asia to Indonesia.

Pyralinae

These species live on dried vegetable matter, and also on dried animal material.

Herculia nigrivitta—larvae destroy dried palm leaves either on the trees or when made into atap and used as a roof covering on buildings; throughout S.E. Asia and Indonesia.

Hypsopygia spp.—larvae feed on dried grasses and damage thatched roofs and haystacks in Europe; one species in India feeds on wax in wasp nests.

Pyralis farinalis (the meal moth)—pest of stored products; now cosmopolitan.

Pyralis manihotalis (gray pyralid)—pest of stored and dried seeds, fruits, tubers, and also dried meats; endemic to Central and South America, but now cosmopolitan.

Pyralis pictalis (brown pyralid)—similar to gray pyralid in habits but native to tropical Asia.

Phycitinae

A large group, mostly tropical, with many pest species; larvae generally specific in choice of food; forewing elongated and without vein R_5.

Anonaepestis bengalella—larvae bore fruits of Annonaceae; S.E. Asia.

Apomyelois dentilinella—larvae predacious on pupae of *Parasa lepida* and *Lymantria serva*; India.

Cactoblastis cactorum (cactus moth)—this Argentine species was introduced into Australia in a spectacularly successful biological control program to control prickly pear (*Opuntia* spp.); also used successfully in India.

Cadra calidella (date moth)—larvae feed on ripening dates; Near East; India.

Cadra cautella (almond/tropical warehouse moth)—larvae feed on dried grains, fruits, beans, and nuts in storage; pantropical, and in warmer regions also (Fig. 12.49).

Cadra figulilella (raisin moth)—larvae feed on dried fruits, dates, cocoa; Near East.

Citripestis eutraphera—larvae bore young mango shoots and fruits; Indonesia.

Citripestis sagittiferella (citrus fruit borer)—larvae bore citrus fruits; Malaysia, Indonesia (Fig. 12.50).

Cryptoblabes angustipennella (webworm)—larvae in panicle of sorghum, millet, castor; India.

Cryptoblabes gnidiella (honeydew moth)—adults are attracted to honeydew on crop plants such as citrus, grapevine, loquat, pomegranate, cotton, maize, sorghum, etc.; often associated with mealybugs, and the larvae may damage the fruits; Mediterranean Region and New Zealand.

Dipha aphidivora—larvae prey on aphids (especially *Ceratovacuna*) on sugarcane and Palmae; Java.

Elasmopalpus lignosellus (lesser cornstalk borer)—larvae bore in stalks of maize, sorghums, wheat, cowpea, beans, peas, turnip, and grasses; southern U.S.A., Central and South America.

Ephestia elutella (warehouse/cocoa moth)—larvae bore dried cocoa beans, grains, fruits, pulses, nuts, coconut, etc.; completely cosmopolitan but less abundant in the hot tropics (Fig. 12.51).

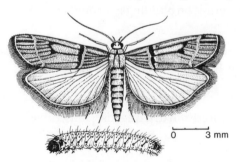

Fig. 12.49 Almond/tropical warehouse moth (*Cadra cautella*) adult and larva (Lepidoptera; Pyralidae; Phycitinae).

Fig. 12.50 Citrus fruit borer (*Citripestis sagittiferella*) infesting pummelo fruit; Malaysia (Lepidoptera; Pyralidae; Phycitinae).

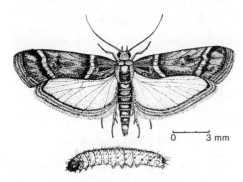 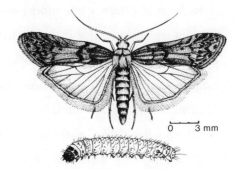

Fig. 12.51 Warehouse/cocoa moth (*Ephestia elutella*) adult and larva (Lepidoptera; Pyralidae; Phycitinae).

Fig. 12.52 Mediterranean flour moth (*Ephestia kuehniella*) (Lepidoptera; Pyralidae; Phycitinae).

Ephestia kuehniella (Mediterranean flour moth)—mostly on meals and flours; cosmopolitan in subtropical regions of the world (Fig. 12.52).

Etiella behrii—larvae damage alfalfa grown for seed in southern Australia, and pigeon pea in the Seychelles.

Etiella zinckenella (pea pod borer)—larvae bore pods of legumes; pantropical and subtropical (C.I.E. Map No. A. 105).

Euzophera bigella (quince moth)—North Africa, Near East and S.W. Asia.

Euzophera perticella—larvae bore stems of Solanaceae (potato, chili, eggplant, etc.); parts of Africa, Near East, India.

Euzopherodes vapidella (citrus stub moth)—on citrus and carob trees, larvae bore under bark; Mediterranean Region.

Homoeosoma vagella (macadamia flower caterpillar)—on macadamia; Australia (Queensland).

Hypsipyla spp.—larvae bore shoots of various forest trees, mahogany, mango, etc.; India, S.E. Asia, Australia.

Indomyrlaea eugraphella (sapota leaf webber)—India.

Leatilia coccidivora—larvae predacious on Coccoidea; North America.

Mussidia pectinicornella—larvae bore in large-podded legumes; S.E. Asia.

Mussidia spp.—larvae tunnel bark of mahogany trees in Malawi, and attack cocoa pods and maize in West Africa.

Phycita diaphana (castor bean moth)—on castor; Near East.

Phycita eulepidella—larvae web flower head and shoots and feed on them; India.

Phycita orthoclina (tamarind fruit borer)—India.

Plodia interpunctella (Indian meal moth)—pest of stored products of vegetable origin; cosmopolitan in warmer regions (Fig. 12.53).

Galleriinae

A small group whose larvae feed on dried material, in nests of Hymenoptera, and in flowers.

Achroia grisella (lesser bee moth)—pest in beehives; cosmopolitan.

Corcyra cephalonica (rice moth)—stored products pest; now cosmopolitan.

Doloessa viridis (green cereal moth)—pest of stored products; pantropical.

Ertzica morosella (pandanus borer)—larvae bore pandanus stems; S.E. Asia.

Galleria mellonella (wax moth)—pest in beehives; cosmopolitan.

Tirathaba mundella (oil palm bunch moth)—pest of oil palm; S.E. Asia.

Tirathaba rufivena (coconut spike moth)—larvae in flower spike of coconut; throughout S.E. Asia.

Tirathaba spp. (flower moths)—larvae web flowers and fruits of coconut, sorghum, coffee, rambutan, durian, etc.; S.E. Asia.

Glaphyriinae

A very small group.

Hellula phidilealis (cabbage webworm)—larvae eat leaves of crucifers; Africa, Central and South America.

Hellula undalis (oriental cabbage webworm)—on cruciferous crops and wild Cruciferae; Africa, S.E. Asia, Near East, Australasia.

Evergestiinae

Another very small group.

Crocidolomia binotalis (cabbage leaf webber/cluster caterpillar)—larvae on crucifers; Africa, India, S.E. Asia, Australia.

Crocidolomia pavonana (cabbage leaf webber/cluster caterpillar)—larvae on crucifers; Africa, India, S.E. Asia, Australia.

Evergestis spp.—larvae of four species regularly damage cruciferous crops; Europe, Asia to Japan, U.S.A., and Canada (Fig. 12.54).

Pyraustinae

The largest group and most abundant in the warmer regions of the world. Most larvae are leaf eaters, but some bore into plant tissues. Many adults are conspicuously marked.

Agathodes spp.—on various ornamental trees (*Erythrina* spp.); S.E. Asia.

Antigastra catalaunalis (sesame webworm)—shoots and pods eaten; Africa, India, and some warmer parts of Europe and Asia (C.I.E. Map No. A. 452) (Fig. 12.55).

Cnaphalocrocis exigua—larvae fold leaves of rice and grasses; India, S.E. Asia, Australasia.

Cnaphalocrocis medinalis (rice leaf roller)—on rice and grasses; Madagascar, India, S.E. Asia, China, Japan, Australia (C.I.E. Map No. A. 212) (Fig. 12.56).

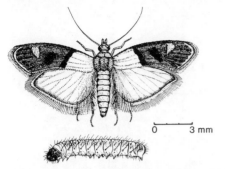

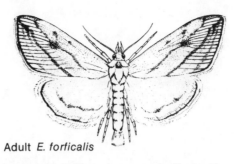

Adult *E. forficalis*

Fig. 12.53 Indian meal moth (*Plodia interpunctella*) adult and larva (Lepidoptera; Pyralidae; Phycitinae).

Fig. 12.54 Cabbageworm moth (*Evergestis* sp.) (Lepidoptera; Pyralidae; Evergestiinae).

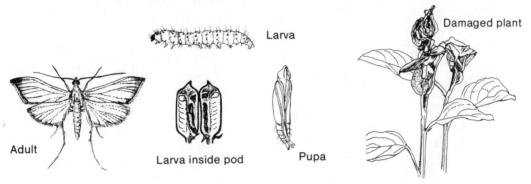

Larva

Adult

Larva inside pod

Pupa

Damaged plant

Fig. 12.55 Sesame webworm (*Antigastra catalaunalis*) (Lepidoptera; Pyralidae; Pyraustinae).

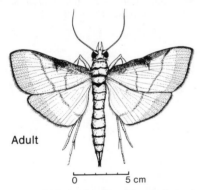

Adult

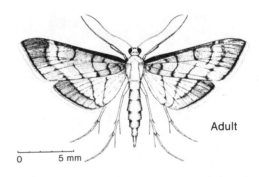

Adult

Fig. 12.56 Rice leaf roller (*Cnaphalocrocis medinalis*) (Lepidoptera; Pyralidae; Pyraustinae).

Fig. 12.57 Maize webworm (*Cnaphalocrocis trapezalis*) (Lepidoptera; Pyralidae; Pyraustinae).

Fig. 12.58 Pumpkin leaf roller (*Diaphania indica*); wingspan 25 mm (1.0 in.); South China (Lepidoptera; Pyralidae; Pyraustinae).

Fig. 12.59 Fig shoot borer (*Glyphodes bivitralis*) (Lepidoptera; Pyralidae; Pyraustinae); wingspan 25 mm (1.0 in.); South China.

Adult

Fig. 12.60   Mung moth (*Maruca testulalis*) (Lepidoptera; Pyralidae; Pyraustinae).

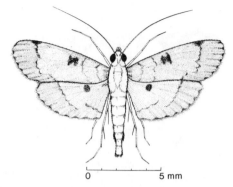

0   5 mm

Fig. 12.61   Banana scab moth (*Nacoleia octasema*) (Lepidoptera; Pyralidae; Pyraustinae).

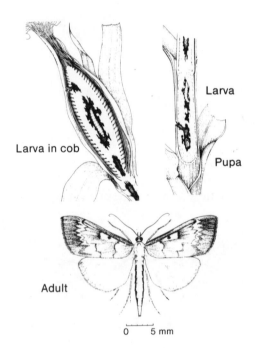

Larva in cob

Larva

Pupa

Adult

0   5 mm

Fig. 12.62   Asian corn borer (*Ostrinia furnacalis*) (Lepidoptera; Pyralidae; Pyraustinae).

Adult
(resting position)

Damaged berry cluster

Fig. 12.63   Coffee berry moth (*Prophantis smaragdina*) (Lepidoptera; Pyralidae; Pyraustinae).

Cnaphalocrocis trapezalis (maize webworm)—on maize, sugarcane, and other cereals; Africa, S.E. Asia, Australasia, Central and South America (Fig. 12.57).

Conogethes punctiferalis (castor capsule borer)—larvae polyphagous on fruits, cotton, maize, etc.; India, S.E. Asia, China, Japan, Australia.

Crypsiptya coclesalis (bamboo leaf roller).

Diaphania indica (pumpkin leaf roller)—larvae spin together leaves of Cucurbitaceae; S.E. Asia and China. (Fig. 12.58).

Eutectona machaeralis (teak skeletonizer)—India, and S.E. Asia.

Glyphodes spp. (shoot borers)—larvae bore shoots on *Ficus*, jackfruit, mulberry; India, S.E. Asia, China, Japan (Fig. 12.59).

Herpetogramma licarsisalis—larvae web and roll leaves of bamboo, rice, and grasses (Gramineae); India and S.E. Asia.

Lamprosema crocodora—larvae roll leaves of coffee; West Africa.

Leucinodes orbonalis (eggplant boring caterpillar)—on cultivated Solanaceae; Africa, India, S.E. Asia.

Loxostege sticticalis—on beet, *Artemesia*, etc.; southern Europe and West Asia; strongly migratory.

Maruca testulalis (mung moth)—larvae eat foliage and pods of beans, peas, other legumes, and leaves of castor, tobacco, rice, etc.; cosmopolitan in warm regions (Fig. 12.60).

Nacoleia octasema (banana scab moth)—damage flowers and fruits of banana, maize, etc.; Indonesia, Australasia (Fig. 12.61).

Omiodes diemenalis—larvae defoliate tobacco, tomato; S.E. Asia.

Omiodes indicata (soybean leaf roller)—larvae defoliate beans, soybean, groundnut, cowpea, and other herbaceous legumes; India, East and West Africa, S.E. Asia, Philippines.

Omphisa anastomosalis (sweet potato stem borer)—larvae bore stem of sweet potato; India, S.E. Asia, China, and Japan.

Ostrinia furnacalis (Asian corn borer)—larvae bore stems of maize and other cereals, and other plants; tropical and warm parts of Asia, Indonesia, Papua New Guinea (C.I.E. Map No. A. 294) (Fig. 12.62).

Ostrinia nubilalis (European corn borer)—in maize and cereals; Europe, North Africa, S.W. Asia, and eastern North America (C.I.E. Map No. A. 11).

Parotis marinata—larvae eat leaves and flowers of cinchona, etc.; S.E. Asia.

Pleuraptya balteata—polyphagous pest of shrubs and trees; S.E. Asia to China and Japan.

Prophantis smaragdina (coffee berry moth)—larvae web and eat berries of coffee; tropical Africa (Fig. 12.63)

Psara ambitalis—larvae web and roll leaves of tobacco, eggplant, and other Solanaceae; India and S.E. Asia.

Spoladea recurvalis (amaranthus leafworm)—larvae polyphagous on a wide range of crops; completely pantropical.

Syllepta derogata (cotton leaf roller)—larvae roll leaves of cotton and other Malvaceae, and some Sterculiaceae; Africa, India, S.E. Asia to China and Japan (C.I.E. Map No. A. 397) (Fig. 12.64).

Syllepta lunalis (grapevine leaf roller)—on grapevine, etc.; India.

Syngamia abruptalis—larvae eat *Mentha,* etc.; India, S.E. Asia.

Tabidia aculealis (sweet potato leaf roller)—S.E. Asia.

Terastia spp. (shoot borers)—destroy shoots of *Erythrina* grown as shade trees for coffee; Indonesia.

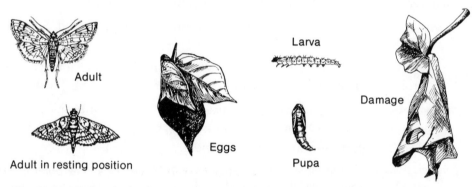

Fig. 12.64 Cotton leaf roller (*Syllepta derogata*) adult; wingspan 25 mm (1.0 in.); Hong Kong (Lepidoptera; Pyralidae; Pyraustinae).

"Eye moths"

A group of Pyraustinae were recorded by Banziger (1968) as feeding on lachrymal fluid from mammals, in Thailand, other parts of S.E. Asia, and India. The species recorded visiting the eyes of mammals include:

Botyodes asialis—pig and elephant.

Botyodes flavibasalis—cow.

Filodes fulvidorsalis—deer and humans.

Glyphodes stolalis—cattle, water buffalo, deer, and pig.

Herpetogramma licarsisalis—cattle.

Pagyda salvalis—water buffalo, pig and elephant.

Pionea aureolalis—cattle, pig, water buffalo, elephant, deer, and humans.

Pionea damastesalis—cattle, goats, deer, humans, elephant, etc.

Pionea flavicinctalis—elephant and cattle.

Typsanodes linealis—mule.

Whether there is any real host preference shown by these moths is not clear; to date, there is no evidence of the accidental transmission of any pathogens, or the causing of conjunctivitis.

PTEROPHOROIDEA

The adult moths have no ocelli; haustellum naked; maxillary palps reduced; and wings split longitudinally into plumes.

Pterophoridae (plume moths)

A widespread and quite large family of moths; lightly built, green or brownish in color, and the long forewing is cleft into two, three, or four plumes, and the hind wing into three. There are no maxillary palps, and the legs are very elongate and bear prominent tibial spurs. Three genera are exceptional in having entire wings. The larvae either feed exposed on flowers and leaves, or else they are internal in stems or fruits. The compositae are attacked more than any other group, although the crop pests are usually feeding on other plants. The caterpillars are long, cylindrical, and with many setae, and prolegs long and thin with uniordinal crochets. Pupae are attached by the cremaster to the foliage, etc., often in a slight cocoon.

Some of the recorded pest species include:

Aciptilia spp.—larvae bore in vines of sweet potato; China.

Exelastis atmosa—larvae eat flowers and seeds on red gram, cajan pea, and other legumes, grapevine, and *Syzygium*; India.

Ochryotica spp. (sweet potato leaf rollers)—roll leaves of sweet potato; Africa, China.

Oxyptilus rugulus (grapevine boring plume moth)—larvae bore into ripening berries on grapevine; India, Australia.

Platyptilia pusilidactyla (lantana plume moth)—larvae feed on flowers and fruits of *Lantana*; Africa, Asia, and America.

Platyptilia spp.—recorded from *Aster, Cissus, Rubus,* grapevine and *Striga*; East Africa and Japan.

Pterophorus periscelidactylus (grapevine plume moth)—larvae web shoots and eat leaves; U.S.A.

Pterophorus spp. (sweet potato plume moths)—larvae bore in the vines of sweet potato and *Ipomoea* species; Europe and Africa.

PAPILIONOIDEA (butterflies)

This, together with the following superfamily (Hesperoidea), constitute what are called the "butterflies". These were at one time regarded as a group (Rhopalocera) of equal systematic rating as the remainder—the moths (Heterocera).

The adults have large wings (often brightly colored) held vertically above the body at rest; slender, elongate antennae with an abrupt terminal club; diurnal in habits; seen feeding from flowers, etc. Many species have chemoreceptors on the tarsi, and there is a reduction of tarsi seen progressively throughout the superfamily.

Nymphalidae (four-footed butterflies)   *5000 spp.*

Both sexes have the anterior legs reduced and useless for walking—at first sight they seem to have only four legs, but the anterior pair can be seen held folded under the prothorax. All adults are diurnal and to be seen generally either feeding on flowers or in the vicinity of the food plant. Some species are notorious for feeding on animal urine and dung, or rotten fruits, and sap exuding from trees. They can be caught in various traps baited with these materials. The group is large and widespread, but best represented in the tropics and warmer regions. There are several quite distinct subfamilies that have in the past been accorded family status—the general consensus of opinion now regards them as subfamilies. Since they are mostly quite distinct they will be regarded separately in this text.

Danainae (monarchs; milkweed butterflies)

A large group of conspicuous butterflies, brightly colored (often aposematically) and slow-flying, and thought to be poisonous to predators. They are tenacious of life and can survive handling and physical injuries that would kill other butterflies—occasionally after having had the thorax inadequately "pinched" (to kill them) adults have been known to fly away and escape. The antennal club is often only slightly pronounced. Males usually have scent glands and expansible hair-pencils. Coloration is usually a mixture of red, yellow, black, or green. Several species of *Danaus* are strongly migratory in parts of Asia, Australia, and the U.S.A.

Most larvae feed on plants containing poisonous substances (often alkaloids), and these poisons are retained in the larval body and render them either poisonous or at least distasteful to insectivores. Most of the food plants belong to the Asclepiadaceae or the Apocynaceae. The caterpillars are usually brightly colored and conspicuous and often have pairs of long fleshy filaments protruding from their backs. The pupa, which hangs from the cremaster attached to plant foliage, is often brilliantly colored gold or silver and is strikingly beautiful.

These butterflies are a conspicuous element of the local (tropical) insect fauna but are at most only minor pests; *Asclepias* species have been cultivated in some areas for stem fibers or for the hairs surrounding the seeds as a kapok substitute, and *Nerium* is a popular and widespread ornamental shrub.

A few minor pest species in the tropics include:

Danaus chrysippus (plain tiger; sodom apple butterfly)—larvae feed on *Solanum* and *Asclepias;* throughout tropical Asia.

Danaus spp. (tigers, etc.)—larvae feed on Asclepiadaceae; Africa and Asia.

Euploea spp. (crows)—larvae feed on *Ficus,* Oleander, *Strophanthus,* etc.; India, S.E. Asia, China.

Ithomiinae

A small Neotropical group closely related to the Danainae.

Satyrinae (browns)

As the name indicates, these butterflies are often brown in color and rather cryptic. Many are essentially forest shade species, but are unusual in that the larval food plants are forest grasses. A number of species are probably best regarded as forest-edge species, and the temperate meadow brown and the heaths also belong to this group. Many adults have eyespots at the tip of the forewings and on the underside of the hind wings.

The larvae tend to be elongate, thin, brown or green in color, and with a peculiarly angled (horned) head; the body is often lined longitudinally. The pupae are similar to those of the Nymphalinae and hang suspended by the cremaster.

Host plants are almost entirely Gramineae and Palmae, being found on sugarcane, grasses, cereals, and cultivated bamboos and palms—damage done by larvae is sometimes of economic consequence. The few pest species include:

Elymnias hypermnestra—larvae on palms; India and S.E. Asia.

Lethe spp. (bamboo browns)—larvae feed on foliage of bamboos; India, China.

Melanitis leda (evening brown; rice butterfly)—larvae feed on rice, maize, sugarcane, millets, and grasses; Africa, India, S.E. Asia, Australasia.

Mycalesis spp. (bush browns)—on sugarcane, rice and grasses; S.E. Asia and China.

Amathusiinae

A small, tropical group, very similar to the "browns", being forest species feeding on monocotyledonous plants, mostly Gramineae and Palmae. The caterpillars are hairy and have a pair of anterior "horns" and a pair of taillike posterior processes. Three species of some interest are:

Amathusia spp.—larvae feed on oil palm, sago, coconut, *Areca,* and banana; S.E. Asia, Indonesia.

Discophora sondaica—gregarious larvae feed on sugarcane in Java, on bamboos in China.

Faunis eumeus (common faun)—larvae on *Pandanus, Phoenix,* and *Smilax;* S.E. Asia, to China.

Morphinae

A small group, exclusively Neotropical; the genus *Morpho* have large wings that are brilliant metallic blue in color; they are extensively collected for ornamental purposes.

Brassolinae

Another Neotropical group of brightly colored butterflies, large in size and with conspicuous eyespots on the underneath of the wings. Many species of *Caligo* are well known in the region.

Acraeinae

The group is mostly African, and most species are placed in the extensive genus *Acraea,* but a few species are Oriental and another few Neotropical. Adults have elongate wings, sparsely scaled and somewhat diaphanous. Neither larvae nor adults are preyed upon by birds and other insectivorous animals—the adults exude a repellant fluid if disturbed. A few species are crop pests of note.

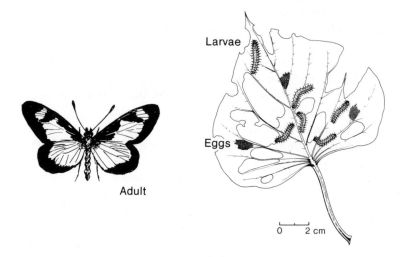

Fig. 12.65 Sweet potato butterfly (*Acraea acerata*) ex Uganda (Lepidoptera; Nymphalidae; Acraeinae).

Acraea acerata (sweet potato butterfly)—larvae eat foliage of *Ipomoea* gregariously; tropical Africa (Fig. 12.65).

Acraea spp.—several species recorded feeding on foliage of tea, cotton, cocoa, *Ficus*, tobacco, *Hibiscus,* etc.; from both West and East Africa.

Heliconiinae

Small- to medium-sized, brightly colored butterflies with long forewings, found only in the Neotropics. They are all thought to possess repellant properties that protect them from predators. Some females are said to feed on pollen which enables them to have a prolonged oviposition period. Few are crop pests.

Heliconius charithonius—larvae feed on foliage of passionflower vine; southern U.S.A.

Nyphalinae (four-footed butterflies, *s.s.*)

By far the largest and most widespread group within the family; the adults have reduced and non-functional fore-legs; large palps, broad anteriorly; male fore-tarsi unjointed, females with 4–5 segments. The larvae are almost always cylindrical and bearing many scoli (tubercules), or else tentacles on the head. The pupae are characteristic and bear rows of tubercules on the abdomen and usually a pointed projection on each side of the head; the pupa is suspended head downwards from the cremaster (with no median girdle). Some of the adults come to baits such as rotting fruits, animal dung, etc., and can be trapped. The adults are generally swift fliers and not easy to catch on the wing.

The painted lady (*Vanessa cardui*) is probably the most widely distributed species in the entire Lepidoptera. It is a minor pest of cultivated plants in most parts of the world but it is seldom abundant anywhere, although it occurs in N.E. Africa in quite large numbers. It is a migratory species and occasionally an influx of migrants can be seen in several parts of the world, including Canada and the U.K.; it seldom survives the winter in Canada and migrates here each year from the U.S.A. and Mexico. Apparently, there is an annual migration from North Africa to western Europe, with some butterflies even reaching Iceland 4000 km/2800 miles away—swarms of vast size have been recorded, sometimes allegedly millions. And there is apparently a return migration southwards in the autumn. In the Mediterranean Region there are apparently 5–6 generations per year, the first usually being on wild plants (thistles and other Compositae) and later ones on crop plants.

A large number of species of Nymphalinae are a dominant element in the local insect fauna of many parts of the world. Many are pests of cultivated plants, though seldom serious pests, including:

Argyreus hyperbius (fritillary)—larvae feed on *Viola* spp; S.E. Asia and China.

Ariadne ariadne (angled castor)—on castor; India, S.E. Asia, and China.

Charaxes bernardus (tawny rajah)—larvae feed on leaves of camphor and other trees; India, S.E. Asia, China.

Ergolis merione—larvae feed on castor; India.

Euthalia spp.—larvae feed on leaves of mango; India, Indonesia, S.E. Asia, China.

Hypolimnas bolina (great eggfly)—larvae eat foliage of sweet potato; S.E. Asia.

Junonia spp.— on sweet potato, rice, various ornamentals, etc.; India.

Kallima inchus (leaf butterfly)—adults suck juices from ripe peaches; India.

Limenitis spp.—larvae feed on *Lonicera*; S.E. Asia and China.

Neptis spp. (sailors)—larvae feed on tree legumes, castor; Africa, tropical Asia through to China.

Polyura spp.—larvae eat the foliage of many different plants, including tree legumes, camphor, cocoa, roses, rambutan, etc.; S.E. Asia, Indonesia, China.

Vanessa cardui (painted lady)—larvae are polyphagous on a wide range of plants, including cucurbits, beet, maize, tobacco, eggplant, grapevine, castor (China), artichoke (Cyprus), various legumes (East Africa and Canada), sunflower (Canada); cosmopolitan except for Australia.

Many species of Nymphalinae (larvae) feed on the foliage of common ornamental and shade trees and bushes, and so are especially abundant in urban and agricultural areas.

Riodinidae (= Nemeobiidae)   *1000 spp.*

The group is characteristic of the Neotropics, but with some species in the U.S.A. and some (*c.* 100 spp.) in Asia. The adults have short, broad, forewings; anterior legs reduced in the male, functional in the female but small in size. Larvae and pupae resemble Lycaenidae generally. The group is not of any known importance agriculturally.

Lycaenidae (blues; coppers; hairstreaks)

A large group of small butterflies, well represented in all regions of the world. The adults are often metallic blue or iridescent brown (coppers) or orange on the upper wing surface but cryptic and brownish underneath. Sexual dimorphism in coloration is common. The antennae are ringed with white, and there is a ring of white scales around each eye. The legs are all functional, but the anterior tarsus of the male is reduced and may be clawless. The hind wings typically bear delicate taillike processes, and there may be an eyespot underneath. The butterflies generally sit in a head-down position when feeding on flowers or resting, and they deceive would-be predators into thinking that the "tails" are antennae and that the eyespot is the actual eye, and that the insect is facing in the other direction.

The larvae are characteristic in that they are onisciform in shape (resembling woodlice), with a broad flattened body whose projecting sides obscure the legs, and the body tapers at each end. Most larvae are green, but some are brown, some are smooth and others hairy. Many species have a dorsal gland posteriorly which yields drops of fluid attractive to attendant ants. Most of the larvae are clearly phytophagous and either eat leaves and flowers, or bore into pods and fruits, but some will eat aphids or other bugs occasionally. Some species are attended by ants for their dorsal gland secretion. Some of these pupate in the nests of ants and a few are even predacious on ants and live their larval life inside ant nests (*Liphyra,* etc.). Another group of species have larvae that are clearly carnivorous and they prey mostly on woolly aphids (Pemphigidae) and mealybugs, but other Coccoidea and Aphidoidea are also eaten. The division between phytophagous and carnivorous species, as already mentioned, is not very precise. An extreme exception is *Maculinea arion* in Europe, where the first four instars are phytophagous but the last instar becomes carnivorous and preys upon ant larvae in the nests of *Myrmica.*

Pupae are smooth and rounded and held attached to the plant foliage by a median silken band and the cremaster; some as mentioned are subterranean.

Within this family, eggs are laid usually singly and stuck firmly on to the plant foliage so that larvae are generally solitary.

Adults are diurnal in habits and conspicuous in their coloration, so that they are an obvious element in most tropical and temperate faunas. Quite a large number of species have been recorded feeding on cultivated plants in the warmer regions of the world, and many different genera are recorded in the literature. The plants most frequently attacked are the Rutaceae (*Citrus,* etc.) and Leguminosae.

A selection of the more notable pests of cultivated plants include:

Baspa melampus—on mango; India.

Catochrysops spp.—larvae feed on flowers and pods of many different legumes (cowpea, pigeon pea, etc.); India, S.E. Asia, China.

Chilades lajus (lime blue butterfly)—on young shoots of *Citrus* and other Rutaceae; India, S.E. Asia, Philippines, China.

Chiliara othona (orchid blue)—larvae bore buds and eat flowers of orchids; Indonesia.

Deudorix epijarbas—larvae bore fruits of litchi, rambutan, pomegranate; India, S.E. Asia to South China.

Euchrysops spp.—on beans, cowpea, etc.; Africa and tropical Asia.

Freyeria putli—on citrus; India.

Jamides bockus (dark cerulean)—recorded on *Citrus* (India) and legumes (South China).

Lampides boeticus (pea blue butterfly)—larvae bore pods of pea and other legumes; cosmopolitan in the Old World (Fig. 12.66).

Lampides elpis—on flowers and fruits of cardamon; India.

Rapala spp.—larvae polyphagous on young leaves, buds, flowers and fruits of apple, guava, litchi, cotton, rambutan; India, S.E. Asia, Australasia, Pacific Islands.

Remelana jangala—larvae damage coffee berries, kapok fruits; India, S.E. Asia, Indonesia, and China.

Talicada nyseus (bryophyllium blue)—larvae damage ornamentals (rockery plants); India, Sri Lanka, Burma.

Tarucus theophastus—on citrus, jujube; India.

Thecla basilides—larvae feed on buds and young fruits of pineapple; Trinidad.

Virachola bimaculata (coffee berry butterfly)—larvae bore berries of coffee; East Africa, Sierra Leone (Fig. 12.67).

Virachola isocrates (pomegranate fruit borer)—polyphagous on apple, citrus, guava, litchi, loquat, peach, pear, plum, pomegranate, etc.; India.

Virachola livia (pomegranate butterfly)—larvae in fruits of pomegranate or seed pod of popinac; tropical Africa, Near East.

Zizeeria otis (lesser grass blue)—larvae feed on a wide range of Leguminosae; pantropical.

Adult ♂   Adult ♀

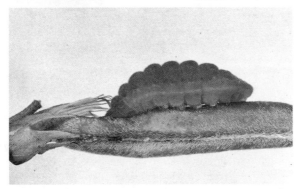

Larva on pod

Fig. 12.66 Pea blue butterfly (*Lampides boeticus*) adult male and female; South China (Lepidoptera; Lycaenidae); and larva of *Lampides boeticus* on a legume pod; Hong Kong (larva ex G. Johnston).

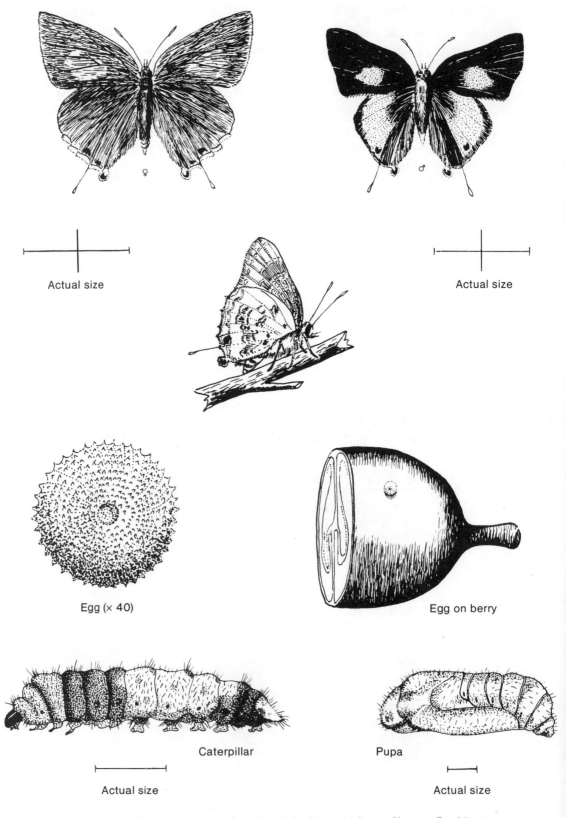

Actual size

Actual size

Egg (× 40)

Egg on berry

Caterpillar

Pupa

Actual size

Actual size

Fig. 12.67 Coffee berry butterfly (*Virachola bimaculata*) ex Kenya (Lepidoptera; Lycaenidae).

Many species of Lycaenidae (larvae) feed on the foliage of clovers and other legumes to be found in grass swards and so the adults are typically seen flying low over grassland seeking out the legume plants, especially in cooler temperate regions.

A few of the better known carnivorous species include:

Feniseca tarquinius—preys on woolly aphids (*Eriosoma,* etc.) in the U.S.A.

Gerydus chinensis—larvae prey on aphids; China.

Liphyra brassolis—in nests of *Oecophylla smaragdina* (red tree ant); Indo-Australian region.

Miletus boisduvali—preys on woolly aphids and coccoids, and attended by black ants; throughout S.E. Asia.

Spalgis epius—preys on a wide range of Coccoidea and woolly aphids; India, S.E. Asia.

Pieridae (white butterflies; yellows; orange-tips, etc.)

Medium-sized butterflies, many of which are extremely abundant, usually either white or yellow in color with orange or black markings. Legs are well developed in both sexes.

The larvae are elongate with body segments divided into annuli and generally many setae. The proleg crochets are bi-or tri-ordinal, and the larvae do not have any osmeteria, horns, or fleshy filaments.

Pupae are suspended in an upright position attached posteriorly by the cremaster resting on a pad of silk, and supported by a central (thoracic) band of silk, and characteristically have a single median cephalic spine; the hindwings are not evident in the pupa externally.

Some species lay eggs singly stuck on to the plant foliage, others lay their eggs in clusters, usually underneath the leaf (see Fig. 12.68). Eggs are tall and spindle-shaped with distinctive longitudinal ribbing. On hatching, the larvae may be gregarious for the first couple of instars and then they generally disperse over the plant.

A few species are strongly migratory—for example, in northern India several species (including *Pieris brassicae, P. boeticus, Eurema blanda*) move up into the cooler regions to avoid the hot season (May to August) and they return with the start of the cooler weather. *Colias* (clouded yellows) are migratory in the Northern Hemisphere; *Pieris rapae* is a regular migrant in North America, flying northwards in warm weather. The two most striking groups of migratory species are probably *Appius* (known to lepidopterists as "albatrosses") and *Catopsila* (common name—"migrants"), both of which can be seen on migration in different parts of Asia in clouds of thousands of individuals. With most species there is a return journey later; most fly northwards from the subtropics in the spring and early summer into more temperate regions, and return southwards in the autumn.

The host plants for the larvae of Pieridae are mostly Cruciferae, Capparidaceae and Leguminosae, with one group (*Delias,* etc.) defoliating the tropical mistletoes (*Loranthus* spp.). The *Pieris* group is extremely damaging to cultivated Cruciferae, but is really a more temperate genus; however these butterflies are found in numbers in North Africa, the Near East and from northern India to South China, so they could be regarded as subtropical as well as temperate; *P. canidia* is certainly abundant in the warm areas of S.E. Asia and the Philippines, and *P. brassicoides* is very abundant in Ethiopia (in the highlands).

The *Delias* species that feed on the leaves of the parasitic "mistletoes" (Loranthaceae) are unusual in that the adults have the undersurface of the wings most strikingly colored whereas the upper surface is quite drab.

Some of the widespread species (of *Pieris,* etc.) are said to occur as a series of geographical subspecies in the extremes of their range. Some species in the tropics occur as distinct wet-season forms (W.S.F.) and dry-season forms (D.S.F.), a phenomenon often observed in the butterflies generally (not just Pieridae). In the tropics, most butterflies are seen in greatest numbers in the dry seasons after the rains have finished. The rainy season provides the flush growth for larval food and by the time the adults emerge the rains are over.

The more important pests of crops and other cultivated plants include the following:

Aporia crataegi (hawthorn white)—larvae on hawthorn and rosaceous fruit trees; Near East.

Appius spp. (albatrosses)—migratory species whose larvae feed on Capparidaceae and Cruciferae; widely distributed throughout Africa and Asia.

Catopsila spp. (migrants)—larvae feed on leaves of shrubs and trees in the Leguminosae; adults are notable migrants; found throughout Asia, as far south as Indonesia and East Africa.

Colias spp. (clouded yellows)—migratory species characteristic of the Northern Hemisphere (Europe, Asia, North America); larvae feed on clovers and other herbaceous legumes; also found in eastern Africa.

Delias spp. (jezebels, etc.)— larvae feed gregariously on leaves of parasitic mistletoes (*Loranthus* spp., etc.; Loranthaceae); S.E. Asia, Australia, China. These could be regarded as being beneficial insects as the larvae severely damage these parasitic plants in coffee and cocoa plantations in S.E. Asia.

Eurema blanda (grass yellow)—larvae eat leaves of tree legumes; India, S.E. Asia, China.

Eurema hecabe (common grass yellow)—larvae on leguminous trees; Old World tropics.

Hebomia glaucippe (great orange tip)—larvae feed on leaves of ornamental/shade trees *Crateva* and *Capparis* in S.E. Asia and China.

Mylothris spp.—on *Loranthus* spp. in Africa (the African equivalent of the Asian *Delias*).

Pieris—the genus is basically a northern temperate one, but some species are found in the warm regions of North Africa, Near East, Middle East, and southern U.S.A. Host plants are all in the Cruciferae and mostly on *Brassica* spp.; some species, e.g. *P. napae,* are to be found more on wild Cruciferae than on crop plants. Many geographical subspecies (races) are known.

Pieris brassicae (large white butterfly)—found throughout Europe, North Africa, Near East, Middle East, India, and now also in Chile (South America) (C.I.E. Map No. A. 25) (Fig. 12.68).

Pieris brassicoides (cabbage white)—on *Brassica* spp.; Ethiopia (Highlands).

Pieris canidia (small white butterfly)—India, S.E. Asia (Fig. 12.69).

Egg mass

Young larvae

Damaged cabbage

Large larvae

Adult ♂

Fig. 12.68 Large white butterfly (*Pieris brassicae*) (Lepidoptera; Pieridae), immature stages and adult male; Skegness, U.K.

Larva

Pupa

Adult ♀

Fig. 12.69 Small white butterfly (*Pieris canidia*) adult female; larva and pupa on nasturtium (Cruciferae); Hong Kong.

Larva

Damaged cabbage leaf

Adult

Fig. 12.70 Small white butterfly (*Pieris rapae*) (Lepidoptera; Pieridae); Cambridge, U.K.

Pieris napi (green-veined white)—larvae mostly on wild Cruciferae; Europe, Asia to Siberia, China, Korea, Japan, North Africa, North America.

Pieris rapae (small white butterfly; imported cabbageworm/U.S.A.)—now cosmopolitan, throughout Europe, Asia, Australia, New Zealand, U.S.A. and Canada (C.I.E. Map No. A. 19) (Fig. 12.70).

Pontia helice—larvae on *Brassica* spp., potato, etc.; eastern Africa.

Pontia protodice (formerly *Pieris protodice*) (southern cabbageworm)—on Cruciferae; southern U.S.A.

Papilionidae (swallowtails)   *600 spp.*

A tropical group of striking beauty, including some of the most beautiful insects in the world. Most are large in size and many have distinctive trailing "tails" on the hind wings, but within the group wing shape is variable. Basic coloration is usually black with markings of red, yellow, blue, or green, sometimes quite iridescent.

Larvae are either smooth or with a series of fleshy dorsal tubercles; or sometimes a raised prominence on segment four. Setae are generally absent. An eversible osmeterium is present on the prothorax. Some of the *Papilio* species have larvae that in their early instars resemble bird droppings (feces) but the later instars assume the green coloration (with bands and eyespots) that characterize the group.

The pupa is somewhat variable in shape, but has two lateral cephalic projections and the hind wings are visible ventrally. The chrysalis is suspended upright, attached posteriorly, and secured by a median silken filament.

Eggs are usually laid singly on the host plant; often they are globular in shape. Adults are remarkable in that there is great sexual dimorphism in many species, and sometimes profound differences in habits. In a few species, both sexes may be polymorphic but in many species only the female is polymorphic. Others have very distinct seasonal forms, and some widespread species occur as distinct geographical subspecies or races. In a few cases the adults apparently mimic certain Danaine butterflies. This is shown by the species *Chilasa clytia* which also has aposematically colored larvae that sit conspicuously on the plant foliage. Another interesting feature of this species is that the chrysalis is a cryptic brown color and shaped to look like a broken twig.

The most striking of these butterflies are now generally regarded as species of special interest and are protected from collectors. As an alternative to field collection, in many countries breeding colonies in captivity have been established to provide specimens for collectors. The collecting of butterflies, and some of the larger moths, has for many years been a popular hobby in many western countries. It is now an extensive commercial business with several well-known companies in Europe and North America offering vast selections of specimens for sale to private collectors. In Papua New Guinea the several species of birdwings (*Ornithoptera* spp.) are now legally protected, but are also reared in captivity for export purposes. Some of the pinned specimens are sold widely to lepidopterists who buy specimens for their collections. Eggs and pupae are also sold, and many now go to the various botanical gardens, zoos, etc. which have displays of living tropical butterflies for public viewing.

A number of species are pests of cultivated plants; although the larvae are found singly, they are large in size; a small *Citrus* bush may have up to a dozen scattered throughout the foliage, and then defoliation may result. Most of the pest species are placed in the large genus *Papilio* and they eat the foliage of *Citrus* and other members of the Rutaceae. Occasionally other plants are recorded as hosts for the citrus swallowtails, such as alfalfa. In temperate regions some *Papilio* larvae feed on Umbelliferae and can be damaging to carrot crops and the like. Throughout the tropics and subtropics cultivated *Citrus* plants are attacked by caterpillars of at least 15–20 different species of *Papilio* in total. In practical terms, each country or region has probably 4–6 different species of *Papilio* attacking the citrus plants; a few species are quite widely distributed, but most are quite local. India has apparently 12 species of *Papilio* recorded from *Citrus*. This is a case where a pantropical crop (*Citrus* spp.) is attacked by autochthonous pest species (*Papilio* spp.) in the different regions of cultivation.

Some of the more important species of swallowtail butterflies include:

Chilasa clytia (common mime)—on *Litsea glutinosa*; S.E. Asia, China.

Graphium agamemnon (tailed green jay)—on *Michelia*, durian, etc.; S.E. Asia, Australasia, China.

Graphium sarpedon (blue triangle)—on camphor, cinnamon; S.E. Asia, Australasia, China.

Lamproptera curius (black and white dragontail)—a species sought by collectors; S.E. Asia, to South China.

Ornithoptera priamus (birdwing)—the largest and most spectacular species of birdwing, sought after by collectors; Papua New Guinea, Australia.

Papilio demodocus (orange dog)—on *Citrus* and Rutaceae; Africa.

Papilio demoleus (lemon butterfly)—on *Citrus* and Rutaceae; Middle East, India, S.E. Asia, China, Australasia (C.I.E. Map No. A. 396) (Fig. 12.71).

Papilio helenus (red helen)—on *Citrus* and other Rutaceae, cultivated and wild; widely distributed throughout S.E. Asia and China (Fig. 12.71).

P. polytes   *P. demoleus*

Fig. 12.71 Citrus butterflies (*Papilio demoleus* right; *P. polytes* left); South China (Lepidoptera; Papilionidae).

Fig. 12.72 Rajah Brook's birdwing (*Trogonoptera brookiana*); Malaya (Lepidoptera; Papilionidae).

Papilio memnon (great mormon)—on *Citrus* and other Rutaceae, cultivated and wild; widely distributed throughout S.E. Asia and China.

Papilio polytes (common mormon)—on *Citrus* and other Rutaceae, cultivated and wild; widely distributed throughout S.E. Asia and China.

Papilio polyxenes (black swallowtail; parsleyworm)—larvae feed on Umbelliferae (carrot, parsley, celery, etc.); Canada, U.S.A. and Central America.

Papilio spp. (10–20 spp.) (citrus swallowtails)—different species present in all tropical and subtropical regions of the world where Rutaceae (and *Citrus*) are found (Fig. 12.71).

Trogonoptera brookiana (Rajah Brook's birdwing)—another beautiful species widely collected and now being extensively reared for sale; Malaysia (Fig. 12.72).

Troides helena (birdwing)—another popular species for collectors; larvae feed on leaves of *Aristolochia* vines; S.E. Asia, Indonesia, China.

HESPERIOIDEA

Antennae dilated apically to form a gradual club which usually ends in a hook; antennal bases widely separated.

Hesperiidae (skippers)

Stout-bodied insects that fly fast and erratically (skipping!), many are diurnal but some crepuscular; small to medium in size. Together with the Papilionoidea they constitute the "butterflies", but in some respects the skippers can be regarded as intermediate between the Papilionoidea and the rest of the Lepidoptera (moths). At rest, the adults, many of which are sun-loving, sit on plant foliage with the wings half-spread (Fig. 12.74).

Larvae feed mostly on the leaves of Gramineae but some other monocotyledenous plants are used; a very few feed on the leaves of legumes. The leaf is either folded over, rolled longitudinally, or cut and rolled into a case, so that the caterpillar is hidden and protected. The larvae are elongate, soft-bodied, tapering at each end, with a small black head capsule, and often coated liberally with white wax. Proleg crochets are triordinal in a circle. The leaf anchorage is achieved with strands of silk.

The pupa is elongate and soft (often waxy), usually enclosed in a slight cocoon in the leaf material but in some cases it is exposed. The cocoon is attached caudally and with a median band of silk.

Eggs are subglobular or hemispherical, flattened underneath and stuck on to the foliage; mostly they have a reticulate pattern, but some are smooth and some ribbed; they are sometimes laid in small groups.

The family is worldwide but the predominance of species is in the tropics. Recorded pest species include:

Ampittia dioscorides (small rice skipper/bush hopper)—on rice and grasses; India, Malaysia, South China.

Borbo spp. (swifts)—on maize, wheat, millets, sugarcane, and grasses; Africa and tropical Asia.

Calpodes ethlius (lesser canna leaf roller)—larvae defoliate arrowroot and *Canna* lilies; West Indies, southern U.S.A.

Erionota thrax (banana skipper)—larvae on leaves of banana and other Musaceae, also on oil palm; S.E. Asia, Indonesia, Philippines.

Erionota torus (banana skipper)—on same hosts as above; India, S.E. Asia, China, and Japan (Fig. 12.73).

Gangara thyrsis—found on all Palmae; India, S.E. Asia to Japan.

Hidari irava (coconut leaf binder)—larvae eat fronds of coconut; S.E. Asia.

Lerodea eufala (rice leaf folder)—larvae fold leaves of rice; U.S.A.

Lotongus calathus (coconut leaf binder)—on coconut; Thailand.

Parnara guttata (rice skipper)—on rice, sugarcane, and other Gramineae; S.E. Asia, Indonesia, to China and Japan (Fig. 12.74).

Parnara spp.—several species recorded on Gramineae; Malaysia, Philippines, and the Solomon Islands.

Eggs   Larva

Pupa   Adult

Fig. 12.73   Banana skipper (*Erionota torus*); South China (Lepidoptera; Hesperiidae).

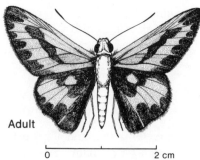

Adult

0   2 cm

Fig. 12.74   Rice skipper butterfly (*Parnara guttata*); wingspan 32 mm (1.28 in.); South China (Lepidoptera; Hesperiidae).

Fig. 12.75   Rice skipper (*Telicota augius*) (Lepidoptera; Hesperiidae).

Pelopidas spp. (leaf folders)—larvae fold leaves of rice, maize, sugarcane, sorghum, bamboos, and grasses; from Africa through the Middle East to India and S.E. Asia.

Rhopalocampta forestana—larvae eat leaves of *Canavalia ensiformis* in Central America.

Suastus gremius—on date palm, coconut; India.

Syrichtus spp.—on *Hibiscus* and some other Malvaceae (not cotton); East Africa.

Telicota augius (rice skipper)—on rice, bamboos, sugarcane; India, S.E. Asia, Papua New Guinea, Australia (Fig. 12.75).

Telicota spp.—larvae on rice, sugarcane, coconut, bamboos; India, Malaysia, Papua New Guinea to China.

Udaspes folus—larvae fold leaves of turmeric; India.

Urbanus proteus (bean leaf roller)—on beans; West Indies, U.S.A.

Zophopetes spp.—larvae eat leaves of Palmae; adults suck juice from citrus fruits on the trees; East and West Africa.

GEOMETROIDEA

Adults with maxillary palps vestigial or absent; tympanal organs in abdomen.

Drepanidae (hook tips; tailed caterpillars)

A small family typical of the Indo-Malayan region, showing considerable diversity of structure but usually with falcate (hooked) forewings at the tip. Eggs are rounded–oval. Larvae are slender without claspers and often have the anal region extended into a taillike protrusion, and some of the anterior segments are enlarged into a hump. Some larvae bear various protruberances along the body. Young larvae usually skeletonize the underside of leaves, but larger instars eat the entire leaf lamina.

A few species are minor crop pests, including:

Cilix glaucata (rosaceous hooktip)—on pome and stone fruits; Near East.

Epicampoptera andersoni (tailed caterpillar)—feeds on the foliage of coffee; tropical Africa (Fig. 12.76).

Epicampoptera marantica (tailed caterpillar)—feeds on the foliage of coffee; tropical Africa.

Oreta extensa (tailed caterpillar)—on coffee foliage; India and S.E. Asia.

Oreta spp.—larvae on various woody plants; S.E. Asia to Japan.

Phyllopteryx elongata—spiny larvae on foliage of coconut and sago palm; S.E. Asia and Indonesia.

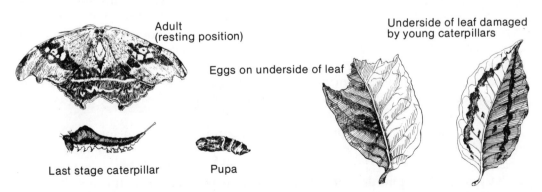

Adult (resting position)

Underside of leaf damaged by young caterpillars

Eggs on underside of leaf

Last stage caterpillar   Pupa

Fig. 12.76 Tailed caterpillar (*Epicampoptera andersoni*) (Lepidoptera; Drepanidae).

Geometridae (carpets/pugs, etc.; loopers)   *12,000 spp.*

A large and ecologically dominating group of moths worldwide. Adults are of slender build with large wings, often held horizontally at rest; nocturnal in habits, to be found resting in foliage during the day; most species are quite small in size, but the pest species are larger. The larvae are the true "looper" caterpillars, elongate and slender and usually with prolegs only on the sixth and the last (tenth) abdominal segments—the terminal pair being usually referred to as claspers. They walk in a looping manner by drawing the posterior segments up to the thorax (forming a body loop) and then the body is extended forwards in the direction of movement, and the process repeated. The larvae are usually green or brown in color and cryptic in habits, spending much of the daytime extended in the foliage resembling a twig (see Fig. 12.80). Pupation is typically underground; but, in a few species, in a cocoon spun in the foliage of the plant.

Some authors regard the group as constituting six definite subfamilies, and others would elevate these groups to separate families.

Eggs are laid in small batches in bark crevices, or on the plant foliage; each female lays from 300 up to 2000 eggs, but a few hundred is average. Some larvae are quite host-specific and may be regarded as monophagous, but some of the more important pest species are both widespread and quite polyphagous. In the larger species the larvae may reach a final length of 4–6 cm (1.6–2.4 in.) and, at this size, host plant defoliation is not uncommon; the most common larval size fully grown is 3–4 cm (1.2–1.6 in.). Larvae are solitary rather than gregarious but there may be many larvae on a single tree. Many can produce silk in their first instar, and tiny larvae dangling from tree foliage by silken threads is a common sight in many countries; using the silk threads there may be dispersal by wind over considerable distances.

A few species of adults are recorded as "eye moths", visiting human eyes, cattle and other large ungulates—the genera concerned are *Hypochrosis, Semiothisa, Pingasa,* and *Problepsis* (Fig. 12.83); found throughout most of S.E. Asia, India, and some parts of Africa.

Despite the ecological dominance of the Geometridae, the number of crop pest species is rather low although there are many species feeding on ornamental and shade trees and woody shrubs. The family is essentially a forest group feeding on trees and understory shrubs. Some species in temperate regions are known as "winter moths", "march moths", etc., and they are adapted for winter emergence; some at the start of the northern winter and others towards the end as the common names indicate. In these species, the female is apterous or with tiny vestigial wings; she climbs up the tree trunk at night and releases her sex pheromones which attract the males for mating. Eggs are laid in bark crevices on the branches and twigs of the trees to be used as hosts, and they hatch in the spring when the buds open. The eating of the young leaves in the spring can be serious and defoliation of fruit and nut trees is not uncommon at this time. This early start in the active life history of these species is probably an adaptation to circumvent the tree's pest-resistance mechanism of tannin and lignin accumulation in the foliage. It typically takes about two weeks for oak, beech, etc. to deposit enough of these chemicals in the young leaves to make them virtually inedible. The eggs of these species are seldom killed by routine winter washes of tar oils and the like, which also tends to increase their importance as orchard pests.

Biston betularia is the peppered moth of Europe of renown for its industrial melanism; it occurs naturally as a polymorphic species with a preponderance of melanistic forms in areas of industrial pollution. Thus, when the adults rest on the bark of birch trees etc., during the day, the dark body is inconspicuous against the sooty bark; the normal pale form can scarcely be seen on a normal pale birch trunk and is very conspicuous on the sooty trunk and therefore is taken by birds. With the heavily industrialized mill-towns in the northwest of England there was a demonstrated development of industrial melanism by these moths. At the turn of the century pale forms predominated but, as the industries developed and pollution increased, the black form became predominant. In recent years the decline of the textile industry in England, combined with pollution control, has resulted in a far cleaner atmosphere, less sooty deposit on the birch trees, and the pale form of *Biston betularia* is again assuming dominance.

Worldwide there is a large number of species, mostly belonging to different genera, and some with quite restricted distributions, so that the literature records many different pests but relatively few are of major importance. Some of the more notable pest species are listed below.

Abraxas spp. (magpie moths)—on currants, other fruit, and nut bushes; Europe, Asia to Siberia, and Japan (Fig. 12.77).

*Alsophila spp. (march moth; fall cankerworms)—on fruit trees and other deciduous trees; Europe, Asia, Canada, U.S.A. (Fig. 12.78).

Ascotis selenaria (giant looper)—polyphagous on coffee, kapok, sweet potato, citrus, soybean, etc.; throughout Africa, and Asia to Japan (Figs. 12.79, 12.80).

Biston betularia (peppered moth)—defoliator of temperate deciduous trees; throughout Europe, Asia and North America.

Biston suppressaria—on tea; India.

Chloroclystis spp. (pug moths)—on a wide range of fruit trees, blackberry, and *Acacia*; Europe, Asia, Near East, Australia.

Ectropis bhurmitra—on cinchona and other trees; India.

Epigynopteryx spp. (3) (coffee loopers)—on coffee; East Africa.

*Erannis spp. (winter moths, etc.)—on deciduous fruit and forest trees; Europe, S.W. Asia, Japan, and the U.S.A. (Fig. 12.81).

Eucylodes spp.—on *Acacia*, cherry, avocado, guava, rose; Australia.

Gelasma illiturata (peach geometrid)—on foliage of peach, etc.; Japan.

Gymnoscelis imperatilis—on mango; India.

Gymnoscelis pumiliata (safflower looper)—polyphagous on a wide range of crops and plants; Central and Southern Europe, West and Central Asia.

Hemerophila spp.—on coffee, cowpea, cotton, sweet potato, cassava, etc.; East Africa.

Hypochrosis spp. (eye moths)—adults come to the eyes of humans and large ungulates; S.E. Asia.

Hyposidra talaca—polyphagous on cocoa, cinchona, coffee, tea, tung, *Derris*, roselle, flax, etc.; throughout many parts of S.E. Asia.

Hyposidra spp.—on cocoa, coffee, castor, and on *Mimosa* and *Sesbania* grown as green manure under rubber, etc.; Africa, India, Indonesia, Australia.

*Species marked thus are belonging to the "winter moth" group.

Adult resting on tree trunk

Adult

Fig. 12.77   Magpie moth (*Abraxas grossulariata*) (Lepidoptera; Geometridae); Cambridge, U.K.

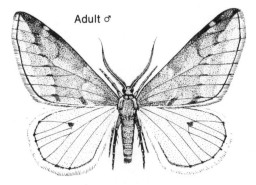

Larva (on gooseberry bush)

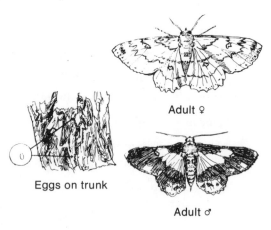

Adult ♀

Eggs on trunk

Adult ♂

Adult ♂

Fig. 12.78   March moth (*Alsophila aescularia*) (Lepidoptera; Geometridae) adult, ex Cambridge, U.K.

Larvae eating leaves

Fig. 12.79   Giant looper (*Ascotis selenaria*) (Lepidoptera; Geometridae) ex Kenya.

Inurois fletcheri (apple fall cankerworm)—on apples, etc.; Japan.

Megabiston plumosaria (tea looper)—larvae on foliage of tea; Japan.

Neocleora spp.—on cocoa, coffee, cinnamon, cotton, *Eucalyptus* and other trees; East and West Africa.

Nychiodes palestinensis (rosaceous looper)—on stone and pome fruits; Near East.

Oenospila spp.—on litchi, mango, tamarind; India, Indonesia.

Operophtera spp. (winter moths)—on fruit and nut trees, and other deciduous trees; Europe, Mediterranean, Asia, to Japan, Canada and the U.S.A. (C.I.E. Map No. A. 69) (Fig. 12.82).

Palaecrita spp. (cankerworms)—on fruit and shade trees; Canada and U.S.A.

Phthonesema spp. (apple/mulberry loopers)—on apple, mulberry, etc.; Japan.

Pingasa spp.—on guava and various Lauraceae; S.E. Asia, China.

Prasinocyma spp.—on castor; East Africa.

Problepsis spp. (eye moths)—adults visit eyes of humans and large ungulates; S.E. Asia to southern China (Fig. 12.83).

Pterotocera sinuosaria (fruit tree looper)—Japan.

Fig. 12.80 A giant looper caterpillar (probably *Ascotis selenaria*) on twig of citrus; body length 48 cm (1.9 in.); Hong Kong (Lepidoptera; Geometridae).

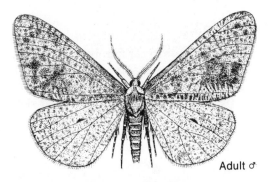

Adult ♂

Fig. 12.81 Mottled umber moth (*Erannis defoliaria*) (Lepidoptera; Geometridae) adult; ex Cambridge, U.K.

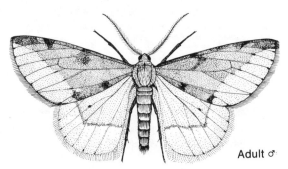

Adult ♂

Fig. 12.82 Winter moth (*Operophtera brumata*) (Lepidoptera; Geometridae) adult; ex Cambridge, U.K.

Fig. 12.83 Eye moth (*Problepsis* sp.) (Lepidoptera; Geometridae); wingspan 25 mm (1 in.); Hong Kong.

Scopula spp. (brassica loopers)—larvae on *Brassica* crops in South China, and on sea lavender in East Africa.

Semiothisa pervolgata (daincha looper)—on *Sesbania* grown as a green manure crop; India.

Semiothisa spp.—on various tree and shade legumes; Africa and India.

Thalassodes spp.—on mango, castor, litchi, rose, etc.; Africa, India, South China.

Theria rupicapraria (early moth)—found in Europe and West Asia.

Zamacra spp. (mulberry/walnut loopers)—on fruit and nut trees; Near East, Japan.

Uranidae

A small but exclusively tropical family, widespread in both the Old and New World. Most are large and slender-bodied and generally resemble butterflies—some are also diurnal in habit; some resemble Geometrids but have caterpillars with normal complement of prolegs. Many adults have tailed hind wings. The nocturnal species are strongly attracted to lights at night. A typical example of the family is the large and spectacular *Lyssa zampa;* this is widespread throughout S.E. Asia and also South China, where it can be seen resting on buildings near house and street lights during the daytime. The larvae are reported to feed on the leaves of *Eugenia* (Myrtaceae). Agriculturally they are not of any importance.

Epiplemidae   *550 spp.*

A small tropical family very closely related to the Uranidae, best developed in Papua New Guinea. The nocturnal adults rest during the daytime with the forewings rather rolled up and the hind wings pressed against the abdomen. The one species, *Leucoplema dohertyi,* is the coffee leaf skeletonizer of tropical Africa (Fig. 12.84).

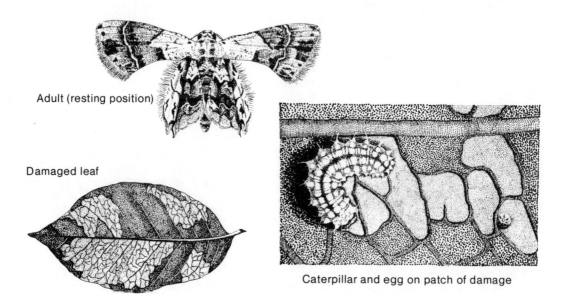

Adult (resting position)

Damaged leaf

Caterpillar and egg on patch of damage

Fig. 12.84   Coffee leaf skeletonizer (*Leucoplema dohertyi*) (Lepidoptera; Epiplemidae).

BOMBYCOIDEA

The previous families all had a frenulum present, but in the Bombycoidea the frenulum is atrophied and usually lost or vestigial; proboscis rarely developed; antennae pectinate, especially in males. Several families are rather small and of little economic interest and they are not mentioned here.

Lasiocampidae (eggars, lappets; tent caterpillars) *1000 spp.*

These moths are quite large, densely scaled, and stout-bodied; proboscis is atrophied, and antennae are bipectinate in both sexes. A widespread group, absent from New Zealand and most abundant in the tropics.

Eggs are smooth, oval, and often laid in a band enclosing a twig (100–200 eggs per band); some females lay up to 2000 eggs each.

The caterpillars usually live gregariously, especially so when young, inside a "tent" of silk on the host plant which is usually a tree or bush. The heavy infestations that are typical may often result in defoliation of the tree. The larvae grow to a large size, up to 10 cm (4 in.) in some species, and they are stout and very bristly; they often have lateral downwardly directed bristle-flanges, and may have dorsal tufts or humps on anterior segments. Many species have bristles that are urticating and may inflict painful wounds on field workers. Some caterpillars have brightly colored spots or bands on the body which make them conspicuous. They are seldom eaten by birds and other vertebrate predators, presumably being protected by the bristles, but they are heavily parasitized by Hymenoptera Parasitica and preyed upon by some bugs.

Pupation is typically inside a firm, oval cocoon made of bristles and silk, and attached to a twig or some solid substrate. Some pupae have two prominent protuberances.

Adults sometimes show distinct sexual dimorphism; in the illustration of *Trabala vishnou* the female shown is large and yellow (Fig. 12.85), whereas the male is somewhat smaller and is green in color. *Lasiocampa* is diurnal and adults fly swiftly in sunshine in Europe and the Mediterranean Region, but most members of the family are nocturnal.

Some of the more important pest species of Lasiocampidae include:

Dendrolemus spp. (pine/oak tent caterpillars)—several species on a range of deciduous and evergreen forest and ornamental trees; Europe, Asia, and North America.

Gastropacha spp. (lappet moths)—on cocoa, tea, pigeon pea, peach, and other plants; S.E. Asia, Indonesia, Philippines, China and Japan.

Lasiocampa trifolii (clover lappet moth)—on clovers, plum, grapevine, etc.; Near East.

Lebeda cognata—on citrus, roses, etc.; Malaysia, Borneo.

Malacosoma indica (Indian lappet moth)—on apple, almond, apricot, pear, cherry, walnut; India.

Malacosoma spp. (8–10) (tent caterpillars)—larvae polyphagous on fruit and forest trees (and some on salt-marsh plants); throughout Europe, parts of Asia, and North America.

Metanastria hyrtaca—larvae on *Cinchona* foliage; Indonesia.

Pachypasa spp.—larvae defoliate *Acacia* and *Pinus,* etc., in South Africa; *P. otus* is the cos silkworm of the Mediterranean Region.

Paralebeda plagifera—found on *Cinchona,* cocoa; India, S.E. Asia, Indonesia.

Suana concolor—larvae on citrus, tea, kapok, cocoa, rambutan, avocado, roses, etc.; India, S.E. Asia, Philippines.

Taragama spp.—on cotton, citrus, mango, rose, etc.; East Africa.

Trabala vishnou (oriental lappet moth)—on guava, almond, pomegranate, *Syzygium, Melastoma,* and *Rhodomyrtus;* India, S.E. Asia to China (Figs. 12.85, 12.86).

Trabala spp.—larvae recorded from guava and a wide range of host plants; S.E. Asia, Indonesia, Borneo, and also in East Africa.

Fig. 12.85 Oriental lappet moth (*Trabala vishnou*) (Lepidoptera; Lasiocampidae); adult female, wingspan 60 mm (2.4 in.); South China.

Fig. 12.86 Caterpillar of oriental lappet moth (*Trabala vishnou*) on leaf of *Rhodomyrtus;* body length 60 mm (2.4 in.); South China.

Bombycidae (silkworm moths)

A small family which now includes the Eupterotidae (according to Richards & Davies, 1977); moderate to large-sized moths, found mostly in the Ethiopian and Oriental regions. Antenna is strongly pectinate, proboscis absent, and a frenulum is present. Larvae are of two types—the *Eupterote* group are stout and bristly and may cause urtication; the *Bombyx* group are smooth and elongate and have a terminal "horn" projecting dorsally. Some larvae live gregariously, but are seldom found in large numbers. The group is of interest for the single species *Bombyx mori*—the Oriental silkworm. This native of China is now dispersed worldwide as the main producer of commercial silk (sericulture). Apparently it is now unknown in the truly wild state, and exists as a series of domesticated races, sometimes regarded as distinct species. The smooth-bodied, tailed larvae feed on leaves of white mulberry (*Morus alba*) and they pupate inside a cocoon of silken thread attached loosely to the leaf surface (Fig. 2.4). Each cocoon is said to be constructed of a single thread of silk about 300 m (330 yd.) in length.

The few species of note in this family are as follows:

Bombyx mori (Oriental silkmoth; silkworm)—native to China; this species feeds on leaves of mulberry (*Morus alba*); now widely cultivated as a source of silk in Japan, China, India, the U.S.A., and other countries (Figs. 2.3, 2.4).

Dreta petola—larvae feed on sugarcane, maize, bamboos, and various grasses; Indonesia.

Eupterote geminata—on banana foliage; India.

Ocinara varians—on foliage of *Ficus,* jackfruit; India, S.E. Asia, China.

Trichola ficicola (banyan silkmoth)—larvae feed on leaves of *Ficus microcarpa;* S.E. Asia to South China.

Saturniidae (emperor moths/giant silkmoths)

The alternative name is giant silkmoths because of the silk produced by the larvae at pupation; several species are used for the production of other types of silk. The family is basically tropical, and it includes the largest species of Lepidoptera known, and probably also the largest insects, with a wingspan of up to 25 cm (10 in.). Most adults have a transparent eyespot near the center of the forewing and the hind wing, and both sexes have bipectinate antennae, with males having the longer rami. They are nocturnal and fly to lights at night—their large size usually causing quite a commotion at domestic lights.

The genus *Attacus* is remarkable for its size and coloration. It occurs as a number of species throughout Africa, and Asia to Japan, Mexico and through Central and South America. Some of the females have a wingspan reaching 25 cm (10 in.)—*Attacus atlas* (Fig. 12.89) and *A. edwardsi* being the largest.

The larvae are distinctive and somewhat specialized; they are stout and smooth and have scoli either well developed or rudimentary. *Attacus* larvae reach a length of 10–15 cm (4–6 in.) when fully grown—such formidable caterpillars eat a great deal of foliage and several larvae may defoliate a small tree quite easily. In S.E. Asia, several species of Saturniidae regularly cause complete tree defoliation.

Pupation takes place inside a firm, dense cocoon spun of silk on a leaf or twig. Several species of Saturniidae produce silk of commercial quality. Most of these large moths only have one generation per year and in cooler regions they overwinter as pupae. Some of the small species will have several generations per year in the tropics.

The following species are of some importance as silkworms:

Antheraea assama (muga silkworm)—polyphagous silkworm of the Oriental region.

Antheraea paphia (tussor silkworm)—as above.

Antheraea pernyi (Chinese oak silkworm)—the producer of shantung silk in China.

Antheraea yamamai (Japanese oak silkworm)—larvae eat *Quercus;* Japan mostly.

Eriogyna pyretorum (giant silkworm moth)—of S.E. Asia and China; larvae feed on camphor.

Samia cynthia (wild silkmoth)—found in India, and throughout S.E. Asia to China; larvae recorded feeding on *Lantana* leaves and *Michelia alba*.

Samia ricini (eri silkworm)—found throughout India and S.E. Asia.

There is some overlap with some species between their being beneficial as a source of silk and the larval damage to some crop plants, as can be seen from the list of crop pest species (mostly minor pests) below. Sericulture is discussed on page 68.

Actias selene (moon moth)—recorded from apple, cherry, pear, walnut, *Betula, Moringa,* etc., often causing complete defoliation; in North India; and from camphor, tallow and liquidamber trees in South China (Figs. 12.87, 12.88).

Antheraea pernyi (Chinese oak silkworm)—damages apple, pear, walnut, etc.; in North India.

Attacus atlas (atlas moth)—on camphor, cinnamon, cashew, coffee, guava, mango, tea, cinchona, citrus, avocado, pepper; in Java recorded from 40 host plants; India, S.E. Asia, Indonesia to China (Figs. 12.88–91). The genus *Attacus* is pan-tropical and occurs as half a dozen different species.

Bunaea spp.—on coffee, guava, *Sapium,* etc.; East Africa.

Caligula simla—on apple, pear, walnut, avocado, mango, cocoa, cinnamon, and other trees; northern India through S.E. Asia to Indonesia.

Citheronia regalis (hickory horned devil)—on foliage of pecan; U.S.A.

Cricula elaezia—larvae feed on leaves of *Cinchona;* Indonesia.

Fig. 12.87 Moon moth (*Actias selene*); adult resting on tree trunk; South China; wingspan 40 mm (5.6 in.) (Lepidoptera; Saturniidae).

Fig. 12.88 Moon moth caterpillar (*Actias selene*); body length 70 mm (2.8 in.); South China (Lepidoptera; Saturniidae).

Fig. 12.89 Atlas moth (*Attacus atlas*); newly emerged adult male; wingspan 210 mm (8.4 in.); South China (Lepidoptera; Saturniidae).

Fig. 12.90 Atlas moth caterpillar (*Attacus atlas*); length 90 mm (3.6 in.); on twig of camphor; South China (Lepidoptera; Saturniidae).

Fig. 12.91   Pupal cocoon of atlas moth (*Attacus atlas*) fastened to leaves of camphor; South China (Lepidoptera; Saturniidae).

Cricula trifenestrata—on mango, cashew, pepper, tea; India.

Epiphora spp.—larvae feed on citrus and many different trees; Africa.

Imbrasia spp.—on tamarind, citrus, and various legume trees; East and West Africa.

Lobobunea spp.—on citrus, mango, guava, roses, *Acacia, Eucalyptus,* etc.; East Africa.

Loepa sikkima—on *Cissus* and *Leea*; S.E. Asia, Borneo, South China.

Nudaurelia spp. (silkworms, etc.)—on tung, cashew, pigeon pea, *Ficus,* cassava, castor, cocoa, *Croton,* mango, *Canna, Acacia, Pinus,* and other trees and ornamentals; East and South Africa.

SPHINGOIDEA

An isolated group containing a single family. Adults have antennae gradually thickened with the apex pointed and usually hooked; proboscis and frenulum both very strongly developed.

Sphingidae (hawk moths/hornworms)   *1000 spp.*

A conspicuous group of large moths, found worldwide but essentially a tropical group; several important pest species are virtually cosmopolitan. The vast majority are nocturnal; the adults fly strongly and fast, they have long, tapering wings and a streamlined body. Most species have a long proboscis—fully extended 25 cm (10 in.) is not uncommon. Many adults usually take nectar from flowers with a long narrow corolla that other moths cannot reach, and, typically, they hover in front of the flower while feeding. A few species, including the humming bird hawk moths (*Macroglossum,* etc.) are diurnal and to be seen feeding in daylight—their wings move so fast as to appear only as a blur. Some species have the wing scaleless and hyaline, either partially or completely.

The larvae are smooth and elongate and with a postero-dorsal horn on the last (eighth) abdominal segment—hence the common name of "hornworms" used in the U.S.A. The larvae are not gregarious but many single eggs may be laid on the same host plant so the bush, or whatever, may literally be covered with large caterpillars and then defoliation can be expected. There are many records of coffee bushes in Sri Lanka, etc., being defoliated by larvae of the coffee hawk moth, and the same species in Hong Kong defoliated a hedge of *Gardenia* bushes. There are five larval instars; in the tropics most species have several generations per year, but in the cooler temperate regions most Sphingidae are univoltine. The larvae are leaf lamina eaters and some are pests, purely on the basis of the quantity of leaf material they consume. Some are quite host-specific; in fact, most species are restricted to a single family of plants for food (oligophagous) but a few are truly polyphagous. The Solanaceae and Convolvulaceae are favored groups for food, but the most popular host plant agriculturally is grapevine with more than a dozen Sphingidae recorded worldwide.

The pupae are long and pointed posteriorly and distinctive in that the long coiled proboscis of the adult is always evident (see Figs. below). Pupation takes place inside an earthen cocoon in the soil, or occasionally in leaf litter on the surface.

A number of species are of importance as pests of cultivated plants, including the following:

Acherontia atropos (death's head hawk moth)—on potato, tomato, tobacco, eggplant, and other Solanaceae (also recorded on olive, sesame, and some other plants); Europe, southern Asia to China and Japan, and the whole of Africa.

Acherontia lachesis (death's head hawk)—larvae more polyphagous and also on mango, grapevine, etc.; India and S.E. Asia.

Acherontia styx (small death's head hawk)—larvae more polyphagous and also on mango, grapevine, etc.; India and S.E. Asia (Fig. 12.92).

Agrius cingulata (sweet potato hornworm)—on sweet potato, etc.; U.S.A., Central and South America.

Agrius convolvuli (sweet potato/convolvulus hawk moth)—on *Ipomoea,* other Convolvulaceae, and on some Leguminosae, sunflower, grapevine, citrus; Europe, Africa, southern Asia to China, Australasia (C.I.E. Map No. A. 451) (Fig. 12.93).

Cephonodes hylas (coffee hawk moth)—on coffee, *Gardenia,* and some other Rubiaceae; Africa, India, S.E. Asia to China, and Australasia (Fig. 12.94). It is not clear whether *C. piceus* is a synonym or whether it is a valid species. There are several species of *Cephonodes* in Asia and Australasia, most to be found feeding on Rubiaceae; adults are diurnal and conspicuous.

Coelonia spp.—on tomato, castor, *Lantana,* etc.; East Africa.

Daphnis hypothous—on various Rubiaceae including *Cinchona;* S.E. Asia to China.

Daphnis nerii (oleander hawk)—larvae on oleander, eggplant, grapevine, *Cinchona, Gardenia,* etc.; Europe, Africa, Asia, and the U.S.A.

Eumorpha spp. (grapevine sphinxes)—on grapevine foliage, etc.; U.S.A.

Hippotion celerio (silver-striped/vine hawk moth)—polyphagous on a wide range of cultivated and wild plants, especially grapevine; southern Europe, Africa, South Asia, Australasia.

Hyles lineata (= *Celerio lineata*) (striped hawk moth)—polyphagous on grapevine, cotton, olive, sweet potato, buckwheat, *Prunus,* etc.; cosmopolitan (C.I.E. Map No. A. 312) (Fig. 12.95).

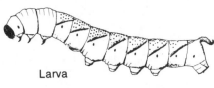

Larva

Fig. 12.92 Death's head hawk moth (*Acherontia styx*); wingspan 100 mm (4 in.); and larva; South China (Lepidoptera; Sphingidae).

Fig. 12.94 Coffee hawk moth (*Cephonodes hylas*) adult, wingspan 65 mm (2.6 in.); Hong Kong; caterpillar on *Gardenia*; South China (ex F. Bascombe) (Lepidoptera; Sphingidae).

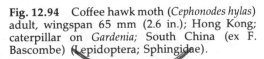

Adult

Fig. 12.93 Sweet potato/convolvulus hawk moth (*Agrius convolvuli*); wingspan 96 mm (3.8 in.); South China (Lepidoptera; Sphingidae).

Larva

Fig. 12.95 Striped hawk moth (*Hyles* sp.); wingspan 68 mm (2.7 in.); Hong Kong (Lepidoptera; Sphingidae).

Fig. 12.96 Pine hawk moth (*Hyloicus pinastri*); wingspan 85 mm (3.4 in.).

Pupa in cocoon

Hyles spp. (sphinx moths)—on a wide range of hosts in the U.S.A. and Asia.

Hyloicus pinastri (pine hawk moth)—on *Pinus;* throughout Europe and Asia (Fig. 12.96).

Langia zeuzeroides—recorded on deciduous fruits; northern India.

Laothoe populi (poplar hawk moth)—on poplar and *Salix;* western Palaearctic.

Leucophlebia lineata (grass hawk moth)—on sugarcane, maize, and grasses; tropical Asia.

Macroglossa belis (hummingbird hawk moth) recorded on citrus; India.

Manduca quinquemaculata (tomato hornworm)—serious pest on tomato, tobacco, potato, and other solanaceous crops; in southern U.S.A., Central and South America (Fig. 12.97).

Manduca sexta (tobacco hornworm)—serious pest on tomato, tobacco, potato, and other solanaceous crops; in southern U.S.A., Central and South America (Fig. 12.97).

Manduca spp. (hornworms)—serious pests on tomato, tobacco, potato, and other solanaceous crops; in southern U.S.A., Central and South America.

Marumba gaschkewitschi (peach hornworm)—on peach, etc.; China and Japan.

Mimas tiliae (lime hawk moth)—on lime, elm, alder, oaks, birch, etc.; Europe, Asia.

Smerinthus spp. (fruit hornworms, etc.)—on various fruit trees and deciduous forest trees; Europe, Asia and U.S.A. (Fig. 12.98).

Sphinx ligustri (privet hawk moth)—on privet, lilac, ash; Palaearctic region.

Sphinx spp. (fruit tree sphinxes, etc.)—on many fruit trees; throughout the U.S.A.

Theretra spp. (grapevine moths)—on grapevine, etc.; southern Europe, India, eastern Asia.

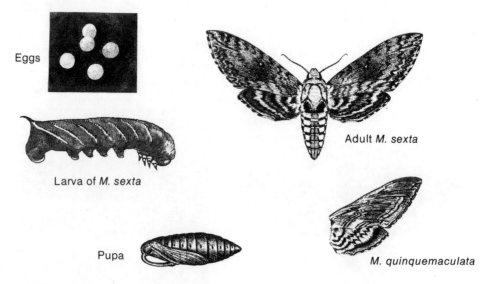

Fig. 12.97 Tomato and tobacco hornworms (*Manduca sexta* and *M. quinquemaculata*) (Lepidoptera; Sphingidae).

Fig. 12.98   Eyed hawk moth (*Smerinthus ocellata*) (Lepidoptera; Sphingidae); wingspan 80 mm (3.2 in.).

NOCTUOIDEA

Notodontidae (prominents, etc.; processionary caterpillars)

Quite large moths, stout-bodied, elongate wings; with a worldwide distribution except for New Zealand, and scarce in Australia. Adults are all nocturnal, sometimes attracted to lights at night but not often, and generally they are seldom seen. The caterpillars generally occur on trees and shrubs, only rarely are they found on herbaceous plants; they are either solitary or gregarious in habit, and when disturbed they assume a characteristic posture with both anterior and posterior parts elevated. The larvae may have spines and fleshy tubercules or humps on the body, and the anal claspers are sometimes modified into tubercules. Generally, the more bizarre larvae are solitary and the more "normal" ones are gregarious. Some of the gregarious species feed nocturnally and during the day they shelter, often in a silken nest, and in the evening they emerge in a procession up the tree trunk, etc., to feed on the tree foliage.

The adult of *Tarsolepis sommeri* is recorded as an "eye moth" frequenting the eyes on humans in Malaysia.

Some species pupate inside a hard spun cocoon stuck to the bark on the trunk of trees, usually in a depression of some sort. In these species, the imago not only has a sharply pointed head-process in order to break through the hard cocoon, but secretes from the mouth a solution of potassium hydroxide for softening the cocoon. Other species pupate in the soil, inside a cocoon.

Some species are of importance as defoliators of trees and other cultivated plants in the warmer parts of the world, and include the following:

Anaphe spp. (African silkworm, etc.)—on coffee and various trees; eastern Africa.

Andraca apodecta—a gregarious species feeding on tea; Indonesia.

Andraca bipunctata (Assam bunch caterpillar)—on tea; Assam and Java.

Datana spp. (handmaid moths, etc.)—on apple, walnut, etc.; U.S.A.

Desmeocraera spp. (guava moth, etc.)—on guava, *Eucalyptus,* etc.; East and West Africa.

Dudusa nobilis—larvae feed on rambutan, etc.; S.E. Asia, Indonesia, and China.

Dynara combusta—on Gramineae (sugarcane, maize, sorghum, rice, and grasses); India, Indonesia, Philippines, and China.

Phalera javana—on Leguminosae; Indonesia.

Phalera raya—on Leguminosae; India, S.E. Asia, Indonesia, and South China.

Phalera spp. (bufftip moth, etc.)—larvae on deciduous fruit, nut, and forest trees; Europe, Asia to Japan.

Scalmicauda sp.—recorded damaging cocoa; West Africa.

Schizura spp. (red-humped caterpillar, etc.)—on apple and orchard trees; U.S.A.

Stauropus alternus (crab caterpillar)—larvae polyphagous on coffee, tea, rambutan, litchi, mango, cowpea, etc.; India, S.E. Asia, Indonesia, Philippines, China.

Tarsolepis sommeri—on rambutan; Indonesia, Malaysia.

Thaumetopoea wilkinsoni (Cyprus processionary caterpillar)—gregarious pest of *Pinus*; Mediterranean Region; other species on trees in eastern Africa.

Ctenuchidae (= Amatidae, Syntomidae) (wasp moths)   *3000 spp.*

A tropical group with a few representatives in temperate regions. Small to medium-sized moths, mostly diurnal, and mostly sluggish in habits, with a well-developed proboscis; they often resemble wasps in their general body coloration and have narrow wings basically black but with clear patches that are scale-free (hyaline). Sometimes this group is included in the Arctiidae as a subfamily.

The larvae are mostly rather short and have verrucae bearing numerous secondary setae. They feed on grasses, and living foliage, or on fallen flowers and leaves. The larval bristles may cause urtication if handled. Generally, the larvae resemble the "woolly bears" of the Arctiidae.

Pupation takes place inside a cocoon of silk and felted hairs, attached to a twig or plant foliage.

A few species are notable as either crop pests, or else because of their local abundance, including:

Balacra spp.—larvae eat dry leaves, bark, etc. on coffee, cotton, etc.; East Africa.

Ceramidia viridis (leaf-cutting caterpillar)—on banana; Central and South America.

Euchromia horsfieldi—on Convolvulaceae; Indonesia.

Euchromia polymena—on sweet potato, etc.; India.

Euchromia spp.—larvae on banana, sweet potato, etc.; Africa.

Metarctia spp.—on maize, grasses (sometimes on roots), sweet potato, etc.; East Africa.

Syntomis alicia (tomato wasp moth)—larvae polyphagous on many plants; eastern Africa.

Syntomis spp. (wasp moths)—on grasses, citrus, *Rhus*, etc.; Africa, S.E. Asia to China (Fig. 12.99).

Fig. 12.99 Wasp moth (*Syntomis polymita*) (Lepidoptera; Ctenuchidae); body length 15 mm (0.6 in.); Hong Kong.

Arctiidae (= Lithosiidae) (tiger moths; woolly bears)   *10,000 spp.*

Adults are small to medium-sized, mostly brightly colored, banded or spotted, many are yellow, white, red, orange, and black in color. Most are nocturnal in habits and some are attracted to lights at night. Some are clearly aposematic in coloration and secrete toxins, and are quite poisonous to would-be predators. There are two distinct subfamilies: the Arctiinae are the "tiger moths", the "buffs", and the "ermines"; the Lithosiinae includes the "footmen" with their long narrow wings, diurnal or crepuscular in habit, and the larvae (only sparsely hairy) feed on lichens on tree trunks and rocks. Generally the adults show great diversity in coloration but a few widespread genera do show some uniformity in color patterns, as mentioned later.

Many species are polyphagous, either on low herbage, or trees and shrubs, and are commonly found in areas of cultivation and recorded feeding on a wide range of cultivated plants. Two genera of special note are *Diacrisia* (buffs), an Old World genus with many species in the tropics, some on trees and shrubs, some on herbaceous crops; and *Spilosoma* (ermine moths), basically a genus of Old World tropics extending into warmer temperate regions. Some pest species have been at different times placed in both of these genera; most feed on herbaceous plants.

Eggs are generally laid on the host plant foliage in groups, often covered with hairs from the female abdomen.

The larvae (especially of Arctiinae) are densely hairy, having many secondary setae on verrucae and the setae may be urticating. As a group, the larvae have few predators but one group of birds worldwide has adapted for feeding on them—the cuckoo family (Cuculidae). The caterpillars are referred to as "woolly bears" and tend to be very active. In temperate regions many species overwinter as larvae and they resume feeding in the spring before they pupate.

Pupation takes place in a cocoon covered with the long larval setae, usually also protected by folded leaves held together with strands of silk, either still on the host plant, or in leaf litter, or under the surface of the soil.

Some of the species recorded as pests of crops and cultivated plants in warmer parts of the world include the following:

Alphaea biguttata (cardamon tiger moth)—on cardamon; India.

Amsacta albistriga (red hairy caterpillar)—a pest of dryland crops (groundnut, sorghum, etc.); southern India.

Amsacta gangis—larvae feed on Gramineae (maize, sugarcane, etc.) and other plants; Indonesia.

Amsacta lactinea (woolly bear)—polyphagous on leek, tea, cucurbits, citrus, soybean and other pulses; South and East Asia to Japan.

Amsacta moorei (red hairy caterpillar)—on cowpea, sorghum, millets, cotton, castor, etc.; India, especially in the north.

Amsacta transiens—larvae are polyphagous on trees, shrubs and herbs; throughout S.E. Asia, Philippines to Japan.

Arctia spp. (tiger moths)—larvae polyphagous in low herbage; Europe and parts of Asia (Fig. 12.100).

Argina cribraria—larvae inside seed pods of *Crotalaria;* India, S.E. Asia, China, and North Australia.

Argina spp.—larvae on *Crotalaria* and other legumes; Africa.

Asura spp.—on bananas, mango, etc.; India.

Diacrisia obliqua (bihar hairy caterpillar)—polyphagous on cabbage, lettuce, pea, jute, cotton, sesame, beans, castor, groundnut, maize, papaya, linseed (flax), sweet potato, mulberry, etc.; India, S.E. Asia, to China.

Diacrisia spp. (tiger moths; buffs)—larvae generally polyphagous on a wide range of crops (coffee, cotton, mulberry, papaya, groundnut, soybean, alfalfa, tobacco, cocoa, *Solanum,* maize and other cereals, and various weeds); throughout Africa and southern Asia.

Ecpantheria icasia—on banana (larvae eat skin of fruits); Central and South America.

Estigmene spp. (saltmarsh caterpillars, etc.)—on Gramineae, groundnut, citrus, etc.; Africa, India, Asia, U.S.A.

Maenas maculifascia—polyphagous, web-spinning larvae found on trees and shrubs (cocoa, castor, etc.); S.E. Asia, Indonesia, Philippines, North Australia.

Ocnogyna loewi (tiger moth)—polyphagous; Near East, Egypt.

Pericallia ricini (castor woolly bear)—polyphagous pest recorded from castor, sweet potato, banana, cucurbits, sesame, okra, cotton, and various ornamentals; India.

Fig. 12.100 Tiger moth (*Arctia caja*) (Lepidoptera; Arctiidae); wingspan 75 mm (3 in.); Gibraltar Point, U.K.

Fig. 12.101 Cinnabar moth caterpillars (*Tyria jacobaeae*) (Lepidoptera; Arctiidae) defoliating a plant of ragwort; Gibraltar Point, U.K.

Pericallia spp. (lantana tiger moth, etc.)—on lantana and various hosts; parts of Africa.

Spilosoma spp. (ermine moths)—on cherry, mulberry, and other trees; Europe, Asia to Japan. (Several African species formerly placed in *Spilosoma* are now moved to *Diacrisia*.)

Teracotona submaculata—larvae recorded feeding on cabbages, beans, beets, lettuce; South Africa.

Tyria jacobaeae (cinnabar moth)—an interesting species that feeds on the ruderal plant pest ragwort; native to Eurasia, now introduced into New Zealand and the U.S.A. in an attempt to control the ragwort weeds (Fig. 12.101).

Uthetheisa pulchella (harlequin moth; sunnhemp hairy caterpillar)—larvae on *Crotalaria* and other legumes, eating leaves and boring pods; Europe, Africa, India, S.E. Asia, Indonesia, and Australia.

Noctuidae (= Agrotidae) *25,000 spp.*

This is the largest family of Lepidoptera and probably the most important family of insects so far as pests of cultivated plants are concerned. The adults are of moderate size, often drab in color, nocturnal in habits; proboscis usually well developed (very rarely atrophied); frenulum always present, and antennal shaft never dilated. They fly to lights at night and usually feed on flower nectar, although the *Plusia* group are crepuscular and may be seen flying in the evenings, or occasionally in daytime. A few species have the proboscis stout enough to pierce unripe and ripe fruits, and even the hide of large vertebrates. Forewing coloration is always drab and cryptic, but the hind wings of many species are brightly colored yellow or red. The intricate pattern of coloration of the forewing is usually a good specific character. There are some instances where two different species are identical morphologically but differ in their genitalia. For example, only recently was the Old World "species" *Spodoptera litura* (a very serious polyphagous crop pest) shown to consist of two sibling species that are allopatric in distribution with *S. littoralis* in Africa and the Mediterranean Region and *S. litura* in tropical Asia and Australia. Some species are regular migrants and annually will fly from the warmer subtropical regions up into temperate countries during the summer.

The larvae are quite large, stout-bodied caterpillars when fully grown, typically smooth, often striped longitudinally, and in some species somewhat variable in color (there are often both green and brown forms). There are usually 5–6 larval instars, but at times of food shortage the life cycle may be shortened and have only five larval instars. Under such crowded conditions cannibalism is not uncommon. Most of the larvae are phytophagous and feed on all parts of the plant body (of flowering plants), but a few species (*Eublemma, Catoblemma,* etc.) are predacious on Coccoidea and some larvae feed on fungi and lichens. Most species have the full complement of four pairs of abdominal prolegs, but in the subfamilies Plusiinae, Catocalinae, Ophiderinae, and Hypeninae, the number of prolegs is reduced to three or two pairs and the larvae are termed "semiloopers". Most larvae are nocturnal and so only feed at night; during the daylight hours they hide, either in the plant foliage or leaf litter. Crops inspected at night by lamplight may reveal the caterpillars climbing all over the aerial parts of the plants.

Eggs are usually more or less spherical and ribbed with fine reticulations, but in some groups they are flattened. There is great variation in the number laid, ranging from about 100 to 2000–3000 in some *Spodoptera*. They are usually laid in batches on the host plant or in its proximity. In the Plusiinae and some other subfamilies it is usual for eggs to be laid singly, although quite a large number may be laid on a single host plant scattered throughout the foliage. Hatching larvae usually first eat the egg shells before starting on plant material. In some species the first instar larvae are gregarious, and they often eat only the lower epidermis of leaves making a large "windowed" area; as they grow they become more solitary and disperse over the plant body.

The pupa is typically cylindrical with a thick integument; the cremaster is usually well developed and often is of a diagnostic appearance. Pupation is often inside an earthen cell in the soil, but may occur in plant debris inside a flimsy, silken cocoon. In temperate regions some species have pupae that occasionally overwinter twice inside the subterranean cocoon. There is no general larval silk production recorded for the Noctuidae, as distinct from the Pyralidae where it is the rule, but a few species do produce a little silk for cocoon formation.

A recent publication on the group by Heath & Emmet (1979) lists 14 separate subfamilies, but most are established on somewhat esoteric grounds; whereas adult noctuids are generally easily recognized as such, with few exceptions they are difficult to place in their respective subfamilies. The Ophiderinae have adults with a short stout proboscis used for piercing fruits (fruit-piercing moths) and their larvae are "semiloopers". As mentioned above, the larvae of Plusiinae (also Ophiderinae, Catocalinae, and Hypeninae) have only three or two pairs of abdominal prolegs and are called "semiloopers" for the way in which they locomote. For a text such as this it does not seem worthwhile considering the different subfamilies separately, except possibly the Plusiinae as the larvae are recognizable and virtually all are simple leafworms.

From an agricultural point of view, it is useful to consider the Noctuidae according to the behavior of the larvae and the nature of the damage inflicted on the plant body. However it should be stressed, both here and preceding the list of noctuid pest species, that the literature is in a state of great confusion regarding names of Noctuidae. The widespread and polyphagous pests generally have long lists of recorded synonyms; the generic limits have continually been reappraised over the years, so some species have previously been placed in several different genera. Also there is a tendency for taxonomists in different parts of the world to hold differing views as to various generic definitions. In addition there will also be the usual misidentifications published regionally to bedevil the general overview. Thus the lists produced below will not satisfy all readers, as this is almost impossible, but hopefully most of the species will be recognizable. Field entomologists should perhaps be reminded that the taxonomists do now rely heavily on microscopic examination of genitalia preparations in the identifications and systematic studies on the Lepidoptera.

In the present state of noctuid taxonomy it appears that there is a large number of "small" genera (i.e., genera each containing only a few species) with the exception of the "large" genera *Achaea, Agrotis, Euxoa, Heliothis* and *Xestia*, etc. The important pest species are more or less equally distributed between the large and the small genera.

Viewing the Noctuidae according to behavior of the larvae and the type of damage, and sometimes including the adult moths, the following categories are obvious. Generalizations are difficult, however, in that different species of the same genus may have quite different habits, both as larvae and adults.

Leafworms. Most larvae live on the foliage of plants and eat the leaf lamina, and they may generally be termed "leafworms". This includes all the "semiloopers" (*Plusia, Trichoplusia, Achaea, Autographa, Anomis, Othreis,* etc.), and many of the species that occur in the following genera—*Acronicta, Aedia, Agrotis* (some), *Alabama, Ceramica, Euxoa,* some *Heliothis, Mamestra, Melanchra, Mythimna, Orthosia, Xanthodes,* etc.

Cereal stalk borers. These larvae bore in the stems of sugarcane, maize, sorghum, rice, other cereals, and various grasses. The main tropical species are *Busseola fusca* and

Sesamia spp.; *Luperina* spp. on more temperate cereals and grasses, with *Mesapamea secalis* and *Apamaea* spp.

Stem borers. In more temperate conditions, the species of *Hydraecia* bore in the stems of potato, hop (vines), sugar beet and other crops, and the leaf petioles of rhubarb in the Holarctic region. Confined to North America, several species of *Papainema* bore into the stems of potato, eggplant, tomato, maize, rhubarb, tobacco, capsicums, cotton and various flowers. Several other species are also recorded boring the stems of maize and the temperate cereals.

Budworms. The larvae of a few species of *Heliothis* and *Anomis* show a tendency to gnaw deep holes into buds on some plants, in addition to leaf eating, and they could be called "budworms".

Bollworms. Because of the importance of cotton as a world crop, and the many caterpillars that bore into developing bolls, it is customary to regard these larvae (Gelechiidae, Pyralidae, Noctuidae) as "cotton bollworms". The main groups of Noctuidae involved are several species of *Earias,* two of *Diparopsis* and several of *Helicoverpa.* Most of these species (except *Diparopsis*) also infest other Malvaceae (okra, *Hibiscus,* etc.).

Fruitworms. Clearly cotton bolls are botanically classed as fruits, but in this category are placed commercial fruits (peach, citrus, etc.), legume pods, seed capsules of flax and kapok, and the like. *Mudaria* spp. are kapok pod borers, *Hadena* spp. bore in the capsule of carnation and other Caryophyllaceae, and *Helicoverpa* caterpillars bore into various soft fruits, legume pods, tomatoes, sweet peppers, maize cobs, etc. *Tiracola plagiata* larvae eat the surface of banana fruits. All of these caterpillars are rather large in size; if the fruit or pod is large, then the caterpillars may live completely inside, but on smaller fruits a hole is made through which the caterpillar pushes its head and thorax in order to eat the seeds and fruit contents. *Helicoverpa* larvae will also graze on maize cobs and may destroy the entire "fruit"; these species and some *Mythimna,* etc., will feed on the head (panicle) of sorghum, rice, and other cereals, and may destroy all the grain.

Armyworms. Some species that are basically leaf eaters show gregarious behavior coupled with occasional population outbreaks (irruptions) and typically the dense mass of caterpillars defoliate an area or crop and may then march gregariously searching for a new feed supply. Sometimes there are distinct coloration changes when the population is truly gregarious, and these are the real armyworms. Crop damage by armyworm swarms can be devastating. A number of species show this swarming and irruption tendency to varying degrees; some are really just gregarious caterpillars feeding collectively but the others are truly armyworms; the most notable belong to the genera *Spodoptera* and *Mythimna. S. exempta* is the notorious African armyworm, generally showing a preference for Gramineae. *S. exigua* is the lesser or beet armyworm of Asia and the U.S.A., and *S. frugiperda* is the polyphagous fall armyworm of the U.S.A. *Mythimna loreyi* is the cereal armyworm and *M. unipuncta* the American or rice armyworm.

Cutworms. These species have larvae, usually large in size, that tend to live in the surface litter or soil and emerge at night to attack seedlings. Typical damage is to cut through the stems of seedlings, often several plants in a row, and just to eat part of each plant. Seedling destruction can be very serious, particularly for transplanted crops such as tobacco and cabbages. The caterpillars are essentially nocturnal and at night may do a certain amount of aerial foliage damage. The main genera concerned worldwide are *Agrotis* and *Euxoa,* which are very closely related; many pest species have been placed in both genera at different times. Additionally, there are some species of *Feltia, Xestia,* and also the spectacular and widespread *Noctua pronuba.*

Rootworms. Some of the cutworm species listed above spend more time underground than others. Under different weather conditions they may also stay underground more, and then the cutworms will feed on the roots of plants and will also tunnel into potato tubers, hollow out the root base of lettuce, etc. Whether these basic behavioral differences reflect different species or a response to different conditions is not too clear.

Coccid predators. Some species of the genera *Autoba, Catoblemma,* and *Eublemma* have larvae that are carnivorous (sometimes facultatively so) and they feed on Coccoidea (*Coccus, Saissetia, Tachardina,* etc.) throughout India and S.E. Asia. In India, some species (especially *E. amabilis*) can be regarded as pests as they prey on the lac insect that is reared as a commercial source of shellac.

Fruit piercers. These are adult moths that pierce ripening fruits to suck the sap. They are important fruit pests in some areas and their feeding leaves a puncture that often becomes infected by bacteria or fungi causing rots. Damaged fruits invariably become infected and then fall prematurely. Most of these species belong to the sub-family Ophiderinae. The group is best represented in West Africa where 20–30 genera of fruit piercers are known, and a wide range of crops are damaged.

Possibly the genus *Achaea* is best known, found as quite a large number of species throughout Africa, tropical Asia, and Australasia. The larvae are "semiloopers" with three pairs of prolegs and some are pests on cocoa, castor, *Acacia,* and many other forest trees; most of the larvae feed on the leaves of forest trees. The larger genera of fruit piercing moths include the following: *Aburina, Achaea, Anomis, Calyptra, Cyligramma, Dermaleipa, Ophiusa, Othreis,* and *Serrodes.* Only some of the species in these genera are fruit piercers, with one or two exceptions. The most northerly records for fruit-piercing moths are probably the *Calyptra* spp. in Japan.

"Eye" moths and bloodsucking moths. The adult moths of a few species of Noctuidae are known to visit the eyes of humans and other large mammals to suck lachrymal secretions. This occurs in the region from India through S.E. Asia to Thailand; the hosts including humans, domestic cattle, buffalo, wild cattle, deer, pigs, horses, donkeys, and tapir. Species of Noctuidae recorded by Banziger (1968) include *Lobocraspis griseifusa, Calyptra minuticornis, C. eustrigata, Mocis frugalis, Arcyophora both-rophora* and *Arcyophora* spp., and *Hypena* spp.

Even more remarkable is the fact that adult moths of *Calyptra eustrigata* in S.E. Asia will actually pierce the skin of buffalo, tapir, cattle and deer to suck the blood.

A number of Noctuidae are strongly migratory and regularly fly from warmer regions to the more temperate parts of Europe, Asia, and North America (Canada especially), where they may have one or two summer generations before being killed by the winter cold. Some of the most important migratory species include *Heliocoverpa armigera, Autographa gamma, Phlogophora meticulosa, Spodoptera exigua, Noctua pronuba, Mythimna unipuncta* and *Trichoplusia ni.*

Some of the more important pest species of Noctuidae are listed below:

Aburina spp. (fruit-piercing moths)—West Africa.

Achaea janata (castor semilooper)—Africa, India, S.E. Asia, Australasia.

Achaea spp. (fruit-piercing moths)—larvae feed on leaves of *Acacia, Euphorbia,* etc.; adults pierce fruits of many kinds; Africa, India, S.E. Asia.

Acronicta spp. (apple/cherry dagger moths)—pests of fruit trees; Japan (a large Holarctic genus with many species).

Aedia leucomelas (sweet potato leafworm)—on sweet potato; Japan.

Agrotis exclamationis (heart and dart moth)—polyphagous cutworm; Europe (Fig. 12.102).

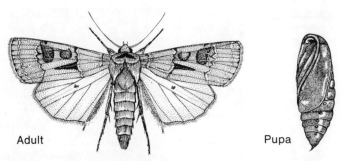

Fig. 12.102 Heart and dart moth (*Agrotis exclamationis*), a common cutworm (Lepidoptera; Noctuidae) and pupa (a typical noctuid pupa); Skegness, U.K.

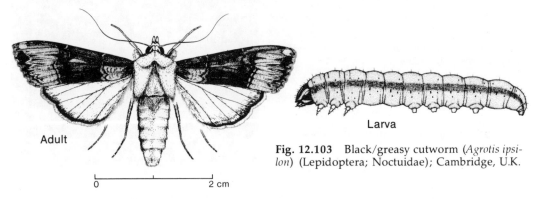

Fig. 12.103 Black/greasy cutworm (*Agrotis ipsilon*) (Lepidoptera; Noctuidae); Cambridge, U.K.

Agrotis ipsilon (black/greasy cutworm)—polyphagous cutworm; cosmopolitan (C.I.E. Map No. A. 261) (Fig. 12.103).

Agrotis munda (cutworm)—polyphagous cutworm; Australia.

Agrotis repleta (armyworm/cutworm)—a polyphagous species; Trinidad.

Agrotis segetum (common cutworm/turnip moth)—polyphagous cutworm; cosmopolitan throughout the Old World (Fig. 12.104).

Agrotis spp. (10–20) (cutworms)—many species, some polyphagous, in all parts of the world, but most abundant in the Holarctic region.

Alabama argillacea (cotton leafworm)—on cotton; U.S.A., Central and South America.

Anomis spp. (fruit piercers, etc.)—larvae eat foliage of citrus, kapok, cola, cocoa, etc.; adults pierce ripening fruits of many types; Africa and tropical Asia.

Anomis (= *Cosmophila*) *erosa* (cotton semilooper)—on cotton, etc.; U.S.A., Central and South America.

Anomis flava (cotton semilooper)—on cotton and other Malvaceae; Africa, Asia to Japan, Australasia (C.I.E. Map No. A. 379) (Fig. 12.105).

Anticarsia gemmatalis (velvetbean caterpillar—on legumes; tropical U.S.A., West Indies.

Anticarsia irrorata (bean caterpillar)—larvae on *Phaseolus* beans; adults pierce fruits; West Africa, Asia, Australasia and Pacific Islands.

Anua spp. (fruit-piercing moths, etc.)—larvae on *Combretum*; adults pierce fruits; West Africa.

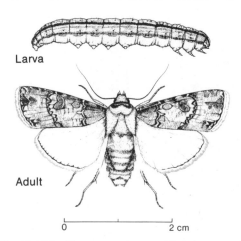

Larva

Adult

0   2 cm

Fig. 12.104 Common cutworm (*Agrotis segetum*) (Lepidoptera; Noctuidae).

Fig. 12.105 Cotton semilooper (*Anomis flava*) feeding on flower of *Hibiscus;* body length 30 mm (1.2 in.).

Apamea spp. (grass root moths)—larvae feed on roots and stem bases of Gramineae; Europe and Asia (a large genus found throughout the Holarctic region).

Argyrogramma signata (tobacco semilooper)—polyphagous; India, S.E. Asia, Australasia.

Autographa californica (alfalfa semilooper)—on alfalfa and legumes; U.S.A.

Autographa gamma (silver-Y moth)—larvae polyphagous semiloopers (adults migratory); Europe, Asia, North Africa (Fig. 12.106). (*Autographa* is a large genus, mostly Holarctic but most species occur in North America.)

Busseola fusca (maize stalk borer)—larvae bore stalks of maize and sorghum; tropical Africa (Fig. 12.107).

Callopistria spp. (fern caterpillars)—larvae feed on ferns; U.S.A. and pantropical.

Calyptra spp. ("eye" moths/fruit-piercing moths, etc.)—adults pierce fruits; some visit eyes; some bloodsucking; Africa, India, S.E. Asia, Japan.

Carea spp.—larvae with swollen thorax and resembling bird droppings, on *Eugenia* and *Eucalyptus;* S.E. Asia to China.

Ceramica pisi (broom moth)—larvae polyphagous on many plants; Europe, Asia (Fig. 12.108).

Ceramica picta (zebra caterpillar)—polyphagous; North America.

Cerapteryx graminis (antler moth; cereal armyworm)—on cereals and Gramineae; Europe, and western Asia (a small genus with only two Palaearctic species) (Fig. 12.109).

Fig. 12.106 Silver-Y moth (*Autographa gamma*); Hong Kong (Lepidoptera; Noctuidae).

Fig. 12.107 Maize stem with hole bored by *Busseola fusca;* Alemaya, Ethiopia (Lepidoptera; Noctuidae).

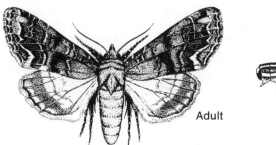

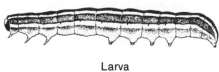

Adult   Larva

Fig. 12.108 Broom moth (*Ceramica pisi*) (Lepidoptera; Noctuidae) adult and caterpillar; ex Cambridge, U.K.

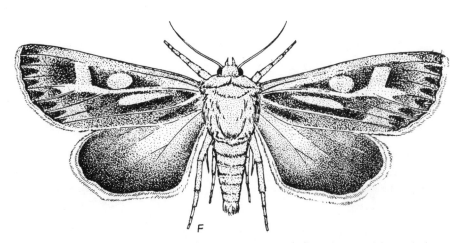

Fig. 12.109 Cereal armyworm (*Cerapteryx graminis*) (Lepidoptera; Noctuidae).

Chrysodeixis chalcites (golden twin-spot; green garden looper)—polyphagous on dicotyledons; Old World tropics and subtropics, now in the U.S.A.

Cucullia spp. (sharks)—a large genus mostly Mediterranean and Near East; many larvae feed on Compositae, and some recorded as pests of lettuce.

Diparopsis castanea (southern red bollworm)—monophagous on *Gossypium* (cotton); Africa south of the Equator.

Diparopsis watersi (northern red bollworm)—on cotton; Africa north of the Equator.

Earias biplaga (spiny bollworm)—on cotton and other Malvaceae, cocoa, etc.; Africa (Fig. 12.110).

Earias insulana (spiny bollworm)—on cotton and Malvaceae; Mediterranean, Africa, Asia (C.I.E. Map No. A. 251) (Fig. 12.110).

Earias vittella (spotted bollworm)—on cotton; S.E. Asia, Australasia (C.I.E. Map No. A. 282).

Earias spp. (4+) (spiny/spotted bollworms)—also on cotton in the Old World tropics, as far north as Japan. (*Earias* is a large genus, well represented in the Old World tropics and subtropics, with a few species polyphagous.)

Eublemma spp.—larvae prey on Coccoidea; other species feed on foliage of Solanaceae, and flowers of Compositae, etc.; Africa and India. (A large cosmopolitan genus, but mostly tropical and subtropical.)

Euxoa nigricans (garden dart)—a polyphagous cutworm; Europe, western Asia. (Fig. 12.111).

Euxoa tritici (white-line dart)—on cereals; Europe and western Asia.

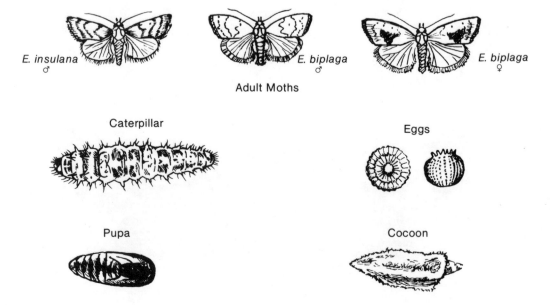

Fig. 12.110   Spiny bollworms (*Earias biplaga* and *E. insulana*) (Lepidoptera; Noctuidae).

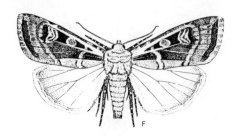

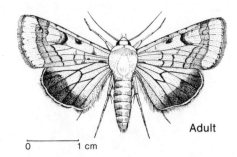

Fig. 12.111 Garden dart moth (*Euxoa nigricans*) (Lepidoptera; Noctuidae).

Fig. 12.112 African/American bollworm (*Helicoverpa armigera*) (Lepidoptera; Noctuidae).

Euxoa spp. (cutworms)—a large genus, well represented in cooler regions but worldwide; 12 species recorded as cutworm pests in U.S.A. and Canada.

Feltia spp. (cutworms)—polyphagous larvae; U.S.A. and Canada.

Gortyna—now regarded as a very small genus, of little importance agriculturally; Palaearctic region.

Hadena spp.—larvae feed in seed capsules of Caryophyllaceae (carnation, etc.); Europe, Africa, Asia. (A well-defined genus with many Holarctic species.)

Helicoverpa armigera (African/American bollworm, etc.)—polyphagous throughout the Old World tropics and subtropics (C.I.E. Map No. A. 15) (Fig. 12.112).

Helicoverpa assulta (Cape gooseberry/oriental tobacco budworm)—on Solanaceae, etc.; Old World tropics.

Helicoverpa punctigera (native budworm)—polyphagous; Australia.

Helicoverpa zea (corn earworm, etc.)—polyphagous on a wide range of crops; U.S.A., Central and South America (C.I.E. Map No. A. 239).

Heliothis—a large genus with many species (80+); some are polyphagous, and some are minor pests. The genus is quite cosmopolitan in distribution. Some 17 species have recently been placed in the new genus *Helicoverpa* including the four important pest species listed above.

Heliothis ononis (flax bollworm)—on flax and other crops; North America.

Heliothis virescens (tobacco budworm)—pest of tobacco, etc.; North, Central, and South America.

Heliothis viriplaca (flax budworm)—on flax, etc.; Japan.

Hydraecia micacea (rosy rustic moth; potato stem borer)—polyphagous stem and root borer; Europe, Asia, and North America. (The genus contains a small number of species of stem borers in the Holarctic region.) (Fig. 12.113).

Hypena humuli (hop semilooper)—larvae eat leaves of hop, etc.; North America. *Hypena* is a very large genus found in Africa and throughout both Holarctic and Oriental regions.

Lacanobia—a diverse genus with numerous species, some Palaearctic and some Holarctic.

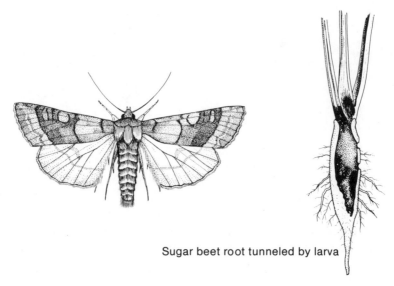

Sugar beet root tunneled by larva

Fig. 12.113 Rosy rustic moth (*Hydraecia micacea*) (Lepidoptera; Noctuidae) and tunneled sugar beet plant; ex Cambridge, U.K.

Larva feeding on lettuce

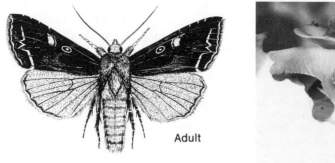

Adult

Fig. 12.114 Tomato moth (*Lacanobia oleracea*) (Lepidoptera; Noctuidae) adult, and caterpillar feeding on lettuce; Skegness, U.K.

Fig. 12.115 Cabbage moth (*Mamestra brassicae*) (Lepidoptera; Noctuidae).

Lacanobia legitima (striped garden caterpillar)—polyphagous; U.S.A.

Lacanobia oleracea (tomato moth)—larvae are polyphagous; Europe, Asia (Fig. 12.114).

Luperina testacea (flounced rustic)—larvae bore in cereal stems and roots; Europe and western Asia; the genus contains many species in the Holarctic region.

Mamestra—a small genus, most found in northern temperate regions.

Mamestra brassicae (cabbage moth)—larvae polyphagous on *Brassica* crops and many others; throughout Europe and Asia (Fig. 12.115).

Mamestra configurata (Bertha armyworm)—polyphagous armyworm species; North America.

Melanchra—a small temperate genus closely related to *Mamestra*.

Melanchra persicariae (dot moth; beet caterpillar)—caterpillars polyphagous; Europe, Asia to China and Japan (called beet caterpillar in Japan) (Fig. 12.116).

Melanchra picta (zebra caterpillar)—also polyphagous in North America.

Mesapamea secalis (common rustic moth)—larvae bore stems of cereals and grasses; Europe, and western Asia.

Mocis spp.—on the foliage of sugarcane, other Gramineae, legumes; Africa, Asia, Indonesia, Australasia, Central and South America.

Mudaria spp. (kapok pod borers)—larvae bore pods of kapok and durian fruits; India, Indonesia, Malaysia.

Mythimna—a large genus with many Holarctic species (Fig. 12.117).

Mythimna loreyi (cereal armyworm)—polyphagous on Gramineae; Africa, Mediterranean, tropical Asia (C.I.E. Map No. A. 275) (Fig. 12.117).

Mythimna separata (oriental armyworm, etc.)—polyphagous on many crops; S.E. Asia, Australasia (C.I.E. Map No. A. 230) (Fig. 12.117).

Larva feeding on geranium foliage

Adult

Fig. 12.116 Dot moth (*Melanchra persicariae*) (Lepidoptera; Noctuidae) adult, and caterpillar feeding on geranium foliage; Skegness, U.K.

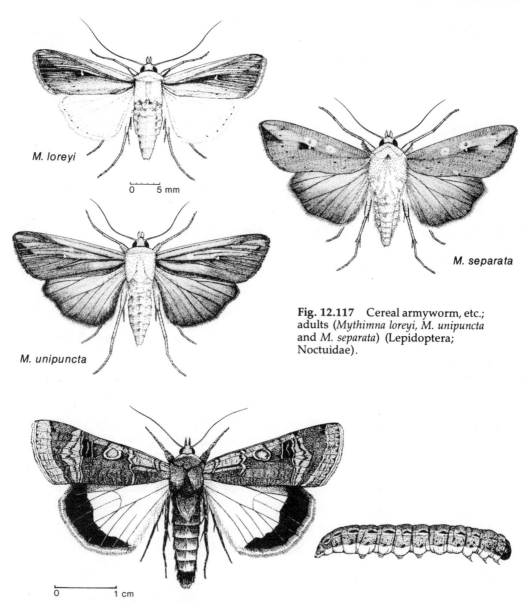

M. loreyi

0   5 mm

M. unipuncta

M. separata

Fig. 12.117   Cereal armyworm, etc.;
adults (*Mythimna loreyi, M. unipuncta*
and *M. separata*) (Lepidoptera;
Noctuidae).

0   1 cm

Fig. 12.118   Large yellow underwing (*Noctua pronuba*) (Lepidoptera; Noctuidae) adult
and larva; Cambridge, U.K.

Fig. 12.119   Fruit-piercing moth (*Othreis fullonia*); wingspan 100 mm (4 in.); Hong Kong
(Lepidoptera; Noctuidae).

Mythimna unipuncta (rice/American armyworm)—polyphagous on cereals and other crops; Europe, parts of Africa, western Asia, U.S.A., Central and South America (C.I.E. Map No. A. 231) (Fig. 12.117).

Noctua pronuba (large yellow underwing)—a polyphagous cutworm; Palaearctic, including North Africa and the Near East (a small Holarctic genus) (Fig. 12.118).

Ophiusa tirhaca (fruit-piercing moth)—adult polyphagous on fruits; S.E. Asia.

Orthosia spp. (drab moths, etc.)—on temperate fruit and forest trees; Europe, Asia (a large Palaearctic genus).

Othreis fullonia (fruit-piercing moth)—adults pierce citrus and other fruits; Africa, tropical Asia, Australasia (C.I.E. Map No. A. 377) (Fig. 12.119).

Othreis spp. (fruit-piercing moths)—larvae feed on Menispermaceae (lianas); adults pierce ripening fruits of many species; Africa and tropical Asia.

Papaipema spp. (stalk borers)—larvae bore stalks of many plants (potato, tobacco, maize, eggplant, rhubarb, tomato, capsicums, flowers, etc.); U.S.A. and Canada.

Penicillaria/Bombotelia spp.—larvae on mango and other trees; Australasia.

Phlogophora meticulosa (angleshades moth)—larvae are polyphagous leaf eaters on many crops; Europe, Mediterranean Region, western Asia; a migrant species going as far north as Scandinavia and Iceland. (A large genus confined to the Holarctic region.)

Plusia festuca (gold-spot moth; rice semilooper)—larvae feed on Gramineae and other monocotyledenous plants in Europe, and on rice in Japan. Many Plusiinae were formerly placed in the genus *Plusia*, which was then very large, but the present interpretation of *Plusia* is that it is a small Holarctic genus.

Rivula spp.—larvae on bamboos and legumes; Japan.

Serrodes spp. (fruit-piercing moths)—a large genus found in Africa and Asia.

Sesamia botanephaga (maize stalk borer)—on maize, etc.; West Africa.

Sesamia calamistis (pink stalk borer)—larvae bore stalks of maize, other cereals, and large grasses; tropical Africa (C.I.E. Map No. A. 414) (Fig. 12.120).

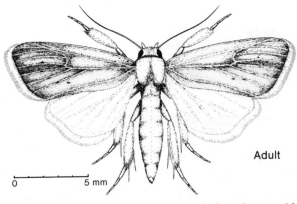

Adult

0   5 mm

Fig. 12.120   Pink stalk borer (*Sesamia calamistis*) (Lepidoptera; Noctuidae).

Sesamia cretica (sorghum borer)—larvae bore sorghum, etc.; Mediterranean, tropical Africa.

Sesamia inferens (purple stem borer)—on cereals and grasses; India, S.E. Asia, tropical Australasia (C.I.E. Map No. A. 237).

Sesamia spp. (stem borers)—several other species recorded boring stems of cereal crops and grasses; throughout Africa.

Spodoptera—this is a large genus, found throughout the tropics and subtropics; several species are strongly migratory; some species show a preference for Gramineae, others not so.

Spodoptera eridania (southern armyworm)—polyphagous pest in southern U.S.A.

Spodoptera exempta (African armyworm)—sporadically serious pest on Gramineae; Africa, parts of Asia, and Australasia (C.I.E. Map No. A. 53) (Figs. 12.121, 12.122).

Spodoptera exigua (lesser/beet armyworm)—polyphagous; southern Europe, Africa, Asia, U.S.A. (C.I.E. Map No. A. 302) (Fig. 12.123).

Spodoptera frugiperda (fall armyworm)—polyphagous; U.S.A. and South America.

Spodoptera littoralis (cotton leafworm)—polyphagous; Africa and Mediterranean Region (C.I.E. Map No. A. 232).

Spodoptera litura (fall armyworm)—polyphagous and gregarious pest; India, Asia, and Australasia (C.I.E. Map No. A. 61) (Fig. 12.124).

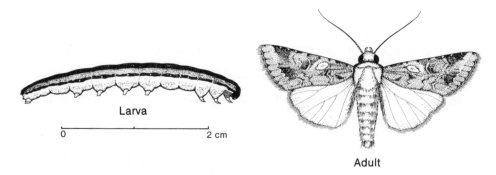

Larva

0   2 cm

Adult

Fig. 12.121   African armyworm (*Spodoptera exempta*) (Lepidoptera; Noctuidae).

Fig. 12.122   African armyworm (*Spodoptera exempta*) (Lepidoptera; Noctuidae) caterpillar found on grass; Alemaya, Ethiopia.

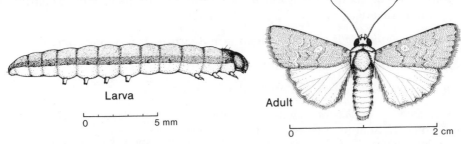

Fig. 12.123 Lesser/beet armyworm (*Spodoptera exigua*) (Lepidoptera; Noctuidae).

Fig. 12.124 Fall armyworm (*Spodoptera litura*); adult, wingspan 35 mm (1.4 in.); Hong Kong (Lepidoptera; Noctuidae); and caterpillar, body length 40 mm (1.6 in.); Hong Kong.

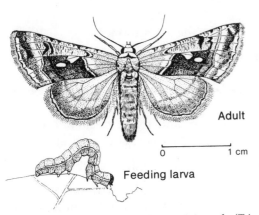

Fig. 12.125 Cabbage semilooper/ni moth (*Trichoplusia ni*) (Lepidoptera; Noctuidae; Plusiinae).

Fig. 12.126 Cotton/semilooper leafworm (*Xanthodes* sp.); Malawi (Lepidoptera; Noctuidae; Plusiinae).

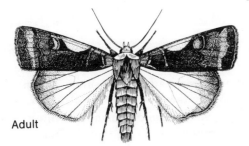

Fig. 12.127 Spotted cutworm (*Xestia c-nigrum*) (Lepidoptera; Noctuidae) moth.

Spodoptera mauritia (paddy armyworm)—pest of Gramineae; parts of Africa, Asia, and Australasia (C.I.E. Map No. A. 162).

Spodoptera ornithogalli (yellow-striped armyworm)—U.S.A. and West Indies.

Spodoptera praefica (western yellow-striped armyworm)—U.S.A.

Tarache spp.—on cotton and other Malvaceae; Africa.

Thysanoplusia orichalcea (slender burnished brass)—larvae polyphagous on dicotyledonous plants; southern Europe, Africa, and tropical Asia. (A Palaearctic and tropical genus with only a small number of species.)

Tiracola plagiata (banana fruit-scarring caterpillar)—larvae damage banana fruits; India, S.E. Asia, and Australasia.

Trichoplusia ni (cabbage semilooper; ni moth)—polyphagous on vegetable and other crops; quite cosmopolitan (except Australia); this genus only has the one species (C.I.E. Map No. A. 328) (Fig. 12.125).

Xanthodes spp. (cotton/semilooper leafworms)—on cotton and other Malvaceae; Africa, Asia to Korea and Japan, Australasia (Fig. 12.126).

Xestia—a large genus found worldwide; many species are polyphagous, some are pests.

Xestia c-nigrum (spotted cutworm)—polyphagous cutworm; Europe, North Africa, Asia, U.S.A., and Canada (Fig. 12.127).

Xylena fumosa (rape caterpillar)—on rape, etc.; Japan. A genus of large noctuids (wingspan of 60 mm/2.4 in. or more) mostly in the Palaearctic region; larvae are large and very polyphagous, and feed both by day and night.

Nolidae (sometimes regarded as a subfamily in the Noctuidae)   *100 spp.*

A small family of worldwide distribution, most to be found in the Old World tropics. The moths resemble grey noctuids, but are generally smaller in size and have tufts of raised scales on the forewing. A few are recorded as pests of cultivated plants.

Meganola spp.—larvae mostly on foliage of oaks and forest trees, but also on Rosaceae; Europe and Asia Minor to North America; also tropical.

Nola sorghiella (sorghum webworm)—a major pest of sorghum in the U.S.A.

Nola spp.—recorded from a range of plants, apple, etc., in Europe and U.S.A., to various ornamentals in Africa; 81 species of *Nola* recorded from Australia; diverse in Asia.

Uraba spp.—larvae defoliate *Eucalyptus* in Australia.

Lymantriidae (tussock moths, etc.)   *2500 spp.*

A group of moderately-sized moths, without a proboscis, without ocelli; usually drab in color; some are white; and the female typically has a thick anal tuft of scale hairs which can be detached and used to cover the egg mass after oviposition. The male has antennae prominently bipectinate to the apex. A few species have the female with reduced wings or even wingless (*Orgyia* spp.); sexual dimorphism is common in this group. The family reaches its greatest development in the Old World tropics.

The larvae are stout-bodied and hairy caterpillars, often densely hairy, and often bearing a few thick dorsal tufts on certain body segments. Some species (especially *Euproctis*) have urticating bristles with barbed spicules, and they cause intense irritation to human skin. Some larvae are very brightly colored. They are general leaf eaters and may cause defoliation; some are quite polyphagous. They feed on trees, both deciduous and evergreen, both forest and fruit trees, and also on some shrubs.

Eggs are laid in groups on the tree trunks or branches, or occasionally on the ground in leaf litter, usually near the female cocoon. In cooler regions, overwintering usually takes place in the egg stage. The egg mass is protected by a layer of bristles.

The larvae are active and spread over the host tree rapidly and may move from tree to tree. Because of the numbers of larvae from each egg mass (100–300) defoliation of the host tree is usual. The caterpillars are active during daylight and very conspicuous in the foliage (Fig. 12.128) and presumably safe from predation.

Pupation takes place in the foliage, often in a loose cocoon of silken threads (Fig. 12.132), and the pupa is characterized by having setae arranged around the scars of the larval verrucae. In some species pupation takes place under loose bark. The silken cocoon incorporates larval setae, and so should not be handled.

Probably the best-known species is *Lymantria dispar* (gypsy moth) of Europe, a serious forest pest on shade and ornamental trees, whose abundant larvae can be dispersed in the first larval instar by air currents and wind. This species occurs naturally throughout the Palaearctic region and was introduced into North America in about 1868 (together with the brown-tail moth) where it has become a very serious pest of forest trees, as well as on shade and ornamentals.

The more important pest species of Lymantriidae are as follows:

Dasychira—a genus with more than 400 species recorded in the Old World; many are polyphagous on a wide range of crops and cultivated plants.

Dasychira basalis—polyphagous on onions, pigeon pea, coffee, cotton, cassava; East Africa.

Dasychira inclusa—polyphagous on cocoa, coffee, citrus, *Ficus,* legumes, etc.; Indonesia.

Dasychira mendosa—polyphagous on fruit trees and shrubs; India, S.E. Asia to Australia.

Euproctis—600 species known in the Old World, a large proportion of which are recorded as pests of cultivated plants; many species are polyphagous.

Euproctis chrysorrhoea (brown-tail moth)—polyphagous on trees and shrubs; Europe, Asia, now in the U.S.A. (C.I.E. Map No. A. 362).

Euproctis flava—on apple, grapevine, mango, pear, plum; India.

Euproctis flexuosa (cinchona tussock moth)—on cinchona plantations in the mountains of Indonesia (Java, etc.).

Euproctis fraterna—polyphagous on castor, cotton, pomegranate, peach, mango, roses, pear, pigeon pea, etc.; India and China.

Euproctis lunata—on apple, grapevine, mango, mulberry, plum, pomegranate; India.

Euproctis oreosaura—on cinchona plantations in lowland areas of Indonesia; also in India and S.E. Asia.

Euproctis producta—polyphagous on pigeon pea, coffee, cotton, citrus, banana, castor, maize, pomegranate, etc.; East Africa.

Euproctis pseudoconspersa (tea tussock moth)—on tea foliage; Japan.

Euproctis (= *Porthesia*) *scintillans*—polyphagous on mango, castor, rose, apple, flax, legumes, jute, etc.; India, S.E. Asia to China.

Euproctis similis (yellow-tail moth)—polyphagous on trees and shrubs; Europe and Asia (C.I.E. Map No. A. 388) (Fig. 12.128).

Euproctis virguncula (tropical yellow-tail moth)—polyphagous on Gramineae, Leguminosae, Malvaceae, and on rubber; India and S.E. Asia.

Lacipa spp.—on a wide range of plants (cotton, millets, *Hibiscus*, etc.); East Africa.

Laelia spp.—larvae feed on Gramineae (rice, sugarcane, millets, etc.), Cyperaceae, and some recorded on *Acacia*; Africa, S.E. Asia to Japan.

Lymantria—150 species recorded from the Old World, and several are recorded as pests of cultivated plants; most are on forest trees.

Lymantria dispar (gypsy moth)—polyphagous on deciduous trees; Europe, Asia, now North America (C.I.E. Map No. A. 26).

Lymantria lapidicola (almond tussock moth)—on a wide range of stone and pome fruits; Asia Minor.

Lymantria monacha (nun moth)—larvae polyphagous on trees; Europe and Asia (C.I.E. Map No. A. 60).

Orgyia—about 60 species recorded mostly in the Holarctic region, and several others are recorded as polyphagous pests of cultivated plants.

Orgyia antiqua (vapourer moth)—polyphagous on trees, shrubs and a few herbaceous plants; Europe, Asia, U.S.A. (Fig. 12.129).

Orgyia postica (oriental tussock moth)—polyphagous on a wide range of cultivated plants (cocoa, coffee, cinchona, tea, cinnamon, pulses, orchids, castor, rubber, citrus, guava, roses, and other flowers), and several conifer trees; India, S.E. Asia to China.

Perina nuda (fig tussock moth)—larvae eat foliage of *Ficus;* India, S.E. Asia to China (Figs. 12.130–132).

Porthesia scintillans (= *Euproctis scintillans*)

Psalis pennatula (hairy rice caterpillar)—larvae found on Gramineae (rice, sugarcane, etc.); Indonesia and Oriental tropics.

Fig. 12.128   Yellow-tail moth caterpillar (*Euproctis similis*) (Lepidoptera; Lymantriidae) on leaf of *Rosa;* Gibraltar Point, U.K.

Adult ♂   Adult ♀

Fig. 12.129 Vapourer moth (*Orygia antiqua*) male and female (Lepidoptera; Lyman-triidae); Cambridge, U.K.

Fig. 12.130 Fig tussock moth (*Perina nuda*); adult female and male; wingspan 35–40 mm (1.4–1.6 in.); Hong Kong (Lepidoptera; Lymantriidae).

Fig. 12.131 Fig tussock caterpillar (*Perina nuda*) on leaf of *Ficus microcarpa;* body length 32 mm (1.28 in.); Hong Kong (Lepidoptera; Lymantriidae).

Fig. 12.132 Pupa of fig tussock moth (*Perina nuda*) on leaf of *Ficus microcarpa;* Hong Kong (Lepidoptera; Lymantriidae).

13

ORDER HYMENOPTERA
sawflies, ants, bees, wasps, etc.
60+: 100,000

This is a large group whose adults range in size from minute (0.2mm/0.008 in. long) to large. On the basis of their structure and behavior they are now regarded as the most highly evolved group of insects. They are anatomically very specialized and in some groups complex social behavior has developed with large societies living communally.

The adults are characterized by having two pairs of membranous wings, with the hind pair smaller and interlocked with the forewings by special hooks. Their mouthparts are basically the biting and chewing type, but are modified in some cases for lapping, and in the bees for sucking fluids (nectar, etc.). The first segment of the abdomen (propodeum) is fused with the metathorax, and in many cases the second (and third) segment is constricted and narrow. An ovipositor is always present, and is modified for sawing, piercing, or stinging. Metamorphosis is complete; the larva in many cases is apodous, and a head is usually more or less well developed. The pupa is adecticous, usually exarate, and mostly enclosed inside a cocoon.

In the social groups the female adult has become structurally changed so that there is polymorphism with caste formation and division of labor within the nest. In its most developed form, social life has evolved into long-lived colonies such as seen in some of the bees (Apidae): reproduction has become the purview of a single large queen, and nest building and foraging is carried out by the large force of usually sterile worker females, who are also responsible for the mass-provisioning of the nest.

Parthenogenesis is practiced in several groups in the order, probably the best known case is in the honeybee where unfertilized eggs become male bees (drones); this also occurs in the social wasps (*Vespa,* etc.).

The group is actually very diversified and overall generalization is difficult. The many species are probably best viewed at the family or superfamily level. A recent book with many illustrations and taxonomic details titled *The Hymenoptera* is now available (Gauld & Bolton, 1988).

SUBORDER SYMPHYTA (sawflies)

The sawflies are distinctive as a group in that the adult has the abdomen broadly attached to the thorax, and no abdominal constriction; the larvae are eruciform (caterpillars) and have 5–6 pairs of abdominal prolegs in addition to thoracic legs; they are phytophagous and eat plant foliage or bore into plant tissues. Many of the larvae are pests of forest trees and cultivated plants. Many larvae are spotted rather than having the more usual striping shown by the Lepidoptera. In some species the larvae are gregarious and feed collectively, causing localized defoliation. Many species are diurnal and can be seen feeding during daylight hours. There are no crochets on the abdominal prolegs. The female sawfly generally has a well-developed ovipositor which is adapted for sawing or boring into plant tissues, including wood. Thus eggs are laid inserted into the plant tissues. All sawflies except the Cephoidea have a pair of small raised bosses on the metanotum (called cenchri) which engage with a roughened patch on the underneath of the forewings to keep them in place at rest. Most adults are small or medium-sized, and many are black, blue, or black and yellow in color, with hyaline wings. There are many more species of sawflies associated with trees and bushes than there are on herbaceous hosts. Pupation takes place inside a woven cocoon either in litter or in the soil.

There is considerable diversity shown by the larvae; most are solitary (after the first instar) and are leaf eaters. The species living inside plant tissues generally have legs reduced, and a few leaf eaters are slug-like with tiny reduced legs and resembling bird droppings. The gregarious leaf eaters behave in a characteristic manner—they feed on the margin of the leaf lamina and when disturbed they stop feeding and elevate the abdomen (see Fig. 13.2). The group is worldwide in distribution but probably best represented in the North Temperate Zone (Holarctic region), where many species eat foliage and bore tree trunks in the deciduous and taiga conifer forests; a few species are even subarctic, eating conifer needles or boring grass stems.

The larval life-styles may be viewed under the following headings:

1. Leaf lamina eaters —solitary: most Tenthredinidae, Diprionidae, etc.
   —gregarious: *Croesus, Hemichroa,* some Diprionidae.

2. Leaf skeletonizers—*Caliroa, Endelomyia* (Caliroini).

3. Leaf rollers—*Blennocampa.*

4. Leaf miners—*Fenusa, Metallus, Schizocerella,* etc.

5. Leaf gall makers—*Pontania* on *Salix* leaves.

6. Shoot and stem borers—in Gramineae; some Cephidae.
   —in woody shrubs; some Cephidae, etc.

7. Fruit borers—*Hoplocampa* (Tenthredinidae).

8. Wood borers (tree trunks and branches)—deciduous trees: *Tremex* (Siricidae).
   —evergreens: many Siricidae.

In the past it was sometimes customary to regard most species as belonging to a single large family, the Tenthredinidae, but now a series of small families are clearly established. Generally the sawflies are, of course, regarded as the more primitive Hymenoptera, both on the basis of anatomy and their biology.

XYELOIDEA

Xyelidae

A single small family, with the most generalized wing venation, and characteristic antennae with a greatly elongate third segment followed by a flagellum; the ovipositor is long. The larvae have legs on all abdominal segments; and the larvae of some of the better known species feed in the staminate flowers of *Pinus*.

MEGALODONTOIDEA

Pamphiliidae (sometimes called "web-spinning" or "leaf-rolling sawflies")

Stout-bodied sawflies, with a short ovipositor, long, thin antennae, and a primitive wing venation. They are sun-lovers and fly rapidly. The larvae have no legs on the abdomen, and they often live gregariously in webs or in rolled leaves. They are mostly recorded in the Holarctic region. A few notable crop pests occur.

> *Acantholyda* spp.—on *Pinus, Picea, Larix, Abies*; eastern Asia and North America.

> *Neurotoma* spp. (web-spinning sawflies)—on foliage of fruit trees and other Rosaceae; Europe, Asia, U.S.A. and Canada.

> *Pamphilius* spp. (many) (blackberry sawflies, etc.)—on many crops; Asia and North America.

Megalodontidae

A small group whose adults have flabellate antennae, and whose larvae live gregariously in silk webs on herbaceous plants; an Asian group.

SIRICOIDEA (wood wasps)

Xiphydriidae   *100 spp.*

A small family, worldwide, but little studied. Host plant associations are not known except for a few Holarctic species whose larvae, with vestigial legs, bore in deciduous angiosperms (elms, beech, *Acer*, etc.).

Siricidae (wood wasps/ horntails)

These are large insects of conspicuous coloration, often black and yellow, or else metallic blue or black. The abdomen terminates in a spine or "horn", short and triangular in the male and elongate in the female. The female has a stout ovipositor projecting backwards and it operates on a boring and drilling mechanism (not sawing as in Tenthredinidae). Eggs are laid singly, deep into the sapwood of trees. The larvae burrow deep into the heartwood of the trees and may cause serious damage; they grow to a length of 5 cm (2 in.) or more and the tunnel is 5–7 mm (0.2–0.28 in.) wide, without any holes to the exterior. Pupation takes place inside a silken cocoon covered with frass inside the larval gallery. It appears that usually weakened or sickly trees are preferred for oviposition. Several species aid in the establishment of fungal pathogens in the tree hosts.

Several species are important pests, but mostly of forest trees rather than crop trees (fruit trees, etc.) in northern temperate forests, including:

> *Sirex cyaneus* (steel-blue wood wasp)—larvae in *Pinus* and spruce; Europe, Asia, North America.

Sirex noctilio (blue wood wasp)—this European species is now well established in Australia and Tasmania where it has caused serious damage to *Pinus radiata* and other important conifers.

Sirex spp. (blue wood wasps)—larvae bore in a wide range of conifers throughout Europe and Asia (from India to Japan) and North America.

Tremex spp.—larvae bore in a wide range of deciduous trees; Europe, Asia, North America.

Urocerus gigas (giant wood wasp)—larvae bore in *Pinus* and firs; Europe and western Asia.

Urocerus spp. (giant horntails)—larvae bore in pines and firs; Japan, U.S.A., and Canada.

ORUSSOIDEA

Orussidae (parasitic wood wasps)

So far as is known these larvae are ectoparasitic upon certain wood-boring beetles, especially Buprestidae. It is conjectured that larvae of *Orussus* also attack larval Siricidae. Most species are, however, tropical or Australian.

CEPHOIDEA

Cephidae (stem sawflies)

A small family of slender-bodied sawflies, mostly black in color, small in size, without cenchri; adults may be found taking nectar from flowers (clovers, etc.); they have fore-tibiae with a single, apical spur. The larvae are apodous and bore in the stems of Gramineae, roses, willows, currants, raspberry and other plants. Pupation takes place inside a cocoon in the stem of the host plant. A few species are pests of importance and are listed below:

Cephus cinctus (wheat stem sawfly)—larvae in stems of grasses, wheat, rye; U.S.A., Canada.

Cephus pygmeus (European wheat stem sawfly)—larvae bore stem under the ear of wheat, barley, rye; Palaearctic (including Near East), now also in Canada and the U.S.A.

Hartigia spp. (blackberry shoot sawfly, etc.)—larvae bore in blackberry, raspberry, rose, etc.; East Asia, U.S.A., and Canada.

Janus spp. (currant stem girdler, etc.)—larvae bore in pith of currants, various woody shrubs and some trees; East Asia, U.S.A.

Syrista similis (rose stem sawfly)—larvae bore stems of roses; Japan.

Trachelus tabidus (black grain stem sawfly)—in stems of wheat and rye; Palaearctic, now also in U.S.A. and Canada.

TENTHREDINOIDEA

A large group, mostly stout-bodied, cenchri present on mesothorax, and fore-tibiae with two apical spurs.

Argidae (rose sawflies, etc.)   *800 spp.*

A worldwide family with adults recognized by the third antennal segment being long and often bifid (Y-shaped). The larvae have 6–8 pairs of abdominal legs. At least one species is a leaf miner, and several species damage ornamental trees (including *Cupressus*) in Australia.

There are a few well-known pests in the genus *Arge;* they are mostly Holarctic in distribution.

> *Arge* spp. (apple/rose/birch sawflies, etc.)—larvae eat foliage of apple, birch, willow trees, and roses, azaleas, etc.; Europe, Asia to Japan, U.S.A., and Canada.

Blasticotomidae   *10 spp.*

A very small European family whose legless larvae bore into the stems of some ferns.

Cimbicidae

A small family of stout-bodied, quite large, sawflies, with strongly clubbed antennae. The larvae feed mostly on leaves of trees and have a characteristic appearance sitting partly curled up on the leaves; they are covered with a waxy powder.

A few pest species are known:

> *Cimbex* spp.—larvae on *Ulmus, Salix, Acer, Populus, Tilia, Alnus,* etc.; Europe, Asia, U.S.A. and Canada.

> *Palaecimbex carinulata* (pear cimbicid sawfly)—on foliage of pears; Japan.

Diprionidae (conifer sawflies)

These are medium-sized sawflies, females with serrate antennae, and males pectinate, with 13 or more segments; they are associated with conifers where the larvae feed externally on the foliage like caterpillars. As a group they are important forestry pests and large tracts of conifer forest may be defoliated in parts of Europe, Asia, and North America. Trees grown as ornamentals or for shade in the temperate regions are often damaged. In some species the larvae are gregarious and very conspicuous; many species are distinctively spotted. Most of the pests are in northern regions; several serious pests are European but have been accidentally introduced into North America over the period 1910–1930 and there have been spectacular population outbreaks with devastating defoliation.

> *Diprion* spp. (pine sawflies)—larvae feed almost entirely on *Pinus* foliage; Europe, Asia, and North America.

> *Gilpinia* spp. (spruce sawflies)—mostly on spruces; Europe, Asia, North America.

> *Monoctenus* spp. (cupressus sawflies)—on *Cupressus;* Europe, Asia, North America.

Neodiprion spp. (15+) (pine sawflies, etc.)—on *Pinus, Picea,* etc.; Europe, Asia, North America.

Pergidae

A large family of diverse habits, but best represented in Australia and South America. Adults have antennae often clubbed or serrate; third segment short; six segments only. Females of *Perga* brood over their young, which are legless and feed gregariously on leaves of *Eucalyptus.* A few pest species are recorded:

Acordulecera spp.—larvae feed on foliage of oaks and hickory; U.S.A. and Central America.

Perga spp. (eucalyptus sawflies)—larvae eat leaves of *Eucalyptus,* sometimes causing defoliation; Australasia.

Phylacteophaga spp. (leaf miners)—larvae make blotch leaf mines on *Eucalyptus* and *Tristania;* Australasia.

Tenthredinidae (typical sawflies)   *4000 spp.*

Adults with nine antennal segments (rarely one or two more), thorax with a definite postscutellum, and fore-tibiae with two apical spurs. The great majority of sawflies still belong to this family, and the group exhibits great diversity. In some species males are rare and parthenogenesis is common. Unfertilized eggs may give rise to either males or females, or even both in a few cases. Adult habits vary, most visit flowers for nectar, but a few are carnivorous and catch small flies and beetles.

Eggs are deposited into plant tissues using the sawlike ovipositor, some into woody shoots, some into soft shoots or leaves. The larvae resemble lepidopterous caterpillars (and are often called "caterpillars") but are mostly spotted; they always have more than the usual lepidopteran four pairs of prolegs, and these are without crochets. The larvae are all phytophagous and feed on almost all orders of flowering plants as well as some ferns. Generally more species are to be found feeding on foliage of trees and shrubs than on herbaceous plants. As already mentioned, there is great larval diversity shown, although the majority are leaf eaters, either solitary or gregarious. Pupation usually takes place inside an oval, silken cocoon in ground litter or in the soil inside an earthen cell.

Within this family is a series of quite well-defined subfamilies and tribes, but for the present purposes these lower taxa will not be used, except to mention that in the tribe Caliroini are the "slug sawflies" whose larvae are green, legless, and covered with black slime—they sit on leaves (pear, rose, etc.) and skeletonize the upper surface with their feeding. The Fenusini are all leaf miners.

There is a large number of recorded pest species; the majority are found in the cooler temperate climates.

Allantus spp. (strawberry/rose/cherry sawflies, etc.)—larvae eat leaves and may tunnel into the pith of shoots on roses and woody hosts; Europe, Japan, U.S.A.

Ametastegia glabrata (dock sawfly)—larvae bore stem and fruits of weed species (*Rumex, Polygonum*), also grapevine, apple fruits, and maize; Europe, North America.

Athalia spp. (cabbage sawflies)—on cultivated Cruciferae—larvae eat leaves; Europe, Africa, India, Asia to Japan (Fig. 13.1).

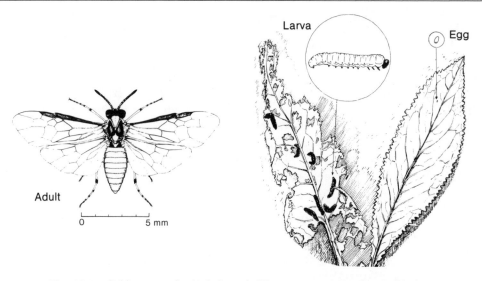

Fig. 13.1   Cabbage sawfly (*Athalia* sp.) (Hymenoptera; Tenthredinidae).

Belennocampa spp. (rose/butternut leafrolling sawflies)—one species on butternut in U.S.A., another on *Rosa* in Europe.

Caliroa spp. (pear slug sawfly, etc.)—larvae skeletonize leaves of pear, chestnut, plum, peach, and oak trees; Europe, Asia to Japan, U.S.A.

Cladius spp. (antler/rose sawflies)-–larvae hole leaves of *Rosa* and strawberry; Europe, Asia to Japan, U.S.A.

Croesus spp. (hazel sawfly, etc.)—gregarious larvae defoliate hazelnut, birch, chestnut, willow, etc.; Europe, Asia, Japan, U.S.A. (Fig. 13.2).

Dolerus spp. (wheat/grass sawflies)—larvae eat stems of wheat, grasses, sedges; Europe, Asia and U.S.A.

Fenusa spp. (birch/elm leaf-mining sawflies)—larvae mine leaves of birch, elm, alder; Japan and U.S.A.

Hemichroa spp. (alder/camphor sawflies)—gregarious larvae eat leaf margins on camphor in China, and alder in Europe and North America.

Hoplocampa spp. (many) (fruit tree sawflies)—many species attack different fruit trees (apple, pear, cherry, plum, etc.) belonging to the Rosaceae; Europe, Asia including Asia Minor, and the U.S.A. (C.I.E. Maps Nos. A. 166–169) (Fig. 13.3).

Metallus spp. (rubus leaf-mining sawflies)—larvae mine in leaves of *Rubus;* Europe, Asia, and North America.

Nematus spp. (currant sawflies)—several different species feed on currant foliage; in Europe and also worldwide; in the U.K. 40 species are recorded, with a similar number in North America feeding on foliage of most common trees.

Pristiphora spp.—many species feed on foliage of fruit and most common trees; Holarctic.

Tenthredo spp.—polyphagous on Rosaceae and many other plants; Holarctic.

Fig. 13.2 Hazel sawfly (*Croesus septentrionalis*); gregarious larvae feeding on leaf of hazel; Gibraltar Point, U.K.

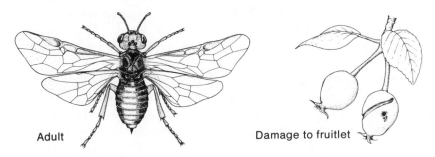

Adult

Damage to fruitlet

Fig. 13.3 Pear sawfly adult (*Hoplocampa brevis*) (Hymenoptera; Tenthredinidae), and damage by apple sawfly larva to a fruitlet; Cambridge, U.K.

SUBORDER APOCRITA

The majority of Hymenoptera are placed in this group; adults all have the abdomen basally constricted (petiolate), and they show extensive specialization in habits, some being social in large nests and complex communities. Larvae are apodous, often parasitic or in nests and provisioned or fed. For many years it was customary to subdivide this group into parasitic forms (Parasitica) and stinging forms (Aculeata), but Richards & Davies (1977) prefer to discard these rather indeterminate taxa as, in practice, the distinctions between the two groups are rather vague both anatomically and biologically. However, the terms are used descriptively to such an extent that perhaps they need to just be mentioned.

(PARASITICA)

The female ovipositor (which is often long) is used for piercing and depositing eggs into host tissues, either animal or plant.

ICHNEUMONOIDEA

A very large group of parasitic wasps having a pterostigma, venation usually complete; long antennae (usually more than 16 segments). The species in this superfamily are all parasitic on either other insects or some other Arthropoda, and the group is of tremendous importance in the natural control of insect pest populations; some species have been very successfully used in biological control projects. There are no known phytophagous species in the Ichneumonoidea.

Ichneumonidae (ichneumon wasps)   *60,000 spp.*

These are moderate in size, although some are small and a few large; many species are known. A few species are hyperparasites, but the vast majority are parasites; hosts are mostly Lepidoptera (caterpillars), then sawflies (Tenthredinidae), and some Coleoptera, some Diptera, some aphids, Neuroptera, and also spiders. Some species are quite host-specific but many appear to be more habitat specific and will parasitize different insects within the same general habitat. Most are found in rank vegetation in wooded areas where water and dew are available—there tend to be only a few species in hotter, drier areas although some species have adapted to such conditions. The group is most abundant in cooler temperate regions. Adults are either diurnal or nocturnal; many can be seen feeding on flowers in the sunshine, particularly Umbelliferae. Many of the nocturnal species such as *Ophion* are yellowish in color.

There is great diversity shown in the different types of parasitism practiced. Most species prefer prey that are exposed, but some specialize in attacking insects concealed inside plant tissues or thick cocoons. Some are endoparasites, with the egg usually being laid in the hemocoel of the host, others are ectoparasites and live inside the pupal cocoon or with the larva concealed in plant tissues. As with the Braconidae, some species are solitary but others are gregarious and several may develop in the same host. In cooler regions the adult females of many species overwinter in a state of hibernation.

The family is subdivided into distinctive subfamilies, some of which show definite host preferences, for example, for exposed lepidopterous caterpillars, sawfly larvae, larvae of wood-boring Coleoptera, etc. The larger species of Ichneumonidae tend to parasitize the larger prey (usually singly) whereas many Braconidae are small and feed on aphids and smaller caterpillars (often gregariously). The hosts of these parasites are invariably killed by the feeding of the wasp larvae. Several species have been used in very successful biological control programs. There is a large number of different genera in this family, and a few of the most abundant and widespread genera are listed below:

Exentrus—important parasites of *Neodiprion* sawflies damaging forests in Europe and North America.

Ichneumon—endoparasites of pupal Lepidoptera, sometimes ovipositing in the larvae.

Metopius—solitary endoparasites of Lepidoptera in early stages but emerging from the pupa.

Ophion—a common European genus (Holarctic) endoparasitic in larval Lepidoptera, especially Noctuidae (orange adults nocturnal and fly to lights at night).

Pimpla—ectoparasites of concealed large lepidopterous larvae or pupae in plant tissues or cocoons.

Rhyssa—ectoparasites on larvae of *Urocerus* deep in the tissues of tree trunks.

Thalessa—also ectoparasites on larval Siricidae in tree trunks.

Fig. 13.4 Typical ichneumonid wasp—female *Xanthopimpla* (Hymenoptera; Ichneumonidae); body length (excluding ovipositor) 13 mm (0.52 in.); Hong Kong.

Tryphon spp.—and closely related genera attach eggs to the host body (sawfly larvae, etc.) by a pedicel, and parasite development is completed inside the host cocoon.

Xanthopimpla—ectoparasites on larval Lepidoptera in concealed situations, including *Chilo*, and various skippers (*Erionota, Parnara*, etc.) in S.E. Asia (Fig. 13.4).

Braconidae (Braconids)   *40,000 spp.*

Most are smaller in size than ichneumons, but the main taxonomic character is that there is no second recurrent vein (2m–cu) in the forewing, and in the hindwing the crossvein r–m is median rather than distal. They parasitize a very wide range of hosts, particularly lepidopterous caterpillars and aphids. In caterpillars, gregarious parasitism is often practiced and 100 braconids of the same species may emerge from a single caterpillar. In caterpillars, some species pupate inside the host body but most (such as *Apanteles*) emerge to pupate on the outside of the dead caterpillar body in tiny white cocoons. Pupation takes place within a cocoon spun of fine silk. In aphids, *Aphidius* pupates inside the nymphal body in a "mummy", but *Praon* larvae emerge under the host body and spin a cocoon which joins the dead aphid to the substrate on a broad pedicel. The age of the aphid parasitized by *Aphidius* is important—when young nymphs are parasitized they fail to reach full size, older nymphs may still reach maturity, but nymphs nearing maturity when attacked often produce a few small offspring before they die.

The subfamilies show very definite host preferences. For example, the Aphidiinae are all internal parasites of aphids (including *Aphidius, Praon, Trioxys*, etc.). In the present state of braconid taxonomy there are two very large genera—*Apanteles* and *Aphidius*—both worldwide in distribution and extremely important as parasites of caterpillars and aphids respectively. Most species appear to be quite host specific, but a few are clearly polyphagous on many different species. Several species have been used very successfully in biological control projects in different parts of the world.

Some of the best-known species of Braconidae encountered by agricultural entomologists are included in the following genera:

Apanteles—many species; found worldwide; endoparasites of caterpillars; most are small black wasps.

Aphidius—many species; worldwide; solitary endoparasites of Aphididae; tiny in size.

Bracon—ectoparasites of concealed larvae of Lepidoptera (such as bollworms).

Chelonus—solitary endoparasites of Lepidopterous larvae; oviposition usually taking place into the eggs.

Dacnusa—solitary endoparasites of larval Diptera; adult wasps emerging from the fly puparium.

Euphorus—parasitize *Helopeltis* bugs in Indonesia (natural parasitism rates on cocoa often reach 50–80%).

Leiophron—endoparasites of Miridae, ovipositing into the body of young nymphs.

Macrocentrus—solitary or gregarious endoparasites of caterpillars.

Meteorus—solitary or gregarious endoparasites of Lepidopterous larvae.

Microgaster—endoparasites, often gregarious, of unconcealed caterpillars.

Opius—small, solitary endoparasites of larval Diptera, especially Agromyzidae and Tephritidae.

Praon—solitary endoparasites of aphids, pupating underneath the bug corpse.

Spathius—external parasites of boring Coleoptera larvae in trees in tropical Asia, and Australasia.

Spinaria—parasitize larvae of Limacodidae.

Trioxys—solitary endoparasites of Aphididae.

There are two other small families placed here (Stephanidae and Megalyridae) whose larvae parasitize larvae of wood-boring beetles.

EVANOIDEA

The three families placed here are united in having an unusual attachment of the petiole just behind the scutellum, but otherwise they are quite different from each other.

Evaniidae (ensign wasps)   *400 spp.*

This small group is worldwide but best represented in the tropics; all known species are parasites of the oöthecae of cockroaches (Blattidae, etc.) and as such are valuable members of domestic habitats and food stores. Most of the species belong to the large genus *Evania* (Fig. 13.5)—they are called "ensign wasps" because of the shape of the small gaster which is flicked up and down as the wasp walks. The body is black, and the long antennae are constantly flickering.

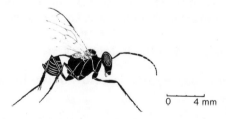

Fig. 13.5   Ensign wasp (*Evania* sp.) (Hymenoptera; Evaniidae); Hong Kong.

Aulacidae   *150 spp.*

Medium-sized wasps with a long ovipositor, which parasitize wood-boring insects; some species of *Aulacus* attack Xiphydriidae, and others Coleoptera (Buprestidae and Cerambycidae). The group is well represented in Australia and the U.S.A., but is basically worldwide.

Gasteruptiidae   *500 spp.*

These wasps have a long "neck" region, a long gaster, long ovipositor, and swollen hind tibiae. They are found worldwide, parasitizing solitary bees, especially those nesting in tree stumps or twigs.

CYNIPOIDEA   *3300 spp.*

A group of small to tiny-sized wasps, particularly renowned as gall makers, and also as parasites and inquilines in galls. Most are dark colored, and many are black. Some species have incredibly complicated life cycles with an alternation of generations between sexual and asexual forms; parthenogenesis is widespread and males of some species are very rare. Taxonomic characters include: wings (when present) without pterostigma and proximal part of costa, venation usually reduced, but cell R1 complete and hind wing with no anal lobe; antennae quite elongate, female with 13 (up to 19) segments, male with 14–15, antennae not geniculate; ovipositor exposed and distinct but not protruding. Between three and six families are placed here but there is little agreement over their classification.

Figitidae   *c. 200 spp.*

A small group, mostly parasitic upon Syrphidae and Neuroptera; most are small in size.

Ibaliidae   *10 spp.*

A very small group of quite large insects; Holarctic in distribution. The main genus is *Ibalia* and these species are endoparasites of Siricidae.

Cynipidae (gall wasps)

This is the main group within the superfamily, but it is composed of several quite distinct subgroups—here regarded as subfamilies. The main taxonomic character is that the gaster has the second and third segments very large and together forming more than half, viewed laterally.

Eucoilinae   *70 genera; 1000 spp.*

One of the largest subfamilies with larvae that are protelean parasites of cyclorrhaphous Diptera in the puparia, and some species are important as parasites of dipterous pests of cereal and vegetable crops in several parts of the world. Three of the main genera are *Eucoila* (parasitoid of *Oscinella frit*), *Kleidotoma,* and *Cothonaspis* (attacks *Delia radicum*). Some species are reared from plant galls.

Charipinae

Two distinct groups are included here in this small subfamily. One group are hyperparasites, preying on *Aphidius* (Braconidae) that have parasitized aphids; the other are primary parasites of Psylloidea.

Cynipinae   *1000 spp.*

These are the true "gall wasps"; most are gall makers on plants, with a very interesting biology; the plants attacked are mainly ornamentals (*Rosa* spp.) or trees of the Fagaceae used either as ornamental/shade trees or in forestry. *Quercus* (oaks) is host to more species than any other plant, but other Rosaceae and some Compositae are attacked. Most species are true gall makers but on the oaks there are many inquilines that also inhabit the galls. Mostly these insects are temperate (Holarctic) and to be found in the northern oak forests. It has been estimated that 86% of these gall wasps occur on *Quercus* and are confined to this genus, with 7% on *Rosa* and the final 7% on 35 genera of angiosperms, especially in the Compositae.

Most of the oak and rose galls are very conspicuous and distinctive, and some are damaging to the host; some destroy the seeds (acorns) of the oaks and can prevent regeneration of natural oak forest. The female wasp lays her egg into the growing host plant tissues using the ovipositor, usually in the spring. It is thought that the gall is actually formed as a result of the feeding activity of the young larva stimulating the cambium tissues—the galls are both quite specific and very distinctive. Most of the galls are unilocular with only one larva per gall, but the rose pincushion gall is multilocular and there may be a couple of dozen larvae in one gall. With many of the oak cynipids there is an alternation of generations, with an asexual (agamic) overwintering generation followed by a bisexual generation in the summer, either on different parts of the same host or on different hosts. This practice is referred to as heterogony. Galls of the asexual generation develop slowly so as to enable the larvae to overwinter, and they emerge in the spring of the following year (but sometimes one or two years later). The insects of the different generations are often quite different in appearance and were initially described in quite different genera.

Some of the more distinctive species regularly encountered are included below:

On *Rosa* spp.

Diplolepis eglanteriae (smooth pea gall)—on the leaves.

Diplolepis nervosa (spiked pea gall)—also on leaves.

Diplolepis rosae (robins's pincushion gall)—on the stem, mostly on wild species of *Rosa*; large multilocular gall, bright red in color.

On *Quercus* spp.

Most galls are on the leaf surface (usually underneath), or in buds, shoots, twigs, acorns, catkins, and on the roots.

Andricus fecundator (artichoke gall)—the gall is a swollen shoot.

Andricus kollari (marble gall)—on twigs of *Q. robur* for overwintering agamic generation (Fig. 13.6); sexual generation in axillary buds of *Q. cerris*.

Fig. 13.6 Marble galls on oak (*Andricus kollari*) (Hymenoptera; Cynipidae), showing adult emergence hole; Alford, U.K.

Fig. 13.7 Acorn gall (*Andricus quercuscalicis*) on *Quercus* (Hymenoptera; Cynipidae); Alford, U.K.

Fig. 13.8 Oak leaf with spangle galls on underside (*Neuroterus quercusbaccarum*) (Hymenoptera; Cynipidae).

Andricus quercuscalicis (acorn gall)—recently introduced into the U.K. from North America; especially damaging for it destroys acorns (Fig. 13.7).

Biorrhiza pallida (oak apple)—sexual generation develops in a swollen (galled) shoot; agamic generation overwinters in root galls.

Cynips spp.—many different leaf galls are formed on *Quercus* spp.

Neuroterus quercusbaccarum (spangle galls)—small, round galls densely clustered under the leaves (asexual generation) and bisexual generation are in "current galls" on male catkins in the spring (Fig. 13.8).

Synergus—spp. and members of the Synerginae are all inquilines to be found inhabiting many different cynipid galls on *Quercus* spp.

Both insects and galls are described in the *Handbook* by Eady & Quinlan (1963), and color illustrations of the galls are in various books and particularly in Darlington (1968).

CHALCIDOIDEA

One of the largest superfamilies in the order, showing considerable adaptive radiation and diversity of life-styles. Most of the species are parasites or hyperparasites of other insects, and of great economic significance in the natural control of many insect pest populations. Many have now been used very successfully in biological control programs throughout the world. A not insignificant number are phytophagous, mostly infesting seeds, but some are gall formers. The groups of insects most parasitized by chalcids are the Lepidoptera, Hemiptera (Homoptera) and Diptera. Coleoptera and other insects are parasitized, as also are some Arachnida. Mostly, it is the larval stage that is parasitized, but some chalcids specialize in pupae or eggs. The egg parasites (*Trichogramma* spp.) are among the smallest of insects and some are reputed to be only 0.2 mm (0.008 in.).

Most species are characterized by having a greatly reduced wing venation, and antennae distinctly elbowed (geniculate). The larvae are very reduced with vestigial head and thirteen body segments; hypermetamorphosis is a frequent occurrence. Pupation usually occurs inside, or else close to, the remains of the host which is killed by the infestation; except for one eulophid species there is never a cocoon constructed.

There are some differences of opinion concerning the classification of this group, but in this work the family arrangement used by Richards & Davies (1977) is followed. Major works on Chalcidoidea are mostly temperate, and include Peck (1963), Peck, Boucek, & Hoffer (1964), and Nikolskaya (1963); also Boucek (1988).

Fourteen major families are recognized here as follows:

Agaonidae (fig wasps)   *1500 spp.*

A strange group that lives in a state of obligatory mutualism with plants of the genus *Ficus* (Moraceae), and it shows extreme sexual dimorphism. Females are relatively "normal" in appearance, but show special modifications for penetrating the ostiole ("eye") of young syconia (figs). The males are wingless, sometimes eyeless, with elongation of the gaster, and totally adapted for a brief hectic life inside the mature syconium. Biologically this association of a family of wasps with a large pantropical genus of plants is unique and is of great basic interest on several counts. For a general account of *Ficus* and fig wasps, and for an extensive bibliography, the papers by Hill (1967), Grandi (1916) and Wiebes (1976) can be consulted.

One point of agricultural interest is that when the edible fig (*Ficus carica*) was introduced into California, in about 1900, fruit development was nonexistent as caprifigs (male form of tree) had not been included, and the wasps vital for fruit pollination were not present. After establishment of caprifig trees and the wasps (*Blastophaga psenes*), the Smyrna fig trees bore crops of fruit. A number of parthenocarpic varieties of edible fig have now been developed so that insect pollination is no longer necessary.

Torymidae   *1500 spp.*

A large family, females often with a long conspicuous ovipositor, mostly parasitic upon gallicolous insects. A few are found in nests of bees and wasps, and some are phytophagous and eat seeds of Rosaceae and Coniferae.

The Sycophaginae are fig wasps and are thought to be either phytophagous and feeding on the ovules inside the syconia, or else parasitic and feeding upon the developing Agaonid larvae. The long ovipositor is used for piercing the wall of the young syconium to deposit eggs singly inside the young ovules. There is a group of large, robust chalcids found in the syconia of banyans (*Ficus* subgenus *Urostigma*); they are a well defined group but their family placing is something of a problem—they were thought to belong to the Torymidae and to constitute the subfamily Epichrysomallinae (Hill, 1967), but present opinion inclines to the Pteromalidae.

Megastigmus is a very large genus found worldwide. Some species have been bred from plant galls made by Diptera or Hymenoptera, but quite a large number are clearly phytophagous and feed on seeds in pine cones, etc., and fruits of various Rosaceae. *Monodontomerus* generally parasitizes caterpillars (Lepidoptera) or other species of Hymenoptera, and some *Torymus* attack gall-making Diptera and Hymenoptera.

Some species of Torymidae of interest to agriculturalists include:

Megastigmus aculeatus (rose torymid)—larvae in fruits of *Rosa;* Europe and Asia.

Megastigmus pinus (pine cone torymid)—larvae in pine cones; Europe.

Megastigmus spp. (seed chalcids)—some 12 species feed on seeds in the cones of many different conifers (up to 15% of seed crops may be destroyed); Europe, Asia, North America.

Torymus druparum (apple seed chalcid)—larvae in apple seeds; Europe, Asia, North America.

Torymus varians (apple seed chalcid)—U.S.A.

The Podagrionidae are often placed in the Torymidae as a subfamily. They are very characteristic in appearance with large swollen hind femora and curved tibiae that reflex against them as they walk. The female has a long ovipositor which is used to pierce the thick oötheca of Mantidae, and the larvae parasitize the eggs. Throughout the warmer parts of the world it appears that each species of mantid has its own species of parasitizing *Podagrion*. The level of parasitism is often very high; parasitized oöthecae have a series of tiny holes through which the adult wasps emerged.

Ormyridae   *60 spp.*

This group is sometimes regarded as a subfamily of Torymidae, and Richards & Davies (1977) finally place it in the Pteromalidae, but it is a distinctive group and might be best viewed separately. In the literature these wasps are regarded as parasites of gall makers, but species of *Ormyrus* are found in syconia of banyan figs and here they appear to be phytophagous. Similarly, in Hong Kong, leaves of *Ficus microcarpa* were often found to be galled; from several rearings of a couple of dozen galled leaves only *Ormyrus* wasps were obtained. *Ormyrus* is a stout-bodied wasp of characteristic appearance with a short ovipositor and gaster with rows of large punctures on the tergites.

Chalcididae   *1500 spp.*

In this group are found some of the largest chalcids. They are stout-bodied, often black in color, and they parasitize Lepidoptera (larvae or pupae) or puparia of Diptera. The group is best represented in South America and other tropical regions. In S.E. Asia, many emperor moth caterpillars (Saturniidae) and pupae are heavily parasitized. This is one of the families important in contributing to the natural control of insect pests on cultivated plants, though little used in biological control programs.

Brachymeria—large, black chalcids that parasitize larvae and pupae of many Lepidoptera (some species are hyperparasites); abundant in S.E. Asia.

Leucospidae

A small group of large black and yellow chalcids which parasitize various bees (*Megachile,* etc.); adults have the forewings folded longitudinally.

Eurytomidae (seed chalcids, etc.) *1100 spp.*

A widespread and abundant group that shows great diversity of life-styles, some being parasitic, others phytophagous on a wide range of plants. Adults are often either black or yellow in color, with a rounded, punctate thorax, often elongate petiole, and rounded gaster; ovipositor concealed. A number of species are regularly reared from galls, and often their biology is not really known, so it is not clear whether they are parasites or gall formers.

Several genera are phytophagous, including *Tetramesa* (= *Harmolita*) and some *Bruchophagus* (*Bruchophagus* may be a subgenus of *Eurytoma*). The genus *Eurytoma* is very large and diverse in habit—some species are phytophagous and feed on the seeds of *Vitis,* plum, etc., others are clearly entomophagous and parasitize Lepidoptera, Diptera, and Coleoptera, and other species reared from galls have unknown biology.

Sycophila (= *Eudecatoma*) is another very large and widespread genus—many species are found inside the syconia of banyan figs, and many others are reared from oak galls and other galls.

Some of the more notable species include the following:

Bruchophagus spp. (clover/legume seed chalcids)—larvae develop in seeds of clovers and other legumes; Europe, Asia, India, U.S.A. and Canada (mostly temperate).

Eurytoma amygdali (almond fruit wasp)—in almond seeds; Near East.

Eurytoma spp.—parasitize Lepidoptera, Diptera, Coleoptera, etc.; worldwide.

Eurytoma spp. (seed chalcids)—found in seeds of grape, plum, apricot, etc.; Holarctic including India.

Sycophila spp.—reared from ovary galls in syconia of *Ficus* (*Urostigma*), oak galls, and other galls.

Tetramesa spp. (wheat/barley/rye jointworms)—many species, some specific to different temperate cereals, others to different grasses; Holarctic in distribution.

Perilampidae *200 spp.*

A small group of stout-bodied, metallic green or black wasps, closely related to the Eurytomidae. Most are hyperparasites, but some are primary parasites of other insects.

Eucharitidae *350 spp.*

Another small group, black or metallic blue or green, somewhat humpbacked in appearance, and with a long petiole. They are a tropical group that, so far as is known, parasitize pupae of ants.

Pteromalidae *3100 spp.*

A very large family, adults very diverse in appearance; they are parasites or hyperparasites and attack virtually all orders of Insecta. The Western European species have been the subject of a review by Graham (1969). Some species have been successfully used in biological control programs in different parts of the world; several species are important parasites of stored-products beetles.

A few of the more important and widespread genera and species are listed below:

Anisopteromalus—attack *Sitophilus* weevils in stored grains and foodstuffs.

Asaphes—mostly hyperparasites of Aphididae.

Cerocephala—parasitize larvae of wood-boring beetles, especially Scolytidae.

Choetospila—parasitize beetles in stored grains and foodstuffs.

Cleonyminae—parasitize wood-boring Coleoptera.

Cyrtogaster—parasitize leaf-mining Diptera (Agromyzidae, etc.).

Dinarmus—attack Bruchidae in seeds of legumes.

Lariophagus—parasitize *Sitophilus* and other grain beetles.

Miscogasterinae—parasitize Diptera.

Nasonia brevicornis—important pupal parasite of calyptrate Diptera, especially Muscidae and Calliphoridae.

Pachyneuron—parasitize Aphididae.

Pteromalus puparum—widespread parasite of pupae of *Pieris* (on Brassicas) and other Lepidoptera.

Scutellista—important parasites of *Ceroplastes, Saissetia,* and other Coccidae.

Systasis—attack gall-making Cecidomyiidae.

Trichomalus—many species; some parasitize *Ceutorhynchus* weevils and other Coleoptera.

Encyrtidae   *3000 spp.*

Another large family of small or tiny wasps (body size 0.2–6.0 mm/0.008–0.24 in.; mostly 1–2 mm/0.04–0.08 in.), but stout-bodied; they have the middle legs modified for jumping. Most live as parasites in the eggs, larvae, or pupae of Homoptera and Lepidoptera. The groups most frequently attacked are aphids, whiteflies, and scale insects. The Encyrtidae are extremely important in their population-controlling effect on many homopterous crop pests, and several species are widely used for controlling scale insects (Coccoidea) on *Citrus* and other crops. Polyembryony is recorded for some species, particularly *Ageniaspis, Copidosoma,* and *Litomastix* on lepidopterous larvae. Some species are hyperparasites, usually on other Encyrtidae.

The group is best represented in the tropical and subtropical regions of the world where most of the host insects are to be found. Some of the more abundant, widespread and important genera of Encyrtidae are as follows:

Ageniaspis—polyembryonic parasites of Micro-lepidoptera (especially *Yponomeuta, Phyllocnistis* and *Phyllonorycter*).

Anagyrus—mostly recorded from mealybugs (Pseudococcidae).

Aphycus—ecto- or endo-parasites of Coccoidea.

Apterencyrtus—parasitic on Diaspididae.

Blastothrix—on many Coccidae and Diaspididae.

Cheiloneurus—recorded from many Coccidae.

Coccidencyrtus—parasitize Coccoidea.

Coccidoxenus—attack many different Coccoidea.

Comperiella—control several important tropical species of Diaspididae.

Copidosoma—polyembryonic parasites of larval Lepidoptera.

Habrolepis—parasitize many different scale insects.

Homalotylus—parasitize Coccinellidae and Chrysomelidae (Coleoptera).

Hunterellus—cosmopolitan parasites of nymphal ticks (Acarina).

Leefmansia—egg parasites of *Sexava* grasshoppers in S.E. Asia and the Pacific Islands.

Litomastix—polyembryonic endoparasites of lepidopterous larvae.

Metaphycus—on Coccidae and Diaspididae.

Microterys—specialize in parasitizing the larger species of Coccoidea.

Oöencyrtus—many important species; they parasitize the eggs of Lepidoptera, Heteroptera, Homoptera, etc.

Psyllaephagus—parasitize nymphs of Psyllidae worldwide.

In some parts of the tropics, local scale insect populations may be parasitized by a large number of different Encyrtidae. Unfortunately *Erencyrtus* is a successful parasite of the lac insect in India and S.E. Asia.

Eupelmidae   *750 spp.*

This family contains a few genera only, but mostly with a very wide host-range.

Anastatus—egg parasites of grasshoppers, bugs (*Nezara,* etc.), and moths.

Eupelmus—shows great diversity of habits; a few are phytophagous, others are parasites or hyperparasites of Coleoptera, Lepidoptera, etc.

Eulophidae   *3000 spp.*

A very large group of mostly small species, with the body weakly sclerotized; tarsi always with four segments; an antennal funicle has at most four segments. Many small genera are found in different parts of the world. A few genera are, however, large, abundant, and widespread, and to be found attacking many different crop pests; they include the following:

Chrysocharis—many species recorded parasitizing leaf miners.

Entedon—many species; they parasitize larvae of Coleoptera in plant stems and twigs (*Scolytus,* etc.).

Euderus—some species attack eggs of Coleoptera; other species parasitize small species of Lepidoptera.

Melittobia—are common ectoparasites on pupae of the Aculeate Hymenoptera (*Bombus, Osmia,* etc.) and some Diptera.

Pediobius—a large genus of parasites and hyperparasites of a wide range of grass-stem dwellers, leaf miners, and pupae of Lepidoptera, Diptera, and Coleoptera.

Tetrastichus—a very large genus, with most species parasitic or hyperparasitic on a wide range of different insects, both on eggs and larvae.

## Aphelinidae									*900 spp.*

Tiny insects, mostly less than 1 mm (0.04 in.) in length, sometimes classified as a subfamily of Eulophidae; they are important parasites of Coccoidea, Aphididae, Aleyrodidae, and occasionally on Cecidomyiidae, eggs of Orthoptera, etc. They are most abundant in the warmer parts of the world. The tiny, fragile body is generally plump, soft, yellow to black in color, and with antennae composed of only a few segments.

Several species have been employed very successfully in biological control programs to control especially armored scales (Diaspididae) on fruit crops. An important monograph on the Aphelinidae of the Mediterranean Basin is by Ferrière (1965).

Some large genera, found widely and abundantly on a wide range of crop pests, include the following:

Aphelinus—endoparasites of Aphididae and Pemphigidae (wooly aphids).

Aphytis—ectoparasites on Diaspididae.

Coccophagus—parasites on mealybugs (Pseudococcidae) and Coccidae.

Encarsia—parasitize Aleyrodidae and some Pemphigidae.

Eretmocerus—ectoparasites on Aleyrodidae.

Marietta—hyperparasites on Coccidae.

Physcus—parasites on Diaspididae (armored scales).

Prospaltella—recorded as all being parasitic on Diaspididae.

## Elasmidae									*200 spp.*

A small family, closely related to the Eulophidae; forewing narrow and marginal vein long. Adults are black and have enlarged hind legs; they parasitize Lepidoptera as larvae, either as primary or secondary parasites.

## Trichogrammatidae									*530 spp.*

These are tiny chalcids, 0.3–1.0 mm (0.012–0.04 in.) in body length, with a broad forewing and three-segmented tarsi. The body is delicate; antennae short, often clubbed, but funicle never with more than two segments. A very important group—they are all egg parasites (mostly of Lepidoptera) and several species are being widely used in very successful biological control programs—and some of these species are commercially available, in both Europe and North America. The group was revised by Doutt & Viggiani (1968)—there is quite a large number of small genera, although most of the B-C agents belong to the genus *Trichogramma.*

Trichogramma—many species, worldwide, in several distinct subgenera; some very specific in their choice of prey/host, and some polyphagous.

Trichogramma japonicum—an important Oriental parasite of eggs of rice and sugarcane stalk borers, found throughout S.E. and East Asia.

Trichogramma minutum—a very important polyphagous egg parasite in the New World; obtainable from several commercial companies in North America.

Other interesting species include:

Hydrophylax aquivolans—the female wasp swims underwater to oviposit in the eggs of Dytiscidae.

Prestwichia aquatica—another freshwater species; it parasitizes eggs of aquatic Heteroptera (*Ranatra, Hygrobia,* and *Notonecta*), and *Dytiscus.*

Mymaridae (fairy flies)   *1300 spp.*

A family of tiny insects (0.2–1.0 mm/0.008–0.04 in.) that are all egg-parasites; adults are mostly black or yellow in color. *Alaptus* has species only 0.2 mm (0.008 in.) in body length. *Caraphractus cinctus* is remarkable in that both sexes swim underwater using their wings—the females oviposit in eggs of water beetles (Dytiscidae). The group is of definite importance as a natural control element for various insect pest populations by destroying eggs. *Anagrus* species parasitize eggs of *Empoasca* and *Hypera* weevils in India and in S.E. Asia.

PROCTOTRUPOIDEA

This group was fairly recently redefined and subdivided into three separate small superfamilies. As now defined, the group is of little importance agriculturally; most of the species parasitize beetles, and a few species attack flies or Chrysopidae.

SCELIONOIDEA

Scelionidae

Many tiny species, worldwide, and all are egg parasites of Lepidoptera, Hemiptera, and Orthoptera, and a few attack spiders. Several species are known to be of agricultural importance in that they regularly parasitize eggs of major crop pests. These include the following:

Ascolus—attack eggs of many Pentatomidae (including *Eurygaster*).

Eumicrosoma—regularly parasitize eggs of *Blissus leucopterus.*

Hadronotus—these species use eggs of Coreidae (*Clavigralla,* etc.).

Scelio—parasitize egg–pods of locusts and grasshoppers (*Valanga, Patanga,* etc.).

Teleonomus—parasitize eggs of Lepidoptera (*Chilo, Scirpophaga,* etc.) and Heteroptera (*Blissus,* etc.); very important in S.E. Asia as egg parasites of rice and sugarcane borers.

Platygasteridae

A large group of tiny chalcids that parasitize Cecidomyiidae. The eggs are usually laid inside the host eggs but they do not develop until the midge larvae hatch.

CERAPHRONOIDEA

A small group, mostly recorded as hyperparasites on chalcids and braconids on aphids and scale insects.

BETHYLOIDEA

Rather a miscellany of insects that are difficult to place taxonomically; they are all parasites, yet some of the families are of little interest agriculturally.

Dryinidae   *850 spp.*

Females are often apterous, but most have distinctive chelate fore-tarsi. They are parasitic upon nymphs of various Homoptera, especially Fulgoridae, Cicadellidae, Cercopidae, and Membracidae. They are endoparasites during larval stages, but eventually a cyst protrudes from the bug body, usually either black or yellow in color; this parasitism causes castration of the host. Many important bug pests (*Sogatella, Idioscopus, Typhlocyba,* etc.) are regularly found with dryinid cysts protruding from the body.

Bethylidae   *2000 spp.*

Small black wasps (rarely metallic), some apterous, that mostly parasitize lepidopterous larvae and Coleoptera. A few notable parasites are as follows:

Cephalonomia—they parasitize various Coleoptera.

Goniozus—in S.E. Asia they destroy larvae of *Pectinophora, Enarmonia,* and *Lamprosema.*

Perisierola nephantidis—in India this wasp parasitizes the coconut caterpillar (*Nephantis serinopa*) and is regularly reared for mass release in coconut groves.

Chrysididae (ruby-tailed wasps; cuckoo wasps)   *3000 spp.*

Bright metallic wasps, usually green or blue, with a bright red posterior; the gaster is convex above and flattened underneath, so that it can be turned under the thorax. A large and widespread group; most species appear to be parasites of Aculeate Hymenoptera, especially Eumenidae and Megachilidae. The genus *Chrysis* contains some 1000 species worldwide, and a few species attack caterpillars of some Limacodidae.

(ACULEATA)

The female ovipositor is usually modified into a sting, used for defensive or predatory purposes.

SCOLIOIDEA

This group is often regarded as being the most primitive of the Aculeata. The ovipositor is modified into a sting for predatory purposes and from the female accessory reproductive glands has developed a poison gland. As now constituted, this is a large and diverse group with several very important families.

Scoliidae   *300 spp.*

Large, hairy wasps including some of the largest Hymenoptera; body coloration is black with red or yellow banding, and wings often darkly metallic. Most are found in the tropics and subtropics. The larvae are ectoparsites of larval Scarabaeidae, or rarely larval Curculionidae. There are two very widespread and abundant genera, mentioned below:

Campsomeris—a large genus, adults medium-sized, males generally orange or yellow with golden setae, females are darker. They prey almost entirely on larvae of Melolonthinae and Rutelinae, but otherwise are not very host specific.

Megacampomeris—male specimens from South China were often found flying into Hong Kong.

Scolia (= *Megascolia*) are large, stout-bodied, hairy wasps that prey on the larvae of Dynastinae; four species are of special note:

Scolia azurea—males and females were captured in South China hunting over a village refuse dump; the female has a distinctive reddish yellow head capsule.

Scolia manilae—used in Hawaii to control *Anomala orientalis* quite successfully.

Scolia oryctophaga—native to Mauritius, it preys on the larvae of rhinoceros beetle (*Oryctes monoceros*).

Scolia ruficornis—introduced into Samoa in 1945 from Zanzibar to control *Oryctes,* and by 1949 was locally established. Another parasite of *Oryctes* is *S. flavifrons.*

Tiphiidae   *1500 spp.*

Another large and widespread family showing great diversity in structure and habits. Most species parasitize various ground-nesting Aculeata, and the Thynninae in South America and Australia attack larvae of Scarabaeidae. The genus *Diamma* parasitizes the species of *Gryllotalpa.*

Mutillidae (velvet "ants")   *5000 spp.*

The females are quite ant-like in appearance (hence the common name), they are apterous, the thorax and propodeum are fused into a single plate; body color is black or reddish with silvery pubescence giving a velvety appearance. Males are winged. They parasitize various other aculeate Hymenoptera, especially *Bombus,* Sphecidae, and Pompiliidae. In Africa some species parasitize the pupae of *Glossina.*

Formicidae (ants)   *15,000 spp.*

Previously the ants were placed in a separate superfamily on their own but they are now regarded as being the most advanced members of the Scolioidea. They are found

throughout the world in considerable abundance. They are probably equally well represented in both tropical and temperate regions, but more species are to be found in
warmer regions. These are polymorphic social insects, most forms are wingless, and
they are characterized by having one or two distinct nodes (swellings) on the petiole
(pedicel). The worker caste often has very large mandibles and they may or may not
have a functional sting. In a few advanced forms the sting is lost, but the defense
mechanism consists of squirting a jet of formic acid. There is tremendous variation in
behavior shown throughout the group, which makes generalization difficult. Most
species are quite small (3–10 mm/0.12–0.4 in. body length) but some of the Australian
bulldog ants (*Myrmecia* spp.) measure 25 mm (1 in.) and they are stout-bodied.

The polypmorphism shown by ants is quite extreme. There are generally three
basic types: male, queen, and worker. The male is the least modified and is winged. The
queen is a fertile female, large in size, antennae and legs somewhat stunted; wings are
lost after mating; abdomen is large and distended by the reproductive organs—she
becomes essentially just an egg-laying machine. Workers are wingless females and
smaller in size, with a small gaster, and reduced ovaries (but not necessarily sterile).
Workers are variable in size, and sometimes dimorphic, without intermediates; then the
large forms are called "soldiers". They usually have a large head capsule and large
mandibles and are usually adapted for a particular function, such as fighting, nest-
guarding, seed crushing, etc. The other smaller "workers" may sometimes be modified
for special functions within the nest or society. In a few Formicinae some workers
become living food stores—their crop is extremely large and it swells the entire gaster
into a spherical shape when filled with nectar or honeydew. These individuals are
known as repletes, or honey-pot ants, and are usually confined to the nest; they
regurgitate crop contents when the nectar is required to feed the community. Honey-
ants (repletes) have been part of the diet of primitive man in many tropical communities
for ages.

Most ant species have a mating ritual similar to that of termites in that, at certain
times of the year, young queens are reared together with males, and on a suitable day
(warm and moist) the young queens and the males take to the air in large numbers. From
a distance a mating swarm resembles a cloud of smoke and mating may take place at a
considerable height. These flights are conspicuous and they are usually accompanied by
swarms of birds that gather to catch the flying insects. The fertilized queens tend to
disperse over long distances from the original nest because of the height of swarming.
The fertilized queens fall to earth after the mating flight, then shed their wings and crawl
off to start new nests.

It appears that the more primitive species of ants have a sting, with a well-
developed poison gland, and the sting is used for killing prey and for defense purposes.
Some of the more advanced forms have lost the sting and have instead an acid-squirting
system opening terminally at the gaster tip. These ants squirt formic acid if disturbed;
sometimes they bite their attacker and squirt formic acid into the wound.

An important characteristic feature of ants connected with their social organization is their ability to cooperate in tasks such as food collection. They can often be seen
collectively carrying large prey, such as a dying earthworm or cicada or else a large piece
of food, back to the nest.

The diet of ants is varied and it is difficult to make generalization, except to say that
most species are opportunistically omnivorous; most will make use of honeydew and
nectar sources, and most will readily accept animal food (meat) if available. The exception being the tribe Attini (leaf-cutting ants) which are fungus feeders. Thus the species
to be regarded as crop pests usually eat bark from woody stems, sometimes killing
seedlings and saplings, and some will collect seeds, but they often kill other insects, or

drive them away by their aggressive behavior. Other species may regularly attend aphids and scale insects for their honeydew, but they may also kill a number of small insects as food. As with many omnivorous species, their diet tends to vary with the season as each type of food will be available at different times in the year. In temperate regions, insect prey are only abundant in the summer and early autumn, and honeydew is only available over the same time period; seeds are generally most available in early autumn, and plant sap is most important in the spring when insects are scarce.

Nest structure varies greatly within the group. In cooler regions, nests are usually hidden or well underground for protection against the cold, but in the tropics many nests are aerial in tree foliage (Fig. 13.16), or in crevices, or in specialized parts of plants (Fig. 13.11). The tropical driver ants are nestless; on the floor of the rain forest where they live, these nomads just rest at night in groups (called bivouacs) where the workers cluster round the queen to protect her. With the nesting species, some smaller nests only contain a few dozen individuals but the largest nests have been estimated to house 100,000–300,000.

In order to establish clear trails from the nest to food and water sources the foraging workers lay pheromone trails. Not all species seem to use pheromone trails, but some important and widespread species do and their trails can be very conspicuous; continual reinforcement of the scent trails make them very distinct and easily followed.

Within the nest the life history of most ants is rather simple—there are no elaborate nest combs built, and eggs are just deposited in a central part of the chamber system. The very first larvae are fed by the queen on special secretions from her salivary glands, and the queen herself at this time lives off her fat body and the atrophying flight muscles. Once the first brood of workers are reared, they take over the duties of food collection and care of the brood and the queen confines herself to egg laying. She is fed directly from the mouth of workers on regurgitated liquid food. The developing larvae are also fed on regurgitated liquid food by the workers. There is usually a certain amount of trophallaxis (reciprocal regurgitation) practiced between workers and workers and larvae. This reciprocal feeding is important when ant colonies are being destroyed by the use of chemical poison baits.

From an agricultural point of view, ants are pests in five main ways:

1. Collecting seeds—either before or after sowing, usually from grasses and cereals, and usually in semiarid areas. These are the granivorous harvester ants (*Messor, Pheidole,* etc.).

2. Sap feeders—chewing bark from woody stems or chewing up seedlings; small fruit trees and ornamentals are girdled just above ground level. The many species in this category include *Solenopsis, Formica, Lasius,* and *Dorylus,* etc. (Fig. 13.17).

3. Leaf cutters—they remove pieces of leaf lamina and take them back to the nest to use for the cultivation of fungal mycelium; these are the tribe Attini (fungus, gardening, or leaf-cutting ants), species of *Atta* and *Acromyrmex;* southern U.S.A., Central and South America (Fig. 13.10).

4. Honeydew farmers—these ants encourage aphid and coccoid infestations by guarding the bugs and protecting them from predators and parasites, in return for receiving the honeydew excreted by the bugs. In extreme cases the ants actually construct small, aerial subnests as shelters for the scale insects which are reared inside, literally a form of farming. Most of the species of *Crematogaster* belong to this category and many other ant species practice this husbandry to a

greater or lesser degree (e.g. *Solenopsis, Pheidole, Polyrachis,* etc.). Subterranean infestations of root aphids or root mealybugs are usually attended by ants which often construct a protective earthern "shell" over the infested roots.

5. Aggressive biting/stinging arboreal ants—they may be nesting in the tree or else just foraging. These species are found in fruit trees, etc., and are alleged to harass field workers and to hinder tree maintenance, and fruit harvesting. *Macromischoides* in Africa and *Oecophylla* spp. in Africa, Asia, and Australasia nest arboreally between leaves, and *Solenopsis* (fire ants) nest underground but· forage aerially. So far as the first two genera are concerned, personal experience indicates that accounts of their aggressiveness are exaggerated. These species of ants may, in fact, be important insect predators in the trees they frequent.

The role of ants as agricultural pests is not at all clear so far as many species are concerned. The Attini are undoubtedly pests in their foliage collection. But most of the others to be found in plantations, orchards, etc., are really omnivorous. They will eat a certain amount of plant material, mainly seeds (especially oily seeds), and sap is taken by removing the bark from woody stems. They will also take nectar and honeydew from bugs, but many are sufficiently predacious to kill numbers of smaller and soft-bodied insects, and some are particularly predatory towards other species of ants. The factors controlling the selection of food material appear to be quite variable and to differ for reasons that are not at all obvious. Some species, although generally omnivorous, do show a predilection for honeydew or plant material. Others do show a preference for animal material in the form of other insects (or meat).

Some of the widespread and abundant species do certainly appear to be more predacious and they prey on capsid bugs, mealy bugs, some scale insects, aphids, and small caterpillars and they are thus definitely beneficial to the farmer. Some other arboreal species are very active and it appears that their high level of physical activity discourages both insect pests feeding and also their oviposition, and they are clearly indirectly beneficial species. However, some aerial species of ants are antagonistic towards others and so the latter may be replaced in orchards. Also, several ground nesters regularly displace some of the beneficial arboreal nesters. In South China, *Oecophylla smaragdina* have been encouraged to nest in citrus orchards (and litchi) for many years and the ants apparently successfully discourage citrus shield bugs.

It appears that any crop-tree/ant situation needs to be carefully examined and evaluated before any control measures are contemplated. The cocoa/ant situation in West Africa has long been studied and for many years the precise relationships were still not clear. Basically the two main cocoa pest complexes are the mealybugs and the capsid bugs. In West Africa, *Oecophylla* is a predator on both mealybugs and capsids, and is also antagonistic to the *Crematogaster* that attend the mealybugs. In this context, the *Oecophylla* are clearly far more beneficial than they are pests. In Java, *Dolichoderus bituberculatus* is a small, black, stingless (but acid-spraying) ant that lives in cocoa where it cultivates a relatively harmless mealybug (i.e., nonvirus vector) but discourages mirid and cocoa pod borer by its large numbers and high level of activity. But the ground-nesting *Anoplolepis longipes* (gramang ant) and *Pheidole,* although largely not predacious, often drive out the *Dolichoderus* as well as *Oecophylla* ants from the cocoa foliage although the precise details of how this happens is not clear.

Present interpretation of the Formicidae regards there as being between 8 and 12 subfamilies (a few of which are rather obscure), and a series of often well-defined tribes. However, to the general entomologist the subfamily taxonomic characters do tend to be somewhat obscure, though there are some quite important basic biological differences.

A few of the more important subfamilies are briefly reviewed below.

The Ponerinae is a large group, showing tremendous diversity, mostly tropical, and carnivorous in diet, feeding on other insects and small invertebrates. They are generally regarded as being primitive—they have well-developed mandibles and an effective sting with which they immobilize their prey. They nest underground in small colonies, and show little variation in size. An unusual member of the Ponerinae is the S.E. Asian jumping ant (*Harpegnathus venator*) which forages singly and is easily recognized by the black body and long mandibles.

The Myrmeciinae are regarded as being equally primitive, but they only occur in Australasia; they include the well-known bulldog ants (*Myrmecia* spp.) that are large and aggressive ants, very abundant, and with a powerful sting. Most species live underground in quite large colonies and are carnivorous. But some species of *Myrmecia* feed themselves on nectar and only hunt insect prey when there is a brood of young to be fed (as do the Sphecoidea).

Dorylinae are army/driver ants, nomadic, predatory, tropical, and somewhat primitive, but also very specialized. They march in long files through the countryside and fiercely attack any animal life they encounter. In the past, two main genera were recognized—*Anomma* in tropical Africa, and *Dorylus* in both Africa and tropical Asia, but nowadays *Anomma* is often regarded as a subgenus of *Dorylus*. The prey is largely other insects and earthworms, but *Anomma* species have been filmed killing lizards, frogs, rodents, and birds, and they are reported to have killed penned livestock as large as pigs. They are stingless but the soldiers have large mandibles and they bite ferociously. The workers are blind and they have a large, blind wingless queen. The males are large, winged, and fly to lights at night and are abundant throughout much of tropical Africa and Asia—they are often called "sausage-flies" (Fig. 13.12) because of the elongate and large gaster. In Ethiopian highlands, an ant identified as *Dorylus* sp. nr *brevinodus* is called the Gojam red ant and it damages root crops, soybean, and has been seen to girdle quite large young trees—presumably for the exuding sap.

The Ecitoniinae has been split off to contain the New World army ants of the genus *Eciton*. These ants have a powerful sting but are usually less aggressive than the African *Anomma*.

The Pseudomyrmecinae is a small group of tropical ants, slender in body form, which specialize in having small nests in hollow twigs, empty galls, and various cavities in plant bodies. Some are referred to as being "obligate plant-ants" as they are only found in very intimate association with certain plants. For example, some *Pseudomyrmex* in Central and South America live inside hollow *Acacia* thorns, and other species have equally mutualistic relationships with particular plants. A recent book on ant-plant relationships is by Beattie (1985). The acacia gall ants of Africa (Fig. 13.11) are members of the genus *Crematogaster* and belong in the subfamily Myrmecine.

The Formicinae is the second largest group of ants, characterized by having only a single node on the petiole (pedicel) and the defensive nature of the sting has been replaced by an acid-projecting system. The group is otherwise extremely diverse both in body form, social organization, and life history. Most species tend to be more phytophagous in diet, and the collection of honeydew and nectar from extrafloral nectaries has become very highly developed. Some species have workers with a very large crop for honeydew storage and these are the "repletes" or "honey-pot ants" already mentioned. The most important genera in the Formicinae are the tropical *Camponotus, Oecophylla,* and *Polyrachis,* and the temperate *Formica* and *Lasius.*

The Myrmecinae is the largest group of ants and the most diverse, generally regarded as being very advanced. The pedicel has two nodes which is characteristic. Diet is varied, ranging from species that are quite carnivorous, with many being omnivorous, to some that are more or less completely phytophagous and live either on seeds or nectar, or on fungi. *Messor* and some species of *Pheidole* are the harvester ants that live in dry savanna and desert areas and feed largely on grass seeds. Important genera in the group include *Crematogaster* with more than 1000 species in the warmer parts of the world; *Tetramorium* with about 300 species in Africa, mostly, but with some "tramp" species cosmopolitan; *Monomorium* with 300 species in the tropics and subtropics and some in warm locations in temperate regions; *Pheidologeton* in tropical Asia with underground nests. The Attini contain the leaf-cutting/gardening/fungus ants—a Neotropical group extending through Central America and the West Indies into southern U.S.A. *Atta* occurs as at least 18 species from southern U.S.A. down to Argentina; the foraging workers cut pieces of leaf material from trees, shrubs, and herbs and carry them back to the nest to make their fungus gardens. They are basically forest species and make large underground nests (10–15 m/30–50 ft. in diameter) to a depth of 2m/6.5 ft.) and may contain up to 2 million ants in a single nest. They are serious polyphagous crop pests in cultivated areas at the edges of forests. *Acromyrmex* have small nests about 1 meter (3.5 ft.) in diameter and one to two meters deep, to be found in disturbed areas. They inhabit urban and agricultural land at a density of up to 60 nests/ha in parts of South America. Defoliation by these ants can be serious.

The last of the more important subfamilies is the Dolichoderinae—small in size, with only a few species, but in some respects regarded as highly evolved. The sting is lost and replaced functionally by an anal gland; the pedicel has a single node. Most species collect nectar and honeydew but are actually quite omnivorous in diet. Several species are important domestic pests in urban habitats and may be called "sugar ants" or "meat ants" according to their dietary preference; the main pest genera are *Tapinoma* and *Iridomyrmex*.

The taxonomy of the Formicidae has undergone major changes in recent years and the literature can be quite confusing with a plethora of names. There is also a large number of "tramp" species that have spread around the world from their original center of origin and they may now be very widespread.

Some of the more notable ants of importance to the agricultural entomologist include the following:

Acromyrmex spp. (leaf-cutting ants)—polyphagous fungus growers; southern U.S.A., Central and South America.

Anomma nigricans—eats oily pericarp of oil palm nuts; West Africa.

Anomma spp. (African driver ants)—carnivorous, nomadic (may be a subgenus of *Dorylus*); tropical Africa.

Atta spp. (18+) (leaf-cutting ants)—polyphagous fungus growers; southern U.S.A., Central and South America (Figs. 13.9, 13.10).

Camponotus spp. (1000 spp.) (carpenter ants, etc.)—variable, most nest underground; cosmopolitan.

Crematogaster spp. (1000 spp.) (cock-tail ants)—including the acacia gall ants (Fig. 13.11) of East and South Africa; most feed largely on honeydew and "farm" Coccoidea and Aphididae.

Dorylus spp. (driver ants, etc.)—most are predacious, nomadic and nestless (some are phytophagous sap feeders); Africa and tropical Asia (Fig. 13.12).

Eciton spp. (New World driver ants)—carnivorous with sting; Central and South America.

Fig. 13.9 Nest of leaf-cutting ant (*Atta* sp.) excavated in Paraguay, South America (Hymenoptera; Formicidae) (ex J. M. Cherrett).

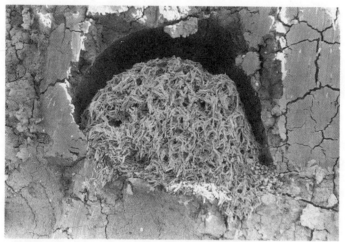

Fig. 13.10 Fungus chamber of *Atta* nest showing garden mixture of grass fragments; Paraguay, South America (Hymenoptera; Formicidae) (ex J. M. Cherrett).

Fig. 13.11 Galls on *Acacia* tree inhabited by acacia gall ant (*Crematogaster* sp.); Kenya (Hymenoptera; Formicidae).

Fig. 13.12 "Sausage-fly"—male *Dorylus* ant; body length 22 mm (0.88 in.); South China (Hymenoptera; Formicidae).

Formica spp.—(wood ant, etc.)—mostly predacious; temperate (Holarctic).

Harpegnathus venator (jumping ant)—solitary, predacious hunter; S.E. Asia.

Iridomyrmex spp. (200 spp.) (house/meat ants)—synanthropic and domestic pests; tropical Asia, Australasia, South America.

Lasius alienus (cornfield ant)—Europe and U.S.A.

Lasius flavus (mound ant)—more of a woodland species; Europe.

Lasius niger (common black ant)—mostly in gardens; an urban pest; Europe.

Macromischoides aculeatus (biting ant)—aggressive predatory species nesting arboreally in coffee bushes, etc.; tropical Africa (Fig. 13.13).

Messor spp. (harvester ants)—collect seeds of grasses and cereals when insect prey scarce; in arid areas of Africa and parts of Asia (Fig 13.14).

Monomorium pharaonis (pharaoh ant)—tiny, warmth-loving, domestic (urban) pest species; cosmopolitan.

Monomorium spp. (300 spp.)—scavengers with omnivorous habits, at most only minor agricultural pests; cosmopolitan.

Myrmecia spp. (bulldog and jumper ants)—large ants in very large colonies, mostly carnivorous; Australasia.

Myrmica spp. (50 spp.)—mostly Holarctic but some species Oriental.

Oecophylla longinoda (African red tree ant)—aerial nests in trees and plantation crops; tropical Africa.

Oecophylla smaragdina (Asian red tree ant)—agricultural status difficult to evaluate, can be a pest and can be beneficial; India, S.E. Asia to China (Fig. 13.15).

Paratrechina longicornis (crazy ant)—urban pest species; U.S.A.

Pheidole spp. (harvester ants, etc.)—live in arid areas and collect seeds of grasses and cereals when insect prey scarce; Africa, Asia.

Pheidologeton spp.—widespread in tropical Asia.

Polyrachis spp. (black tree ants, etc.)—many species in Old World tropics making aerial carton or silk nests (Fig. 13.16).

Pseudomyrmex spp. (acacia ants, etc.)—some species live in hollow thorns on *Acacia* trees; Central and South America.

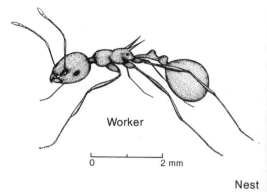

Worker

0   2 mm

Nest

Fig. 13.13 Biting ant (*Macromischoides aculeatus*) worker; Uganda (Hymenoptera; Formicidae).

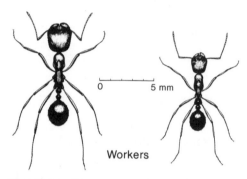

0   5 mm

Workers

Fig. 13.14 Harvester ant (*Messor barbarus*) (Hymenoptera; Formicidae).

Fig. 13.15 Asian red tree ant (*Oecophylla smaragdina*); body length *c.* 10 mm (0.4 in.); South China (Hymenoptera; Formicidae).

Fig. 13.16 Aerial carton nest of black tree ant (*Polyrachis dives*); Hong Kong; nest diameter *c.* 15 cm (6.0 in.) (Hymenoptera; Formicidae).

Solenopsis geminata (fire ant)—omnivorous agricultural pest, also attacking field workers; nest is underground; cosmopolitan (not Europe) (Fig. 13.17).

Solenopsis saevissima (imported fire ant)—agricultural pest species; U.S.A.

Solenopsis xyloni (southern fire ant)—California; feeds on almond kernels while the nuts are drying on the ground.

Solenopsis spp. (fire/thief ants)—about 250 species known, omnivorous, usually nesting underground; tropical U.S.A., Central and South America (also in Europe).

Tapinoma spp. (60 spp.) (house/sugar ants)—urban species, nocturnal in behavior, quite important pests; cosmopolitan (worldwide).

Tetramorium spp. (300 spp.)—worldwide now; most abundant in Africa, but from Old World tropics, now introduced into the New World.

Tetraponera spp. ("plant ants", etc.)—some species live in intimate association with many different plants; Africa, Asia, and Australasia.

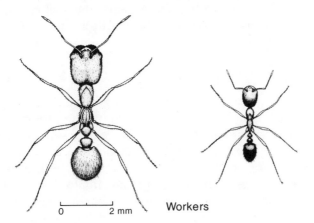

0   2 mm   Workers

Fig. 13.17 Fire ant workers (*Solenopsis geminata*) (Hymenoptera; Formicidae).

POMPILOIDEA

A large and homogeneous group of wasps, usually black in color, but some African species brown, with long legs, ovipositor hidden, and the adults are noticeably active and agile. Most species prey on spiders, and the group is somewhat more abundant in the tropics.

Pompilidae (spider-hunting wasps)   *4000 spp.*

These wasps have very long hind legs, and the abdomen has no definite petiole. Most are fossorial and all are predatory and provision their nests with paralyzed spiders. The group is not of much importance directly to agriculture but the wasps will be frequently encountered. The most spectacular species are *Pepsis* which prey on the giant tarantula spiders in South America.

Rhopelosomatidae

A very small family containing a few species, nocturnal in habit, widespread, but not found in Australia. The Nearctic species of *Rhopalosoma* are apparently ectoparasites of crickets, and they have no nest.

VESPOIDEA

These are what are generally regarded as "wasps"; adults have the forewing folded longitudinally at rest, and eyes usually emarginate on the inner edge. The present interpretation of this group is to have three closely related families that show interesting basic differences in their life-styles.

Eumenidae (potter wasps)   *3000 spp.*

The potter wasps are solitary and predacious, and to be found in both tropical and temperate regions. Adults have bifid claws, and often have an elongate but thick petiole (but in *Odynerus* the petiole is short). The true potter wasps (*Eumenes* spp.) build beautiful, small urn-shaped mud nests stuck on to a solid substrate or a twig and provisioned with paralyzed small caterpillars. Clearly, these species are an important part of the natural control complex of predators. Species of *Odynerus* inhabit keyholes, shelf-holes in bookcases, and similar small crevices in domestic premises. Other species dig holes in the earth or wood (tree stumps, etc.) and often partition the tunnel distally into a series of compartments divided by mud walls. The prey are almost invariably small caterpillars (Lepidoptera), but some species use larvae of Tenthredinidae, and others Chrysomelidae. A single egg is laid inside the cell, usually attached to the wall by a filament. Once the nest is sealed the adult wasp pays it no further attention.

Vespidae (true wasps; social wasps)   *800 spp.*

Mostly stout-bodied, yellow and black, or reddish in color; large in size (up to 20–30 mm/0.8–1.2 in. in body length); claws simple; petiole usually short. They are mostly social in habits and they build large nests of papery consistency either aerially (Fig. 13.18) or underground. In the tropics most nests are aerial or in tree or rock cavities, but in the cooler temperate regions most are underground or in cavities. The group is basically Holarctic but extends throughout Asia into India and down into Papua New Guinea. Most colonies, even in the tropics, only last for one season. In temperate countries the colony dies with the onset of winter, yet the young, fertilized queens hibernate overwinter in sheltered locations. In the spring the young queen emerges from hibernation and builds an initial nest containing about a dozen cells. The nest is made of wood scrapings mixed with saliva (called wasp-paper) and is applied as a series of overlapping arcs; the final nest is often large and either spherical or elongated. Internally it consists of a series of horizontal paper combs with ventral openings, suspended from a central pillar. Externally it is surrounded by an envelope of paper several layers thick. A large nest may contain up to seven combs and 12,000 cells, and each cell may be used twice or three times in a single season. On average most nests of *Vespa* in the summer contain about 5000 wasps. Males are usually only produced in late summer (from unfertilized eggs) when needed to fertilize the new young queens.

Wasps are basically carnivorous and feed on a wide range of insects and other invertebrates, but they also seek nectar, honeydew, and will bite holes in ripe fruit to suck the sugary sap. Their role in agriculture is complex. They clearly kill many crop pests and occasionally assist in pollination (*Vespa analis* with mango and kapok in Java), but they are greatly feared by field workers in orchards and plantations for their aggressiveness and painful stings; and finally they damage ripe fruits by biting holes to suck the sap. In many areas they severely damage ripening grapes. In some countries (e.g. Java) they are thought to be definitely pests but in others, such as China, they are encouraged in orchards as part of a biological control program and are regarded as being definitely beneficial.

The largest colonies are made by the Vespinae (*Vespa*, etc.) and the combs are enclosed within the outer envelope. The large tropical genus *Polistes* (paper wasps) make a single comb, usually horizontal and open, hanging in vegetation from a central pedicel; these are always only annual nests, and the queen is scarcely distinguishable from the workers. They are fiercely aggressive and have a very painful sting, and the aerial nest is usually in a bush and so easily disturbed. But *Polistes* wasps have been encouraged to nest in or near fruit orchards in parts of China where they kill many caterpillars and also lepidopterous pupae. *Polistes* species are mostly yellow with brown markings and they tend to be more slender than the *Vespa* species. *Vespula* usually make nests underground, but most nests of *Dolichovespula* are aerial. Some of the *Vespa/Vespula* species are occasionally domestic pests when they construct their nest inside buildings (houses, sheds, garages, etc.) or in gardens.

The main genera of concern to the agricultural entomologist are as follows:

Balanogaster spp.—build tiny nests hanging from the eaves of buildings in the tropics.

Dolichovespula—Europe, temperate Asia, Canada, and the U.S.A.

Polistes (paper wasps)—many species, most basically quite similar in appearance and habits; completely pantropical, and some also to be found in Europe and temperate U.S.A. (Fig. 13.18).

Vespa (tropical wasps)—many species are recorded throughout the warmer parts of Asia and North Africa (Fig. 13.19).

Vespa crabro (European hornet)—this large species, often the object of great alarm, nests underground and preys on other wasps; Europe, Asia, now in North America.

Vespula (temperate wasps; yellow jackets in the U.S.A.)—these are found throughout Europe, temperate Asia, Canada and the U.S.A.

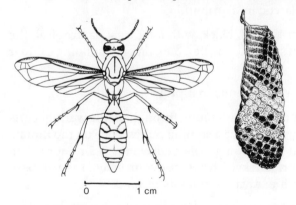

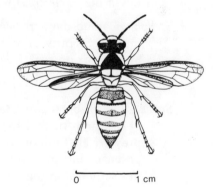

Fig. 13.18 Paper wasp (*Polistes olivaceous*) and nest; South China—some species have the nest comb held horizontally (Hymenoptera; Vespidae).

Fig. 13.19 Oriental wasp (*Vespa bicolor*); South China (Hymenoptera; Vespidae).

SPHECOIDEA

As now constituted, this group contains only a single family including many solitary wasps and the mud-daubers. The hind wing has enclosed cells and an anal lobe; usually a distinct and narrow petiole; male antennae with 13 segments, female with 12.

Sphecidae (sand wasps; mud-daubers, etc.)   *7700 spp.*

Most species are fossorial but some build mud-cells and all are predacious and provision their nests with paralyzed insects or spiders. The groups most used as prey include caterpillars (Lepidoptera), Hemiptera, Orthoptera, and spiders. The prey are stung and paralyzed, and the provisioned nest is then sealed and usually left. Most female sphecids lay less than a dozen eggs as the level of maternal care is sufficient to ensure a very high survival rate of larvae. Some species are recorded as using thrips, mayflies, springtails, cockroaches, bees, and even other wasps.

There is a series of distinct subfamilies that some authorities believe should be elevated to family status. Since these wasps are all predacious, and many use insects that are pests of one sort or another as their prey, they do have some importance as part of the local natural enemy complex.

A few examples of widespread and well-known Sphecidae are listed below, and some are illustrated:

Ammophila (sand wasps)—these are well-known species that live in sand dunes in Europe and provision their nests with caterpillars.

Ampullex (cockroach wasps)—metallic species with short wings; live in the tropics and prey on cockroaches in urban situations.

Bembix—gregarious Palaearctic species that also occur in Australia; they burrow in the ground in dense colonies; the nest is left unsealed and the females show maternal care and feed the larvae daily (progressive provisioning); most species prey on Diptera, often adult Tabanidae being most frequently taken.

Crabro—a very large genus of small, dark (often yellow/orange banded) wasps, they prey on Diptera and make nests in soil burrows, rotten wood, or plant stems.

Ectemnius fossorius—a wood burrower in South China that provisions its nest with hover flies. *Ectemnius* is Holarctic in distribution.

Sceliphron—mud-daubers; slender-bodied, black, with long legs; pantropical; the mud nest is provisioned with spiders and often built on or in buildings, under bridges, or on rock faces.

Sphecius (cicada-killers)—a New World genus that preys on cicadas.

Sphex (sand wasps)—large, dark-bodied wasps found throughout the warmer parts of Asia and Australasia; they often have a shining pubescence on the thorax; prey are all Tettigoniidae. In Hong Kong, *Sphex fulvohirtus* is a minor urban pest of quality grassland or turf (cricket pitches, golf courses, etc.) as groups of nests are found together with excavated mounds of sand.

Trypoxylon—make mud nests in rock cavities usually, but may use other natural cavities, beetle tunnels in trees, etc.; the prey is always small spiders and several are usually sealed in each mud cell.

The family is widespread and abundant and of great interest from various biological points of view, yet is only of peripheral concern to applied entomologists.

APOIDEA (bees, *sensu lato*) *2000 spp.*

These are the solitary and the social bees. The vast majority of species are solitary, but, of course, the social species are conspicuous and well known. The body is generally stout, short, and hairy, and the petiole is obvious; the hind tarsi more or less broadened and used for carrying collected pollen; mouthparts are modified so that the glossae are elongated into a tonguelike suctorial proboscis, but functional mandibles are still retained. Adult diet is mostly a combination of nectar from flowers and pollen, the former supplying carbohydrates and the latter proteins. The larvae are fed on the same diet except that the nectar is regurgitated. The social species store foodstuffs in comb cells in the nest for redistribution to larvae and workers during the dormant winter season; see page 62 on "apiculture".

This entire group is of paramount importance agriculturally as these are the species responsible for most of the pollination of crops and other plants. For the entomophilous plants (insect-pollinated plants) there are some flies and some Lepidoptera that are pollinators of certain types of flowers, but the vast majority are pollinated by bees; see page 53 on "pollination".

Colletidae (plasterer bees; membrane bees)

A somewhat primitive group of solitary bees with short "tongues", making simple nests in soil, holes in wood, hollow plant stems, etc.; the tunnel is lined with a salivary secretion which dries into a thin transparent film—hence the common name. The family is very well represented in Australia and South America.

The genus *Colletes* is Holarctic (as also is *Hylaeus*) with a large number of species. The nest tunnel is filled with a number of cells (2–8 usually), each filled with a mixture of pollen and honey and containing a single egg. *Colletes daviesanus* makes horizontal tunnels in the mortar (if not too hard) of the brickwork of houses in Europe and the U.K., and occasionally a particular house is plagued with a large number of tunnelling bees.

Halictidae (mining bees; sweat bees)

Another large group of bees, very abundant in Australia, generally widespread and common in most parts of the world. Its species are difficult to distinguish; bees are smallish and often metallic in appearance. *Nomia* seems to be the main genus in the New World tropics, and *Halictus* more abundant in slightly cooler areas. Nests are underground and vary in size; some species are quite solitary, but others are subsocial, and some fully social in small colonies. Some species of *Halictus* are attracted to people perspiring heavily and are called "sweat bees".

Andrenidae (burrowing bees; digger bees)

Medium-sized bees that burrow into the ground, making a vertical tunnel with lateral branches, each ending in a single cell. They are predominantly Holarctic, but some species occur elsewhere. *Andrena* is the main genus, with many species in Europe, Asia and North America, but none in Australia. These bees are solitary but often nest in colonies, and several females may use a common nest entrance.

Megachilidae (leaf-cutting bees and mason bees)

A very large group of bees, worldwide and abundant, none of which are social; they are mostly solitary bees but a few are parasitic in the nests of other bees. Adults are stout-bodied and characteristic with a large head well developed in the genal and occipital regions, and with two equally long submarginal cells in the forewing. The females that gather pollen have the pollen brushes (scopa) under the abdomen (rather than on the hind legs). There is a large number of subfamilies and tribes used by taxonomists.

Chalicodoma—builds nests of soil particles mixed with saliva, resin, wax, and chewed leaves (or paper), stuck on to a wall or hard substrate; the nest usually contains several cells and they are plastered over with the same material into rather an amorphous lump; some species are referred to as mason bees.

Megachile—includes the widespread and well-known leaf-cutting bees—they nest underground mostly, some dig their own tunnel but many prefer to use existing burrows. They construct a series of cells in the soil or rotten wood out of a series of cut leaf fragments. The leaves from which the nest material is collected present a very characteristic appearance with a series of almost circular holes around the periphery (Fig. 13.20). Leaf damage may often be extensive and unsightly but is seldom sufficient to be regarded as defoliation.

Osmia—also called mason bees; the group is Holarctic and abundant. They generally nest in existing hollows in wood, stone, empty galls, or plant stems. *Osmia cornifrons* in Japanese fruit orchards is provided with artificial nest sites to encourage large local populations to ensure adequate pollination of the fruit trees; the project seems to have been very successful (see page 55).

Anthophoridae (cuckoo bees, mining bees, etc.) *4000 spp.*

These bees have long tongues, and usually three submarginal cells in the forewing; they are stout-bodied, often very hairy. Some are black in color, but a very common species of *Anthophora* in S.E. Asia has distinctive blue-green banding on the gaster (Fig. 13.21). They are often to be seen feeding at flowers. They are solitary species but sometimes nest gregariously in the soil. *Anthophora* is widespread throughout Europe, Asia and the U.S.A.; in North America the common genus is *Hemesia*. *Nomada* and several other genera are parasitic upon various ground-nesting bees.

Xylocopidae (carpenter bees)

These are mostly large, 2–3.5 cm/0.8–1.4 in. long) hairy bees, very stout, with dark wings and the top of the gaster bare and shiny (Fig. 13.22). Many species are black, but some have a gray or brown pubescence and some African species are bright rufous. These bees are commonly seen at flowers where they make a very loud buzzing sound with their wings. They have very large mandibles, in addition to a long tongue, and they excavate long tunnels in dead tree trunks or branches, in house and structural timbers, as well as telegraph poles. Throughout Africa and tropical Asia, where buildings are constructed with broad overhanging eaves to allow for runoff during torrential or monsoonal rains, *Xylocopa* bees tunnel the exposed part of the wood beams. The tunnels are almost wide enough to push a finger into, and extend for 20–30 cm (8–12 in.) or more. One species occurs as far north as Paris, France. *Xylocopa iridipennis* is the bamboo carpenter bee of China and S.E. Asia, and it nests inside the internode of, usually, a dead bamboo (Fig. 13.23); only the larger species of bamboo are selected for use by this bee. *Ceratina* are the small carpenter bees, only about 6 mm (0.24 in.) in length; hairless; metallic blue-green; they nest in the pith of various shrubs and are well represented in Africa and the U.S.A.

Fig. 13.20 *Bougainvillea* leaves damaged by *Megachile* (leaf-cutting bee); Bird Island, Seychelles (Hymenoptera; Megachilidae).

Fig. 13.21 Cuckoo bee (*Anthophora* sp.); body length 14 mm (0.6 in.); Hong Kong (Hymenoptera; Anthophoridae).

Fig. 13.22 Bamboo carpenter bee (*Xylocopa iridipennis*); body length 26 mm (1 in.); South China (Hymenoptera; Xylocopidae).

Fig. 13.23 Nest entrance in bamboo stem of *Xylocopa iridipennis* (Hymenoptera; Xylocopidae).

Apidae (honey bees; bumble bees, etc.)   *1000 spp.*

These are the true social bees that have corbiculae (pollen baskets) on the hind tibiae of the worker caste. There are a few parasitic species. There are four subfamilies that are very distinct.

The Euglossinae are South American, large, solitary, and they make small nests of mud or resin; this includes the spectacular orchid bees.

The Bombinae are the bumble bees; large, stout, hairy, and mostly conspicuously colored black and yellow or red, with white bands (Alford, 1975). They are abundant in the Holarctic region, but tend to be confined to higher altitudes in hotter countries. Some species are now introduced into New Zealand to pollinate the red clover crops, but, to date, attempts to introduce them into Australia have been unsuccessful. *Bombus* in temperate regions has a life cycle similar to that of *Vespa*, in that the colony (small in this case) is annual and the young fertilized queens hibernate overwinter. In the spring, the queen constructs a small nest underground and tends the first brood, after which the workers take over foraging duties and the queen stays in the nest to lay eggs. The nests are small and usually contain only a couple of dozen workers; the nest is underground, often in a vacated mouse hole, at the end of tunnel of about half a meter (1–2 ft.), and is in the center of a hollow ball of fine grass or moss. A few species are known as "carder bees" and they make surface nests of plaited grass and moss in the grass or ground foliage. They are very important pollinators for many crops, particularly red clover and some other forage legumes, where the heavy body of the bee is required to open the flower properly for pollination to be effected.

The Meliponinae include about 250 species in the tropics—most in South Africa, but also in other parts of Africa, and Australia, and parts of Asia. They are tiny in size (some less than 3 mm/0.12 in. long) and often called "mosquito bees" or "stingless bees". They nest in crevices, both natural and in buildings (toilet overflow pipes, etc.), and the entrance is usually conspicuous as a projecting funnel of cerumen (wax from the abdominal terga mixed with resin or soil). *Melipona,* confined to South America, are almost as large as honeybees. *Trigona* are tiny and common throughout the more southern parts of tropical Africa, as well as parts of S.E. Asia and northern Australia.

The Apinae include the famous and ubiquitous honeybee (*Apis mellifera*) which has probably been studied more than any other species of insect, and has been used by humans since prehistoric times. Apiculture is the rearing and managing of bee colonies and is conducted in virtually every country in the world. On pages 62–68 some detail of the art of apiculture is given, together with various illustrations. The original object of apiculture was for the production of honey, and to some extent also wax. But, in recent years, the use of managed honeybee colonies for crop pollination has become very important (see pages 53–62).

There are four species of *Apis,* native to southern Asia; these are as follows:

Apis cerana (eastern honeybee)—*A. c. indica* is the Indian honeybee.

Apis dorsata (giant/rock honeybee).

Apis florea (little honeybee).

Apis mellifera (the honeybee)—thought to have originated in eastern Europe or western Asia and now is worldwide as a series of subspecies and races; cultivated since prehistoric times.

The nest is perennial and may last for many years in a hollow tree or crevice in rocks, for winter provisioning is practiced so that a reduced population can survive in the nest overwinter or through a pronounced dry season. Polymorphism occurs with the queen being larger than the workers, and winged males (drones) are produced annually solely for the purpose of mating with the young queens. Colony fission is the method of spread of the species; annually the old queen, and sometimes a young queen also, will leave the colony with a swarm of several hundred (or a few thousand) workers to found new colonies. In temperate regions the storage of honey is a major feature of the colony economy but in the tropics most of the colony energy goes into swarm formation. A large colony contains some 50,000–80,000 workers. Honeybees are totally phytophagous and feed on a mixture of nectar and pollen which is collected by the foraging workers and stored in the honeycombs inside the nest, although some is fed directly from mouth to mouth to other workers and the larvae in the nest.

Honeybee management basically involves the suppression of part of the normal life history so as to minimize swarm development and to maximize honey collection. But it should be stressed that, even after thousands of years of apiculture, the honeybees are not domesticated—they are wild insects that have been induced to nest in hives for human convenience, where, with skill, they can to some extent be managed.

14

CLASS ARACHNIDA

There are other Arthropods of some importance in the broad agricultural sense. Myriapoda are divided mainly into two large groups: the predacious Chilopoda (centipedes) that are predators of soil insects, and the few large species (such as *Scolopendra*) hazardous to humans in domestic dwellings. The phytophagus/scavenging Diplopoda (millipedes) do occasionally damage soft-stemmed seedlings in protected cultivation.

The Crustacea include terrestrial woodlice that damage ripening fruits and soft-stemmed plants occasionally, and river rice is damaged by feeding shrimps of different types, and freshwater and mangrove crabs.

It is not really appropriate to include these minor pests in this book, but the Arachnida need a brief mention.

Within the Class Arachnida are several orders that include scorpions, chelifers, harvestmen, etc., that are all predacious and feed mostly on other smaller arthropods, mainly insects, and they all contribute to the general overall natural predator complex responsible for controlling many insect populations. It is very obvious that spiders eat insects, but, in the past, in Western literature it was usually assumed that predation by spiders was not overly important in controlling insect pest populations. However, the Chinese have a long tradition of biological control practices in their agriculture, and some successful biocontrol programs involving the use of spiders were started in 1975 (Hill, 1987). In the last decade, work in Canada and at I.R.R.I. with rice has shown that insect pest predation by spiders can be of great importance and should never be neglected in future I.P.M. programs.

ORDER ARANEIDA (spiders)   *many families*

Several different classifications are being used in different parts of the world, but the main families appear to be fairly well defined, as follows:

GNAPHOSIDAE (nocturnal hunting spiders)—these spiders roam vegetation at night seeking their prey, actively; generally webless.

CLUBIONIDAE (as above)

THOMISIDAE (crab spiders)—diurnal species that lurk in flower heads awaiting their prey; they may kill many pollinating insects.

SALTICIDAE (jumping spiders)—diurnal hunters that jump onto their prey.

LYCOSIDAE (wolf spiders)—mostly nocturnal, fast running spiders; several large, tropical, domestic species are common; bite painfully if handled.

ARGIOPIDAE (orb web spiders)—large, vertical webs built, often quite spectacular, such as those by *Nephila* spp.; prey trapped in the sticky web.

LINYPHIIDAE ("money" spiders)—the largest family with many species, many small, webs in foliage, of many sizes and shapes; young dispersed by wind on gossamer threads.

ORDER ACARINA (mites and ticks)

In the field of agricultural/applied entomology it is customary to include the ticks and mites because they are arthropods and clearly share some of the characteristics of the insects. However, in some universities and research establishments Acarology is regarded as a distinct discipline. In practical terms, most agricultural entomologists are actually working as applied zoologists. It is really necessary to view the crop pest spectrum as a whole and to include mammals, birds, molluscs, as well as insects and mites. Nematoda are a problem in that they are very important, and often underrated pests of cultivated plants, mostly invisible to the unaided eye, and so their field infestations are assessed initially on the basis of symptoms, as with plant diseases. This is why the study of nematodes was formerly a part of plant pathology. Now nematology is generally regarded as a branch of agricultural zoology, but a separate discipline in its own right. Most entomologists are expected to have some basic knowledge of nematology but not detailed expertise. In a book of this nature, it is thought that the Acarina must be included but space problems do not permit the inclusion of Mollusca and vertebrate pests. Books with a crop approach clearly need to have these other pests included as the pest complex must be viewed in its entirety; for example see Hill & Waller (1988).

Classification of the Insecta is now fairly stable, but there is some variation from country to country, as already mentioned. However, the classification of the Acarina by comparison is quite chaotic; it varies from source to source quite considerably. In this book a somewhat simplified system is employed—and the Order Acarina is divided into seven distinct groups regarded as suborders which together contain more than 200 different families and many thousands of species. It is only possible here to view the groups very superficially, for more detailed information the sources listed in the bibliography should be consulted, e.g. Jeppson, Keifer & Baker (1975); Evans, Sheals & Macfarlane (1961); Smith (1973); Harwood & James (1979); and Baker & Wharton (1964).

It must be remembered that the vast majority of species are cryptozoic and scavenging, or predacious, and that only a relatively few species are important to the agricultural entomologist.

SUBORDER MESOSTIGMATA

The largest group; most are parasitic or predacious, well armored, and with the usual four pairs of legs. It should perhaps be pointed out that the first instar in the Acarina has only six legs and is called a larva; the next instar has eight legs and is a nymph. There is a total of 66 families in this group but most are small and rather obscure.

Parasitidae

These are small, predatory mites found in litter and decaying vegetation mostly. They feed on tiny insects and other small soil arthropods.

Dermanyssidae (red mites, etc.)

Medium-sized parasitic species, most occur on wild birds and rodents, but some attack livestock and will also feed on humans if presented with the opportunity. A few important species are mentioned below:

> *Dermanyssus gallinae* (poultry red mite)—major pest of domestic fowl; heavy infestations can kill the birds; also vectors of various virus diseases; can feed on humans; cosmopolitan.

> *Ornithonyssus bacotic* (tropical rat mite)—attacks humans as well as rodents; pantropical.

> *Ornithonyssus bursa* (tropical fowl mite)—pest of all domesticated fowl in the tropical parts of the world.

Phytoseiidae (predatory mites)

These are relatively large predatory mites, long-legged and active, usually to be found on plant foliage. They prey naturally on Tetranchidae (spider mites) and other phytophagous mites. Several species have now been used for years in biological control programs both in greenhouses in temperate regions (U.S.A., Europe, and now Asia) and outdoors in orchards and plantations in warmer regions. A number of species are widely available commercially for use in biological control or I.P.M. programs. Most of the exploited species belong to the three genera listed below:

> *Amblyseius*
> *Phytoseiulus*   A number of species names are used in the literature and have been moved from genus to genus, so it appears that the precise limits of some species and genera are not too clear.
> *Typhlodromus*

SUBORDER IXODIDES (ticks)

These are large species, parasitic (mostly on Vertebrata), with a well-developed, piercing hypostome armed with rows of recurved spines; once the hypostome is fully embedded in the flesh of the host it is immovable and can only be withdrawn by a drop of internal turgor pressure. Most of the hosts are mammals and birds, including humans.

Argasidae (soft ticks)

The integument is soft and distendable and the mouthparts are somewhat hidden ventrally. The group are cosmopolitan and feed on a very wide range of hosts (e.g. mammals, birds, snakes, turtles, etc.). Typically they live in the nest or lair of the host. Often they feed intermittently and may drop off the body of the host between feeds. Many species feed at night and so their presence is often not easily detected as they are not seen during daylight hours. Some species are recorded as having great longevity and under starvation conditions (absence of a host) have been recorded to survive for years.

A few important species that attack domestic livestock include:

Argas persicus (fowl tick)—abundant on a wide range of domesticated birds
throughout the warmer parts of the world; causes anemia and is a vector of
several disease organisms.

Argas spp.—occur on birds, bats, lizards, and small insectivores; worldwide.

Ornithodorus spp.—about 90 species are known, worldwide; several attack
livestock (birds and mammals) in parts of the U.S.A. and tropical Asia; other
species are vectors of several medical and veterinary diseases.

Ixodidae (hard ticks)

These ticks still have a very distendable integument and after a full feed they swell
dramatically to the size of a pea seed, but they have a hard dorsal skeletal plate (shield)
and hence are regarded as "armored"—the shield or scutum extends dorsally over the
whole body surface in the male, but only over the anterior part in the female. The mouthparts are clearly visible in dorsal view at the anterior end of the body.

Eggs are laid in batches of up to 18,000 in crevices or under stones and the female
then dies. The emerging larvae are called "seed ticks" and they are tiny and active, and
climb to the tips of the foliage where they await a passing host. In savanna or grassland
areas the larval ticks can be seen as small brown clusters on the tips of blades of grass.
When touched by a passing host the tiny ticks seize hold of the pelt or plumage (or
clothing). Once firmly established on the host the larval ticks take in a blood meal and
when fully engorged usually dop off onto the ground to molt. Most ticks are classified
biologically according to the number of hosts required to complete the life cycle—as
one-host, two-host, or three-host ticks. The one-host ticks stay on the same host for the
whole life cycle, molting taking place on the body of the host. Two-host ticks have the
larvae and young nymphs staying on the same host and molting there, but the full-
grown nymph drops off onto the ground for its final molt and the young adult tick has to
find a new host. The most common type is the three-host tick where each stage in the life
cycle, larva, nymph, and adult, has a separate host. At each stage the tick takes in so much
blood from the host that it swells enormously (thanks to the soft flexible integument). In
many common species the fully engorged adult tick is the size of a pea seed.

On the vertebrate host, the preferred sites for attachment are locations where the
host is unable to scratch itself or otherwise dislodge the ticks. In and around the ears are
favored sites, also the groin area and around the anus, and some species use the orbital
rim. The main effect of a tick infestation is anemia; with a heavy infestation the host loses
a great deal of blood—it has been estimated that a cow may lose up to 100 kg (200 lb.) of
blood in a single season (Harwood & James, 1979). Dairy cattle show a drop in milk yield
of up to 25%, and weight gain in calves is noticeably retarded by tick infestations. In
Australia, beef losses due to tick infestations were recently (*c.* 1977) estimated as U.S.
$25 million annually. In addition to loss of blood there is usually some toxicosis caused
by the salivary secretions of the ticks, and there may also be paralysis. Many species of
ticks are the natural vectors of viruses, parasitic protozoa, spirochetes, rickettsias, and
the like, causing diseases in cattle and humans that are very debilitating and may be fatal.
Parts of the savanna/grassland regions of Africa, Asia, Australia, North and South
America are rendered unsuitable for cattle rearing owing to the presence of large
numbers of ticks transmitting very serious diseases.

In the past, cattle-tick control consisted almost entirely of regular dipping in an
acaricidal solution. In recent years, the development of systemic acaricides, which are

either injected by syringe or ingested in food, has facilitated chemical control considerably.

Some species are quite host-specific, but others are quite polyphagous and in such cases the abundance of local, wild reservoir hosts can be important. In parts of the U.S.A. picnickers and campers often pick up disease-carrying ticks from vegetation, which have originated on local deer, or other large mammals. Some of the more important genera and species of ticks of importance to humans include:

Amblyomma—about 100 species; large tropical ticks with long mouthparts. They are three-host ticks, mostly found in Africa, tropical Asia, Central and South America. There is a very wide host range, from amphibians, reptiles to birds, rodents and large mammals. Immatures of many species (called "seed" ticks) will attack humans, but adults are less likely to attack. Various diseases are transmitted among cattle by these ticks.

Boöphilus—five species are known, widespread throughout the tropics and important as cattle pests in both Old and New Worlds. They are small in size, especially when unfed, and easily overlooked; they often pass quarantine inspections. Although important on cattle, they are also widely found on deer, horses, donkeys, sheep, and goats.

Dermacentor (30+ spp.)—ornate, mostly three-host ticks but includes some one-host species; very important in temperate North America, spreading cattle diseases, and also attacking humans and dogs; immatures regularly found on wild rodents and lagomorphs.

Haemaphysalis (150 spp.)—worldwide; small in size and sexes similar; probably originating in South Asia; many different hosts are recorded.

Hyalomma (21 spp.)—large ticks without ornamentation; adapted for survival in cold, heat, and arid conditions; thought to be native to Central Asia; the group shows great variation and adaptability.

Ixodes (250 spp.)—the largest genus, wordwide, on a wide range of hosts. On humans the bite may be quite toxic, causing pain, fever, vomiting, and sometimes they cause tick-paralysis; spirochetes and other parasites are transmitted in mammals.

Ixodes ricinus (castor bean tick)—found in Europe and West Asia; quite polyphagous on a wide range of hosts; other species equally as important in other parts of Asia and North America.

Nosomma monstrosum (the only species)—an important pest of cattle, buffalo, humans, and other animals in the region of India and S.E. Asia.

Rhipicephalus (c. 63 spp.)—an Old World group, mainly found in Africa, seldom recorded from either birds or reptiles, but otherwise on many different hosts.

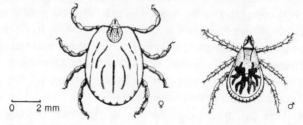

0   2 mm   ♀   ♂

Fig. 14.1   Brown dog tick (*Rhipicephalus sanguineus*); female and male (Acarina; Ixodidae).

Rhipicephalus sanguineus (brown dog tick)—now probably the most widespread tick species, completely worldwide in warmer regions; on a wide range of hosts but with seemingly a preference for dogs; vector of several major canine diseases and some transmissible to humans (Fig. 14.1).

SUBORDER TROMBIDIFORMES (= Prostigmata)

A diverse group of both animal and plant parasites; small to tiny in size, a few larger predatory species. Some 35 families are placed here but only about six are of any importance agriculturally. In some groups there is morphological reduction and the number of legs can be reduced to two pairs (usually the anterior two pairs remain).

Eriophyidae (gall/rust/blister mites)

These are minute mites that gall a very wide range of wild and cultivated plants. They are only visible under a hand lens, and a microscope is really needed to see any detail. The body is elongate, tapering, and rather wormlike, with only two pairs of small anterior legs. They feed by rasping and piercing the surface tissues with chelicerae modified into stylets. Air enters the cells when the epidermal cells are emptied of sap and gives the surface a silvery appearance (by reflection), and as the cells die the overall effect is of bronzing. When the mites damage young tissues, the subsequent growth and expansion of these tissues lead to the development of galls or other characteristic damage. In some cases the effect of the feeding of the young mites is to produce very specific galls at precise locations—these species can, of course, be identified in the field by shape and location of the galls on specific trees and bushes. The association between plants and eriophyid mites is very close, and most mite species are quite host-specific and their damage to the plant is often characteristic. The most usual damage by these mites is more drastic than the silvering and bronzing characteristic of the Tetranychidae; leaf-folding, leaf-rolling, leaf-cupping are practiced but the development of erinea is the most common effect. An erinium is a warty outgrowth of dense hairlike structures from the damaged epidermis, usually on the lower surface of the leaves, and evident on the upper surface as pale sunken patches. Examination of the erinium by microscope will reveal the tiny maggot-like mites located between the outgrowths that form the erinium (Figs. 14.3–14.5). Some species inhabit buds, and the unopened buds enlarge and in this swollen, closed condition they are known as "big-buds". A characteristic species on plums produces a series of small, bead-like galls around the leaf periphery.

Eriophyid females in temperate regions do not lay resistant eggs (as do Tetranchyidae, etc.) to overwinter but the females hibernate; similarly in the tropics the females may aestivate to avoid a prolonged dry season. Eggs of Eriophyidae are so minute as to be very seldom seen by field entomologists.

A few species are recorded as transmitting viruses causing plant diseases, and others are often found with fungal spores on their bodies.

The identification of gall mites is a very specialized task and only successfully completed by acarologists with techniques and equipment not usually to be found in the average field laboratory. However, with the literature now available, if the host-plant identity is known, and the damage symptoms (gall, etc.) observed, it is usually possible to arrive at the identity of the causal mite, providing that the host is a cultivated plant. Gall mites from wild hosts will be returned by most acarologists most of the time without a specific determination.

A number of mites belonging to what some authorities regard as the Eriophyoidea are particularly damaging to plants on two counts. Firstly, they inject a systemic and persistent toxin in their saliva which acts as a form of growth regulator and produces distortion, and/or tissue necrosis in the host plant—the damage often is similar to the symptoms of some types of plant virus; damage is typically progressive and irreversible. Secondly, they can transmit plant viruses causing a number of different diseases. Unfortunately, the two different sets of symptoms are often sufficiently similar to be indistinguishable, so it is usually not clear just how extensive virus transmission may be. But, a number of different mites are definitely known to transmit different viruses causing diseases on cereals, grasses, various shrubs, and some trees.

There are several dozen gall mites of agricultural importance, although it should be stressed that many actually cause little crop loss. However, the damage symptoms may be quite spectacular so their recognition is of importance. Forest tree seedlings and saplings may suffer growth retardation as a result of heavy leaf infestation (see Fig. 14.5). A few species do, however, cause serious damage and some have been recorded killing the host plant. The total list of phytophagous species recorded from crop plants of all types according to Jeppson, Keifer & Baker (1975) is very extensive, but the records of many species appear to be few in number. Some of the more important species are listed below:

Abacarus hystrix (grain rust mite)—on perennial grasses, oats, barley, wheat, and rice (also transmits rye grass mosaic virus); more or less Holarctic in distribution.

Acalitus gossypii (cotton blister mite)—causes growth distortion, erinea, etc., and can destroy whole plants; U.S.A., Central and South America.

Acalitus phloeocoptes (almond and plum bud gall mite)—gall bud edges, and almond trees can be killed (plum usually recovers); Europe, Asia Minor, U.S.A.

Acaphylla theae (pink tea rust mite)—scarify both surfaces of tea leaves; can retard bush growth; prefers Assam varieties to Chinese; found throughout southern Asia to China and Japan.

Aceria spp. (5+)—from different fruit trees; India, South America.

According to the new checklist by Wood (1989) many of the pest species long regarded as belonging to *Eriophyes* should now be placed in *Aceria* and some other genera.

Aceria (= *Eriophyes*) *sheldoni* (citrus bud mite)—on citrus, especially lemon, distorts buds, twigs, and fruits; records from scattered locations in Mediterranean, Africa, India, Java, Australia, U.S.A. and South America (Fig. 14.2).

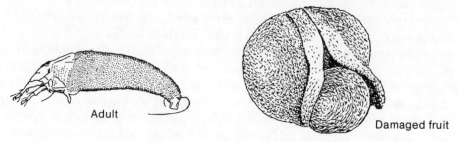

Adult

Damaged fruit

Fig. 14.2   Citrus bud mite (*Aceria sheldoni*) (Acarina; Eriophyidae).

Aculops lycopersici (tomato russet mite)—attacks tomato, potato, tobacco, and petunia; on tomato bronze stem which cracks, plant withers and dies; almost worldwide now.

Aculops pelekassi (pink citrus rust mite)—distorts and bronzes foliage and fruit of citrus and also on new growth; Europe, Asia, Florida and Brazil.

Aculus cornutus (peach silver mite)—leaves of peach and almond silvered; tree devitalized and fruit fall prematurely; now worldwide.

Calacarus carinatus (purple tea mite)—mature tea leaves bronzed; during drought, tea bushes defoliated; Assam teas more susceptible than Chinese; India; S.E. Asia, Japan, Australasia, and U.S.A.

Calepitrimerus vitis (grape rust mite)—rust leaves and distort shoots; Japan.

Cecidophyopsis ribis (currant big-bud mite)—infested black currant buds swell and later die; vector of reversion disease; Europe and North America.

Colomerus vitis (grape leaf blister mite)—vine leaves with erinea; worldwide.

Diptacus gigantorhynchus (big-beaked plum mite)—plum leaves distorted; cosmopolitan.

Eriophyes caryae (pecan leaf-roll mite)—U.S.A.

Eriophyes erineus (Persian walnut blister mite)—erinea under leaves; Europe, Asia, and U.S.A. (Fig. 14.3).

Eriophyes ficus (fig bud mite)—attacks fig buds and transmits fig mosaic virus; U.S.A.

Eriophyes litchi (litchi mite)—leaves galled and rolled; India, Pakistan, China, Hawaii.

Eriophyes lycopersici (tomato erinium mite)—erinea on leaves; pantropical.

Eriophyes mangiferae (mango bud mite)—Near East, S.E. Asia, U.S.A. and South America.

Eriophyes oleae (olive bud mite)—leaves and fruits distorted; Mediterranean Region.

Eriophyes pistaciae (pistachio bud mite)—distort leaves and flower stalks; Near East.

Fig. 14.3 Persian walnut blister mite (*Eriophyes erineus*) (Acarina; Eriophyidae); Louth, U.K.

Fig. 14.4 Erinea on leaf of *Shefflera octophylla* produced by a species of *Eriophyes;* Hong Kong (Acarina; Eriophyidae).

Eriophyes rossettonis (cashew bud mite)—South America.

Eriophyes tulipae (wheat curl mite)—on grasses, cereals, and Liliaceae; distort foliage, and vector of several viruses; Europe, Asia, Australasia, U.S.A.

Eriophyes spp.—on many fruit trees, bushes, and forest trees; worldwide (Figs. 14.4, 14.5).

Eriophyes spp.—on grasses, cereals and sugarcane; worldwide.

Phyllocoptella avellaneae (filbert big-bud mite)—swollen, unopened buds die; Europe, Asia, Australia, and North America (Fig. 14.6).

Phyllocoptruta oleivora (citrus rust mite)—citrus fruit skins thickened and scarified; pantropical (Fig. 14.7).

Phytoptus pyri (pear leaf blister mite)—buds attacked, and small erinea on leaves of pear and apple; Europe, Asia, South Africa, U.S.A. (Fig. 14.8).

Phytoptus similis (plum leaf bead gall mite)—small, pouch (bead) galls around leaf edge, on plum, damson, and blackthorn; Europe.

Fig. 14.5 Leaf of *Celtis* attacked by gall mites (*Eriophyes* sp.)—some leaves totally distorted; Alemaya, Ethiopia (Acarina; Eriophyidae).

Fig. 14.6 Filbert big-bud mite (*Phyllocoptella avellaneae*); Alford, U.K.

Fig. 14.7 Citrus rust mite (*Phyllocoptruta oleivora*) ex Uganda (Acarina; Eriophyidae).

Damaged fruit

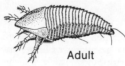

Adult

Fig. 14.8 Pear leaf blister mite (*Phytoptus pyri*) infested leaves; Skegness, U.K.

Tarsonemidae (tarsonemid mites)

Small plant mites, slightly elongate, with short legs, but not so reduced anatomically as the Eriophyidae. A few are important crop pests, others are free-living, some scavenging, and some parasitic on insects. The plant feeders tend to attack young leaves, buds, and young shoots, and their feeding causes growth deformation and a reduction of crop yield. Some of the pest species appear to be quite polyphagous and attack a wide range of cultivated plants. Some important species are listed below:

Polyphagotarsonemus latus (broad mite; yellow tea mite)—truly polyphagous, on tea, cotton, coffee, jute, mango, tomato, potato, beans, peppers, avocado, etc.; on field crops in warmer regions but in greenhouses in cooler temperate regions of Europe and U.S.A.; cosmopolitan but records somewhat scattered to date (Fig. 14.9).

Stenotarsonemus ananas (pineapple tarsonemid)—specific to pineapple, attack fruits and cause segment rotting; Australia and Hawaii.

Stenotarsonemus bancrofti (sugarcane stalk mite)—specific to cane, infests cane tops causing scarification; Africa; S.E. Asia (not India), North, Central and South America.

Stenotarsonemus laticeps (bulb scale mite)—on *Narcissus* and other Amaryllidaceae; feed on bulb scales and distort flowers and leaves; Europe and U.S.A.

Stenotarsonemus spirifex (oat spiral mite)—on oats, etc., cause deformed blind spike; Europe.

Tarsonemus pallidus (strawberry/cyclamen mite)—on cyclamen, watercress, strawberry; leaves and flowers distorted, plant stunted; Europe, Asia, North America.

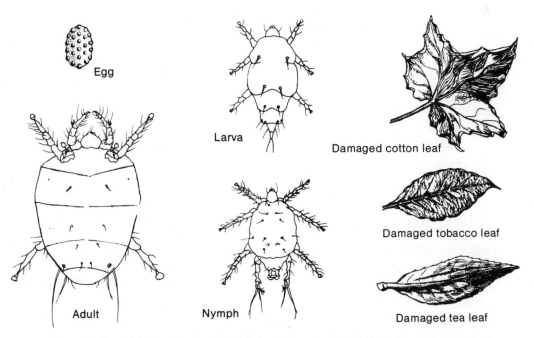

Fig. 14.9 Broad (yellow tea) mite (*Polyphagotarsonemus latus*) (Acarina; Tarsonemidae).

Tetranychidae (spider mites)

These mites by comparison with the previous families are large—they can just be seen with the unaided eye! Either red or green in color, they are quite long-legged and relatively active, found in plant foliage, usually on leaves, and there may be silk webbing over the foliage, especially in heavy infestations when webbing can be very extensive. They lay large, globular, red eggs on the foliage, usually singly. Sexual dimorphism is usually obvious. By piercing epidermal cells with their stylets and removing the sap, they give the damaged leaves a silvery or bronzed appearance, and, if foliage is heavily attacked, the leaves wilt and may die. In temperate regions the overwintering eggs are laid on twigs and spurs in crevices and scars. The group is renowned for the resistance to pesticides shown by so many species. One experiment recorded that *Panonychus* mites actually thrived on a diet that was supplemented with D.D.T.

Many species are known throughout the world, and there are both tropical and temperate pest species. Some are polyphagous; many have a definite host preference. Many species also have a definite preferred location on the host plant, but will move elsewhere when populations are high. Some prefer the upper surface of leaves, others the underside, others prefer fruits, and occasionally young twigs; and a polyphagous species may prefer different locations on different host plants.

Damage to crop plants is often difficult to evaluate with any precision, but is often more pronounced during dry periods or drought when damaged leaves may be shed. On fruit and other trees, damage is particularly difficult to assess. On many field crops, severely damaged plants may die. Fruits are often marked by discoloration so that there is a decrease in quality and value. A few species are recorded as virus vectors.

A very large number of Tetranychidae are recorded from cultivated plants, especially ornamental shrubs and flowers, but the more serious crop pests are restricted to a few dozen species, some of which are listed below:

Bryobia—this genus is characterized by having a rounded body with setae broadly clavate and lightly serrate which gives them a distinctive appearance.

Bryobia cristata (grass-pear bryobia mite)—on grasses, herbaceous plants and occasionally fruit trees; Europe, North Africa, Japan, Australia, New Zealand.

Bryobia praetiosa (clover mite)—polyphagous on herbaceous plants and cereals (seldom seen on trees); Europe, Asia, Africa, Australia, North and South America.

Bryobia rubrioculus (apple and pear bryobia/brown mite)—multivoltine temperate species, overwintering as eggs, on all fruit and nut trees; Europe, Asia, South Africa, Japan, Australia, North and South America.

Eotetranychus—these mites are slender-bodied, yellow or greenish with dark spots along each side of the body; most feed underneath the leaves in small colonies.

Eotetranychus boreus (apricot spider mite)—on apricot, etc.; Japan, U.S.A.

Eotetranychus hicoriae (pecan leaf scorch mite)—on pecan, chestnuts, oaks; U.S.A.

Eotetranychus lewisi (Lewis' spider mite)—on citrus, papaya, olive, clovers; U.S.A., Central America.

Eotetranychus sexmaculatus (six-spotted spider mite)—on citrus, avocado, camphor and other trees and shrubs; India, China, Japan, New Zealand, and U.S.A.

Eotetranychus willametti (Willamette mite)—serious pest of grapevine in California; also on apple, pear, elms, oak, etc., in other parts of the U.S.A.

Eutetranychus banksi (Texas citrus mite)—on citrus, almond, fig, castor, croton, and other plants; North, Central and South America.

Eutetranychus orientalis (oriental mite)—on citrus, pear, grapevine, walnut, quince, cotton, squash, frangipani, etc.; Africa, India, S.E. Asia (Fig. 14.10).

Eutetranychus ssp.—several species recorded on fruit trees; Africa and India.

Mononychellus tanajoa (green cassava mite)—an endemic minor pest in South America; recently accidentally introduced into Uganda and now spread through Central and West Africa—rated a very serious pest of cassava and is doing serious damage.

Oligonychus—a large genus, in appearance very like *Tetranychus;* one group of species infests fruit, nut, and forest trees, worldwide; another group of species infests a wide range of Gramineae including sugarcane; in both tropical and temperate regions.

Oligonychus afrasiaticus (date spider mite)—major date pest; Africa, Near East.

Oligonychus coffeae (red coffee/tea mite)—on tea, coffee, jute, rubber, citrus, cashew, mango, grapevine, oil, palm, camphor, mulberry, etc.; pantropical.

Oligonychus indicus (sugarcane leaf mite)—on sugarcane foliage; India and U.S.A.

Oligonychus mangiferus (mango spider mite) on mango, cotton, peach, pear, quince, pomegranate, grapevine, *Rosa,* etc.; pantropical.

Oligonychus orthius (sugarcane spider mite)—on foliage of sugarcane; Japan.

Oligonychus oryzae (rice spider mite)—on rice; India.

Oligonychus pratensis (Banks grass mite)—on grasses and cereals; Africa, North, Central and South America.

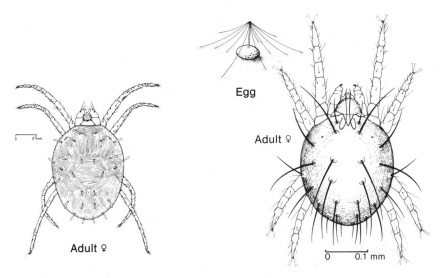

Fig. 14.10 Oriental mite (*Eutetranychus orientalis*); adult female; Hong Kong (Acarina; Tetranychidae).

Fig. 14.11 Citrus red spider mite (*Panonychus citri*); adult and egg (Acarina; Tetranychidae).

Oligonychus punicae (avocado brown mite)—on avocado, pomegranate, grapevine, etc.; tropical Asia, Central and South America.

Oligonychus shinkajii (rice spider mite)—on rice foliage; Japan.

Oligonychus yothersi (avocado red mite)—on avocado, apple, mango, litchi, grapevine, camphor, eucalyptus, etc.; southern U.S.A., Central and South America.

Panonychus citri (citrus red spider mite)—very serious on citrus, also on almond, pear, rose, castor, and evergreen ornamentals; tropicopolitan (Fig. 14.11).

Panonychus ulmi (fruit tree red spider mite)—on deciduous, temperate fruit trees, also elm, rose, chestnut, grapevine, raspberry, etc.; cosmopolitan.

Panonychus spp.—several species found on a wide range of fruit bushes, legumes, herbs, and flowers; found throughout the world.

Petrobia apicalis (legume mite)—on peas, clovers, onions; Portugal and the U.S.A.

Petrobia harti (oxalis spider mite)—on *Oxalis,* clovers, sugarcane and citrus; worldwide.

Petrobia latens (brown wheat mite)—on wheat and other small grains, grasses, onions, strawberry, carrot, cotton, lettuce, alfalfa, etc.; Europe, Africa, Asia (India to Japan), Australia, and the U.S.A.

Schizotetranychus asparagi (asparagus spider mite)—on asparagus fern in greenhouses in Europe and U.S.A.; also on pineapple in Hawaii and Central America.

Schizotetranychus celarius (bamboo spider mite)—on leaves of bamboos, wild *Ficus,* rice, sugarcane; China, Japan, southern U.S.A.

Schizotetranychus spp.—on citrus, rice, sugarcane, etc.; Africa, India, and South America.

Tetranychus—a large genus, many species are recorded on a very wide range of crops in both tropical and temperate regions of the world; some species are restricted to a single country, others widely distributed; a few are oligophagous, but most are polyphagous. The northern species tend to be green and the tropical ones tend to be red in color; they usually feed on the undersurface of leaves, on broad-leaved plants, and usually produce copious webbing which covers the foliage.

Tetranychus cinnabarinus (carmine spider mite)—totally polyphagous through the warmer regions of the world; completely pantropical (Fig. 14.12).

Tetranychus evansi (solanum spider mite)—on Solanaceae and some other plants; Mauritius, southern U.S.A., and South America.

Tetranychus ludeni—polyphagous on a wide range of herbs and shrubs; Africa, Australia, southern U.S.A., Central and South America.

Tetranychus marianae (solanum spider mite)—on Solanaceae and cotton; S.E. Asia, Australia, U.S.A., Central and South America.

Tetranychus neocaledonicus (vegetable spider mite)—polyphagous (recorded from 110 host plants); worldwide in tropics and subtropics.

Tetranychus urticae (two-spotted spider mite)—temperate and totally polyphagous; worldwide in cooler regions.

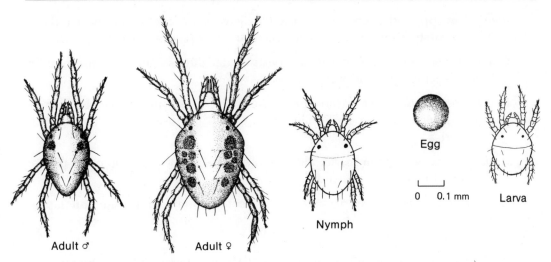

Fig. 14.12 Carmine spider (tropical red spider) mite (*Tetranychus cinnabarinus*); all stages (Acarina; Tetranychidae).

Tenuipalpidae (= Phytoptipalpidae, etc.) (false spider mites)

This rather small group has been known by several different names historically and by region. These mites are reddish, slow-moving, mostly found on the undersurface of leaves but some feed on the bark of plants, some in flower heads, or under the leaf sheath of Gramineae. Most species are not of economic importance. A few species listed below are agricultural pests; damage symptoms are silvering and bronzing of leaves, scaly bark on twigs and speckling of citrus fruits.

Brevipalpus californicus (scarlet tea mite)—polyphagous on a wide range of crops and cultivated plants; a serious pest of tea in S.E. Asia; probably cosmopolitan in the warmer parts of the world but records to date are scattered.

Brevipalpus lewisi (citrus flat mite)—on citrus, walnut, pomegranate, grapevine, and many ornamentals; Near East, Australia, U.S.A.

Brevipalpus obovatus (privet mite)—on privet, citrus, and more than 50 different genera of ornamental plants; totally cosmopolitan.

Brevipalpus phoenicis (red crevice tea mite; scarlet mite)—on tea, citrus, passion fruit, coffee, rubber, apple, guava, fig, olive, walnut, grapevine, and more than 50 genera of ornamentals; probably cosmopolitan throughout the warmer regions of the world (Fig. 14.13).

Fig. 14.13 Red crevice tea (scarlet) mite (*Brevipalpus phoenicis*); adult female (Acarina; Tenuipalpidae).

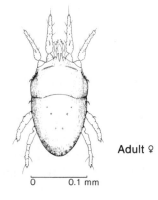

Brevipalpus spp.—other species recorded on leaves and bark of olive, fruit and nut trees, orchids, and many ornamentals; in all parts of the world.

Cenopalpus spp.—several species are pests of deciduous fruit trees and nuts in southern Europe and western Asia.

Dolichotetranychus floridanas (pineapple false spider mite)—monophagous only on pineapple; S.E. Asia, Philippines, Japan, Hawaii, Florida, Central and South America.

Raoiella indica (date palm scarlet mite)—on date palm, coconut, and other palms; Egypt, India, S.E. Asia, and Central America.

Tenuipalpus zhizhilashviliae (persimmon false spider mite)—on persimmon; Japan.

Penthaleidae

A group of soft-bodied mites, rather specialized in several respects, containing a few pest species.

Hylotydeus destructor (red-legged earth mite)—on annual, broad-leaved plants and grasses (clovers, wheat, tobacco, tomato, potato, beet, peas, pastures, and ornamentals—especially seedlings); infested plants may be killed; Africa and Australasia.

Penthaleus major (winter grain mite)—scorch and brown foliage of small grains, grasses, legumes, cotton, vegetables, flowers, etc.; worldwide.

Siteroptes ceralium (grass and cereal mite)—on cereals and grasses causing "silver-tip"; found throughout Europe.

Demodicidae (hair follicle mites)

These are tiny and wormlike in appearance; they live on the skin and in the hair follicles of various mammals including humans. With people they tend to live in the hair follicles of the face. They are minute, 0.1–0.4 mm (0.004–0.016 in.) long, and all stages live together. Host specificity is pronounced. Usually little harm seems to be done to the host but occasionally damage can be serious; in dogs fatalities have been recorded. The most notable species are probably:

Demodex bovis (cattle follicle mite)—cause swollen nodules in the skin.

Demodex canis (dog hair follicle mite; red mange)—infestation often associated with bacteria (*Staphylococcus*) and can cause skin lesions, hair loss, and even death, but most dogs show few symptoms; cosmopolitan.

Demodex folliculorum (human hair follicle mite)—reported to occur on most humans, especially in the hair follicles on the face; cosmopolitan (Fig. 14.14).

Other species attack pigs, cats, goats, horses, deer, and rodents.

Trombiculidae (chiggers; redbugs; etc.)

These mites are unique in that the nymphs and adults are free-living (in the tropics and southern U.S.A.) but the larvae are skin parasites of humans and various mammals and other animals. They are reputedly the most irritating (itching) ectoparasites known.

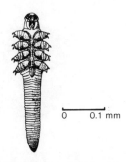

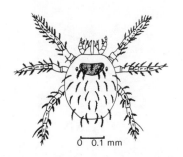

Fig. 14.14 Human hair follicle mite (*Demodex folliculorum*) (Acarina; Demodicidae).

Fig. 14.15 Chigger—larva of *Trombicula* mite (Acarina; Trombiculidae).

Wild birds and rodents are the natural hosts but larvae of *Trombicula* will attack humans and cause dermatitis. Some species transmit *Rickettsia* causing scrub typhus. Turkeys, chickens, quail, horses, and other livestock may be attacked, and economic losses are recorded. Several other genera are recorded as important pests (Harwood & James, 1979). Fig. 14.15 shows a typical "chigger".

SUBORDER HYDRACHNELLE

Hydrachnidae (water mites)

These are the large, red or black, plump, long-legged predatory mites to be found in ponds, streams, and reservoirs around the world.

SUBORDER SARCOPTIFORMES

Another group generally free-living or parasitic, showing rather diverse habits biologically.

Acaridae (Astigmata *partim*)

A rather diverse group of mites, mostly free-living, but some are pests of stored foodstuffs and a few are phytophagous and damage growing plants and bulbs. Some other astigmata are tiny mites in foodstores and are human pests when they infest lungs or intestines, or cause dermatitis. It is thought that the dermatitis is caused by the biting of the tiny mites, although there may be some contact allergy involved. In the wild some of these species occur naturally in bird nests and animal lairs or dens underground. Some of the species listed below may be more suitably placed in other families.

Acarus siro (flour mite)—this cosmopolitan pest of domestic premises and foodstores can be quite important as a pest of foodstuffs, despite its miniscule size (Fig. 14.16). It can cause "bakers' itch" by dermal irritation.

Caloglyphus spp.—live on damp grains, nuts, and decaying fruits.

Glycyphagus domesticus (house mite)—causes "grocers' itch".

Rhizoglyphus echinopus (bulb mite)—lives in bulbs of many species, onions, mushrooms, and on a wide range of plants; cosmopolitan.

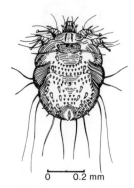

Fig. 14.16 Flour mite (*Acarus siro*) (Acarina; Acaridae).

Fig. 14.17 Itch and mange mite (*Sarcoptes scabiei*) (Acarina; Sarcoptidae).

Rhizoglyphus spp. (cheese mites)—several species are involved.

Tyrophagus dimiatus—found under the leaf sheaths of grasses and cereals, especially after attack by frit fly.

Tyrophagus putrescentiae—causes "grocers' itch" and "copra itch".

Sarcoptidae (itch and mange mites)

Tiny, white, hemispherical mites with reduced legs and claws often absent; the posterior legs usually do not extend beyond the body margin. They make definite tunnels in the skin in which the females lay their eggs. The larvae make their way on to the skin surface where development takes place. The skin burrows may reach a length of 3 cm (1.2 in.). A few species are quite serious pests, and are cosmopolitan in their distribution.

Cnemidocoptes spp. (bird skin mites)—sometimes placed in a separate family; these mites burrow in the skin of the legs of domestic fowl and wild gallinaceous birds, causing "scalyleg"; some species cause feather-shedding.

Sarcoptes scabiei (human itch mite; scabies mite; mange mite)—this mite burrows in the soft skin between the fingers, the bend of elbow and knee, shoulder blades, and around the penis or breasts (Fig. 14.17). On cats and dogs the infestation is referred to as "mange" and there may be extensive hair-fall. *Sarcoptes scabei* occurs as a series of distinct varieties (races) that attack other hosts; var. *bovis* on cattle; var. *equi* on horses, etc.

Psoroptidae (scab mites, etc.)

These mites generally live at the base of hairs where they pierce the skin to feed and cause inflammation (scab mange); their legs are not so reduced as mange mites. *Chorioptes* and *Psoroptes* occur as several different species and attack cattle and domesticated livestock. *Otodectes* (Fig. 14.18) is the carnivore ear mite and to be commonly found on cats and dogs.

The widespread house dust mite (*Dermatophagoides pteronyssinus*) responsible for asthma and respiratory/allergic disorders in man is sometimes placed in the family Pyroglyphidae.

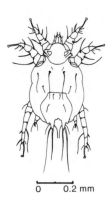

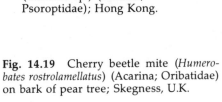

Fig. 14.18 Carnivore ear mite (*Otodectes* sp.) (Acarina; Psoroptidae); Hong Kong.

Fig. 14.19 Cherry beetle mite (*Humerobates rostrolamellatus*) (Acarina; Oribatidae) on bark of pear tree; Skegness, U.K.

SUBORDER ORIBATEI

Some 35 families are grouped here; most are saprozoic and live in soil and leaf litter and are of ecological interest partly because of their sheer abundance. Many species are rounded and well sclerotized hence their common name of beetle mite or armadillo mite. Members of seven families are now known to act as vectors for various tapeworms in different parts of the world; the sheep tapeworm (*Moniezia*) is transmitted by these mites.

Oribatidae (beetle mites)

These rounded, well-sclerotized soil and litter mites are very abundant in most parts of the world. Occasionally they are encountered agriculturally. In Europe some-times clusters of large, bright red mites are found sheltering in crevices on the bark of fruit (unsprayed) and ornamental trees (Fig. 14.19) where they also resemble an insect egg mass—their appearance generally causes considerable concern. This species is the cherry beetle mite (*Humerobates rostrolamellatus*), sometimes placed in the family Mycobatiae; the mites are quite large, about 1 mm (0.04 in.) long, dark red in color, and have been reported feeding on ripe and splitting fruits of cherry and plum. Their normal diet is reported to be algae and lichens growing on the tree bark.

APPENDIX
Notes on Collection, Preservation, and Identification of Insect Specimens

IDENTIFICATION

A major problem facing the field entomologist is to be reasonably certain of the identity of the insect under concern. In many temperate countries there are now text-book series, field guides, and handbooks available for the identification of the local flora and fauna. And many workers have access to local museums and insect collections, so it is usually not too difficult to identify most pest species. But, in most tropical countries there is usually an absence of such reference books and good museum collections are few and scattered. The tropical entomologist is often isolated and with limited reference facilities, and he usually has a very rich insect fauna to contend with. However, most Department of Agriculture research stations, and most universities/colleges do have a named collection of local insect (and other) pests, and a collection of local publications in their library together with the major international publications. Thus the experienced entomologist is usually able to identify most of the insects regularly encountered on the local crops.

Yet, with some groups of insects and most mites, particularly gall midges, Agromyzidae, muscoid flies, some aphids, Coccoidea, and many Microlepidoptera, for example, the taxonomic criteria are quite esoteric. As previously mentioned, infestations are often composed of several closely related species. In some cases insect genitalia have to be dissected and where chaetotaxy is important (some fly larvae, scale insects, some caterpillars), the insect often has to be cleared, stained, and made into a slide mount for detailed microscopic examination. Many such insects can only be identified by experts at one of the major international museums with extensive reference collections. With some groups there may be only a single world expert (e.g., Agromyzidae), with others, such as many Microlepidoptera, there will be regional experts in Europe and North America, but in the tropics most such insects cannot be identified, although some may be placed in the appropriate genus.

Generally, it is not feasible to identify larval stages of most Diptera, Coleoptera, and some Lepidoptera, and these insects have to be reared through to the adult stage. There are, of course, some exceptions as some groups have been extensively studied, such as the tropical Cerambycidae, and monographs on the larval stages have been published. It is always strongly recommended that when collecting a pest species for identification a series of individuals be taken (10–20 probably, depending upon size and abundance); if they are larvae then some should be fixed and some should be reared. If possible, a photograph of the living insects on the host should be taken. The sample should be carefully labeled and recorded (ideally, each sample being given a unique number for recording), preserved appropriately, and some sent away for identification while the remainder are stored locally for future reference. Insects sent away for determination may be retained at that institution if they are of particular interest.

There are now several excellent publications dealing with the general collection and preservation of insect specimens, and some of the more useful sources of information are listed below:

1. Amateur Entomologists' Society, England. They publish a series of *Leaflets,* most of which are concerned with the collecting of different groups of insects, and also a series of *Handbooks* for Lepidopterists, Coleopterists, Dipterists, Hymenopterists, etc.—these are about 100–150 pages each. Most of these are available commercially from E. W. Classey, Ltd., P.O. Box 93, Faringdon, Oxon, SN7 7DR, England.
2. Cogan, B. H., & K. G. V. Smith (1974). *Instructions for Collectors No. 4a Insects* (5th edition) (British Museum, Natural History: London) p. 169.
3. McNutt, D. N. (1976). *Insect Collecting in the Tropics* (C.O.P.R.: London) p. 68. Copies are available from the publisher, free to workers in Commonwealth and aided countries, otherwise £1.30; please note that C.O.P.R. became T.D.R.I. and then O.D.N.R.I., and finally moved to a new location at Chatham, Kent, as part of the newly formed N.R.I. (Natural Resources Institute).
4. Oldroyd, H. (1958). *Collecting, Preserving and Studying Insects* (Hutchinson: London) p. 327.
5. Walker, A. K., & T. K. Crosby (1988). *The Preparation and Curation of Insects* (D.S.I.R. Information Series 163) (D.S.I.R.: Wellington) p. 92.

In the British Isles, for faunistic studies, there is the series of *Handbooks for the Identification of British Insects*—the first was published in 1957 and now about half the groups are covered and a few new books are published each year (now published by the Natural History Museum Publications); they can be bought from Classey (as above). Each handbook serves both as an introduction to that group and as an identification manual.

For pest species, it is hoped that this text by Hill serves as an introduction to all the groups of importance and indicates the scope and importance of each group; the illustrations show the main features of some species. Recognizing the need for such manuals, the C.I.E. International have started a series of *C.I.E. Guides to Insects of Importance to Man;* No. 1. Lepidoptera was published in 1987; No. 2. Thysanoptera in 1989; No. 3. Coleoptera in 1990; and forthcoming guides include Diptera, Hymenoptera, and Hemiptera.

It should perhaps be stressed that most determinations are made by comparison with previously researched and named specimens (ultimately with the type specimens). There is general agreement that in the Third World there should be established at least regional museums or centers of reference which would provide an identification service locally for the region, and some such centers are being established. Most specimens could be sent to a local or regional center for identification. Failing that, it may be necessary to contact one of the major international institutions that have worldwide collections. Good national or regional collections and identification facilities are established in most of the larger countries throughout the world. Recently a regional center for eastern Africa was established in Nairobi, Kenya. Each national museum usually only accommodates local enquiries. As a last resort it may be necessary to send specimens abroad for identification to the two institutions listed below. It should be stressed that in either case it is necessary to receive prior approval before dispatching any specimens. In the case of the U.S.D.A. there is a form to be completed and submitted with the insect specimens, and with the I.I.E. there will now be payment required for non-member countries. The two addresses are:

1. The Director
   C.A.B. International Institute of Entomology
   c/o The Natural History Museum
   Cromwell Road, South Kensington
   London, SW7 5BD
   ENGLAND

   (Preference may be given to member countries of the Commonwealth Agricultural Bureaux International.)

2. The Chairman (Dr. Lloyd Knutson)
   Insect Identification and Beneficial Insect Introduction Institute
   U.S. Department of Agriculture
   Building 003, Room 1
   Beltsville Agricultural Research Center—West
   Beltsville, Maryland 20705
   U.S.A.

   (Preference may be given to institutions within the U.S.A.; form N.E.R.—625 (Jul. 83) to be completed and submitted with specimens.)

Parcels sent to either of these institutions should be sent by "Air Parcel Post" (or by surface post) and *not* by "Air Freight" since this service involves heavy charges for customs clearance and delivery from the airport which also results in a considerable delay before actual delivery.

KILLING

For specimens to be pinned or preserved dry, a killing bottle should be used; ethyl acetate is the recommended killing agent as it leaves the insect relaxed—potassium cyanide is seldom used now because of the dangers. Other volatile liquids can be used, such as chloroform, ether, petrol, etc., but they are less effective. For larvae or soft-bodied insects to be preserved in fluids, very hot water is the best, or else the preservative fluid; the hot water method is by far the best. With large beetles, hot water killing is very effective. For large moths, hypodermic injection with formalin (40%) is best, but they must be mounted immediately.

LABELING

Specimens without collecting data are of very limited value scientifically, and most museums will not spend their time to identify them. All specimens should be labeled as soon as possible after collection. The data should include the place of collection, date, and name of collector clearly stated, e.g., "Kabanyolo, Kampala, UGANDA—12 April (iv). 1967—D. S. Hill." In remote districts, the latitude and longitude (or recording grid reference) should be used rather than the name of the village which probably does not appear on any maps. For crop pests, the name of the host plant is vitally important, whether it is a crop plant or weed or wild host, and preferably the international name or its Latin name rather than a local name (only an Indonesian would know that Gandrung refers to sorghum!). Notes such as "eating cocoa leaves", "boring sugarcane stem", etc., should be added when relevant. It should be clearly established that the insect is actually feeding on the plant and that it is a true host plant and not just a temporary resting site. If an extensive collection is being made it is worthwile having collecting labels printed. For pinned specimens (and for small tubes) the labels need to be small and inconspicuous, such as the sample shown below.

| HONG KONG: |
| . 199 |
| ex |
| leg. D.S. Hill |
| Zoology Dept., H.K.U. |

Length about 25 mm (1 in.),
depth about 12 mm (0.48 in.)

For specimens stored in alcohol, preferably in tubes/bottles such as Macartney bottles or Wheaton snap-cap bottles, a larger label some 5 cm (2 in.) square can be used; a sample label is shown below.

| ORDER: |
| FAMILY: |
| . . . . |
| . . . . |
| HONG KONG: |
| . . 199 |
| ex |
| leg. D.S. Hill |
| ZOOLOGY DEPT., H.K.U. |

"Ex" refers to the host (plant), or wherever collected; "leg." is a Latin abbreviation meaning "collected by"; such labels are always placed *inside* the tube, and the information typed or written in Indian ink or soft pencil.

DRIED SPECIMENS

Most of the larger adult insects have a stout exoskeleton and they are traditionally pinned and dried. Only stainless steel insect pins should be used, especially in the humid tropics—other pins soon rust and break. If pinning is not convenient, specimens can be stored dry in small pill boxes or paper envelopes. If placed in a box, some soft tissue paper should be packed to prevent movement; *never* use cotton-wool as insect claws become entangled. After pinning or packing, the specimens should be protected from mold and insect pests with liberal amounts of naphthalene or some other preservative.

PRESERVATION IN LIQUIDS

Soft-bodied adult insects (termites, aphids, mealybugs, etc.) and all insect larvae (caterpillars, grubs, maggots) should be preserved in liquid, preferably 70–80% alcohol, with 5% added glycerine. Bottles/tubes should not be filled more than halfway with specimens to avoid excessive dilution of the preservative. Avoid using formaldehyde if possible, but a 3–4% solution could be used as a last resort. In emergency, substitute preservatives that could be used include gin, rum, vodka (undiluted), and methylated spirits (slightly diluted). The I.I.E. "Instructions for Users of the Identification Service" indicates the best method of preservation for the different groups of insects, and likewise the U.S.D.A. "Identification Request—Information/Instructions" indicates their preferences in these respects.

PACKING AND DISPATCH OF SPECIMENS FOR IDENTIFICATION

For pinned specimens, the small postal boxes commercially available are to be preferred; the pins must be pushed deep into the cork, otherwise many of them come adrift during postal handling. Large insects must be prevented from rotating on their pins by extra pins touching each side of the body. Boxes of pinned insects should ideally be packed inside another box, with at least 5 cm (2 in.) of thin wood shavings ("excelsior"), polystyrene chips, or something equivalent, packed all around sufficiently firmly to absorb all shock and to prevent any movement. Unless the boxes are packed very carefully, the more fragile specimens are unlikely to survive the mailing and will arrive totally fragmented.

For specimens in fluid in corked tubes it is necessary to push the corks very firmly into the tubes/vials, but screw-capped tubes are to be preferred. The tubes should be wrapped individually in cotton-wool, tissue paper, or polystyrene sheet, so that the glass surfaces do not touch. Care has to be taken with corked tubes if they are to be sent by air, for, unless they are very tight, they might pop out at high altitude in unpressurized cargo holds: corked tubes may be dipped in hot wax for greater security.

LIVING SPECIMENS

Occasionally it is necessary to send living insects to the taxonomist, but this should only be done at their specific request. Live insects should be sent in a suitable strong container of cardboard, wood or tin, with small holes to permit ventilation and prevent condensation. In general, larvae should be supplied with an adequate supply of food material; if their food is of a succulent nature it will probably decompose in transit and it may be preferable to pack them loosely in damp moss. Larvae need to be kept moist but will die if conditions are wet. Noctuid caterpillars, and some other insects, are often cannibalistic if kept under crowded conditions. Most adult insects survive several days without food, but, in a dry climate, some damp moss or cotton-wool should be placed in the container with them, together with enough soft tissue paper or muslin to prevent their being thrown about during transit.

FURTHER INFORMATION

For more detailed information concerning the collecting, preservation, storage, and identification of insect specimens, the sources listed on page 572 are recommended.

GLOSSARY

Acephalous. Headless; larva without a distinct head capsule; applies to certain Diptera.

Adecticous. Pupa with immovable mandibles; most common type.

Aestivation. Dormancy of individuals during a period of heat/drought; usually occurs during larval or pupal stages.

Agamic. Parthenogenetic reproduction; reproduction without mating.

Alatae. Winged insects; usually refers to winged aphids.

Alienicole. Refers to Aphididae; parthenogenetic viviparous females, usually on the secondary host plant, and often differing from the other forms.

Allochthonous. Not an aboriginal species; exotic; an introduced species. The opposite is autochthonous.

Allopatric. Having separate and mutually exclusive areas of geographical distribution. The opposite is sympatric.

Amphipneustic. Respiratory system in which only the first (thoracic) and the last one or two pairs of spiracles (abdominal) are functional; usually refers to insect larvae.

Anemophilous. Plants which are pollinated by the wind.

Anholocycly. Refers to Aphididae; species without winged sexual forms, reproduction parthenogenetic and viviparously, hosts herbaceous, and no alternation of generations.

Aniline. Synthetic dyes produced from oily liquid (aniline) extracted from coal tar.

Annulate. Ringed; usually refers to narrow ringlike segments in the funicle of the antenna (see diagram of insect antenna).

Annulus. Narrow, ringlike segments in the antenna, follow after the pedicel. See G.1.

Antenna. One of the pair of appendages borne on the insect head, bearing many sensillae and functioning as a sense organ. See G.1, G.5, G.6.

Aphidophagous. Aphid-eating; predator that feeds largely on aphids (for example, lacewing larva).

Apiculture. The commercial rearing of honeybees for bee products or for pollination purposes.

Apneustic. Insect larva (usually aquatic Diptera) without functional spiracles, but still possessing an internal tracheal system.

Apodous. Legless; without feet; larva without thoracic legs.

Aposematic. Conspicuous insect coloration used to frighten away enemies or to warn predators that it is unpalatable/poisonous.

Apterae. Wingless insects; usually refers to wingless forms of Aphididae.

Apterous. Wingless; without wings. The opposite is macropterous, etc.

Aquaculture. Water cultivation of plants or breeding (rearing) of animals.

Arista. Antennal bristle characteristic of higher Diptera; represents the reduced antennal flagellum.

Asymptote. Point in the growth of a population at which numerical stability or equilibrium is reached.

Autochthonous. Aboriginal; native; indigenous; species thought to have evolved where they are found.

Autocide. Self-destruction; control of an insect pest population by the sterile male technique.

Benthic. Living on the lake bottom or sea bed.

Benthos. Flora and fauna of the lake bottom or sea bed.

Biotype. Type of plant or animal; group of individual insects of similar appearance and equal genotype.

Bivoltine. Animals having two generations per year.

Branchial. Refers to gills (branchiae).

Caecae. Saclike or tubular structures open at only one end (blind), often opening from the mid-intestine.

Campodeiform. Insect larva that resembles the dipluran genus *Campodea;* usually agile, predacious forms with a slender, mobile body, thoracic legs, and well-developed sense organs.

Cantharidin. Chemical found in the body of adult Meloidae and Oedemeridae; mostly found in the elytra, with urticating properties and a reputation as an aphrodisiac.

Carnivorous. Flesh-eating; predacious. The opposite is herbivorous.

Caudal. Refers to a tail or tail-like structure at the end of the abdomen.

Cellulase. Enzyme capable of breaking down or digesting plant cellulose.

Cenchri. Special lobes on the metanotum of adult sawflies (Hymenopetera, Symphyta) which engage on rough areas of the undersides of the forewings to hold them in place.

Cephalopharangeal. Refers to a piece of skeleton in some larval Diptera; held withdrawn into the thoracic segments but the mouth-hooks protrude during feeding.

Cerci. Paired appendages at the distal end of the abdomen, usually many segmented; most obvious in more primitive insects. See G.5, G.6.

Cerumen. A mixture of wax and plant resin used by some bees for nest construction.

Chaetotaxy. The arrangement and nomenclature of the setae (bristles or chaetae) on part of the exoskeleton of insects, usually adult Diptera and larval Lepidoptera.

Chelicerae. The pincer-like first pair of head appendages found in Arachnida and Merostomata.

Chemoreceptors. Sense organs having structures sensitive to certain chemicals and chemical properties.

Chitin. A complex and chemically resistant polysaccharide forming a major part of insect cuticle.

Clavus. A club; antennal club, usually the terminal three segments; also the anal area of the hemelytron (forewing) in the Heteroptera. See G.1.

Cloacal. Refers to the rectum or cloaca at the termination of the intestine; a common chamber into which opens both rectum and genital ducts (some Coleoptera and Lepidoptera).

Coccidophagous. Coccid-eating; predators/parasites that feed largely on Coccidoidea.

Cochineal. Red dye produced from the dried bodies of certain scale insects (*Dactylopius* spp.).

Colony fission. The breaking up of an insect aggregation and dispersal of part to form a new colony.

Comb. 1. A layer of nest cells crowded in a distinctive regular arrangement, found in social Hymenoptera. 2. A serrated structure bearing a resemblance to a hair comb.

Commensalism. A type of symbiosis where two species live together, one benefits and the other appears neither benefited nor harmed.

Conidia. Asexual fungal spores.

Coprophilous. Living in dung (for example, dung beetles,), or decaying vegetable matter resembling excrement.

Corbicula. Pollen basket on the hind leg of some bees.

Costa. Wing vein; the first longitudinal vein running along the anterior edge and terminating before the apex; usually reduced in the hind wing; also any elevated ridge rounded at its crest.

Coxa. The basal segment of a leg, articulated to the insect body at the sternum.

Cremaster. Spines, usually hooked, on the posterior tip of the pupa in Lepidoptera; often used for attachment purposes.

Crepuscular. Active before sunrise and at evening twilight or dusk.

Crescentric. Crescent-shaped.

Cryptobiosis. Survival of an organism during a dry period whereby metabolic activity is reversibly ceased and activity stops; some Chironomidae in dried-up pools in Africa.

Cryptozoic. Fauna concealed under stones, tree bark, in crevices, or otherwise hidden in darkness.

Ctenidia. Comb-like structures on any part of an insect body.

Cu. Cubitus; the sixth longitudinal wing vein with two primary branches. See wing diagram.

Cubito-anal. Region of wing between the cubitus and anal veins.

Cuneus. Part of the forewing in some Heteroptera; a triangular postero-lateral part of the corium and separated from it by the costal fracture.

Cyclorrhapha. A group of Diptera, regarded as highly evolved; characterized by the puparium inside which the adult develops.

Detritivore. Animal feeding upon organic detritus or of animal or plant origin; saprophage.

Diapause. Spontaneous state of dormancy during development in many insects; usually physiologically induced and broken to avoid a long period of inclement weather, usually winter.

Diatom. Tiny, unicellular, freshwater green alga.

Dimorphism. Having two different life forms; differing sexes, seasonal forms, or geographical forms.

Dipterous. Two-winged; having characters of the Diptera.

Diurnal. Active during the daytime. The opposite is nocturnal. See also crepuscular.

Dorsum. The upper surface of the insect body.

Ecdysis. Molting or shedding the outer layer the body surface ("skin").

Eclosion. Emergence of the adult insect from the cocoon, puparium, or cuticle of the pupa.

Ectoparasite. An organism that lives parasitically on the surface of its host, usually in the fur or feathers of mammals or birds; most insects in this category are blood suckers.

Elytra. The hardened protective forewings of Coleoptera, usually meeting in the mid-dorsal line of the body at rest.

Endoparasite. An organism that lives parasitically inside the body of the host organism or under the skin surface.

Endysis. Developing a new outer layer.

Entognathous. Having mouthparts inserted in a pocket on the face of the insect; internal mouthparts. The opposite is ectognathous.

Entomology. The branch of zoology that refers to the study of the Class Insecta.

Entomophagous. Insect-eating; organisms that feed more or less exclusively on insects.

Entomophilous. Plants that are pollinated by insects; insect-loving.

Epizoötic. Outbreak of disease in an animal (insect) population at the same time; a group phenomenon.

Erinea. Outgrowths of abnormal hairs on leaves which is produced by certain gall-mites (Eriophyidae), accompanied by some leaf distortion.

Eruciform. Larva resembling a caterpillar (polypod) in appearance; applies to larval Lepidoptera and some sawfly larvae.

Eucephalous. An apodous larva with a well-developed head capsule (for example, Diptera, Nematocera).

Euryhaline. Usually aquatic organisms adaptable to a wide range of salinity. The opposite is stenohaline.

Euryhygric. Organisms adaptable to a wide range of atmospheric humidity. The opposite is stenohygric.

Eurythermal. Organisms adaptable to a wide range of temperatures. The opposite is stenothermal.

Eurytopic. Having a wide geographical distribution, and tolerant of a wide range of ecological habitats. The opposite is stenotopic.

Exarate. Sculptured; an excavated surface; a pupa in which legs and antennae are free from the body and the abdomen is movable.

Exuvium. Cast-off skin (cuticle) of an arthropod after molting.

Femur. The third, and usually largest, segment of the insect leg.

Filaria. A type of small, thin, nematode parasite; usually blood parasites in vertebrate hosts.

Flocculent. Woolly; like tufts of wool.

Foreceps. The claspers or pincers at the apex of the abdomen in some insects; most evident as the male cerci in earwigs (Dermaptera); part of the male genitalia in many different insects used either for copulation or for defense purposes.

Fore-coxae. Coxae of the forelegs (most anterior pair).

Forewing. One of the anterior pair of wings. See G.6.

Fossorial. In the habit of or adapted for digging or burrowing.

Frass. Usually a mixture of plant fragments and insect excrement produced by plant-boring insects, most commonly wood-boring insects.

Frenulum. The spines or bristles on the base of the costal vein in the hindwing of Lepidoptera which serve to unite the wings during flight.

Fundatrigenia. Daughter of female aphid living on the primary host plant.

Fundatrix. Apterous, parthenogenetic, viviparous female aphid emerging from the over-wintering egg in the spring, on the primary host plant.

Fungivorous. Feeding on fungi.

Funicle. The central and major part of geniculate antenna between the pedicle and the club. See G.1.

Galea. The outer two-jointed lobe of the insect maxilla; in adult Lepidoptera the two are joined and enlarged to form the proboscis. See G.4.

Gallicolous. Living in a plant gall either as the producer or inquiline.

Gallinaceous. A game bird; member of the Order Galliformes.

Gaster. The abdomen of adult Hymenoptera Apocrita, located behind the waist or petiole and always excluding the propodeum (first abdominal segment).

Gena. The front cheek region of the insect head.

Geniculate. Elbowed; abruptly bent; refers to the antennae of weevils (Coleoptera) and Chalcidoidea (Hymenoptera).

Habit, Habitus. The general appearance of an organism; facies.

Haematophagous. Blood-sucking; blood-feeding.

Hair pencil. Found in some male Lepidoptera; a scent brush for dispersing pheromones, protrusible and usually located at the end of the abdomen.

Haltere. Modified hindwing in adult Diptera; a balancing organ.

Haustellate. Mouthparts adapted for sucking. The opposite is mandibulate.

Haustellum. General term for a proboscis. In adult Diptera it is the distal part of the proboscis representing the prementum.

Hematozoa. Certain blood-inhabiting parasitic Protozoa.

Hemelytra. Anterior wings of Heteroptera with basal part thickened and distal part membranous.

Hemicephalous. Larvae of certain Diptera (Tipulidae and Brachycera) with a reduced head capsule which can be withdrawn into the thorax.

Hemocoel. The general body cavity of insects in which the blood flows.

Herbiphagy. Herbivorous; phytophagous; feeding on plants.

Heterogony. Reproduction with alternation of generations, often involving sexual and asexual forms.

Hindwings. The posterior pair of wings attached to the insect metathorax.

Holoptic. Compound eyes extended dorsally so that they are contiguous along the midline (dragonflies and many male Diptera).

Honeydew. A watery sugar solution, usually saturated, excreted from the anus of some sap-sucking Homoptera (Aphidoidea, Coccoidea, some Fulgoroidea).

Hydrofuge. Water repelling.

Hypermetamorphosis. Insect development with successive larvae having different body forms, more or less having two metamorphoses during development (some Coleoptera, some Diptera, etc.).

Hypostoma. Region of mouthparts in some Diptera; part of the subgena behind the mandible.

Imago. Imagine; the adult, sexually mature insect.

Inquiline. An animal living in the home or nest of another and sharing its food; usually refers to an insect that lives in the nest of termites or social Hymenoptera, or in a hymenopterous plant gall.

Insolation. Sunlight; solar radiation.

Instar. Stage between molts in insect nymphs or larvae, numbered to designate the various stages.

Intercalary. Inserted between others; extra wing veins inserted between the main longitudinal veins (as in Ephemeroptera and others).

Irruption. Local population explosion; irregularly large population increase.

Jugal. Lobe at the base of the hindwing in some Lepidoptera and Trichoptera.

Jugate. Possessing a jugum wing-coupling apparatus as in some Lepidoptera.

Jugum. Basal area of insect wing in some Lepidoptera, set off by the jungal fold; a basal wing lobe.

Keratin. The inert exoskeletal protein of vertebrate animals forming skin, hair, feathers, claws, nails, scales, etc.; similar in structure and function to the arthropod chitin.

Kleptoparasite. An animal that steals its food from another species.

Labella. Paired lobes at the end of the proboscis in some adult Diptera and derived from the labial palps.

Labrum. The upper lip of insects; adjoins the clypeus in front of the mouth. See G.4, G.6.

Lachrymal. Referring to tears in the Mammalia.

Lacinia. The inner lobe of the insect maxilla bearing a brush of setae or spines; used for food particle handling. See G.4.

Larva. An immature insect that leaves the egg in a very early stage of morphological development while differing greatly from the adult form (Insecta, Holometabola); metamorphosis occurs during development.

Larviparous. An insect that gives birth to living young instead of laying eggs.

Larviposition. The deposition of larvae by the adult female of some Diptera.

Lentic habitat. Pond or lake; swamp; a region of standing water.

Life-cycle. The succession of stages through which an organism passes to reach maturity.

Lignified. Woody; plant tissues impregnated with lignin.

Littoral. The shore zone between high-water mark and low-water, at the edge of lakes and on the seashore; shore zone; also encompasses the flora and fauna living there.

Logistic curve. Sigmoid-shaped growth curve applicable to both individuals and populations of organisms plotted graphically.

Lotic habitat. Running waters; streams and rivers.

Lunule. A crescentic mark (sclerite) on the face of some adult Diptera; a paired structure on some Hymenoptera.

Macroclimate. General climate prevailing over a large area, considered as a unit. The opposite is microclimate.

Macrofauna. The larger members of a particular fauna. Open to different levels of interpretation for they may be larger vertebrates or alternatively the larger insects of a habitat. The opposite is microfauna.

Macropterous. With long or large wings. The opposite of apterous.

Maculate. Spotted; usually refers to insect wings. Opposites are immaculate, hyaline.

Mallophagan. Refers to a member of the Order Mallophaga.

Mandibles. Jaws; the first pair of jaws in the Insecta; stout and serrated in chewing insects or needlelike and piercing in sucking insects. See G.5, G.6.

Mandibular. Pertaining to the mandibles. See G.3, G.4.

Mariculture. The rearing and commercial cultivation of marine organisms.

Maxillary. Part of or belonging to the maxillae of the mouthparts. See G.3, G.4, G.5.

Meiofauna. Tiny animals that move through the substratum without displacing the particles; in sand usually species 50μ–3 mm in length.

Melanism. Dark-pigmented forms; being black in color. The opposite is albinism.

Mesoclimate. Climate of a localized area, such as a valley or hillside, that differs from the macroclimate.

Mesonotum. Notum or dorsal plate of the mesothorax.

Mesoscutellum. Scutellum of the mesothorax.

Mesothorax. The middle and second segment of the thorax, bearing the forewings and middle legs.

Metamorphosis. The change in body form during insect development; either complete (drastic change) or incomplete (slight change).

Metanotum. Tergum (notum; dorsal plate) of the metathorax.

Metapneustic. Respiration when the larva has only the last pair of abdominal spiracles open; usually refers to Diptera.

Metathorax. The third part of the thorax, bearing the hindwings and posterior pair of legs. See G.6.

Microclimate. Climate of a microhabitat; typically more constant and less variable than either the general climate of an area or a mesoclimate.

Microfauna. The fauna of a microhabitat or used as a general term for microscopic animals.

Microfilariae. A group of very small, thin, parasitic nematodes, usually blood-inhabiting; often transmitted among the vertebrate hosts by blood-sucking Diptera.

Microtrichia. Tiny cuticular setae on the surface of the wings, especially in adult Diptera.

Migrantes. Adult, winged, parthenogenetic viviparous female aphids, developing on the primary host and then dispersing to secondary hosts.

Moniliform. Antenna resembling a bead necklace.

Monolectic. Refers to bees that use pollen from only a single plant species.

Monophagous. Feeding on only one type of food; monotrophic; insects feeding on only one genus or species of plants.

Multivoltine. Having several generations per year.

Mutualism. Symbiosis; relationship between two organisms when both benefit.

Mycangia. Pocket-shaped receptacles used to carry symbiotic fungi (for example, Coloptera, Scolytidae).

Mycelium. Vegetative structure of a fungus composed of a network of fine, filamentous hyphae.

Mycetome. The ovary follicle cells through which some aphid eggs are infected by symbiotic fungi.

Mycetophagous. Fungivorous; feeding on fungi, on fungal mycelium.

Myiasis. Infestation and injury of animal or human body by living larvae of some Diptera; often termed invasion of body tissues.

Myrmecodomatia. Structures in higher plants that seem to have evolved, through mutualistic association, to serve as dwelling places for ants.

Myrmecophile. An insect that lives in the nest of ants; a form of inquiline.

Necrosis. Death of part of a plant or animal.

Nekton, Necton. Organisms that swim actively in water.

Nemathelminthes. Systematic interpretation which places the Platyhelminthes and Nematoda into a single Phylum.

Neuston. Organisms that live on the surface film of water bodies.

Nocturnal. Animals active, and food seeking, at night only.

Nymph. Immature stage of hemimetabolous insects which bears close resemblance to the adult.

Obtect. Concealed; covered; refers to pupa where wings and legs are appressed to the body, abdomen immovable.

Occipital. Refers to the back part of the head, the occiput.

Ocellus. Simple eye of adult insects on the top of the head, occurring singly, or in a small group (2–3); or simple eye at the side of the head on some insect larvae, usually several; also a colored spot on the wing of Lepidoptera. See G.6.

Oligolectic. Bees that gather pollen from only a few closely related species of flowers.

Oligophagous. Having a restricted range of food; phytophagous insects that feed on only one family of plants.

Oligopod. Insect larva with fully developed thoracic legs and segmented abdomen.

Omnivorous. Eating both plant and animal food materials.

Onisciform. Insect larva resembling a woodlouse, being depressed and broadly ovate.

Oötheca. Eggcase; a mass of eggs enclosed in a container secreted by the female accessory glands, best shown in the Dictyoptera.

Operculum. A lid-like structure; cover. Many different specific uses in the Class Insecta.

Organochlorine. A group of insecticides also referred to as the chlorinated hydrocarbons, including DDT, HCH, dieldrin, and others.

Osmeterium. A fleshy, tubular structure, eversible through a prothoracic slit found on the larvae of some Papilionidae and producing a strong odor.

Ostiole. A small opening or hole; the external opening of certain glands (as in many Heteroptera).

Oviparous. Egg-laying; reproduction by producing eggs.

Oviposition. Deposition of eggs; the process of egg-laying.

Ovipositor. Structure used for egg deposition; usually an elongation of the terminal abdominal segments (oviscapt) or the appendages (typical ovipositor). See G.6.

Oviscapt. Functional ovipositor produced by elongation of terminal abdominal segments rather than the appendages; applies to some Diptera and Trichoptera.

Ovoviparous. Forms producing an egg with a definite shell hatching inside the maternal body.

Palp, Palpus. Segmented tactile structure born on the maxillae and labium and used in the feeding process. See G.6.

Parasitoid. A species of parasite where only one stage (usually larva) is parasitic; the other (usually adult) is free-living.

Parasitism. One organism living on or within another, to its own advantage in food or shelter; a form of symbiosis.

Parenchyma. Soft and thin-walled plant cells which form the pith of plant stems, leaf mesophyll, and other plant parts.

Parthenogenesis. Development of eggs without fertilization.

Pecten. Comb; any comb-like structure or organ; can refer to many different structures in different groups of insects.

Pectinate. Having a comb-like structure; usually applied to some claws, antennae, and sometimes leg spines, with teeth-like serrations.

Peduncle. Stalk or petiole; a stalk-like structure supporting another structure, such as a flower stalk.

Pelagic. Living in the open sea or the open water of large freshwater lakes.

Periphyton. Organisms in water attached to or clinging to rocks, twigs, and other submerged structures.

Peripneustic. Respiratory system with one metathoracic and eight abdominal pairs of spiracles; most insect larvae have this type.

Petiolate. Stalked; some Diptera and Hymenoptera Apocrita have slender segments between the thorax and abdomen.

Pharate. Cloaked; living inside the cuticle of a previous instar; usually occurs as a pharate pupa or pharate adult insect.

Pheromone. A chemical secreted by a special gland used for communication within the species, usually eliciting a specific behavioral response.

Phloeophagy. Feeding on the phloem tissues in the inner bark of tree trunks.

Phoresy. A relationship where one organism is carried on the body of a larger organism; symbiotic transportation.

Phorids. Flies belonging to the family Phoridae.

Phylogenetic. Relating to the evolutionary history of a group of organisms.

Phytophagous. Herbivorous; feeding on or in plants, or on vegetable matter.

Phytotelemata. Plant structures inhabited by or used by insects, ranging from pitchers, special galls, to flooded leaf axils.

Phytotoxic. Poisonous to plants.

Planarium. Flatworm; a free-living member of the Platyhelminthes, Turbellaria; either terrestrial or aquatic in habits.

Plankton. Tiny marine or freshwater organisms floating and drifting with the surrounding water.

Plastron. A film of air on the outer surface of the body of some aquatic insects and used in respiration.

Plumose. Feathery; usually refers to some antennae, such as found in male mosquitoes; and feathery setae.

Poikilothermous. "Cold-blooded"; animals with little or no control of their body temperature, which usually varies with that of the ambient medium. The opposite is homeiothermous.

Polydemic. Refers to species occurring in several different habitats/areas.

Polyembryony. The production of several or many embryos from a single egg; characteristic of some parasitic insects.

Polylectic. Bees that gather pollen from a wide range of flowers.

Polyphagous. Eating various types of food; insects that feed on plants belonging to several or many different families.

Porrect. Projecting forward; head or antennae that project anteriorly and horizontally.

Postnotum. An intersegmental, dorsal thoracic skeletal plate; the dorsal sclerite behind the scutellum in adult Diptera.

Postscutellum. Postnotum in adult Diptera.

Prestomal. Refers to the distal end of the proboscis in some adult Diptera; teeth on the labella of some blood-sucking Muscidae.

Proboscis. An extended mouthpart structure used for feeding; for example, rostrum, and others.

Prognathous. Jaws anterior on a head held horizontally.

Prolegs. The abdominal appendages of caterpillars used for locomotion.

Pronotum. Dorsal skeletal plate of the prothorax.

Propalis. A resinous substance collected by honeybees from tree buds and used as a cement for nest damage repair.

Propodeum. The first abdominal segment attached to the end of the thorax in adult Hymenoptera Apocrita.

Prothorax. The first (anterior) segment of the thorax. See G.6.

Protolean. A parasite parasitoid; parasitic when immature; free-living as adult.

Protozoa. A phylum of unicellular animals, most free-living, some parasitic.

Pseudopod. False leg; pseudopodium; locomotory protuberances on the legless larvae of some Muscidae.

Pseudopupa. The larva of certain Coleoptera in a quiescent condition preceding the true pupal stage.

Pseudotracheae. The ridged grooves on the ventral surface of the labella of some adult Diptera; used for absorbing liquid food materials.

Psyllidophagous. A parasite or predator that feeds on Psyllidae; host-specific to the Psyllidae.

Psyllose. Infestation symptoms of cotton plants after attack by Cotton Psyllids.

Pterostigma. A pigmented thickened spot near the anterior apex of the wings of Odonata and Mecoptera, and on the forewings of Hymenoptera and Psocoptera.

Ptilinum. An eversible sac that can be thrust out of a fissure on the face of adult Schizophora (Diptera) to push off the cap of the puparium at the time of adult emergence.

Pulvillus. The bladder-like structure between the claws of the pretarsus of most insects; used for gripping the substratum.

Pupa. The inactive stage of insects, having complete metamorphosis between the larva and the adult, when metamorphosis occurs.

Puparium. The exuvium of the third (last) larval instar in higher Diptera within which the pupa develops.

Pupate. To become a pupa.

Pygiduim. The tergum of the last visible segment on the insect abdomen.

Raptorial. Predatory; adapted for seizing prey.

Reduviid. A member of the Reduviidae (Hemiptera, Heteroptera).

Reticulate. Net-like; meshed; a network of lines.

Reticulo-punctate. A descriptive term for cuticle sculpturing; a combination of a network of lines with a series of regular punctures.

Rostrum. Proboscis in the Hemiptera; the snout-like anterior extension of the head in the Curculionoidea.

Royal jelly. Secretion from the hypopharyngeal glands of worker honeybees; used for feeding larvae in the production of queens.

Ruderal. Plants growing in artificial habitats created by people (waste ground, roadsides, and others); in some sense considered weeds; secondary pioneers.

Rufous. Reddish colored.

Saltatorial. Adapted for, or used in, jumping as a means of locomotion.

Saprophagous. Scavenging; feeding on dead or decaying animal or plant materials.

Saprozoic. Animal living on or in dead or decaying matter.

Scape. The first or basal segment of the insect antenna. See G.1.

Scarabaeiform. Oligopod larva; a white grub of the Scarabaeidae with a fat, fleshy C-shaped body and sluggish behavior.

Scarification. Scarring; appearing scratched; often applied to leaf surfaces after feeding by thrips or spider mites.

Sclerite. Any piece of skeletal cuticle bounded by either sutures or membranes.

Sclerotize. Hardening of the body cuticle.

Scoli. Tubercules in the form of spiny projections on the body of some insect larvae (some Lepidoptera and Coccinellidae).

Scopa. A brush; a mass of bristles on the hind leg (tibia) of adult bees; abdominal fringe of long scales on some male Lepidoptera.

Scutellum. A shield-shaped plate on the thoracic dorsum of many adult insects.

Scutum. The shield of ticks; a shield-shaped dorsal plate on the thorax, occupying most of the dorsum of adult Diptera.

Sericulture. The commercial rearing of silkworms and silk production.

Setae. Bristles; macrotrichia; sclerotized hairlike projections of skeletal cuticle. Note: in strict usage never called "hairs".

Sexuales. Sexually reproductive males and females of Aphidoidea and Coccoidea.

Sexuparae. In Aphidoidea the alate parthenogenetic females that produce males and sexual females.

Shellac. Commercial lac produced as a red resinous substance from the scales of female lac insects.

Sigmoid. S-shaped; the S-shaped growth curve representing development of either an individual organism or a population plotted graphically.

Siliqua. The fruiting body of Crucifera; a long pod divided in two by a false septum.

Spanandrous. A shortage of males.

Spermaphagy. Seed-eating.

Spicule. A small needlelike spine.

Spiracle. One of the external openings of the tracheal system. See G.5.

Stenohygric. Adapted for life in a small range of relative humidity. The opposite is euryhygric.

Stenotopic. Restricted to living in a single type of habitat. The opposite is eurytopic.

Sternite. Part of any sternum on the ventral part of a body segment. See G.5.

Stridulate. Sound production by rubbing two rough surfaces together.

Style. Small, pointed, fingerlike process, non-articulated; many different references in insect anatomy, often to parts of the male genitalia.

Stylet. A small style; needlelike structure; often refers to part of the mouthparts in Diptera and other orders, used for piercing or sucking.

Subcorticolous. Beneath the bark of trees.

Sublittoral. Shore zone below the littoral; shallow water zone at sea or lake edge.

Suctorial. Sucking; adapted for sucking.

Syconium. A receptacular fruit; for example, fig, the fruit of *Ficus* spp.

Symbiosis. The intimate, dependent relationship between two different organisms; now usually regarded as consisting of three basic kinds: commensalism, mutualism, and parasitism.

Sympatric. Two or more species having the same, or overlapping, areas of geographical distribution. The opposite is allopatric.

Synanthropic. Associated with humans or living in human dwellings.

Tarsus. The last (distal) segment of the leg bearing the pretarsus and claws; usually composed of 4–5 segments. See G.6.

Tegmina. The hardened leathery forewings in Dictyoptera and Orthoptera.

Teneral. A newly emerged adult insect before the cuticle is fully hardened and colored.

Tentiform. Shaped like a tent.

Tergite. A dorsal sclerite.

Tergum. The dorsal surface of any body segment.

Termitophilous. Living in the nest of termites; a nest inquiline.

Thigmotactic. Contact-loving; insects that inhabit crevices; also refers to insects that live very close together—touching.

Thorax. The middle part of the insect body between head and abdomen; composed of three segments each of which bears a pair of articulated legs.

Tibia. The fourth leg segment between femur and tarsus. See G.6.

Tillering. The process when a grass or cereal plant develops secondary shoots.

Tortricid. A member of the family Tortricidae (Lepidoptera).

Triungulin. The active, tiny, first-instar larva of Meloidae and some others; campodeiform and predacious; undergoes hypermetamorphosis.

Trochanter. A leg segment between coxa and femur.

Trophallaxis. Reciprocal feeding; the practice in some social insects of exchanging alimentary liquids.

Trypanosome. A protozoan blood parasite, usually transmitted from host to host by biting flies. The most notorious are probably the species of *Trypanosoma* that cause "sleeping sickness" in people in Africa.

Tubercule. A small, rounded protuberance.

Univoltine. Having only one generation per year.

Urogomphi. Paired processes on the end of the abdomen in some larval Coleoptera.

Urticating. Stinging; nettling; setae or chemicals that cause skin irritation, either mechanical or chemical.

Vermiform. Wormlike; shaped like a worm, being elongate and thin.

Verruca. A wartlike prominence on some caterpillars, usually bearing tufts of setae.

Vesicle. A small sac or bladder, sometimes extensible.

Vestigial. Small and imperfectly developed; degenerate and non-functional.

Viviparae. Female aphids that reproduce viviparously.

Viviparity. The bearing of live young instead of eggs.

Warble. The swelling on the host body caused by the larva of a warble fly.

Xerophyte. Plant adapted for survival in desert or physiologically dry soil.

Xylomycetophagy. Cultivation and feeding upon symbiotic fungi growing in woody tree tissues; for example, some Scolytidae.

Xylophagy. Feeding on xylem tissues.

See also the updated edition of *The Torre-Bueno Glossary of Entomology* compiled by Nichols in 1989, consisting of 840 pages.

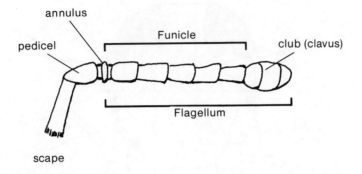

G.1 Insect antenna.

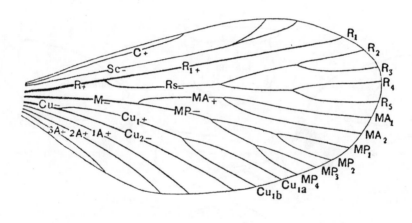

G.2 A theoretical primitive insect wing showing basic venation. Veins marked + protrude upwards (convex). Veins marked — protrude ventrally (concave). A = Anal series, C = Costa, Cu = Cubital series, M = Median (anterior and posterior), R = Radial series, Sc = Subcosta.

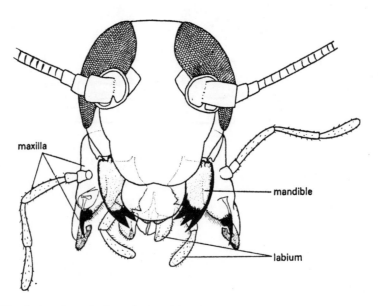

G.3 Head of cockroach, frontal view. Mouthparts are shown *in situ*.

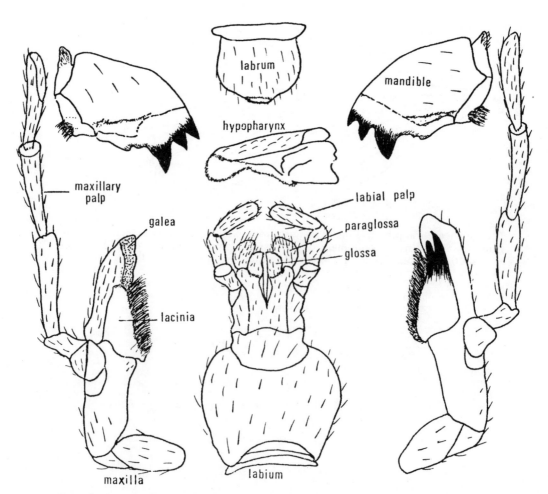

G.4 Cockroach mouthparts. Illustrates a primitive, unspecialized insect with biting and chewing mouthparts.

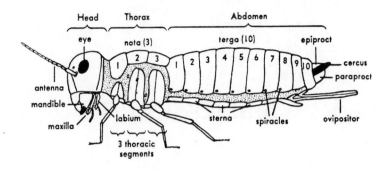

G.5 Main body parts of a generalized primitive insect.

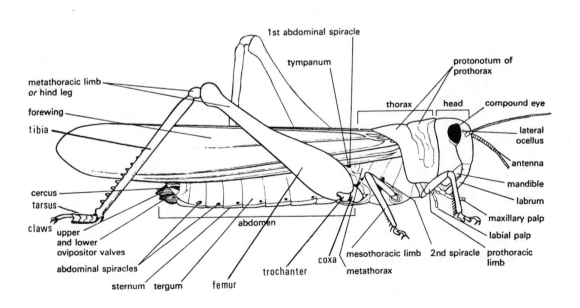

G.6 Female grasshopper (Orthoptera; Acrididae), lateral view. A common unspecialized insect.

BIBLIOGRAPHY

It is not feasible to compile a complete bibliography on this broad topic for it would be too enormous to publish as part of a general textbook. The list included here contains most of the major sources of information that were used in the production of the text. Emphasis has been placed on regional checklists, regional handbooks and textbooks, review articles, and other publications thought to be of particular use to students seeking details or further information concerning either a group of insects or the pests of a particular crop. There is a concentration on insect and mite pests of cultivated plants, and crops post-harvest, as this emphasis is present throughout the book. Many of the more specialized texts referred to can usually be regarded as a source of references on that topic.

A.D.A.S. (M.A.F.F., U.K.). For many years, as part of the advisory service to British agriculture, a series of illustrated leaflets were published (by H.M.S.O.), mostly on specific pests. Since the recent reorganization (1980s) the service is greatly curtailed and only a few publications are still being produced. Similar small publications are produced by U.S.D.A. and also in most states in the U.S.A., and also in many other countries for local use. See Gratwick, 1992.

Alford, D. V. 1975. *Bumblebees.* Davis-Poynter: London.

––––––– 1984. *A colour atlas of fruit pests* (their recognition, biology and control). Wolfe Publishing: London.

––––––– 1991. *A colour atlas of pests of ornamental trees, shrubs and flowers.* Wolfe Publishing: London.

Allee, W. C., A. E. Emerson, O. Park, T. Park, and K. P. Schmidt. 1955. *Principles of animal ecology.* Saunders: Philadelphia.

Allen, W. A., and E. G. Rajotte. 1990. "The changing role of extension entomology in the I.P.M. era." In *Ann. Rev. Entomol.,* 35, 379–397.

Anathakrishnan, T. N. 1979. "Biosystematics of Thysanoptera." In *Ann. Rev. Entomol.,* 24, 159–183.

Andrewartha, H. G., and L. C. Birch. 1961. *The distribution and abundance of animals.* 2nd ed. Univ. Chicago Press: Chicago.

Angus, A. 1962. Annotated list of plant pests, diseases and fungi in northern Rhodesia, recorded at the Plant Pathology Laboratory, Mount Makulu Research Station, parts 1 to 7, and supplement. (Cyclostyled) *c.* 600 pp.

Annecke, D. P., and V. C. Moran. 1982. *Insects and mites of cultivated plants in South Africa.* Butterworths: London.

Anon. 1964. *A handbook on Arabica coffee in Tanganyika.* Tanzania Coffee Board: Lyamungu, Tz.

––––––– 1965. *An atlas of coffee pests and diseases.* Coffee Res. Found.: Ruiru, Kenya.

––––––– 1965. *A handbook for sisal planters.* Tanganyika Sisal Growers Assoc.

Anti-locust Research Centre. 1966. *The locust handbook.* C.O.P.R.: London (2nd ed., 1989; available from N.R.I., Chatham, U.K.; also many other publications on grasshoppers, locusts, and other Orthoptera).

Askew, A. A. 1971. *Parasitic insects.* Heinemann: London.

Avidoz, Z., and I. Harpas. 1969. *Plant pests of Israel.* Israel Univ. Press: Jerusalem.

Axtell, R. C., and J. J. Arends. 1990. "Ecology and management of arthropod pests of poultry." In *Ann. Rev. Entomol.,* 35, 101–126.

Baker, E. W., and A. E. Pritchard. 1960. "The tetranychoid mites of Africa." *Hilgardia,* 29, 455–574.

Baker, E. W., and G. W. Wharton. 1964. *An introduction to acarology.* Macmillan: New York.

Balachowsky, A. S., ed. 1966 *Entomologie appliquée à l'agriculture,* vols. 1–8. Masson: Paris.

Balachowsky, A. S., and L. Mesnil. 1935. *Les insectes nuisibles aux plantes cultivées,* 2 vols. Min. Agric.: Paris.

Banerjee, B. 1981. "An analysis of the effect of latitude, age and area on the number of arthropod pests of tea." In *J. Appl. Ecol.,* 18, 339–342.

Banziger, H. 1968. "Preliminary observations on a skin-piercing bloodsucking moth, *Calyptra eustrigata* (Hamps.) (Lep., Noctuidae) in Malaya." In *Bull. Ent. Res.,* 59, 159–163.

Barbosa, P., and M. A. Wagner. 1989. *Introduction to forest and shade tree insects.* Academic Press: London.

Bardner, R., and K. E. Fletcher. 1974. "Insect infestations and their effects on the growth and yield of field crops: a review." In *Bull. Ent. Res.,* 64, 141–160.

Barnes, H. F. 1946–69. *Gall midges of economic importance.* Crosby Lockwood: London.

_____ 1946. Vol. 1. *Root and vegetable crops.*

_____ 1946. Vol. 2. *Fodder crops.*

_____ 1948. Vol. 3. *Fruit.*

_____ 1948. Vol. 4. *Ornamental plants and shrubs.*

_____ 1951. Vol. 5. *Trees.*

_____ 1949. Vol. 6. *Miscellaneous crops.*

_____ 1956. Vol. 7. *Cereal crops.*

_____ 1969. Vol. 8. *Miscellaneous* (W. Nijveldt).

Bateman, M. 1972. "The ecology of fruit flies." In *Ann. Rev. Entomol.,* 17, 493–518.

Batra, S. W. T. 1984. "Solitary bees." In *Sci. American,* 250 (2), 86–93.

"Bayer." 1968. *Bayer crop protection compendium,* 2 vols. Bayer Co.: Leverkusen.

Beardsley, J. W., and R. H. Gonzales. 1975. "The biology and ecology of armoured scales." In *Ann. Rev. Entomol.,* 20, 47–73.

Beattie, A. J. 1985. *The evolutionary ecology of ant–plant mutualisms.* Cambridge Univ. Press: Cambridge.

Beaver, R. A. 1977. "Bark and ambrosia beetles in tropical forests." In *Biotrop. Spec. Pub. No. 2,* 133–147.

Becker, P. 1974. *Pests of ornamental plants.* M.A.F.F. Bull., No. 97, H.M.S.O.: London.

Beirne, B. P. 1967. *Pest management.* Leonard Hill: London.

_____ 1971. "Pest insects of annual crops in Canada. 1. Lepidoptera; 2. Diptera; 3. Coleoptera." In *Mem. Ent. Soc. Canada,* No. 78.

_____ 1972. "Pest insects of annual crops in Canada. 4. Hemiptera–Homoptera; 5. Orthoptera; 6. Other Groups." In *Mem. Ent. Soc. Canada,* No. 85.

Bell, W. J., and R. T. Carde. 1984. *Chemical ecology of insects.* Assoc. Book Publ.: London.

Bell, T. R. D., and F. B. Scott. 1937. *The fauna of British India,* vol. 5, *moths.* Taylor & Francis: London.

Bellotti, A., and A. van Schoonhoven. 1978. "Mite and insect pests of cassava." In *Ann. Rev. Entomol.,* 23, 39–67.

_____ 1978. *Cassava pests and their control.* C.I.A.T.: Colombia.

Beroza, M., ed. 1970. *Chemicals controlling insect behavior.* Academic Press: New York.

Bevan, D. 1987. *Forest insects* (a guide to insects feeding on trees in Britain). For. Comm. Hdbk. 1. H.M.S.O.: London.

Birch, M. C., and K. F. Haynes. 1982. *Insect pheromones.* Studies in Biology No. 147, Edward Arnold: London.

Blackman, R. L., and V. F. Eastop, eds. 1984. *Aphids on the world's crops.* John Wiley: Chichester.

Bleszynski, S. 1970. "A revision of the world species of *Chilo* Zincken (Lep., Pyraliae)." In *Bull. Br. Mus. Nat. Hist. (B),* 25.

Bodenheimer, F. S. 1951. *Citrus entomology in the Middle East.* W. Junk: The Hague.

_____ 1951. *Insects as human food.* W. Junk: The Hague.

Bodenheimer, F. S., and E. Swirski. 1957. *The Aphidoidea of the Middle East.* Weizman Sci. Press: Jerusalem.

Bohlen, E. 1978. *Crop pests in Tanzania and their control.* Verlag Paul Parey: Berlin & Hamburg.

Boller, E. F., and R. J. Prokopy. 1976. "Bionomics and management of *Rhagoletis.* " In *Ann. Rev. Entomol.,* 21, 223–246.

Borror, D. J., and D. M DeLong. 1971. *An introduction to the study of insects.* 3rd ed. Holt, Rinehart and Winston: New York.

Bostanian, N. J., L. T. Wilson, and T. J. Dennehy. 1990. *Monitoring and integrated management of arthropod pests of small fruit crops.* Intercept: Andover, U.K.

Bottrell, D. R. 1979. *Integrated pest management.* Council on Environmental Quality. U.S. Govt. Printing Office: Washington.

Boucek, Z. 1988. *Australasia Chalcidoidea.* A biosystematic revision of genera of 14 families with a reclassification of the species. CABI: Wallingford.

Bournier, A. 1977. Grape insects. In *Ann. Rev. Entomol.,* 22, 355–376.

Box, H. E. 1953. *List of sugarcane insects.* C.I.E.: London.

Brader, L. 1979. "Integrated pest control in the developing world." In *Ann. Rev. Entomol.,* 24, 154–255.

Bradley, J. D., W. G. Tremewan, and A. Smith. 1973. *British tortricoid moths Cocylidiae and Tortricidea: Tortricinae.* Ray Soc. No. 147, The Ray Soc.: London.

Brown, A. W. A., and R. Pal. 1971. *Insecticide resistance in arthropods.* 2nd ed. W.H.O.: Geneva.

Brown, E. S. 1954. "The biology of the coconut pest *Melittomma insulare* (Col., Lymexylonidae), and its control in the Seychelles." In *Bull. Ent. Res.,* 45, 1–66.

Brown, E. S., and C. F. Dewhurst. 1975. "The genus *Spodoptera* (Lepidoptera, Noctuidae) in Africa and the Near East." In *Bull. Ent. Res.,* 65, 221–262.

Brown, F. G. 1968. *Pests and diseases of forest plantation trees: an annotated list of the principal species occurring in the British Commonwealth.* Clarendon Press: Oxford.

Brown, K. W. 1967. *Forest insects of Uganda, an annotated list.* Govt. Printer: Entebbe.

Buczacki, S. T., and K. M. Harris. 1981. *Collins guide to the pests, diseases and disorders of garden plants.* Collins: London.

Burn, A. J., T. H. Coaker, and P. C. Jepson. 1987. *Integrated pest management.* Academic Press: London.

Busvine, J. R. 1966. *Insects and hygiene.* 2nd ed. Methuen: London.

_____ *Recommended methods for measurement of pest resistance to pesticides.* F.A.O. Pl. Prod. Prot. Paper—21, F.A.O.: Rome.

Butani, D. K. 1979. *Insects and fruits.* Periodical Expert Book Agency: New Delhi.

Buyckx, E. J. E. 1962. *Précis des maladies et des insectes nuisibles rencontrées sur les plantes cultivées au Congo, au Rwanda et au Burundi.* I.N.E.A.C.

C.A.B. 1951–1989. *Distribution maps of insect pests,* series A, agricultural. Nos. 1–510, C.I.E.: London.

_____ 1961–67. *C.I.B.C. technical bulletins.* 1–8, C.A.B.: London.

_____ 1980. *Perspectives in world agriculture.* C.A.B.: Slough.

_____ 1981. *List of research workers in the agricultural sciences in the Commonwealth and in the Republic of Ireland.* 4th ed. C.A.B.: London.

C.A.B.I. 1992. *Quarantine pests for Europe.* C.A.B.I. & E.P.P.O.

(C.A.B. is now Commonwealth Bureaux of Agriculture International, and located at Wallingford, Oxford, U.K.; many specialist publications.)

Caltagirone, L. E. 1981. "Landmark examples in classical biological control." In *Ann. Rev. Entomol.,* 26, 213–232.

Cameron, P. J., R. L. Hill, J. Bain, and W. P. Thomas, eds. 1989. *A review of biological control of invertebrate pests and weeds in New Zealand 1874–1987.* C.I.B.C. Tech. Comm. No. 10, C.A.B.I.B.C. & D.S.I.R.

Campion, D. G. 1972. "Insect chemosterilants: a review." In *Bull. Ent. Res.,* 61, 577–635.

Caresche, L., G. S. Cotterell, J. E. Peachey, R. W. Rayner, and Jacques-Felix. 1969. *Handbook for phytosanitary inspectors in Africa.* OAU/STRC: Lagos.

Carey, J. R. 1989. "The multiple decrement life table: a unifying framework for cause-of-death analysis in ecology." In *Oecologia,* 78, 131–137.

Carey, J. R., and R. V. Dowell. 1989. "Exotic fruit fly pests and California agriculture." In *California Agric.*, May–June, 38–40.

Carpenter, J. B., and H. S. Elmer. 1978. *Pests and diseases of the date palm.* U.S.D.A.: Agric. Hdbk. No. 527.

Caswell, G. H. 1962. *Agricultural entomology in the tropics.* Edward Arnold: London.

Cavalloro, R., ed. 1983. *Fruit flies of economic importance.* Proc. C.E.C./I.O.B.C. Int. Symp.; Athens, Nov. 1982. Balkema: Rotterdam.

Chalfant, R. B., R. K. Jansson, D. R. Seal, and J. M. Schalk. 1990. "Ecology and management of sweet potato insects." In *Ann. Rev. Entomol.*, 35, 157–180.

Chandler, A. C., and C. P. Read. 1961. *Introduction to parasitology* (with special reference to the parasites of man). 10th ed. John Wiley: New York.

Chapman, R. F. 1970. *The insects—structure and function.* English Univ. Press: London.

_____ 1974. *Feeding in leaf-eating insects.* Oxford Biology Reader 69, Oxford Univ. Press: Oxford.

Cheng, T. C. 1967. *The biology of animal parasites.* Saunders: Philadelphia.

Cherrett, J. M., J. B. Ford, I. V. Herbert, and A. J. Probert. 1971. *The control of injurious animals.* English Univ. Press: London.

Cherrett, J. M., and D. J. Peregrine. 1976. "A review of the status of the leaf-cutting ants and their control." In *Ann. Appl. Biol.*, 84, 128–133.

Cherrett, J. M., and G. R. Sager. 1977. *Origins of pest, parasite, disease and weed problems.* 18th Symp. Brit. Ecol. Soc. Blackwell: Oxford.

Chiang, H. C. 1973. "Bionomics of the northern and western corn rootworms." In *Ann. Rev. Entomol.*, 18, 47–72.

_____ 1978. "Pest management in corn." In *Ann. Rev. Entomol.*, 23, 101–123.

Chinnery, M. 1986. *Collins guide to the insects of Britain and Western Europe.* Collins: London.

C.I.E. Guide to insects of importance to Man.

1987. 1. *Lepidoptera.* C.A.B. Int. Inst. Ent. & Brit. Mus. Nat. Hist.

1989. 2. *Thysanoptera.* C.A.B. Int. Inst. Ent. & Brit. Mus. Nat. Hist.

1990. *Coleoptera.* Int. Inst. Ent. & The Nat. Hist. Mus.

C.I.P. 1983. *Major potato diseases, insects and nematodes.* C.I.P.: Lima, Peru.

Clark, L. R., P. W. Geier, R. D. Hughes, and R. F. Morris. 1967. *The ecology of insect populations in theory and practice.* Methuen: London.

Clausen, C. P. 1940. *Entomophagous insects.* 1st ed. McGraw-Hill: New York.

Cloudsley-Thompson, J. L. 1986. "Cochineal." In *Antenna*, 10 (2), 70–72.

Conway, G. R. 1972. *Pests of cocoa in Sabah, Malaysia.* Bull. Dept. Agric., Malaysia.

_____ 1972. "Ecological aspects of pest control in Malaysia." In J. Milton, ed. *The careless technology; ecological aspects of international development.* Nat. Hist. Press.

Conway, G. R., and E. B. Tay. 1969. *Crop pests in Sabah, Malaysia and their control.* St. Min. Agric. Fish., Sabah, Malaysia.

C.O.P.R. 1978. *Pest control in tropical root crops.* P.A.N.S. Manual No. 4, C.O.P.R.: London.

_____ 1981. *Pest control in tropical grain legumes.* C.O.P.R.: London.

_____ 1982. *The locust and grasshopper manual.* C.O.P.R.: London, later O.D.N.R.I. (now T.R.I.)

_____ 1983. *Pest control in tropical tomatoes.* C.O.P.R.: London.

Corbet, S. 1987. "More bees make better crops." In *New Scientist*, 23 July, 40–43.

Coulson, R. N. 1979. "Population dynamics of bark beetles." In *Ann. Rev. Entomol.*, 24, 417–444.

Coulson, R. N., and J. A. Witter. 1984. *Forest entomology* (ecology and management). Wiley & Sons: New York.

Cramer, H. H. 1967. *Plant protection and world crop production.* Bayer Pflanzenschutz: Leverkusen.

Crane, E., and P. Walker. 1983. *The impact of pest management on bees and pollination.* Int. Bee Res. Assoc., T.D.R.I.: London.

Cranham, J. E. 1966. *Insect and mite pests of tea in Ceylon and their control.* Tea Res. Inst. Ceylon: Talawakelle.

_____ 1966. "Tea pests and their control." In *Ann. Rev. Entomol.*, 11, 491–510.

Crawley, M. J. 1989. "Insect herbivores and plant population dynamics." In *Ann. Rev. Entomol.*, 34, 351–364.

Crowe, T. J. 1967. "Common names for agricultural and forestry insects and mites in East Africa." In *E. Afr. Agric. For. J.*, 33, 55–63.

_____ 1967. *Cotton pests and their control.* Dept. Agric.: Nairobi, Kenya.

Crowe, T. J., and G. M. Shitaye. 1977. *Crop pest handbook.* 3rd ed. I.A.R: Addis Ababa.

Crowe, T. J., G. M. Tadesse, and T. Abate. 1977. *An annotated list of insect pests in Ethiopia.* I.A.R.: Addis Ababa.

Crowson, R. A. 1968. *The natural classification of the families of Coleoptera.* Classey: Hampton.

_____ 1981. *The biology of the Coleoptera.* Academic Press: London.

C.S.C.P.R.C. 1977. *Insect control in the People's Republic of China.* Nat. Acad. Sci.: Washington.

C.S.I.R.O. 1973. *Scientific and common names of insects and allied forms occurring in Australia.* C.S.I.R.O. Bull. 287.

_____ *See also* Mackerras.

Darlington, A. 1968. *The pocket encyclopedia of plant galls in colour.* Blandford Press: London.

Davidson, R. H., and L. M. Peairs. 1966. *Insect pests of farm, garden and orchard.* John Wiley: New York.

Davies, R. G. 1988. *Outlines of entomology.* 7th ed. Chapman & Hall: London.

Dean, G. J. 1979. "The major pests of rice, sugarcane and jute in Bangladesh." In *P.A.N.S.*, 25, 378–385.

DeBach, P. 1964. *Biological control of insect pests and weeds.* Chapman & Hall: London.

_____ 1974. *Biological control by natural enemies.* Cambridge Univ. Press: Cambridge.

De Long, D. 1971. "The bionomics of leafhoppers." In *Ann. Rev. Entomol.*, 16, 179–210.

De Lotto, G. 1967. "The soft scales (Homoptera, Coccidae) of South Africa, I." In *S. Afr. J. Agric. Sci.*, 10, 781–810.

Delucchi, V. L., ed. 1976. *Studies in biological control* (I.B.P.-9). Cambridge Univ. Press: Cambridge.

Dent, D. 1991. *Insect pest management.* C.A.B.I.: Wallingford.

Dobie, P., C. P. Haines, R. J. Hodges, and P. T. Prevett. 1984. *Insects and arachnids of tropical stored products—their biology and identification* (a training manual). T.D.R.I.: Slough.

Dolling, W. R. 1991. *The Hemiptera.* Nat. Hist. Museum Publications, Oxford Univ. Press: Oxford.

Doutt, R. L., and G. Viggiani. 1968. "The classification of the Trichogrammatidae (Hymenoptera: Chalcidoidea)." In *Proc. Calif. Acad. Sci.*, 35, 477–586.

Drew, R. A. I., G. H. S. Hooper, and M. A. Bateman. 1978. *Economic fruit flies of the South Pacific region.* Dept. Pri. Indust.: Queensland.

Duffey, E. A. J. 1957. *African timber beetles.* Brit. Mus. Natural History: London.

Eady, R. D., and J. Quinlan. 1963. *Handbooks for the identification of British insects. Hymenoptera Cynipoidea.* Vol. 8, Part I (a). Roy. Ent. Soc. Lond.: London.

Eastop, V. F. 1958. *A study of the Aphididae (Homoptera) of East Africa.* Col. Res. Pub. No. 20, H.M.S.O.: London.

_____ 1961. *A study of the Aphididae (Homoptera) of West Africa.* Brit. Mus. Natural History: London.

_____ 1966. "A taxonomic study of Australian Aphidoidea (Homoptera)." In *Aust. J. Zool.*, 14, 399–592.

Ebbels, D. L., and J. E. King. 1979. *Plant health.* Blackwell: Oxford.

Ebeling, W. 1959. *Subtropical fruit pests.* 2nd ed. Univ. Calif. Press: California.

_____ 1975. *Urban entomology.* Univ. Calif., Div. Agric. Sci.: California.

Edwards, C. A., and G. W. Heath. 1964. *Principles of agricultural entomology.* Chapman & Hall: London.

Elkinton, J. S., and A. M. Liebhold. 1990. "Population dynamics of gypsy moth in North America." In *Ann. Rev. Entomol.*, 35, 571–596.

Emden, H. F. van, ed. 1972. *Insect/plant relationships.* Symp. Roy. Ent. Soc. Lond., No. 6. Blackwell: Oxford.

_____ ed. 1972. *Aphid technology.* Academic Press: London.

_____ 1974. *Pest control and its ecology.* Inst. of Biol. Studies in Biology, No. 50. Edward Arnold: London.

_____ 1989. *Pest control.* 2nd ed. IOB New Studies in Biology. Edward Arnold: London.

Emden, H. F. van, V. F. Eastop, R. D. Hughes, and M. J. Way. 1969. "The ecology of *Myzus persicae.*" In *Ann. Rev. Entomol.*, 14, 197–270.

Emmet, A. M. 1979. *A field guide to the smaller British Lepidoptera*. Brit. Ent. and Nat. Hist. Soc.: London.

Entwhistle, P. F. 1972. *Pests of cocoa*. Longmans: London.

E.P.A. 1976. *List of insects and other organisms*. 3rd ed. Parts 1, 2, 3, and 4. E.P.A.: Washington.

Evans, G. O., J. G. Sheals, and D. Macfarlane. 1961. *The terrestrial Acari of the British Isles*. Vol. 1. *Introduction and biology*. Brit. Mus. Natural History: London.

Evans, J. W. 1952. *Injurious insects of the British Commonwealth*. C.I.F.: London.

F.A.O. (Rome) produce a series of publications; some are conference proceedings, and others are F.A.O. Plant Production and Protection Papers—a selected few titles are listed below.

F.A.O. 1966. *Proceedings of the F.A.O. Symposium on Integrated Pest Control (11–15 October 1965)*, 1, 2, 3. F.A.O.: Rome.

_____ 1974. *Proceedings of the F.A.O. Conference on Ecology in Relation to Plant Pest Control. Rome, Italy, 11–15 December 1972*. F.A.O.: Rome.

_____ 1979. *Guidelines for integrated control of maize pests*. F.A.O. Pl. Prod. Prot. Paper—18. F.A.O.: Rome.

_____ 1979. *Elements of integrated control of sorghum pests*. F.A.O. Pl. Prod. Prot. Paper—19. F.A.O.: Rome.

F.A.O./C.A.B. 1971. *Crop loss assessment methods*. F.A.O. Manual on the Evaluation of Losses by Pests, Diseases and Weeds. Supplement 1. 1973, 2. 1977, 3. 1981.

Feakin, S. D., ed. 1971. *Pest control in bananas*. P.A.N.S. Manual No. 1, 2nd ed. C.O.P.R.: London.

_____ ed. 1973. *Pest control in groundnuts*. P.A.N.S. Manual No. 2, 3rd ed. C.O.P.R.: London.

_____ ed., 1976. *Pest control in rice*. P.A.N.S. Manual No. 3, 2nd ed. C.O.P.R.: London.

Fenemore, P. G. 1984. *Plant pests and their control*. Revised ed. Butterworths: London.

Ferrière, C. 1965. *Faune de L'Europe et du Bassin Mediterranéen. 1. Hymenoptera Aphelinidae*. Masson: Paris.

Ferro, D. N., ed. 1976. *New Zealand insect pests*. Lincoln Univ. Coll. Agric.: Canterbury.

Ferron, P. 1978. "Biological control of insect pests by entomogenous fungi." In *Ann. Rev. Entomol.*, 23, 409–442.

Fichter, G. S. 1968. *Insect pests*. Paul Hamlyn: London.

Finch, S. 1989. "Ecological considerations in the management of *Delia* pest species in vegetable crops." In *Ann. Rev. Entomol.*, 34, 117–137.

Fitt, G. P. 1989. "The ecology of *Heliothis* species in relation to agroecosystems." In *Ann. Rev. Entomol.*, 34, 17–52.

Fletcher, J. T., P. F. White, and R. H. Gaze. 1989. *Mushrooms—pest and disease control*. Intercept: Andover, U.K.

Fletcher, W. W. 1974. *The pest war*. Blackwell: Oxford.

Forsyth, J. 1966. *Agricultural insects of Ghana*. Ghana Univ. Press: Accra.

Fox Wilson, G. 1960. *Horticultural pests—detection and control*. 2nd ed. Crosby Lockwood: London.

Free, J. B. 1970. *Insect pollination of crops*. Academic Press: London.

Free, J. B., and I. H. Williams. 1977. *The pollination of crops by bees*. Apimondia, Bucharest, & Int. Bee Res. Assoc., U.K.

Frolich, G., and W. Rodewald. 1970. *General pests and diseases of tropical crops and their control*. Pergamon Press: London.

Gauld, I., and B. Bolton, eds. 1988. *The Hymenoptera*. Brit. Mus. Natural History & O.U.P.: London.

Gay, F. J., ed. 1966. *Scientific and common names of insects and allied forms occurring in Australia*. Bull. No. 285, C.S.I.R.O.: Melbourne.

Geier, P. W. 1966. "Management of insect pests." In *Ann. Rev. Entomol.*, 11, 471–490.

Geier, P. W., L. R. Clark, D. J. Anderson, and H. A. Nix. 1973. *Insects: studies in population management*. Ecol. Soc. Australia, Memoir 1: Canberra.

Gentry, J. W. 1965. *Crop pests of northeast Africa–southwest Asia*. Agric. Hdbk. No. 273, U.S.D.A.: Washington.

Gerling, D. 1990. *Whiteflies: their bionomics, pest status and management*. Intercept: Andover, U.K.

Getz, W. M., and A. P. Gutierrez. 1982. "A perspective of systems analysis in crop production and insect pest management." In *Ann. Rev. Entomol.,* 27, 447–466.

Gilbert, P., and C. J. Hamilton. 1990. *Entomology—a guide to information sources.* 2nd ed. Mansell: London.

Glass, E. H. (Coordinator). 1975. *Integrated pest management: rationale, potential, needs and implementation.* Ent. Soc. Amer., Special Publ. 75–2.

Goater, B. 1986. *British pyralid moths.* Harley Books: Colchester.

Good, R. 1953. *The geography of the flowering plants.* 2nd ed. Longmans: London.

Graham, M. W. R. de V. 1969. "The Pteromalidae of north-western Europe (Hymenoptera: Chalcidoidea)." In *Bull. B.M. (N.H.), Ent. Suppl.,* 16, 1–908.

Gram, E., P. Bovien, and C. Stapel. 1969. *Recognition of diseases and pests of farm crops.* 2nd ed. Blandford Press: London.

Grandi, R. 1961. "The hymenopterous insects of the superfamily Chalcidoidea developing within the receptacles of figs. Their life-history, symbioses and morphological adaptations." In *Boll. Ist. Ent. Univ. Bologna,* 26, 1–13.

Gratwick, M., ed. 1989. *Potato pests.* A.D.A.S. Ref. Book 187. H.M.S.O.: London.

———— 1992. *Crop pests in the U.K.* Collected edition of M.A.F.F. leaflets. Chapman & Hall: London.

Gray, B. 1972. "Economic tropical forest entomology." In *Ann. Rev. Entomol.,* 17, 313–354.

Greathead, D. J. 1971. *A review of biological control in the Ethiopian region.* Tech. Commun. C.I.B.C., No. 5, C.A.B.: London.

Grist, D. H., and R. J. A. W. Lever. 1969. *Pests of rice.* Longmans: London.

Gruys, P., and A. K. Minks, eds. 1979. *Integrated control of pests in the Netherlands.* Pudoc: Netherlands.

Hainsworth, E. 1952. *Tea pests and diseases.* Heffers: Cambridge.

Hanson, H. C. 1963. *Diseases and pests of economic plants of Central and South China, Hong Kong and Taiwan (Formosa).* Amer. Inst. Crop Ecology: Washington.

Harcourt, D. G. 1969. "The development and use of life tables in the study of natural insect populations." In *Ann. Rev. Entomol.,* 14, 175–196.

Hare, J. D. 1990. "Ecology and management of the Colorado potato beetle." In *Ann. Rev. Entomol.,* 35, 81–100.

Harris, K. M. 1962. "Lepidopterous stem borers of cereals in Nigeria." In *Bull. Ent. Res.,* 53, 139–171.

———— 1966. "Gall midge genera of economic importance (Diptera: Cecidomyiidae). Part 1: Introduction and subfamily Cecidomyiinae: supertribe Cecidomyiidi." In *Trans. Roy. Ent. Soc. Lond.,* 118, 313–358.

———— 1968. "A systematic revision and biological review of the cecidomyiid predators (Diptera: Cecidomyiidae) on world Coccoidea (Hemiptera: Homoptera." In *Trans. Roy. Ent. Soc. Lond.,* 119, 401–494.

———— 1970. "The sorghum midge." In *P.A.N.S.,* 46, 36–42.

Harris, K. M., and E. Harris. 1968. "Losses of African grain sorghums to pests and diseases." In *P.A.N.S.,* 44, 48–54.

Harris, W. V. 1969. *Termites as pests of crops and trees.* C.I.E.: London.

———— 1971. *Termites, their recognition and control.* 2nd ed. Longmans: London.

Harwood, R. F., and M. T. James. 1979. *Entomology in human and animal health.* Macmillan: New York.

Hassan E. 1977. *Major insect and mite pests of Australian crops.* Ento Press: Queensland.

Hassell, M. P., and T. R. E. Southwood. 1978. "Foraging strategies of insects." *Ann. Rev. Ecol. Syst.,* 9, 75–98.

Headley, J. C. 1972. "Economics of agricultural pest control." In *Ann. Rev. Entomol.,* 17, 273–286.

Heath, J., and A. M. Emmet, eds. 1976– . *The moths and butterflies of Great Britain and Ireland:*

———— 1976. Vol. 1. *Micropterigidae—Heliozelidae.* Curwen Press: London.

———— 1985. Vol. 2. *Cossidae—Heliodiniae.* Harley Books: Colchester.

———— 1989. Vol. 7. *(Part 1) Hesperiidae—Nymphalidea.* Harley Books: Colchester.

———— 1979. Vol. 9. *Sphingidae—Noctuidae (Part I).* Curwen Press: London.

———— 1983. Vol. 10. *Noctuidae (Part II) and Agaristidae.* Harley Books: Colchester.

Hercules Powder Co. 1960. *Cotton insect pests.* Hercules Powder Co.: Wilmington, U.S.A.

Hering, E. M. 1951. *Biology of the leaf miners.* Dr. W. Junk: Gravenhage.

Herren, H. R. 1981. "IITA's role and actions in controlling the cassava mealybug in Africa." In *IITA Research Briefs,* 2 (4), 1–8.

Hill, D. S. 1967. "Figs (*Ficus* spp.) and fig-wasps (Chalciodoidea)." In *Journal Natural History,* 1, 413–434.

_____ 1982. *Hong Kong insects.* Vol. 2. Urban Council: Hong Kong.

_____ 1983. *Agricultural insect pests of the tropics and their control.* 2nd ed. Cambridge Univ. Press: Cambridge.

_____ 1987. *Agricultural insect pests of temperate regions and their control.* Cambridge Univ. Press: Cambridge.

_____ 1989. *Catalogue of crop pests of Ethiopia.* 1st ed. Alemaya Univ. Agric. Bull. No. 1.

_____ 1990. *Pests of stored products and their control.* Belhaven Press: London.

Hill, D. S., P. Hore, and I. W. B. Thornton. 1982. *Insects of Hong Kong.* H. K. Univ. Press: Hong Kong.

Hill, D. S, and J. M. Waller. 1982. *Pests and diseases of tropical crops.* Vol. 1: *Principles and methods of control.* (I.T.A.S.). Longmans: London.

_____ 1988. *Pests and diseases of tropical crops.* Vol. 2: *Field handbook.* (I.T.A.S.). Longmans: London.

Hinckley, A. D. 1963. *Trophic records of some insects, mites and ticks in Fiji.* Dept. Agric. Fiji, Bull. No. 45.

Hinton, H. E., and A. S. Corbet. 1955. *Common insect pests of stored food products.* 3rd ed. Econ. Ser., No. 15, Brit. Mus., Natural History: London. (For 7th ed. see Mound, 1989.)

Hodgson, C. J. 1970. "Pests of citrus and their control." In *P.A.N.S.,* 16, 647–666.

Hodkinson, I. D., and M. K. Hughes. 1982. *Insect herbivory* (Outline Studies in Ecology). Chapman & Hall: London.

Holt, V. M. 1988. *Why not eat insects?* 2nd ed.

Huffaker, C. B., ed. 1971. *Biological control.* Plenum: New York.

Huffaker, C. B., J. A. McMurtry, and M. van de Vrie. 1969. "The ecology of tetranychid mites and their natural control." In *Ann. Rev. Entomol.,* 14, 125–174.

Hughes, A. M. 1961. *The mites of stored food.* M.A.F.F. Tech. Bull. No. 9. H.M.S.O.: London.

Hussey, N. W., W. H. Read, and J. J. Hesling. 1969. *The pests of protected cultivation.* Edward Arnold: London.

I.C.R.I.S.A.T. 1985. *Proceedings of the international sorghum entomology workshop, 15–21 July, 1984, Texas A & M University, College Station, Tex., U.S.A.* I.C.R.I.S.A.T.: Patancheru, India.

I.I.T.A. 1981. *Annual report, 1981.* I.I.T.A.: Ibadan.

Imms, A. D. 1960. *A general textbook of entomology.* 9th ed. revised by O. W. Richards and R. G. Davies. Methuen: London.

Ingram, W. R., J. R. Davies, and J. N. McNutt. 1970. *Agricultural pest handbook (Uganda).* Govt. Printer: Entebbe.

I.R.R.I. 1967. *The major insect pests of the rice plant.* Johns Hopkins Press: Baltimore, Md., U.S.A.

Jameson, J. D. 1970. *Agriculture in Uganda.* 2nd ed. Oxford Univ. Press: London.

Japan Plant Protection Assoc. 1980. *Major insect and other pests of economic plants in Japan.* Jap. Pl. Prot. Assoc.: Tokyo.

Jeppson, L. R., H. H. Keifer, and E. W. Baker. 1975. *Mites injurious to economic plants.* Univ. California Press: Berkeley.

Jepson, W. F. 1954. *A critical review of the world literature on the lepidopterous stalk borers of tropical graminaceous crops.* C.I.E.: London.

Johnson, W. T., and H. H. Lyon. 1976. *Insects that feed on trees and shrubs* (an illustrated practical guide). Cornell Univ. Press: Ithaca.

Jones, F. G. W., and M. Jones. 1974. *Pests of field crops.* 2nd ed. Edward Arnold: London.

Kalshoven, L. G. E. 1981. *Pests of crops in Indonesia.* Revised and translated by P. A. van der Laan and G. H. L. Rothschild. Ichtiar Baru: Van Noeve, Jakarta.

Kettle, D. S. 1990. *Medical and veterinary entomology.* 2nd ed. C.A.B.I.: Wallingford.

Kevan, P. G., and H. G. Baker. 1983. "Insects as flower visitors and pollinators." In *Ann. Rev. Entomol.,* 28, 407–453.

Kilgore, W. W., and R. L. Doutt. 1967. *Pest control—biological, physical and selected chemical methods.* Academic Press: New York.

King, A. B. S., and J. L. Saunders. 1984. *The invertebrate pests of annual food crops in Central America.* O.D.A.: London.

Kiritani, K. 1979. "Pest management in rice." In *Ann. Rev. Entomol.,* 24, 279–312.

Kirkpatrick, T. W. 1966. *Insect life in the tropics.* Longmans: London.

Klages, K. H. W. 1942. *Ecological crop geography.* Macmillan: New York.

Kloet, G. S., and W. D. Hincks. 1964–1978. *A check list of British insects.* 2nd ed.—revised. Roy. Ent. Soc. of Lond.: London.

_____ 1964. Part 1: Small Orders and Hemiptera.

_____ 1972. Part 2: Lepidoptera.

_____ 1977. Part 3: Coleoptera and Strepsiptera.

_____ 1978. Part 4: Hymenoptera.

_____ 1976. Part 5: Diptera and Siphonaptera.

(The above are parts of the Handbooks for the Identification of British Insects.)

Krantz, G. W., and E. E. Lindquist. 1979. "Evolution of phytophagous mites (Acari)." In *Ann. Rev. Entomol.,* 24, 121–158.

Kranz, J., H. Schmutterer, and W. Koch, eds. 1979. *Diseases, pests and weeds in tropical crops.* John Wiley: Chichester.

Krishna, K., and F. M. Weesner. 1969–1970. *Biology of termites.* Vol. 1, 1969. Vol. 2, 1970. Academic Press: New York.

Kumar, R. 1984. *Insect pest control* with special reference to African agriculture. Edward Arnold: London.

Labeyrie, V. 1981. *The ecology of Bruchids attacking legumes (pulses).* Proc. Symposium, 1980.

Laffoon, J. L. 1960. "Common names of insects—approved by the Entomological Society of America." In *Bull. Ent. Soc. Amer.,* 6, 175–211.

Lamb, K. P. 1974. *Economic entomology in the tropics.* Academic Press: London.

Lamb, R. J. 1989. "Entomology of oilseed *Brassica* crops." In *Ann. Rev. Entomol.,* 34, 211–229.

Lange, W. H., and L. Bronson. 1981. "Insect pests of tomatoes." In *Ann. Rev. Entomol.,* 26, 345–371.

Lange, W. H., *et al.* 1970. *Insects and other animal pests of rice.* Circ. 555, Calif. Agric. Expt. Sta. Ext. Service.

Lashomb, J. H., and R. A. Casagrande, eds. 1990. *Advances in potato pest management.* Academic Press: New York.

Lee, K. E., and T. G. Wood. 1971. *Termites and soils.* Academic Press: London.

Le Pelley, R. H. 1959. *Agricultural insects of East Africa.* E. Afr. High Comm.: Nairobi.

_____ 1968. *Pests of coffee.* Longmans: London.

Lepesme, ... 1947. *Les insectes des palmiers.* Lechevalier: Paris.

Leston, D. 1970. "Entomology of the cocoa farm." In *Ann. Rev. Entomol.,* 15, 273–294.

Leuschner, K., E. Terry, and T. Akinlosutu. 1980. *Field guide to identification and control of cassava pests and diseases in Nigeria.* Manual Series No. 3, I.I.T.A.: Ibadan.

Lewis, T. 1973. *Thrips: their biology, ecology and economic importance.* Academic Press: London.

Libby, J. L. 1968. *Insect pests of Nigerian crops.* Res. Bull. 269, Univ. Wisconsin: Madison, Wis., U.S.A.

Lincoln, R. J., G. A. Boxshall, and P. F. Clark. 1982. *A dictionary of ecology, evolution and systematics.* Camb. Univ. Press: Cambridge.

Long, W. H., and S. D. Hensley. 1972. "Insect pests of sugarcane." In *Ann. Rev. Entomol.,* 17, 149–176.

Lozano, J. C., *et al.* 1981. *Field problems in cassava.* C.I.A.T.: Colombia.

Mackauer, M., and L. E. Ehler. 1990. *Critical issues in biological control.* Intercept: Andover, U.K.

Mackerras, I. M., ed. C.S.I.R.O. 1969. *The insects of Australia.* Melbourne Univ. Press: Victoria.

Madge, R. B., R. G. Booth, and M. Cox. 1990. *I.I.E. guides to insects of importance to man.* No. 3, *Coleoptera.* I.I.E.: London.

Madsen, H. F., and J. C. Arrand. 1977. *The recognition and life history of the major orchard insects and mites in British Columbia.* Min. of Agric. Victoria, B.C.

Madsen, H. F., and C. V. G. Morgan. 1970. "Pome fruit pests and their control." In *Ann. Rev. Entomol.,* 15, 295–320.

Maeta, Y., and T. Kitamura. 1980. "Effects of delaying release on the reproduction of *Osmia (Osmia) cornifrons* (Rad.) in apple orchards." In *XVI Int. Congress of Entomology— Abstracts.* Kyoto, Japan.

M.A.F.F. (U.K.). In the past, a series of major texts were published by H.M.S.O. (see Hughes, 1961) as *Technical Bulletins,* and there was a more extensive series of paperbacked *Bulletins* (e.g. Bulletin 162—Sugar Beet Pests; p. 113). But since the reorganization of M.A.F.F. these publications have been mostly discontinued and no longer available.

Marmorosch, K., and K. F. Harris, eds. 1979. *Leafhopper vectors and plant disease agents.* Academic Press: New York.

Marshall. A. T. 1970. "External parasitism and blood-feeding by the lepidopterous larva *Epipyrops anomala* Westwood." In *Proc. Roy. Ent. Soc. Lond.* (A), 45, 137–140.

Martin, E. C., and S. E. McGregor. 1973. "Changing trends in insect pollination of commercial crops." In *Ann. Rev. Entomol.,* 18, 207–226.

Massee, A. M. 1954. *The pests of fruit and hops.* 3rd ed. Crosby Lockwood: London.

Matthews, G. A. 1984. *Pest management.* Longmans: London.

May, R. M., ed. 1976. *Theoretical ecology—principles and applications.* Blackwell: Oxford.

McCallan, E. 1959. "Some aspects of the geographical distribution of insect pests." In *J. Ent. Soc. S. Afr.,* 22, 3–12.

McNutt, D. N. 1976. *Insect collecting in the tropics.* C.O.P.R.: London.

Metcalf, C. L., W. P. Flint, and R. L. Metcalf. 1962. *Destructive and useful insects.* McGraw-Hill: New York.

Metcalf, R. L., ed. 1957–68. *Advances in pest control research.* Vols. 1–8. Interscience: London & New York.

_____ 1980. "Changing role of insecticides in crop protection." In *Ann. Rev. Entomol.,* 25, 219–256.

Metcalf, R. L., and W. H. Luckmann, eds. 1975. *Introduction to insect pest management.* John Wiley: New York.

Miller, D. R., and M. Kosztarab. 1979. "Recent advances in the study of scale insects." In *Ann. Rev. Entomol.,* 24, 1–27.

Moran, V. C. 1983. "The phytophagous insects and mites of cultivated plants in South Africa: patterns and pest status." In *J. Appl. Ecol.,* 20, 439–450.

Morton, A. 1989. "Thailand's million-dollar moth." In *New Scientist,* 25 November, 48–52.

Mound, L. A. 1965. "An introduction to the Aleyrodidae of Western Africa." In *Bull. Brit. Mus. Nat. Hist., Ent.,* 17, 113–160.

_____ ed. 1989. *Common insect pests of stored food products.* 7th ed. Brit. Mus. Nat. Hist., Econ. Series No. 15, Brit. Mus. (NH): London.

Mound, L. A., and S. H. Halsey. 1978. *Whiteflies of the world: a systematic catalogue of the Aleyrodidae (Homoptera) with host plant and natural enemy data.* John Wiley: London.

Mumford, J. D., and G. A. Norton. 1984. "Economics of decision making in pest management." In *Ann. Rev. Entomol.,* 29, 157–174.

Munro, J. W. 1966. *Pests of stored products.* Hutchinson: London.

Nat. Acad. Sci. U.S. 1969. *Principles of plant and animal pest control.* Vol. 3, Insect-pest management and control. Nat. Acad. Sci. U.S., No. 1695: Washington, D.C.

Nayar, K. K., T. N. Ananthakrishnan, and B. V. David. 1976. *General and applied entomology.* Tata McGraw-Hill: New Delhi.

Needham, J. G. 1959. *Culture methods for invertebrate animals.* Dover: New York.

Nichols, S. W., ed. 1989 (updated ed., 1992). *The Torre-Bueno glossary of entomology.* New York Ent. Soc.: New York.

Nickel, J. L. 1979. *Annotated list of insects and mites associated with crops in Cambodia.* S.E.A.R.C.A.: Philippines.

Nikolskaya, M. N. 1962. *The chalcid fauna of the U.S.S.R.* Russian ed. 1952. Israel Prog. for Scientific Translations: Jerusalem.

Nishida, T., and T. Toru. 1970. *Handbook of field methods for research on rice stem borers and their natural enemies.* Blackwell: Oxford.

Nixon, G. E. J. 1951. *The association of ants with aphids and coccids.* C.I.E.: London.

Nobel, E. R., and G. A. Nobel. 1961. *Parasitology (the biology of animal parasites).* Lea & Febiger: Philadelphia.

Nye, I. W. B. 1960. *The insect pests of graminaceous crops in East Africa.* Colonial Res. Studies No. 31. H.M.S.O.: London.

O'Connor, B. A. 1969. *Exotic plant pests and diseases.* South Pac. Comm.: Noumea, New Caledonia.

O.D.N.R.I. Several institutions (five in total—A.L.R.C., C.O.P.R., T.D.R.I., T.P.I., L.R.D.C.) have been amalgamated under the Overseas Development Natural Resources Institute, now called the Natural Resources Institute. This is the scientific unit of the British Overseas Development Administration, now housed at Chatham Maritime in Kent; a great many scientific publications are produced.

Oldfield, G. N. 1970. "Mite transmission of plant viruses." In *Ann. Rev. Entomol.,* 15, 343–380.

Oldroyd, H. 1958. *Preserving and studying insects.* Hutchinson: London.

_____ 1964. *The natural history of flies.* Weidenfeld & Nicholson: London.

_____ 1968. *Elements of entomology.* Weidenfeld & Nicholson: London.

Ordish, G. 1967. *Biological methods in crop pest control.* London.

_____ 1976. *The constant pest.* Peter Davies: London.

Ostmark, H. E. 1974. "Economic insect pests of bananas." In *Ann. Rev. Entomol.,* 19, 161–176.

Painter, R. H. 1951. *Insect resistance in crop plants.* Univ. Press Kansas: Lawrence.

_____ 1958. "Resistance of plants to insects." In *Ann. Rev. Entomol.,* 3, 267–290.

Palmer, J. M., L. A. Mound, and G. J. du Heaume. 1989. *C.I.E. guides to insects of importance to man.* No. 2, *Thysanoptera.* C.I.E.: London.

Papaj, D. R., and R. J. Prokopy. 1989. "Ecological and evolutionary aspects of learning in phytophagous insects." In *Ann. Rev. Entomol.,* 34, 315–350.

Parkin, E. A. 1956. "Stored product entomology (the assessment and reduction of losses caused by insects to stored foodstuffs)." In *Ann. Rev. Entomol.,* 1, 223–240.

Pathak, M. D. 1968. "Ecology of common insect pests of rice." In *Ann. Rev. Entomol.,* 13, 257–294.

Pearson, E. O. 1958. *The insect pests of cotton in tropical Africa.* C.I.E.: London.

Peck, O. 1963. "A catalogue of the Nearctic Chalcidoidea (Insecta: Chalcidoidea)." In *Canadian Entomologist Suppl.,* 30, 1092.

Peck, O., Z. Boucek, and A. Hoffer. 1964. "Keys to the Chalcidoidea of Czechoslovakia (Insecta: Hymenoptera)." In *Mem. Ent. Soc. Canada,* No. 34, 1–120.

P.E.S.T.D.O.C. 1974. *Organism thesaurus.* Vol. 1, *Animal organisms.* Ciba-Geigy, Basle & Derwent Pub.: London.

Peterson, A. 1953. *Entomological techniques.* Edwards Bros.: Ann Arbor, Mich., U.S.A.

Pfadt, R. E. 1962. *Fundamentals of applied entomology.* Macmillan: New York.

Pinhey, E. C. G. 1960. *Hawkmoths of Central and Southern Africa.* Longmans: London.

_____ 1968. *Introduction to insect study in Africa.* Oxford Univ. Press: London.

Pirone, P. P. 1978. *Diseases and pests of ornamental plants.* 5th ed. John Wiley: New York.

Popham, W. L., and D. G. Hall. 1958. "Insect eradication programs." In *Ann. Rev. Entomol.,* 3, 335–354.

Price Jones, D., and M. E. Solomon. 1974. *Biology in pest and disease control.* 13th Symp. Brit. Ecol. Soc. Blackwell: Oxford.

Pritchard, G. 1983. "Biology of Tipulidae." In *Ann. Rev. Entomol.,* 28, 1–22.

Proctor, M., and P. Yeo. 1973. *The pollination of flowers.* The New Naturalist, No. 54. Collins: London.

Prokopy, R. J., and E. D. Owens. 1983. "Visual detection of plants by herbivorous insects." In *Ann. Rev. Entomol.,* 28, 337–364.

Proverbs, M. D. 1969. "Induced sterilization and control of insects." In *Ann. Rev. Entomol.,* 14, 81–102.

Pschorn-Walcher, H. 1977. "Biological control of forest insects." In *Ann. Rev. Entomol.,* 22, 1–22.

Prescott, J. M., et al. 1986. *Wheat diseases and pests: a guide for field identification.* C.I.M.M.Y.T.: Mexico.

Purseglove, J. W. 1968. *Tropical crops. Dicotyledons,* vols. 1 and 2. Longmans: London.

_____ 1972. *Tropical crops. Monocotyledons,* vols. 1 and 2. Longmans: London.

Rabb, R. L., and F. E. Guthrie. 1970. *Concepts of pest management.* Proc. of a conference held at N.C. State Univ., Raleigh, N.C., 25–27 March 1970. N.C. State Univ.: Raleigh, N.C.

Rabb, R. L., F. A. Todd, and H. C. Ellis. 1976. "Tobacco pest management." In Apple and Smith, *Integrated pest management,* pp. 71–106. Plenum Press: New York.

Radcliffe, E. B. 1982. "Insect pests of potato." In *Ann. Rev. Entomol.,* 27, 173–204.

Rajamohan, N. 1976. "Pest complex of sunflower—a bibliography." In *P.A.N.S.,* 22, 546–563.

Rao, B. S. 1965. *Pests of Hevea plantations in Malaya.* Rubber Res. Inst.: Kuala Lumpur.

Rao, G. N. 1970. "Tea pests in Southern India and their control." In *P.A.N.S.,* 16, 667–672.

Reddy, D. B. 1968. *Plant protection in India.* Allied Publ.: Bombay.

Richards, O. W., and R. G. Davies. 1977. *Imms' general textbook of entomology.* 10th ed. Vol. 1. *Structure, physiology and development.* Vol. 2, *Classification and biology.* Chapman & Hall: London.

Ripper, W. E., and L. George. 1965. *Cotton pests of Sudan.* Blackwell: Oxford.

Roelofs, W. L., ed. 1979. *Establishing efficacy of sex attractants and distruptants for insect control.* Ent. Soc. Amer.: Washington, D.C.

Rose, G. 1963. *Crop protection.* 2nd ed. Leonard Hill: London.

Russell, G. E. 1978. *Plant breeding for pest and disease resistance.* Butterworths: London.

_____ ed. *See* "Intercept Publications".

Schal, C., and R. L. Hamilton. 1990. "Integrated suppression of synanthropic cockroaches." In *Ann. Rev. Entomol.,* 35, 521–551.

Schmutterer, H. 1969. *Pests of crops in Northeast and Central Africa.* G. Fischer: Stuttgart.

Schneider, D. 1961. "The olfactory sense of insects." In *Drapco Report,* 8, 135–146.

Schoohoven, L. M. 1968. "Chemosensory basis of host plant selection." In *Ann. Rev. Entomol.,* 13, 115–136.

Schreck, C. E. 1977. "Techniques for the evaluation of insect repellants: a critical review." In *Ann. Rev. Entomol.,* 22, 101–119.

Sclater, P. L. 1858. "On the general geographical distribution of the members of the Class Aves." In *J. Linn. Soc. Lond.,* 2, 130.

Scopes, N., and M. Ledieu, eds. 1983. *Pest and disease control handbook.* 2nd ed. B.C.P.C. Pub.: London.

Seymour, P. R. 1979. *Invertebrates of economic importance in Britain* (formerly M.A.F.F. *Tech. Bull.* No. 6). H.M.S.O.: London.

Sharples, A. 1936. *Diseases and pests of the rubber tree.* Macmillan: London.

Shillito, J. F. 1971. "The genera of Diopsidae (Insecta: Diptera)." In *Zool. J. Linn. Soc.,* 50, 287–295.

Shorey, H. H. 1976. *Animal communication by pheromones.* Academic Press: New York.

Short, L. R. T. 1963. *Introduction to applied entomology.* Longmans: London.

Singh, J. P. 1970. *Elements of vegetable pests.* Vora & Co.: Bombay.

Singh, S. R., H. F. van Emden, and T. A. Taylor, eds. 1978. *Pests of grain legumes: ecology and control.* Academic Press: London.

Singh, S. R., and H. F. van Emden. 1979. "Insect pests of grain legumes." In *Ann. Rev. Entomol.,* 24, 255–278.

Skaife, S. H. 1953. *African insect life.* Longmans: London.

Skinner, B. 1988. *Colour identification guide to the moths of the British Isles (Macrolepidoptera).* Viking: London.

Smit, B. 1964. *Insects in Southern Africa—how to control them.* Oxford Univ. Press: South Africa.

Smith, K. G. V., ed. 1973. *Insects and other arthropods of medical importance.* Brit. Mus. Nat. Hist.: London.

_____ 1989. *An introduction to the immature stages of British flies: Diptera larvae, with notes on eggs, puparia and pupae.* (Handbooks for the Identification of British Insects, vol. 10, part 14.) Roy. Ent. Soc. Lond.: London.

Smith, K. M. 1951. *Agricultural entomology.* 2nd ed. Cambridge Univ. Press: London.

Soulsby, E. J. L. 1968. *Helminths, arthropods and protozoa of domesticated animals.* 6th ed. Baillier, Tindall & Cassell: London.

Southgate, B. J. 1979. "Biology of the Bruchidae." In *Ann. Rev. Entomol.,* 24, 449–473.

Southwood, T. R. E., ed. 1968. *Insect abundance, a symposium.* Roy. Ent. Soc. Lond.: London.

_____ 1972. "The insect/plant relationship—an evolutionary perspective." In *Insect/plant relationships,* ed. H. F. van Emden, pp. 3–30.

_____ 1977. "The relevance of population dynamic theory to pest status." In *Origins of pest, parasite, disease and weed problems,* pp. 35–54. Blackwell: Oxford.

_____ 1978. *Ecological methods (with particular reference to the study of insect populations).* 2nd ed. Chapman & Hall: London.

Spencer, K. A. 1973. *Agromyzidae (Diptera) of economic importance* (vol. 9 of *Series Entomologica*). W. Junk: The Hague.

_____ 1990. *Host specialization in the world Agromyzidae (Diptera).* Series Ent., vol. 45. Kluwer Acad. Pubs.: Dordrecht, Netherlands.

_____ (Many regional faunistic monographs on the Agromyzidae.)

Spradberry, J. P. 1973. *Wasps.* Sidgwick & Jackson: London.

Stapley, J. H., and F. C. H. Gayner. 1969. *World crop protection.* Vol. 1, *Pests and diseases.* Vol. 2, *Pesticides.* K. A. Hassell.

Stern, V. M., R. F. Smith, R. van den Bosch, and K. S. Hagen. 1959. "The integrated control concept." In *Hilgardia,* 29, 81–101.

Stinner, B. R., and G. J. House. 1990. "Arthropods and other invertebrates in conservation-tillage agriculture." In *Ann. Rev. Entomol.,* 35, 299–318.

Stinner, R. E., C. S. Barfield, J. L. Stimac, and L. Dohse. 1983. "Dispersal and movement of insect pests." In *Ann. Rev. Entomol.,* 28, 319–336.

Strong, D. R., J. H. Lawton, and T. R. E. Southwood. 1984. *Insects on plants: community patterns and mechanisms.* Blackwell: Oxford.

Stroyan, H. L. G. 1961. "Identification of aphids living on citrus." In *F.A.O. Pl. Prot. Bull.,* 9, 30.

Swaine, G. 1961. *Plant pests of importance in Tanganyika.* Min. Agric. Bull. No. 13, Min. Agric.: Tanganyika.

Sweetman, H. L. 1958. *Principles of biological control.* W. C. Brown: Iowa.

Talhouk, A. M. S. 1969. *Insects and mites injurious to crops in Middle Eastern countries.* Mon. zur angew. Ent., Nr 21, Paul Parey: Hamburg.

T.D.R.I. Information Service Annotated Bibliographies.

      1981. No. 1. Insect Pests of Pre-harvest Wheat and Their Control in the Developing World: 1975–80. Compiled by E. Southam and P. Schofield.

      1983. No. 2. *Heliothis* Dispersal and Migration. Compiled by N. W. Widmer and P. Schofield. T.D.R.I.: London.

Teetes, G. L., *et al.* 1983. *Sorghum insect identification handbook.* Info. Bull. No. 12, I.C.R.I.S.A.T.: Patancheru, India.

Theberge, R. L., ed. 1984. *Common African pests and diseases of cassava, yam, sweet potato and cocoyam.* I.I.T.A.: Ibadan, Nigeria.

Thomas, R. T. S. 1962. Checklist of pests on some crops in West Irain (New Guinea). Bull. Econ. Aff., Agric. Series No. 1, Dept. Econ. Affairs: Hollandia.

Thompson, 1943–82. *A catalogue of the parasites and predators on insect pests.* C.A.B. (C.I.B.C.): London. Section 1, Parasite Host Catalogue; parts 1–2. Section 2, Host Parasite Catalogue; parts 1–5. Section 3, Predator Host Catalogue. Section 4, Host Predator Catalogue.

Tinsley, T. W. 1979. "The potential of insect pathogenic viruses as pesticidal agents." In *Ann. Rev. Entomol.,* 24, 63–87.

Trought, T. E. T. 1965. *Farm pests.* Blackwell: Oxford.

Tsedeke, A., ed. 1987. *A review of crop protection research in Ethiopia.* Inst. Agric. Res.: Addis Ababa.

Turnipseed, S. G., and M. Kogan. 1976. "Soybean entomology." In *Ann. Rev. Entomol.,* 21, 247–282.

Tuttle, D. M., and E. W. Baker. 1968. *Spider mites of Southwestern United States and a revision of the Family Tetranychidae.* Univ. Arizona Press.

U.S.A. (In the U.S.A. there is a range of publications dealing with insects and entomology, ranging from taxonomic monographs, checklists, to faunal studies, and field handbooks. A few examples are listed below.)

_____ 1951. (Muesebeck, C. F. W., *et al.*) *Hymenoptera of America North of Mexico (Synoptic catalogue).* Agric. Mon. No. 2. Supplements, U.S. Govt. Printing Off.: Washington, D.C.

_____ 1965. (Stone, A., *et al.*) *A catalogue of the Diptera of America North of Mexico.* Agric. Hdbk. No. 276; U.S.D.A. Supplements, U.S. Govt. Printing Off.: Washington, D.C.

_____ 1971– . *The moths of America North of Mexico.* (Many volumes.)

University of California Statewise Integrated Pest Management Project.

_____ 1983. IPM for rice (Pub. No. 3280).

_____ 1984. IPM for cotton (Pub. No. 3305).

_____ 1986. IPM for potatoes (Pub. No. 3316).

_____ 1984. IPM for citrus (Pub. No. 3303).

U.S.D.A. 1952. *The yearbook of agriculture (1952) insects* (U.S.D.A.). U.S. Govt. Printing Off.: Washington, D.C.

_____ 1966. *The yearbook of agriculture (1966) protecting our food* (U.S.D.A.) U.S. Govt. Printing Off.: Washington, D.C.

Uvarov, B. 1966. *Grasshoppers and locusts.* Cambridge Univ. Press: Cambridge.

Villani, M. G., and R. J. Wright. 1990. "Environmental influences on soil macroarthropod behaviour in agricultural systems." In *Ann. Rev. Entomol.,* 35, 249–269.

Visser, J. H. 1986. "Host odour perception in phytophagous insects." In *Ann. Rev. Entomol.,* 31, 121–144.

Waters, W. E., and R. W. Stark. 1980. "Forest pest management: concept and reality." In *Ann. Rev. Entomol.,* 25, 479–509.

Watson, M. A., and R. T. Plumb. 1972. "Transmission of plant-pathogenic viruses by aphids." In *Ann. Rev. Entomol.,* 17, 425–452.

Watson, T. F., L. Moore, and G. W. Ware. 1975. *Practical insect pest management.* W. H. Freeman & Co.: San Francisco.

Watt, A. D., ed. 1990. *Population dynamics of forest insects.* Intercept: Andover, U.K.

Weber, N. A. 1972. "Gardening ants: the Attines." In *Mem. Amer. Phil. Soc.,* 92, 1–146.

Werner, F. G. 1982. *Common names of insects and related organisms.* Ent. Soc. Amer.: Washington, D.C.

Wheatley, P. E., and T. J. Crowe. 1967. *Pest handbook (the recognition and control of the more important pests of agriculture in Kenya).* Govt. Printer: Nairobi.

Whellan, J. A. 1964. "Some Rhodesian pests of entomological interest." In *Span,* 7, 3.

Wiebes, J. T. 1976. "A short history of fig wasp research." In *Gdns' Bull.* (Singapore), XXIX, 207–232.

Williams, D. J. 1969. "The family-group of the scale insects (Hemiptera: Coccoidea)." In *Bull. Br. Mus. Nat. Hist., Ent.,* 32, 315–341.

_____ 1970. "The mealybugs (Homoptera, Coccoidea, Pseudococcidae) of sugarcane, rice and sorghum." In *Bull. Ent. Res.,* 60, 109–188.

_____ 1971. "Synoptic discussion of *Lepidosaphes* Shimer and its allies with a key to genera (Homoptera, Coccoidea, Diaspididae)." In *Bull. Ent. Res.,* 61, 7–11.

Williams, J. R., J. R. Metcalfe, R. W. Mungomery, and R. Mathes, eds. 1969. *Pests of sugarcane.* Elsevier: London and New York.

Wilson, F. 1960. *A review of the biological control of insects and weeds in Australia and Australian New Guinea.* Techn. Comm. No. 1, C.I.B.C.: Ottawa.

Wood, A. M. 1989. *Insects of economic importance: a checklist of preferred names.* C.A.B.I.: Wallingford.

Wood, B. J. 1968. *Pests of oil palms in Malaysia and their control.* Inc. Soc. of Planters: Kuala Lumpur.

_____ 1971. "The importance of ecological studies to pest control in Malaysian plantations." In *P.A.N.S.,* 17, 411–416.

Wyniger, R. 1962. *Pests of crops in warm climates and their control.* Acta Tropica. Supp. 7.

_____ 1968. *Control measures.* 2nd ed. Verlag Recht Ges.: Basel, Switzerland.

Yasumatsu, K., and T. Toru. 1968. "Impact of parasites, predators, and diseases on rice pests." In *Ann. Rev. Entomol.,* 13, 295–324.

Young, W. R., and G. L. Teets. 1977. "Sorghum entomology." In *Ann. Rev. Entomol.,* 22, 193–218.

Yunus, A., and A. Balasubramanian. 1975. *Major crop pests in Peninsular Malaysia.* Min. Agric. and Rur. Dev.: Malaysia.

INTERCEPT SCIENTIFIC PUBLICATIONS

A new publishing house started in 1983 by Emeritus Professor G. E. Russell (Agricultural Biology, Newcastle University, U.K.) is designed to produce reviews of research and recent developments. A major publication is

Agricultural Zoology Reviews:

Vol. 1. 1986.
Vol. 2. 1987.
Vol. 3. 1989.

Collections of related articles from these reviews are then published as books, with Professor Russell as editor. For example:

Russell, G. E., ed. 1989. *Biology and population dynamics of invertebrate crop pests.*
Russell, G. E., ed. 1989. *Management and control of invertebrate crop pests.* Intercept: Andover, U.K.

Titles by other authors/editors are also being published. This publisher is clearly of particular interest to agricultural entomologists worldwide.

INDEX OF COMMON NAMES

INDEX OF SCIENTIFIC NAMES